T0213167

Second Edition

Problems
and
Solutions
on
MECHANICS

Major American Universities Ph.D. Qualifying Questions and Solutions - Physics

ISSN: 1793-1487

Published

Problems and Solutions on Mechanics (Second Edition)
    edited by Swee Cheng Lim, Choy Heng Lai and Leong Chuan Kwek
    (NUS, Singapore)

Problems and Solutions on Thermodynamics and Statistical Mechanics
    edited by Yung-Kuo Lim (NUS, Singapore)

Problems and Solutions on Optics
    edited by Yung-Kuo Lim (NUS, Singapore)

Problems and Solutions on Electromagnetism
    edited by Yung-Kuo Lim (NUS, Singapore)

Problems and Solutions on Mechanics
    edited by Yung-Kuo Lim (NUS, Singapore)

Problems and Solutions on Solid State Physics, Relativity and Miscellaneous Topics
    edited by Yung-Kuo Lim (NUS, Singapore)

Problems and Solutions on Quantum Mechanics
    edited by Yung-Kuo Lim (NUS, Singapore)

Problems and Solutions on Atomic, Nuclear and Particle Physics
    edited by Yung-Kuo Lim (NUS, Singapore)

Problems and Solutions on Optics (Second Edition)
    edited by Swee Cheng Lim, Choy Heng Lai and Leong Chuan Kwek
    (NUS, Singapore)

Major American Universities Ph.D. Qualifying Questions and Solutions - Physics

Second Edition

Problems
and
Solutions
on
MECHANICS

Editors

Swee Cheng Lim

Choy Heng Lai
National University of Singapore, Singapore

Leong Chuan Kwek
*CQT, National University of Singapore and
NIE and EEE, Nanyang Technological University, Singapore*

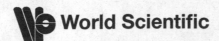 World Scientific

EW JERSEY · LONDON · SINGAPORE · BEIJING · SHANGHAI · HONG KONG · TAIPEI · CHENNAI · TOKYO

Published by

World Scientific Publishing Co. Pte. Ltd.

5 Toh Tuck Link, Singapore 596224

USA office: 27 Warren Street, Suite 401-402, Hackensack, NJ 07601

UK office: 57 Shelton Street, Covent Garden, London WC2H 9HE

British Library Cataloguing-in-Publication Data
A catalogue record for this book is available from the British Library.

Major American Universities Ph.D. Qualifying Questions and Solutions - Physics
PROBLEMS AND SOLUTIONS ON MECHANICS
Second Edition

ISBN 978-981-121-340-3 (hardcover)
ISBN 978-981-121-445-5 (paperback)
ISBN 978-981-121-341-0 (ebook for institutions)
ISBN 978-981-121-342-7 (ebook for individuals)

For any available supplementary material, please visit
https://www.worldscientific.com/worldscibooks/10.1142/11642#t=suppl

Typeset by Diacritech Technologies Pvt. Ltd.
Chennai - 600106, India

Printed in Singapore

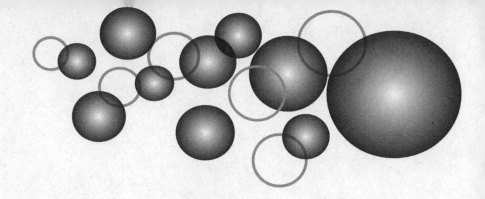

Preface

This is the second edition of a former popular series on Problems and Solutions in various topics of physics ranging from Mechanics, Electromagnetism, Optics, Atomic, Nuclear and Particle Physics, Thermodynamics and Statistical Mechanics, Quantum Mechanics, to Solid State Physics, Relativity and Miscellaneous topics. We have greatly expanded the volumes. Each volume is divided into several subtopics, and there are new problems and solutions for many of them. In total, there are several thousand problems and solutions that almost cover an entire undergraduate physics course. There are different types of questions: qualitative, quantitative, fill-in-the-blanks, and even multiple-choice. These questions are chosen at the level of the PhD Qualifying Examinations at American universities. However, these questions can also serve as an excellent resource for Physics Competitions, like the International Physics Olympiad or the regional Physics Olympiads.

We believe a good grounding in problems solving is necessary for studying physics, and therefore this compendium is an invaluable resource for the preparation of graduate study in physics or competitions. It is suggested that the student attempt the problems first, and consult the solution only afterwards. Where applicable, the solutions are accompanied by figures that illustrate how to set up the problem.

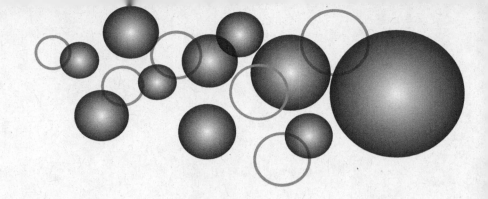

Contents

Part I
Newtonian Mechanics

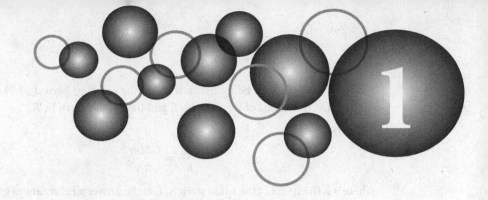

DYNAMICS OF A POINT MASS (1001–1108)

1001

A body weighs W on Earth. A man on a moving elevator weighs the body using a spring balance and finds that its weight is $w < W$. Describe the possible motion of the elevator.

Sol: $W = mg$ and $w = mg'$, where g' is the effective acceleration in the elevator. Since $w < W$, $g' < g$, which means there's an extra acceleration downward in addition to Earth's gravity.

Thus, the motion of the elevator is either decelerating while moving upward or accelerating in the downward direction.

1002

An orbiting space station is observed to remain always vertically above the same point on the earth. Where on earth is the observer? Describe the orbit of the space station as completely as possible.

(Wisconsin)

Sol: The observer must be on the equator of the earth. The orbit of the space station is a large circle in the equatorial plane with center at the center of the earth. The radius of the orbit can be figured out using the orbiting period of 24 hours* as follows. Let the radius of the orbit be R and that of the earth be R_0.

We have

$$\frac{mv^2}{R} = \frac{GMm}{R^2},$$

where v is the speed of the space station, G is the universal constant of gravitation, m and M are the masses of the space station and the earth respectively, giving

$$v^2 = \frac{GM}{R}.$$

As

$$mg = \frac{GMm}{R_0^2},$$

we have

$$GM = R_0^2 g.$$

Hence

$$v^2 = \frac{R_0^2 g}{R}.$$

For circular motion with constant speed v, the orbiting period is

$$T = \frac{2\pi R}{v}.$$

Hence

$$\frac{4\pi^2 R^2}{T^2} = \frac{R_0^2 g}{R}$$

and

$$R = \left(\frac{R_0^2 T^2 g}{4\pi^2}\right)^{\frac{1}{3}} = 4.2 \times 10^4 \text{ km}.$$

1003

A block of mass m is pushed at constant velocity v along a horizontal surface with coefficient of kinetic friction μ_k. A bullet of mass M moving with

---

* For a more accurate calculation, the orbiting period should be taken as 23 hours 56 minutes and 4 seconds.

a velocity V hits the block and embeds in it. The combination moves a distance d before returning to its initial velocity. Obtain an expression for d. What happens if $m = M$?

Sol: For the body to initially move with a constant velocity, a force must be applied to counter the frictional force, i.e., $f = \mu_k mg$. When the mass is embedded in the bullet, the force exerted on the combination is

$$F = (m + M)a = \mu_k mg - \mu_k(m + M)g$$

Hence

$$a = -\frac{\mu_k Mg}{m + M}$$

From the law of conservation of momentum

$$mv + MV = (m + M)v_{comb}$$

Hence

$$v_{comb} = \frac{mv + MV}{(m + M)}$$

With this as the initial velocity of the combination and v as the final velocity

$$v^2 = \left(\frac{mv + MV}{m + M}\right)^2 - 2\frac{\mu_k mg}{m + M}d$$

Solving the above equation

$$d = \frac{M(V - v)[2mv + M(V + v)]}{2\mu_k mg(m + M)}$$

When $m = M$

$$d = \frac{V^2 + 2vV - 3v^2}{4\mu_k g}$$

1004

A cord passing over a frictionless pulley has a 9 kg mass tied on one end and a 7 kg mass on the other end (Fig. 1.1). Determine the acceleration and the tension of the cord.

(Wisconsin)

Sol: Neglecting the moment of inertia of the pulley, we obtain the equations of motion

$$m_1\ddot{x} = m_1g - F$$

and

$$m_2\ddot{x} = F - m_2g.$$

Hence the tension of the cord and the acceleration are respectively

$$F = \frac{2m_1m_2g}{m_1 + m_2} = 77.2 \text{ N}$$

and

$$\ddot{x} = \frac{(m_1 - m_2)g}{m_1 + m_2} = \frac{2g}{16}$$
$$= 1.225 \text{ m/s}^2.$$

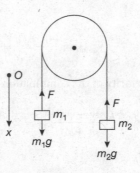

Fig. 1.1

1005

Consider a block on an inclined plane. The coefficient of kinetic friction between the block and the plane is μ_k. When the block slides down the plane, a force F is required to make the block slide at a constant velocity. When the block slides up the plane, a greater force nF is required to maintain a constant velocity. If the angle of the plane is θ, show that the angle of inclination for such a condition depends only on μ_k and n. Show mathematically that it is impossible for the two forces to be the same.

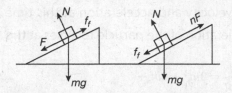

Fig. 1.2

Sol: Let the angle of inclination in each case be θ. In each case $f_f = \mu_k mg \cos \theta$.

In the first case, the force equations along to the plane are

$$F = \mu_k mg \cos \theta - mg \sin \theta \tag{1}$$

In the second case,

$$nF = \mu_k mg \cos \theta + mg \sin \theta \tag{2}$$

Dividing one by the other and simplifying

$$\theta = \tan^{-1}\left(\mu_k \frac{n-1}{n+1}\right),$$ which depends only on n and μ_k.

If the forces are equal $n = 1$ and $\theta = 0$, which indicates a horizontal plane that contradicts the assumption of an inclined plane.

1006

A person of mass 80 kg jumps from a height of 1 meter and foolishly forgets to buckle his knees as he lands. His body decelerates over a distance of only one cm. Calculate the total force on his legs during deceleration.

(Wisconsin)

Sol: The person has mechanical energy $E_1 = mg(h + s)$ just before he lands. The work done by him during deceleration is $E_2 = fs$, where f is the total force on his legs. As $E_1 = E_2$,

$$f = \frac{mgh}{s} + mg = \left(\frac{80 \times 1}{0.01} + 80\right)g = 8080g \text{ N}.$$

1007

A body moves in a curved path according to the equation $x = ut^3 + bt^2 + ct + d$. Determine the conditions that must be imposed on the constants to ensure that the velocity changes direction only once. Also determine an expression for the velocity and acceleration at this time in terms of *b* and *c*. Show that the acceleration of the particle reverses at the same time.

Sol: Velocity, $v = \dfrac{dx}{dt} = 3ut^2 + 2bt + c$

Since the velocity is a 2nd degree polynomial, there is only one stationary point and that point cannot be an inflection point. At stationary points, $v = \dfrac{dx}{dt} = 0$

or $3ut^2 + 2bt + c = 0$

or $t = \dfrac{-2b \pm \sqrt{4b^2 - 12uc}}{6u} = \dfrac{-b \pm \sqrt{b^2 - 3uc}}{3u}$

The velocity will change direction only once if

$$b^2 = 3uc \qquad \text{and} \qquad t = -\frac{b}{3u}$$

Substituting for *u* and *t* in the equation for *v*

Velocity, $v = b^2 t^2 + 2btc + c^2 = (bt + c)^2$

or, $v = \left(\left(b \times \dfrac{-b}{3u}\right) + c\right)^2 = \left(-\dfrac{b^2}{3u} + c\right)^2$

Acceleration, $a = \dfrac{d^2x}{dt^2} = 6ut + 2b$

Substituting *t* in the equation for *a*

$a = 6u\left(-\dfrac{b}{3u}\right) + 2b = 0$. The acceleration of the particle will change direction when

$$a = \frac{d^2x}{dt^2} = 0$$

or $6ut + 2b = 0$

or $t = -\dfrac{b}{3u}$

1008

Consider a rotating spherical planet. The velocity of a point on its equator is V. The effect of rotation of the planet is to make g at the equator 1/2 of g at the pole. What is the escape velocity for a polar particle on the planet expressed as a multiple of V?

(*Wisconsin*)

Sol:　Let g and g' be the gravitational accelerations at the pole and at the equator respectively and consider a body of mass m on the surface of the planet, which has a mass M. At the pole,

$$mg = \frac{GMm}{R^2},$$

giving

$$GM = gR^2.$$

At the equator, we have

$$\frac{mV^2}{R} = \frac{GMm}{R^2} - mg' = mg - \frac{mg}{2} = \frac{mg}{2}.$$

Hence $g = 2V^2/R$.

If we define gravitational potential energy with respect to a point at infinity from the planet, the body will have potential energy

$$-\int_{\infty}^{R} -\frac{GMm}{r^2}dr = -\frac{GMm}{R}.$$

Note that the negative sign in front of the gravitational force takes account of its attractiveness. The body at the pole then has total energy

$$E = \frac{1}{2}mV^2 - \frac{GMm}{R}.$$

For it to escape from the planet, its total energy must be at least equal to the minimum energy of a body at infinity, i.e. zero. Hence the escape velocity v is given by

$$\frac{1}{2}mv^2 - \frac{GMm}{R} = 0,$$

or

$$v^2 = \frac{2GM}{R} = 2gR = 4V^2,$$

i.e.

$$v = 2V.$$

1009

A small mass m rests at the edge of a horizontal disk of radius R; the coefficient of static friction between the mass and the disk is μ. The disk is rotated about its axis at an angular velocity such that the mass slides off the disk and lands on the floor h meters below. What was its horizontal distance of travel from the point that it left the disk?

(*Wisconsin*)

Sol: The maximum static friction between the mass and the disk is $f = \mu mg$. When the small mass slides off the disk, its horizontal velocity v is given by

$$\frac{mv^2}{R} = \mu mg.$$

Thus

$$v = \sqrt{\mu Rg}.$$

The time required to descend a distance h from rest is

$$t = \sqrt{\frac{2h}{g}}.$$

The the horizontal distance of travel before landing on the floor is equal to

$$vt = \sqrt{2\mu Rh}.$$

1010

A body of mass m is travelling with a velocity v m/s in positive x direction when it collides with a massive stationary object and is turned anticlockwise through an angle of 45°. It continues to move with the same speed. If the impact lasted 1 ms, determine the acceleration of the body, the force it experienced, and impulse of the force.

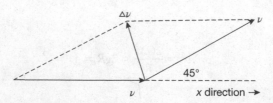

Fig. 1.3

Sol: $|\Delta v| = \sqrt{v^2 + v^2 + 2v^2 \cos 135°} = v\sqrt{2 - \sqrt{2}} = 0.765 \, v \, \text{m/s}$ at an angle of $(45° + 67.5°) = 112.5°$ counterclockwise from the x direction.

Acceleration: $a = 0.765 \, v \times 10^3 \, \text{m/s}^2$ in a direction $112.5°$ counterclockwise from the x direction.

Impulse, $I = F\Delta t = 0.765 \, mv \, \text{Ns}$

1011

Assume all surfaces to be frictionless and the inertia of pulley and cord negligible (Fig. 1.4). Find the horizontal force necessary to prevent any relative motion of m_1, m_2 and M.

(*Wisconsin*)

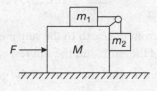

Fig. 1.4

Sol: The forces f_1, F and mg are shown in Fig. 1.5. The accelerations of m_1, m_2 and M are the same when there is no relative motion among them. The equations of motion along the x-axis are

$$(M + m_1 + m_2)\ddot{x} = F,$$
$$m_1\ddot{x} = f_1.$$

As there is no relative motion of m_2 along the y-axis,

$$f_1 = m_2 g.$$

Combining these equations, we obtain

$$F = \frac{m_2(M + m_1 + m_2)g}{m_1}.$$

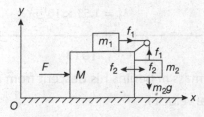

Fig. 1.5

1012

The sun is about 25,000 light years from the center of the galaxy and travels approximately in a circle with a period of 170,000,000 years. The earth is 8 light minutes from the sun. From these data alone, find the approximate gravitational mass of the galaxy in units of the sun's mass. You may assume that the gravitational force on the sun may be approximated by assuming that all the mass of the galaxy is at its center.

(Wisconsin)

Sol: For the motion of the earth around the sun,

$$\frac{mv^2}{r} = \frac{Gmm_s}{r^2},$$

where r is the distance from the earth to the sun, v is the velocity of the earth, m and m_s are the masses of the earth and the sun respectively.

For the motion of the sun around the center of the galaxy,

$$\frac{m_s V^2}{R} = \frac{Gm_s M}{R^2},$$

where R is the distance from the sun to the center of the galaxy, V is the velocity of the sun and M is the mass of the galaxy.

Hence

$$M = \frac{RV^2}{G} = \frac{R}{r}\left(\frac{V}{v}\right)^2 m_s.$$

Using $V = 2\pi R/T$, $v = 2\pi r/t$, where T and t are the periods of revolution of the sun and the earth respectively, we have

$$M = \left(\frac{R}{r}\right)^3 \left(\frac{t}{T}\right)^2 m_s.$$

With the data given, we obtain

$$M = 1.53 \times 10^{11} m_s.$$

1013

An Olympic diver of mass m begins his descent from a 10 meter high diving board with zero initial velocity.

a. Calculate the velocity V_0 on impact with the water and the approximate elapsed time from dive until impact (use any method you choose). Assume that the buoyant force of the water balances the gravitational force on the diver and that the viscous force on the diver is bv^2.

b. Set up the equation of motion for vertical descent of the diver through the water. Solve for the velocity V as a function of the depth x under water and impose the boundary condition $V = V_0$ at $x = 0$.

c. If $b/m = 0.4$ m^{-1}, estimate the depth at which $V = V_0/10$.

d. Solve for the vertical depth $x(t)$ of the diver under water in terms of the time under water.

(Wisconsin)

Sol:

a.
$$V_0 = \sqrt{2gh} = \sqrt{2 \times 9.8 \times 10} = 14 \text{ m/s}.$$

The time elapsed from dive to impact is

$$t = \frac{V_0}{g} = \frac{14}{9.8} = 1.43 \text{ s}.$$

b. As the gravitational force on the diver is balanced by the buoyancy, the equation of motion of the diver through the water is

$$m\ddot{x} = -b\dot{x}^2,$$

or, using $\ddot{x} = \dot{x}d\dot{x}/dx$,

$$\frac{d\dot{x}}{\dot{x}} = -\frac{b}{m}dx.$$

Integrating, with $\dot{x} = V_0$ at $x = 0$, we obtain

$$V \equiv \dot{x} = V_0 e^{-\frac{b}{m}x}.$$

c. When $V = V_0/10$,

$$x = \frac{m}{b}\ln 10 = \frac{\ln 10}{0.4} = 5.76 \text{ m}.$$

d. As $dx/dt = V_0 e^{-\frac{b}{m}x}$,

$$e^{\frac{b}{m}x}dx = V_0 dt.$$

Integrating, with $x = 0$ at $t = 0$, we obtain

$$\frac{m}{b}(e^{\frac{b}{m}x} - 1) = V_0 t,$$

or

$$x = \frac{m}{b}\ln\left(1 + bV_0\frac{t}{m}\right).$$

1014

The combined frictional and air resistance on a bicyclist has the force $F = aV$, where V is his velocity and $a = 4$ newton-sec/m. At maximum effort, the cyclist can generate 600 watts propulsive power. What is his maximum speed on level ground with no wind?

(Wisconsin)

Sol: When the maximum speed is achieved, the propulsive force is equal to the resistant force. Let F be this propulsive force, then

$$F = aV \quad \text{and} \quad FV = 600 \text{ W}.$$

Eliminating F, we obtain

$$V^2 = \frac{600}{a} = 150 \text{ m}^2/\text{s}^2$$

and the maximum speed on level ground with no wind

$$v = \sqrt{150} = 12.2 \text{ m/s}.$$

1015

A pendulum of mass m and length l is released from rest in a horizontal position. A nail a distance d below the pivot causes the mass to move along the path indicated by the dotted line. Find the minimum distance d in terms of l such that the mass will swing completely round in the circle shown in Fig. 1.6.

(Wisconsin)

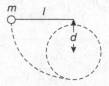

Fig. 1.6

Sol: Take the mass m as a point mass. At the instant when the pendulum collides with the nail, m has a velocity $v = \sqrt{2gl}$. The angular momentum of the mass with respect to the point at which the nail locates is conserved during the collision. Then the velocity of the mass is still v at the instant after the collision and the motion thereafter is such that the mass is constrained to rotate around the nail. Under the critical condition that the mass can just swing completely round in a circle, the gravitational force is equal to the centripetal force when the mass is at the top of the circle. Let the velocity of the mass at this instant be v_1, and we have

$$\frac{mv_1^2}{l-d} = mg,$$

or

$$v_1^2 = (l-d)g.$$

The energy equation

$$\frac{mv^2}{2} = \frac{mv_1^2}{2} + 2mg(l-d),$$

or

$$2gl = (l-d)g + 4(l-d)g$$

then gives the minimum distance as

$$d = \frac{3l}{5}.$$

1016

A mass m moves in a circle on a smooth horizontal plane with velocity v_0 at a radius R_0. The mass is attached to a string which passes through a smooth hole in the plane as shown in Fig. 1.7. ("Smooth" means ffictionless.)

a. What is the tension in the string?

b. What is the angular momentum of m?

c. What is the kinetic energy of m?

d. The tension in the string is increased gradually and finally m moves in a circle of radius $R_0/2$. What is the final value of the kinetic energy?

e. Why is it important that the string be pulled gradually?

(Wisconsin)

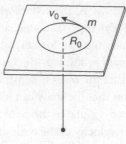

Fig. 1.7

Sol:

    a. The tension in the string provides the centripetal force needed for the circular motion, hence $F = mv_0^2/R_0$.

    b. The angular momentum of the mass m is $J = mv_0R_0$.

    c. The kinetic energy of the mass m is $T = mv_0^2/2$.

    d. The radius of the circular motion of the mass m decreases when the tension in the string is increased gradually. The angular momentum of the mass m is conserved since it moves under a central force. Thus

$$mv_0R_0 = mv_1\left(\frac{R_0}{2}\right),$$

or

$$v_1 = 2v_0.$$

The final kinetic energy is then

$$T_1 = \frac{mv_1^2}{2} = \frac{m(2v_0)^2}{2} = 2mv_0^2.$$

    e. The reason why the pulling of the string should be gradual is that the radial velocity of the mass can be kept small so that the velocity of the mass can be considered tangential. This tangential velocity as a function of R can be calculated readily from the conservation of angular momentum.

1017

When a 5000 lb car driven at 60 mph on a level road is suddenly put into neutral gear (i.e. allowed to coast), the velocity decreases in the following manner:

$$V = \frac{60}{1 + \left(\frac{t}{60}\right)}\ \text{mph,}$$

where t is the time in sec. Find the horsepower required to drive this car at 30 mph on the same road.

Useful constants: $g = 22$ mph/sec, 1 H.P. = 550 ft.lb/sec, 60 mph = 88 ft/sec.

(Wisconsin)

Sol: Let $V_0 = 60$ mph, then

$$\frac{t}{60} = \frac{V_0}{V} - 1.$$

Hence

$$\frac{dV}{dt} = \frac{-V^2}{60V_0},$$

and the resistance acting on the car is $F = mV^2/(60V_0)$, where m is the mass of the car. The propulsive force must be equal to the resistance F' at the speed of $V' = 30$ mph in order to maintain this speed on the same road. It follows that the horsepower required is

$$P' = F'V' = \frac{mV'^3}{60V_0} = 37500 \, \frac{\text{mph}^2.\text{lb.}}{\text{s}}$$

$$= \frac{37500}{g} \frac{\text{mph}^2.\text{lb wt}}{\text{s}} = \frac{37500}{22} \text{mph.lb wt}$$

$$= \frac{37500}{22} \cdot \frac{88}{60} \frac{\text{ft.lb wt}}{\text{s}}$$

$$= 2500 \, \frac{\text{ft.lb wt}}{\text{s}} = 4.5 \text{ H.P.}$$

Note that pound weight (lb wt) is a unit of force and 1 lb wt $= g$ ft.lb/s^2. The horsepower is defined as 550 ft.lb wt/s.

1018

The bob (mass $= m$) of a conical pendulum of length l moves in a horizontal circle such that the string makes an angle θ with the vertical. Obtain an expression for the angular speed and show that it is independent of m. If the tension in the string is doubled, determine the change in the angular speed and θ and hence describe the subsequent motion.

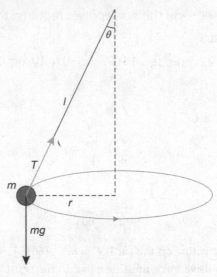

Fig. 1.8

Sol: Resolving the forces vertically and horizontally,

$$T \cos \theta = mg$$

$$T \sin \theta = m\omega^2 r = m\omega^2 l \sin \theta$$

Solving these two equations, we get $\omega = \left(\dfrac{g}{l \cos \theta} \right)^{\frac{1}{2}}$ (independent of m)

When the tension is doubled

$$2T \cos \theta' = mg$$

and

$$\cos \theta' = \frac{\cos \theta}{2}$$

The new angular velocity is $\omega' = \left(\dfrac{2g}{l \cos \theta} \right)^{\frac{1}{2}}$

If the tension is doubled, the bob rotates with an angular velocity that is $\sqrt{2}$ times greater. Since $\cos \theta'$ decreases, the angle increases and the bob rotates with a larger radius.

1019

A particle of mass m is subjected to two forces: a central force $\mathbf{f}_1$ and a frictional force $\mathbf{f}_2$, with

$$\mathbf{f}_1 = \frac{\mathbf{r}}{r} f(r),$$

$$\mathbf{f}_2 = -\lambda \mathbf{v} \qquad (\lambda > 0),$$

where **v** is the velocity of the particle. If the particle initially has angular momentum $\mathbf{J}_0$ about $r = 0$, find its angular momentum for all subsequent times.

(Wisconsin)

Sol: Write out the equations of motion of the particle in polar coordinates:

$$m(\ddot{r} - r\dot{\theta}^2) = f(r) - \lambda\dot{r},$$
$$m(2\dot{r}\dot{\theta} + r\ddot{\theta}) = -\lambda r\dot{\theta},$$

or

$$\frac{1}{r}\frac{d(mr^2\dot{\theta})}{dt} = -\lambda r\dot{\theta}.$$

Letting $J = mr^2\dot{\theta}$, we rewrite the last equation as follows:

$$\frac{dJ}{dt} = \frac{-\lambda J}{m}.$$

Integrating and making use of the initial angular momentum $\mathbf{J}_0$, we obtain

$$\mathbf{J} = \mathbf{J}_0 e^{-\frac{\lambda}{m}t}.$$

1020

a. A spherical object rotates with angular frequency ω. If the only force preventing centrifugal disintegration of the object is gravity, what is the minimum density the object must have? Use this to estimate the minimum density of the Crab pulsar which rotates 30 times per second. (This is a remnant of a supernova in 1054 A.D. which was extensively observed in China!)

b. If the mass of the pulsar is about 1 solar mass ($\sim 2 \times 10^{30}$ kg or $\sim 3 \times 10^5 M_{earth}$), what is the maximum possible radius of the pulsar?

c. In fact the density is closer to that of nuclear matter. What then is the radius?

(CUSPEA)

Sol:

a. Consider the limiting case that the Crab pulsar is just about to disintegrate. Then the centripetal force on a test body at the equator of the Crab pulsar is just smaller than the gravitational force:

$$\frac{mv^2}{R} = mR\omega^2 \leq \frac{GmM}{R^2},$$

or

$$\frac{M}{R^3} \geq \frac{\omega^2}{G},$$

where m and M are the masses of the test body and the Crab pulsar respectively, R is the radius of the pulsar, v is the speed of the test body, and G is the gravitational constant. Hence the minimum density of the pulsar is

$$\rho = \frac{M}{\frac{4}{3}\pi R^3} \geq \frac{3\omega^2}{4\pi G} = \frac{3(2\pi \times 30)^2}{4\pi \times 6.7 \times 10^{-11}} \sim 1.3 \times 10^{14} \text{ kg/m}^3.$$

b. As $\dfrac{3M}{4\pi R^3} \geq \rho_{\min}$,

$$R \leq \left(\frac{3M}{4\pi\rho_{\min}}\right)^{\frac{1}{3}} = \left(\frac{6 \times 10^{30}}{4\pi \times 1.3 \times 10^{14}}\right)^{\frac{1}{3}} = 1.5 \times 10^5 \text{ m} = 150 \text{ km}.$$

c. The nuclear density is given by

$$\rho_{\text{nuclear}} \approx \frac{m_p}{4\pi R_0^3/3},$$

Where m_p is the mass of a proton and is approximately equal to the mass m_{H} of a hydrogen atom. This can be estimated as follows:

$$m_p \approx m_{\text{H}} = \frac{2 \times 10^{-3}}{2 \times 6.02 \times 10^{23}} = 1.7 \times 10^{-27} \text{ kg}.$$

With

$$R_0 \approx 1.5 \times 10^{-15} \text{ m},$$

we obtain

$$\rho_{\text{nuclear}} \approx 1.2 \times 10^{17} \text{ kg/m}^3.$$

If $\rho = \rho_{\text{nuclear}}$, the pulsar would have a radius

$$R \approx \left(\frac{6 \times 10^{30}}{4\pi \times 1.2 \times 10^{17}}\right)^{\frac{1}{3}} \approx 17 \, \text{km.}$$

1021

Two weightless rings slide on a smooth circular loop of wire whose axis lies in a horizontal plane. A smooth string passes through the rings which carries weights at the two ends and at a point between the rings. If there is equilibrium when the rings are at points 30° distant from the highest point of the circle as shown in Fig. 1.9, find the relation between the three weights.

(UC, Berkeley)

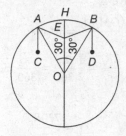

Fig. 1.9

Sol: Assume the string is also weightless. As no friction is involved, the tensions in the segments AC and AE of the string must be the same. Let the magnitude be T. For the ring A to be at rest on the smooth loop, the resultant force on it must be along AO, O being the center of the loop; otherwise there would be a component tangential to the loop. Hence

$$\angle OAE = \angle OAC = \angle AOE = 30°.$$

The same argument applies to the segments BD and BE. Then by symmetry the point E at which the string carries the third weight must be on the radius HO, H being the highest point of the loop, and the tensions in the segments BD and BE are also T.

Consider the point E. Each of the three forces acting on it, which are in equilibrium, is at an angle of 120° to the adjacent one. As two of the forces have magnitude T, the third force must also have magnitude T. Therefore the three weights carried by the string are equal.

1022

Calculate the ratio of the mean densities of the earth and the sun from the following approximate data:

θ = angular diameter of the sun seen from the earth = $\dfrac{1°}{2}$.

l = length of 1° of latitude on the earth's surface = 100 km.

t = one year = 3×10^7 s.

g = 10 ms^{-2}.

(*UC, Berkeley*)

Sol: Let r be the distance between the sun and the earth, M_e and M_s be the masses and R_e and R_s be the radii of the earth and the sun respectively, and G be the gravitational constant. We then have

$$\frac{GM_eM_s}{r^2} = M_er\omega^2,$$

$$\frac{2R_s}{r} = \frac{1}{2}\frac{2\pi}{360} = \frac{\pi}{360} \text{ rad,}$$

i.e.

$$r = \frac{720R_s}{\pi}.$$

The above gives

$$\frac{GM_s}{(720R_s/\pi)^3} = \omega^2,$$

or

$$\frac{GM_s}{R_s^3} = \left(\frac{720}{\pi}\right)^3\left(\frac{2\pi}{3 \times 10^7}\right)^2.$$

For a mass m on the earth's surface,

$$\frac{GmM_e}{R_e^2} = mg,$$

giving

$$\frac{GM_e}{R_e^3} = \frac{g}{R_e} = \frac{g}{\left(\frac{360 \times 100}{2\pi}\right)} = \frac{g\pi}{18 \times 10^3}.$$

Hence

$$\frac{\rho_e}{\rho_s} = \frac{g\pi}{18 \times 10^3}\left(\frac{720}{\pi}\right)^{-3}\left(\frac{2\pi}{3 \times 10^7}\right)^{-2} = 3.31.$$

1023

A parachutist jumps at an altitude of 3000 meters. Before the parachute opens she reaches a terminal speed of 30 m/sec.

  a. Assuming that air resistance is proportional to speed, about how long does it take her to reach this speed?

  b. How far has she traveled in reaching this speed?

      After her parachute opens, her speed is slowed to 3 m/sec. As she hits the ground, she flexes her knees to absorb the shock.

  c. How far must she bend her knees in order to experience a deceleration no greater then $10g$? Assume that her knees are like a spring with a resisting force proportional to displacement.

  d. Is the assumption that air resistance is proportional to speed a reasonable one? Show that this is or is not the case using qualitative arguments.

(UC, Berkeley)

Sol:

  a. Choose the downward direction as the positive direction of the x-axis. Integrating the differential equation of motion

$$\frac{dv}{dt} = g - \alpha v,$$

where α is a constant, we obtain

$$v = \frac{g}{\alpha}(1 - e^{-\alpha t}).$$

This solution shows that v approaches its maximum, the terminal speed g/α, when $t \to \infty$.

  b. Integrating the above equation, we obtain

$$x = \frac{gt}{\alpha} + \frac{ge^{-\alpha t}}{\alpha^2}.$$

Thus $x \to \infty$ as $t \to \infty$. This means that when the parachutist reaches the terminal speed she has covered an infinite distance.

c.   As her speed is only 3 m/s, we may neglect any air resistance after she hits the ground with this speed. Conservation of mechanical energy gives

$$\frac{k\xi^2}{2} = mg\xi + \frac{mv^2}{2},$$

where ξ is the distance of knee bending and v is the speed with which she hits the ground, considering the knee as a spring of constant k. Taking the deceleration $-10g$ as the maximum allowed, we have

$$mg - k\xi = -10mg,$$

i.e.

$$\xi = 11mg/k.$$

The energy equation then gives

$$\xi = \frac{v^2}{9g} = \frac{3^2}{9 \times 9.8} = 0.102 \text{ m}.$$

d.   We have seen that if the air resistance is proportional to speed, the time taken to reach the terminal speed is ∞ and the distance traveled is also ∞. However, the actual traveling distance is no more than 3000 m and the traveling time is finite before she reaches the terminal speed of 30 m/s. Hence the assumption that air resistance is proportional to speed is not a reasonable one.

1024

A satellite in stationary orbit above a point on the equator is intended to send energy to ground stations by a coherent microwave beam of wavelengh one meter from a one-km mirror.

a.   What is the height of such a stationary orbit?

b.   Estimate the required size of a ground receptor station.

(*Columbia*)

Sol:

a.   The revolving angular velocity ω of the synchronous satellite is equal to the spin angular velocity of the earth and is given by

$$m(R + h)\omega^2 = \frac{GMm}{(R + h)^2}.$$

Hence the height of the stationary orbit is

$$h = \left(\frac{GM}{\omega^2}\right)^{\frac{1}{3}} - R = 3.59 \times 10^4 \, \text{km},$$

using $G = 6.67 \times 10^{-11} \, \text{Nm}^2\text{kg}^{-2}$, $M = 5.98 \times 10^{24} \, \text{kg}$, $R = 6.37 \times 10^4 \, \text{km}$.

b. Due to diffraction, the linear size of the required receptor is about

$$\frac{\lambda h}{D} = 1 \times \left(\frac{3.59 \times 10^4}{1}\right) = 3.59 \times 10^4 \, \text{m}.$$

1025

Two blocks of mass m_1 and m_2 are connected through a pulley as shown. The coefficients of static friction for both surfaces are μ_{s1} and μ_{s2}, respectively. The angles θ_1 and θ_2 are greater than α_1 and α_2, the angles at which the masses m_1 and m_2 just begin to slide down the incline if unattached. Determine the condition for $\frac{m_2}{m_1}$ which will make the mass m_2 pull m_1 in terms of the angles $\theta_1, \theta_2, \alpha_1,$ and α_2.

(Columbia)

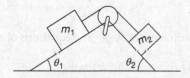

Fig. 1.10

Sol: Let α_1 and α_2, respectively, be the angles at which the m_1 and m_2 begin to slide down their respective planes if not fettered by the string.

The force equations for the two masses, which would make them move down their respective planes, are

$$F_1 = m_1 g \sin\theta_1 - \mu_{s1} m_1 g \cos\theta_1 - T$$

and

$$F_2 = m_2 g \sin\theta_2 - \mu_{s2} m_2 g \cos\theta_2 - T$$

m_2 will pull m_1 if $F_2 > F_1$. Using $\mu_{s1} = \tan\alpha_1$ and $\mu_{s2} = \tan\alpha_2$

$$\frac{m_2}{\cos\alpha_2}(\sin\theta_2\cos\alpha_2 - \sin\alpha_2\cos\theta_2) > \frac{m_1}{\cos\alpha_1}(\sin\theta_1\cos\alpha_1 - \sin\alpha_1\cos\theta_1)$$

$$\frac{m_2}{m_1} > \frac{\cos\alpha_2\sin(\theta_1 - \alpha_1)}{\cos\alpha_1\sin(\theta_2 - \alpha_2)}$$

1026

A particle of mass m is constrained to move on the frictionless inner surface of a cone of half-angle α, as shown in Fig. 1.11

a. Find the restrictions on the initial conditions such that the particle moves in a circular orbit about the vertical axis.

b. Determine whether this kind of orbit is stable.

(*Princeton*)

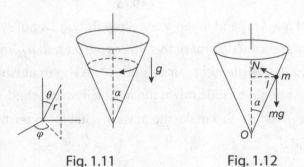

Fig. 1.11   Fig. 1.12

Sol:

a. In spherical coordinates (r, θ, φ), the equations of motion of the particle are

$$m(\ddot{r}\,\dot{\theta}^2 - r\dot{\varphi}^2 \sin^2\theta) = F_r,$$
$$m(r\ddot{\theta} + 2\dot{r}\dot{\theta} - r\dot{\varphi}^2 \sin\theta\cos\theta) = F_\theta,$$
$$m(r\ddot{\varphi}\sin\theta + 2\dot{r}\dot{\varphi}\sin\theta + 2r\dot{\theta}\dot{\varphi}\cos\theta) = F_\varphi.$$

As the particle is constrained to move on the inner surface of the cone,

$$\theta = \text{constant} = \alpha.$$

Then $\dot{\theta} = 0$, $F_r = -mg\cos\alpha$, and Eq. (1) becomes

$$m(\ddot{l} - l\dot{\varphi}^2 \sin^2\alpha) = -mg\cos\alpha, \tag{2}$$

where l is its distance from the vertex O (see Fig. 1.12). For motion in a circular orbit about the vertical axis, $\dot{l} = \ddot{l} = 0$. With $l = l_0$, Eq. (2) becomes

$$l_0\dot{\varphi}^2 \sin^2\alpha = g\cos\alpha. \tag{3}$$

The right-hand side of Eq. (3) is constant so that $\dot{\varphi} = $ constant $= \dot{\varphi}_0$, say. The particle has velocity v_0 tangential to the orbit given by $v_0 = l_0\dot{\varphi}_0 \sin\alpha$. Equation (3) then gives

$$v_0^2 = gl_0\cos\alpha,$$

which is the initial condition that must be satisfied by v_0 and l_0.

b. Suppose there is a small perturbation acting on the particle such that l_0 becomes $l_0 + \Delta l$, $\dot{\varphi}_0$ becomes $\dot{\varphi}_0 + \Delta\dot{\varphi}$. Equation (2) is now

$$\frac{d^2(l_0 + \Delta l)}{dt^2} - (l_0 + \Delta l)(\dot{\varphi}_0 + \Delta\dot{\varphi})^2 \sin^2\alpha = -g\cos\alpha,$$

or

$$\Delta\ddot{l} - 2l_0\,\dot{\varphi}_0\,\Delta\dot{\varphi}\sin^2\alpha - \Delta l\,\dot{\varphi}_0^2\sin^2\alpha = l_0\,\dot{\varphi}^2\sin^2\alpha - g\cos\alpha,$$

where $\Delta\ddot{l}$ is shorthand for $d^2(\Delta l)/dt^2$, by neglecting terms of orders higher than the first order quantities Δl and $\Delta\dot{\varphi}$. As the right-hand side of this equation vanishes on account of Eq. (3), we have

$$\Delta\ddot{l} - 2l_0\,\dot{\varphi}_0\,\Delta\dot{\varphi}\sin^2\alpha - \Delta l\,\dot{\varphi}_0^2\sin^2\alpha = 0. \tag{4}$$

There is no force tangential to the orbit acting on the particle, so there is no torque about the vertical axis and the angular momentum of the particle about the axis is constant:

$$mlv\sin\alpha = ml^2\,\dot{\varphi}\sin^2\alpha = \text{constant} = k, \text{ say,}$$

or

$$l^2\dot{\varphi} = \frac{k}{m\sin^2\alpha}. \tag{5}$$

Substituting $l = l_0 + \Delta l$, $\dot{\varphi} = \dot{\varphi}_0 + \Delta\dot{\varphi}$ into Eq. (5) and neglecting terms of the second order or higher, we have

$$l_0\Delta\dot{\varphi} + 2\Delta l\,\dot{\varphi}_0 = 0. \tag{6}$$

Eliminating $\Delta\dot{\varphi}$ from Eqs. (4) and (6), we obtain

$$\Delta\ddot{l} + (3\dot{\varphi}\theta_0^2\sin^2\alpha)\Delta l = 0.$$

As the factor in brackets is real and positive, this is the equation of a "simple harmonic oscillator." Hence the orbit is stable.

1027

Three point particles with masses m_1, m_2 and m_3 interact with each other through the gravitational force.

   a. Write down the equations of motion.

   b. The system can rotate in its plane with constant and equal distances between all pairs of masses. Determine the angular frequency of the rotation when the masses are separated by a distance d.

   c. For $m_1 \gg m_3$ and $m_2 \gg m_3$, determine the stability condition for motion of the mass m_3 about the stationary position. Consider only motion in the orbital plane.

(*MIT*)

Sol: Take the center of mass C of the system as the origin of coordinates and let the position vectors of m_1, m_2, m_3 be $\mathbf{r}_1$, $\mathbf{r}_2$, $\mathbf{r}_3$ respectively as shown in Fig. 1.13. Denote

$$\mathbf{r}_{ij} = \mathbf{r}_i - \mathbf{r}_j \quad (i, j = 1, 2, 3).$$

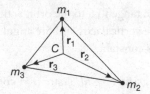

Fig. 1.13

   a. The motion of the ith particle is given by

$$m_i \ddot{\mathbf{r}}_i = -\sum_{j \neq i}^{3} \frac{G m_i m_j}{r_{ij}^3} \mathbf{r}_{ij},$$

or

$$\ddot{\mathbf{r}}_i = -\sum_{j \neq i}^{3} \frac{G m_j}{r_{ij}^3} \mathbf{r}_{ij} \quad (i = 1, 2, 3). \tag{1}$$

   Note that the minus sign is to indicate that the forces are attractive.

   b. With the given condition $r_{ij} = d$, Eq. (1) is rewritten as

$$\ddot{\mathbf{r}}_i = \frac{G}{d^3} \sum_{j \neq i}^{3} m_j (\mathbf{r}_j - \mathbf{r}_i)$$

$$= \frac{G}{d^3} \left[-\sum_{j \neq i}^{3} m_j \mathbf{r}_i + \sum_{j \neq i}^{3} m_j \mathbf{r}_j \right]$$

$$= \frac{G}{d^3} \left[-\sum_{j \neq i}^{3} m_j \mathbf{r}_i - m_i \mathbf{r}_i + \sum_{j=1}^{3} m_j \mathbf{r}_j \right]$$

$$= \frac{G}{d^3} \left[-\mathbf{r}_i \sum_{j=1}^{3} m_j + \sum_{j=1}^{3} m_j \mathbf{r}_j \right]$$

$$= -\frac{GM}{d^3} \mathbf{r}_i,$$

where $M = m_1 + m_2 + m_3$. Note that the choice of the center of mass as origin makes $\sum m_j \mathbf{r}_j$ vanish. Thus the force on each particle points towards the center of mass of the system and is a harmonic force. With d constant, the system rotates about C with angular frequency

$$\omega = \sqrt{\frac{GM}{d^3}}.$$

c. For $m_3 \ll m_1$ and $m_3 \ll m_2$, the equation of motion of either of the masses m_1 and m_2 can be written as

$$\ddot{\mathbf{r}}_i = -\frac{G(m_1 + m_2)}{d^3} \mathbf{r}_i$$

$$= -\frac{G(m_1 + m_2)}{\mathbf{r}_{12}^3} \mathbf{r}_i, \quad i = 1, 2.$$

With the distance between m_1 and m_2 constant, the system rotates about its center of mass with a constant angular frequency

$$w = \sqrt{\frac{G(m_1 + m_2)}{\mathbf{r}_{12}^3}}.$$

Use a rotating coordinate frame with origin at the center of mass of the system and angular frequency of rotation ω and let the quantities $\dot{\mathbf{r}}$, $\ddot{\mathbf{r}}$ refer to this rotating frame. Considering the motion of particle m_3 in the laboratory frame, we have

$$m_3 (\ddot{\mathbf{r}}_3 - \omega^2 \mathbf{r}_3) = -\frac{Gm_3 m_1}{\mathbf{r}_{31}^3} \mathbf{r}_{31} - \frac{Gm_3 m_2}{\mathbf{r}_{32}^3} \mathbf{r}_{32} - 2m_3 \omega \times \dot{\mathbf{r}}_3,$$

or

$$\ddot{\mathbf{r}}_3 = -\frac{Gm_1}{\mathbf{r}_{31}^3}\mathbf{r}_{31} - \frac{Gm_2}{\mathbf{r}_{32}^3}\mathbf{r}_{32} + \omega^2\mathbf{r}_3 - 2\omega \times \dot{\mathbf{r}}_3.$$

If m_3 is stationary, $\ddot{\mathbf{r}}_3 = \dot{\mathbf{r}}_3 = 0$ and the above becomes

$$\frac{Gm_1}{\mathbf{r}_{31}^3}(\mathbf{r}_1 - \mathbf{r}_3) + \frac{Gm_2}{\mathbf{r}_{32}^2}(\mathbf{r}_2 - \mathbf{r}_3) + \omega^2\mathbf{r}_3 = 0.$$

With $m_1, m_2 \gg m_3$, $\sum m_j r_j = 0$ gives $m_1 r_1 \approx -m_2 r_2$ and the above becomes

$$-G\left(\frac{m_1}{\mathbf{r}_{31}^3} + \frac{m_2}{\mathbf{r}_{32}^3}\right)\mathbf{r}_3 + G\left(\frac{m_1}{\mathbf{r}_{31}^3} - \frac{m_1}{\mathbf{r}_{32}^3}\right)\mathbf{r}_1 + \omega^2\mathbf{r}_3 = 0.$$

This relation shows that $\mathbf{r}_3$ is parallel to $\mathbf{r}_1$ and thus the stationary position of m_3 lies on the line joining m_1 and m_2. At this position, the attractions of m_1 and m_2 are balanced.

Consider now a small displacement being applied to m_3 at this stationary position. If the displacement is along the line joining m_1 and m_2, say toward m_1, the attraction by m_1 is enhanced and that by m_2 is reduced. Then m_3 will continue to move toward m_1 and the equilibrium is unstable. On the other hand, if the displacement is normal to the line joining m_1 and m_2, both the attractions by m_1 and m_2 will have a component toward the stationary position and will restore m_3 to this position. Thus the equilibrium is stable. Therefore the equilibrium is stable against a transverse perturbation but unstable against a longitudinal one.

1028

A wheel of radius R is connected with a belt to another wheel of radius r, $R = 3r$.

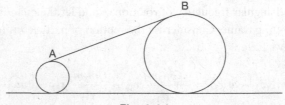

Fig. 1.14

The system starts from rest and moves with a linear acceleration a m/s². After time t, determine the velocity of the system and the angular velocity of both wheels in rev/s. Consider two particles placed at A and B on top of each wheel at the start of the motion. The particles move with the wheel without falling off. If $r = 1$ m, $a = 2$ m/s² and the system starts from rest, determine the displacement of the two particles after 10 s. (Assume the particles A and B stay on the wheel.).

Sol:

$$v = at$$

$$\omega_R = \frac{v}{R} = \frac{at}{R}\frac{\text{rad}}{\text{s}} = \frac{at}{2\pi R}\text{ rev/s} \quad \text{or} \quad \frac{at}{6\pi r}\text{ rev/s}$$

$$\omega_r = \frac{v}{r} = \frac{at}{r}\frac{\text{rad}}{\text{s}} = \frac{at}{2\pi r}\text{ rev/s}$$

Linear distance travelled in 10 s: $s = \frac{1}{2}at^2 = \frac{1}{2} \times 2\text{ m/s}^2 \times (10\text{ s})^2 = 100\text{ m}$

Angular distance covered by A on the smaller wheel: $\theta = \frac{s}{r} = \frac{100\text{ m}}{1\text{ m}} = 100\text{ rad}$

Total number of rotation $= \frac{100}{2\pi} = 15.9$.

The final angle that the position vector of A has covered is $0.9 \times 360° = 324°$. It has a magnitude of $2r\sin18° = 0.618$ m and makes an angle of 162° with its original direction measured clockwise.

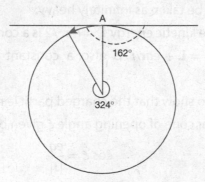

Fig. 1.15

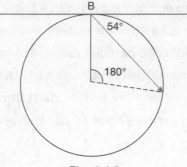

Fig. 1.16

For the larger wheel,

$$\theta = \frac{s}{R} = \frac{s}{3r} = \frac{100}{3} = 33.3 \text{ rad}$$

Total number of rotation $= \dfrac{33.3}{2\pi} = 5.3$.

The final angle that the position vector of B has covered is $0.3 \times 360° = 108°$. It has a magnitude of $2R \cos 36° = 2 \times 3r \times \cos 36° = 4.854$ m and makes an angle of $54°$ wh its original direction measured clockwise.

1029

Point charge in the field of a magnetic monopole.

The equation of motion of a point electric charge e, of mass m, in the field of a magnetic monopole of strength g at the origin is

$$m\ddot{\mathbf{r}} = -ge\frac{\dot{\mathbf{r}} \times \mathbf{r}}{\mathbf{r}^3}.$$

The monopole may be taken as infinitely heavy.

a. Show that the kinetic energy $T = m\dot{\mathbf{r}}^2/2$ is a constant of the motion.

b. Show that $\mathbf{J} = \mathbf{L} + eg\mathbf{r}/r$ is also a constant of the motion, where $\mathbf{L} = m\mathbf{r} \times \dot{\mathbf{r}}$.

c. Use part (b) to show that the charged particle moves on the surface of a right circular cone of opening angle ξ given by

$$\cos \xi = \frac{eg}{|\mathbf{J}|},$$

with **J** as its symmetry axis (see Fig. 1.17). [Hint: Consider **r · J**.]

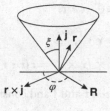

Fig. 1.17

Define a new variable **R** by

$$\mathbf{R} = \frac{1}{\sin \xi} \hat{\mathbf{J}} \times (\mathbf{r} \times \hat{\mathbf{J}}) = \frac{1}{\sin \xi}[\mathbf{r} - \hat{\mathbf{J}}(\mathbf{r} \cdot \hat{\mathbf{J}})],$$

where $\hat{\mathbf{J}} = \mathbf{J}/|\mathbf{J}|$. **R** lies in the plane perpendicular to **J**, but with $|\mathbf{R}| \equiv R = |\mathbf{r}|$ so that R may be obtained by rotating **r** as shown in the figure. You may use the fact that $m\mathbf{R} \times \dot{\mathbf{R}} = \mathbf{J}$.

d. Find the equation of motion for **R**.

e. Solve the equation of motion part (d) by finding an effective potential $V_{\text{eff}}(R)$, and describe all possible motions in **R**.

(MIT)

Sol:

a. $\dfrac{dT}{dt} = \dfrac{d}{dt}\left(\dfrac{1}{2}m\dot{r}^2\right) = m\dot{\mathbf{r}} \cdot \ddot{\mathbf{r}} = \dot{\mathbf{r}} \cdot \left(-ge\dfrac{\dot{\mathbf{r}} \times \mathbf{r}}{r^3}\right) = 0.$

Hence T is a constant of the motion.

b. $\dot{\mathbf{j}} = \dfrac{d}{dt}\left(m\mathbf{r} \times \dot{\mathbf{r}} + \dfrac{eg\mathbf{r}}{r}\right)$

$\qquad = m\mathbf{r} \times \ddot{\mathbf{r}} + m\dot{\mathbf{r}} \times \dot{\mathbf{r}} + \dfrac{eg\dot{\mathbf{r}}}{r} + \left(-\dfrac{eg\mathbf{r}}{r^2}\right)\dfrac{\mathbf{r} \cdot \dot{\mathbf{r}}}{r}$

$\qquad = m\mathbf{r} \times \ddot{\mathbf{r}} + \left[\dfrac{eg\dot{\mathbf{r}}}{r} - \dfrac{eg\mathbf{r}(\mathbf{r} \cdot \dot{\mathbf{r}})}{r^3}\right]$

$\qquad = -ge\dfrac{\mathbf{r} \times (\dot{\mathbf{r}} \times \mathbf{r})}{r^3} + ge\dfrac{\mathbf{r} \times (\dot{\mathbf{r}} \times \mathbf{r})}{r^3} = 0.$

Hence **J** is a constant of the motion. Note that in the above we have used

$$\dot{r} \equiv \frac{d}{dt}(\mathbf{r} \cdot \mathbf{r})^{\frac{1}{2}}$$

$$= \dot{\mathbf{r}} \cdot \frac{\mathbf{r}}{r},$$

$$\mathbf{r} \times (\dot{\mathbf{r}} \times \mathbf{r}) = \dot{\mathbf{r}}(\mathbf{r} \cdot \mathbf{r}) - \mathbf{r}(\mathbf{r} \cdot \dot{\mathbf{r}}).$$

c. Let ξ be the angle between **r** and **J** and consider

$$\mathbf{r} \cdot \mathbf{J} = r|\mathbf{J}| \cos \xi = \mathbf{r} \cdot \left(m\mathbf{r} \times \dot{\mathbf{r}} + \frac{eg\mathbf{r}}{r} \right) = egr.$$

As

$$\cos \xi = \frac{eg}{|\mathbf{J}|} = \text{constant},$$

the charged particle moves on the surface of a right circular cone of opening angle ξ.

d. As **J** and ξ are constants of the motion, we have, using

$$\mathbf{r} \times \dot{\mathbf{r}} = \frac{\mathbf{L}}{m}, \quad \mathbf{L} = \mathbf{J} - eg\frac{\mathbf{r}}{r}, \quad m\ddot{\mathbf{r}} = -ge\frac{\dot{\mathbf{r}} \times \mathbf{r}}{r^3},$$

$$m\ddot{\mathbf{R}} = \frac{m}{\sin \xi}\hat{\mathbf{J}} \times (\ddot{\mathbf{r}} \times \hat{\mathbf{J}})$$

$$= \frac{1}{\sin \xi}\hat{\mathbf{J}} \times \left[\frac{ge}{mr^3}\left(\mathbf{J} - eg\frac{\mathbf{r}}{r} \right) \times \hat{\mathbf{J}} \right]$$

$$= \frac{1}{\sin \xi}\hat{\mathbf{J}} \times \left[-\frac{g^2 e^2}{mr^4}\mathbf{r} \times \hat{\mathbf{J}} \right]$$

$$= -\frac{e^2 g^2}{mr^4}\mathbf{R}.$$

This is the equation of motion for **R**.

e. Let ϕ be the angle between **R** and a fixed axis in the plane of **R** and $\mathbf{r} \times \hat{\mathbf{J}}$. The above equation can be written as

$$m(\ddot{R} - R\dot{\varphi}^2) = -\frac{e^2 g^2}{mR^3}, \tag{1}$$

$$m(R\ddot{\varphi} + 2\dot{\varphi}\dot{R}) = 0. \tag{2}$$

Equation (2) can be written as

$$m(R^2\ddot\varphi + 2R\dot R\dot\varphi) = \frac{d}{dt}(mR^2\dot\varphi) = 0.$$

Hence

$$mR^2\dot\varphi = \text{constant}.$$

As

$$\begin{aligned}\mathbf{R} \times \dot{\mathbf{R}} &= R\mathbf{i}_R \times (\dot R\mathbf{i}_R + R\dot\varphi\mathbf{i}_\varphi) \\ &= R^2\dot\varphi\mathbf{i}_R \times \mathbf{i}_\varphi,\end{aligned}$$

$$mR^2\dot\varphi = |m\mathbf{R} \times \dot{\mathbf{R}}| = J.$$

Equation (1) can then be written as

$$m\ddot R = -\frac{e^2g^2}{mR^3} + \frac{J^2}{mR^3} = -\frac{d}{dR}V_{\text{eff}}(R), \tag{3}$$

with

$$\begin{aligned}V_{\text{eff}}(R) &= \frac{1}{2mR^2}(J^2 - e^2g^2) \\ &= \frac{e^2g^2}{2mR^2}\left(\frac{1}{\cos^2\xi} - 1\right) \\ &= \frac{e^2g^2}{2mR^2}\tan^2\xi = \frac{K}{R^2},\end{aligned}$$

where $K = e^2g^2\tan^2\xi/2m$. Using $\ddot R = \dot R d\dot R/dR = d\dot R^2/2dR$, Eq. (3) can be integrated to give

$$\frac{1}{2}m\dot R^2 + \frac{K}{R^2} = E,$$

where E is a constant. We then have

$$\frac{dR}{d\varphi} = \frac{\dot R}{\dot\varphi} = \pm\frac{mR^2}{J}\sqrt{\frac{2}{m}\left(E - \frac{K}{R^2}\right)}.$$

Integrating, we obtain

$$\begin{aligned}\pm\left(\tan^{-1}\sqrt{\frac{E}{K}R^2 - 1} - \tan^{-1}\sqrt{\frac{E}{K}R_0^2 - 1}\right) &= \frac{\sqrt{2mK}}{J}(\varphi - \varphi_0) \\ &= (\varphi - \varphi_0)\sin\xi,\end{aligned}$$

which gives the trajectory of the tip of **R**. Note that if $J \gg eg$ the motion is unbounded whatever the initial state, and if $J < eg$ the motion is bounded when $E < 0$ and unbounded when $E \geq 0$.

1030

Paris and London are connected by a straight subway tunnel (see Fig. 1.18). A train travels between the two cities powered only by the gravitational force of the earth. Calculate the maximum speed of the train and the time taken to travel from London to Paris. The distance between the two cities is 300 km and the radius of the earth is 6400 km. Neglect friction.

(MIT)

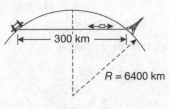

Fig. 1.18

Sol: Define x, h, r as in Fig. 1.19 and assume the earth to be a stationary homogeneous sphere of radius R. Taking the surface of the earth as reference level, the gravitational potential energy of the train at x is

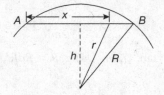

Fig. 1.19

$$V = \int_R^r \frac{GmM}{R^3} r\, dr = \frac{GmM}{2R^3}(r^2 - R^2),$$

where m, M are the masses of the train and the earth respectively. Conservation of mechanical energy gives, as the train starts from rest at the earth's surface,

$$\frac{mv^2}{2} + \frac{GmM(r^2 - R^2)}{2R^3} = 0,$$

or

$$v^2 = \frac{g(R^2 - r^2)}{R},$$

where $g = GM/R^2$ is the acceleration of gravity at the earth's surface. As

$$r^2 = h^2 + (150 - x)^2 = (R^2 - 150^2) + (150 - x)^2 = R^2 - 300x + x^2,$$

$$v^2 = \frac{gx(300 - x)}{R}.$$

v is maximum when $x = 150$ km:

$$v_{\max} = \sqrt{\frac{9.8 \times 150 \times 150 \times 1000}{6400}} = 185.6 \text{ m/s}.$$

The time from London to Paris is

$$T = \int_0^{300} \frac{dx}{v} = \int_0^{300} \sqrt{\frac{R}{g}} \frac{dx}{\sqrt{x(300 - x)}}$$

$$= \int_0^1 \sqrt{\frac{R}{g}} \frac{dt}{\sqrt{t(1 - t)}} = \pi \sqrt{\frac{R}{g}} = 42.3 \text{ min}.$$

1031

Three fixed point sources are equally spaced about the circumference of a circle of diameter a centered at the origin (Fig. 1.20). The force exerted by each source on a point mass of mass m is attractive and given by $\mathbf{F} = -k\mathbf{R}$, where $\mathbf{R}$ is a vector drawn from the source to the point mass. The point mass is placed in the force field at time $t = 0$ with initial conditions $\mathbf{r} = \mathbf{r}_0$, $\dot{\mathbf{r}} = \mathbf{v}_0$.

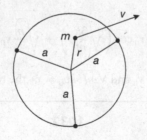

Fig. 1.20

a. Define suitable coordinates and write an expression for the force acting on the mass at any time.

b. Use Newton's second law and solve the equation of motion for the initial conditions given above, namely, find $\mathbf{r}(t)$ in terms of $\mathbf{r}_0$, $\mathbf{v}_0$ and the parameters of the system.

c. Under what conditions, if any, are circular orbits a solution?

(MIT)

Sol:

a. Let $\mathbf{r}_1$, $\mathbf{r}_2$, $\mathbf{r}_3$ be the position vectors of the three fixed point sources. As they are equally spaced on a circle, we have

$$\mathbf{r}_1 + \mathbf{r}_2 + \mathbf{r}_3 = 0.$$

The force acting on the particle m is

$$\mathbf{F} = -k(\mathbf{r} - \mathbf{r}_1) - k(\mathbf{r} - \mathbf{r}_2) - k(\mathbf{r} - \mathbf{r}_3) = -3k\mathbf{r}.$$

b. The equation of the motion of the point mass is

$$m\ddot{\mathbf{r}} + 3k\mathbf{r} = 0,$$

with the general solution

$$\mathbf{r}(t) = \mathbf{a} \cos\left(\sqrt{\frac{3k}{m}}t\right) + \mathbf{b} \sin\left(\sqrt{\frac{3k}{m}}t\right),$$

$\mathbf{a}, \mathbf{b}$ being constant vectors.

Using the initial conditions $\mathbf{r}(0) = \mathbf{r}_0$, $\dot{\mathbf{r}}(0) = \mathbf{v}_0$, we find

$$\mathbf{a} = \mathbf{r}_0, \quad \mathbf{b} = \sqrt{\frac{m}{3k}}\mathbf{v}_0,$$

and hence

$$\mathbf{r}(t) = \mathbf{r}_0 \cos\left(\sqrt{\frac{3k}{m}}t\right) + \sqrt{\frac{m}{3k}}\mathbf{v}_0 \sin\left(\sqrt{\frac{3k}{m}}t\right).$$

c. It is seen that if $\mathbf{r}_0 \perp \mathbf{v}_0$ and $\sqrt{m/3k}v_0 = r_0$, the trajectory is a circle.

1032

A phonograph turntable in the xy plane revolves at constant angular velocity ω around the origin. A small body sliding on the turntable has location $\mathbf{x}(t) = (x(t), y(t), 0)$. Here x and y are measured in an inertial frame, the lab frame. There are two forces in the lab frame: an elastic force of magnitude $k|\mathbf{x}|$

towards the origin, and a frictional force $-c(\dot{\mathbf{x}} - \mathbf{v})$, where c is a constant and $\mathbf{v}$ is the velocity of the turntable at the body's location.

a. If the body is observed to stay at a fixed off-center point on the turntable (i.e. it is at rest with respect to the turntable), how big is k?

b. Assume k has the value you found in (a). Solve for $\mathbf{v}(t) = \dot{\mathbf{x}}\,(t)$ with general initial conditions.

c. In (b), find $\mathbf{x}(t)$. Describe $\mathbf{x}(t)$ in words and/or with a rough sketch.

(UC, Berkeley)

Sol:

a. The body has angular velocity ω around the origin so that $m\omega^2|\mathbf{x}| = k|\mathbf{x}|$, giving $k = m\omega^2$.

b. In the lab frame the equation of motion for the small body is

$$m\ddot{\mathbf{x}} = -k\mathbf{x} - c(\dot{\mathbf{x}} - \mathbf{v})$$
$$= -m\omega^2\mathbf{x} - c(\dot{\mathbf{x}} - \omega \times \mathbf{x}).$$

Let $x, y, \dot{x}, \dot{y}, \ddot{x}, \ddot{y}$ be the coordinates, velocity and acceleration components in the rotating frame attached to the turntable. In the lab frame we have

$$\dot{\mathbf{x}} = (\dot{x} - y\omega)\mathbf{i} + (\dot{y} + x\omega)\mathbf{j},$$
$$\ddot{\mathbf{x}} = (\ddot{x} - 2\dot{y}\omega - x\omega^2)\mathbf{i} + (\ddot{y} + 2\dot{x}\omega - y\omega^2)\mathbf{j},$$
$$-k\mathbf{x} = -kx\mathbf{i} - ky\mathbf{j},$$
$$-c(\dot{\mathbf{x}} - \omega \times x) = -c\dot{x}\mathbf{i} - x\dot{y}\mathbf{j}.$$

Note that in the above we have used $\omega \times \mathbf{i} = \omega\mathbf{j}$, $\omega \times \mathbf{j} = -\omega\mathbf{i}$. The equation of motion in the lab frame is then written as

$$m(\ddot{x} - 2\dot{y}\omega - x\omega^2) = -kx - c\dot{x}, \qquad (1)$$
$$m(\ddot{y} - 2\dot{x}\omega - y\omega^2) = -ky - c\dot{y}. \qquad (2)$$

Multiplying Eq. (2) by $i \equiv \sqrt{-1}$, adding it to Eq. (1) and setting $z = x + iy$, we obtain

$$m\ddot{z} + (2m\omega i + c)\dot{z} = 0.$$

Integrating once we find

$$\dot{z} = \dot{z}_0 e^{-ct/m} e^{-i2\omega t}, \qquad (3)$$

namely,

$$\dot{x} = [\dot{x}_0 \cos(2\omega t) + \dot{y}_0 \sin(2\omega t)]e^{-ct/m}, \qquad (4)$$

$$\dot{y} = [-\dot{x}_0 \sin(2\omega t) + \dot{y}_0 \cos(2\omega t)]e^{-ct/m}. \tag{5}$$

By directly integrating Eqs. (4) and (5) or by integrating Eq. (3) and then using $z = x + iy$, we obtain

$$x = x_0 + \frac{m(c\dot{x}_0 + 2m\omega\dot{y}_0)}{c^2 + 4m^2\omega^2}$$

$$- \left[\frac{m(c\dot{x}_0 + 2m\omega\dot{y}_0)}{c^2 + 4m^2\omega^2} \cos(2\omega t) - \frac{m(2m\omega\dot{x}_0 - c\dot{y}_0)}{c^2 + 4m^2\omega^2} \sin(2\omega t) \right] e^{-ct/m}, \tag{6}$$

$$y = y_0 - \frac{m(2m\omega\dot{x}_0 - c\dot{y}_0)}{c^2 + 4m^2\omega^2}$$

$$+ \left[\frac{m(2m\omega\dot{x}_0 - c\dot{y}_0)}{c^2 + 4m^2\omega^2} \cos(2\omega t) + \frac{m(c\dot{x}_0 + 2m\omega\dot{y}_0)}{c^2 + 4m^2\omega^2} \sin(2\omega t) \right] e^{-ct/m}. \tag{7}$$

In the above, $\dot{x}_0$, $\dot{y}_0$ are the components of the velocity of the small body at $t = 0$ in the rotating frame.

c. Equations (6) and (7) imply that, for the body on the turntable, even if x, y may sometimes increase at first because of certain initial conditions, with the passage of time its velocity in the turntable frame will decrease and the body eventually stops at a fixed point on the turntable, with coordinates $((x_0 + m(c\dot{x}_0 + 2m\omega\dot{y}_0))/(c^2 + 4m^2\omega^2), (y_0 - m(2m\omega\dot{x}_0 - c\dot{y}_0))/ (c^2 + 4m^2\omega^2))$.

1033

A nonlinear oscillator has a potential given by

$$U(x) = \frac{kx^2}{2} - \frac{m\lambda x^3}{3}, \quad \text{with } \lambda \text{ small.}$$

Find the solution of the equation of motion to first order in λ, assuming $x = 0$ at $t = 0$.

(*Princeton*)

Sol: The equation of the motion of the nonlinear oscillator is

$$\frac{md^2x}{dt^2} = \frac{-dU(x)}{dx} = -kx + m\lambda x^2.$$

Neglecting the term $m\lambda x^2$, we obtain the zero-order solution of the equation

$$x_{(0)} = A\sin(\omega t + \varphi),$$

where $\omega = \sqrt{k/m}$ and A is an arbitrary constant. As $x = 0$ at $t = 0$, $\varphi = 0$ and we have

$$x_{(0)} = A\sin(\omega t).$$

Suppose the first-order solution has the form $x_{(1)} = x_{(0)} + \lambda x_1$. Substituting it in the equation of motion and neglecting terms of orders higher than λ, we have

$$\ddot{x}_1 + \omega^2 x_1 = x_{(0)}^2$$
$$= \frac{A^2}{2}[1 - \cos(2\omega t)].$$

To solve this equation, try a particular integral

$$x_1 = B + C\cos(2\omega t).$$

Substitution gives

$$-3\omega^2 C\cos(2\omega t) + \omega^2 B = \frac{A^2}{2} - \frac{A^2}{2}\cos(2\omega t).$$

Comparison of coefficients gives

$$B = \frac{A^2}{2\omega^2}, \quad C = \frac{A^2}{6\omega^2}.$$

The homogeneous equation

$$\ddot{x}_1 + \omega^2 x_1 = 0$$

has solution

$$x_1 = D_1\sin(\omega t) + D_2\cos(\omega t).$$

Hence we have the complete solution

$$x_{(1)} = (A + \lambda D_1)\sin(\omega t) + \lambda\left[\frac{A^2}{2\omega^2} + D_2\cos(\omega t) + \frac{A^2}{6\omega^2}\cos(2\omega t)\right].$$

The initial condition $x = 0$ at $t = 0$ then gives

$$D = -\frac{2A^2}{3\omega^2}$$

and

$$x_{(1)} = A'\sin(\omega t) + \frac{\lambda A^2}{\omega^2}\left[\frac{1}{2} - \frac{2}{3}\cos(\omega t) + \frac{1}{6}\cos(2\omega t)\right],$$

where A' is an arbitrary constant. To determine A' and A, additional information such as the amplitude and the velocity at $t = 0$ is required.

1034

A defective satellite of mass 950 kg is being towed by a spaceship in empty space. The two vessels are connected by a uniform 50 m rope whose mass per unit length is 1 kg/m. The spaceship is accelerating in a straight line with acceleration 5 m/sec^2.

a. What is the force exerted by the spaceship on the rope?

b. Calculate the tension along the rope.

c. Due to exhaustion, the crew of the spaceship falls asleep and a short circuit in one of the booster control circuits results in the acceleration changing to a deceleration of 1 m/sec^2. Describe in detail the consequences of this mishap.

(SUNY, Buffalo)

Sol:

a.
$$F = (m_{\text{rope}} + m_{\text{satellite}}) \cdot a$$
$$= (950 + 50) \times 5 = 5 \times 10^3 \, \text{N}.$$

b. Choose the point where the rope is attached to the satellite as the origin and the x-axis along the rope towards the spaceship. The tension along the rope is then

$$F(x) = (m_{\text{satellite}} + m_{\text{rope}}(x)) \cdot a$$
$$= [950 + 1 \times (50 - x)] \times 5$$
$$= 5 \times 10^3 - 5x \, \text{N}.$$

c. After the mishap, the spaceship moves with an initial velocity v_0 and a deceleration of 1 m/s^2, while the satellite moves with a constant speed v_0. After the mishap, the two vessels will collide at a time t given by

$$v_0 t = 50 + v_0 t - \frac{a}{2} t^2,$$

or

$$t = \sqrt{\frac{100}{1}} = 10 \, \text{s}.$$

1035

A ball of mass M is suspended from the ceiling by a massless spring with spring constant k and relaxed length equal to zero. The spring will break if it is extended beyond a critical length l_c ($l_c > Mg/k$). An identical spring hangs below the ball (Fig. 1.21). If one slowly pulls on the end of the lower spring, the upper spring will break. If one pulls on the lower spring too rapidly, the lower spring will break. The object of this problem is to determine the force $F(t)$ which, when applied to the end of the lower spring, will cause both springs to break simultaneously.

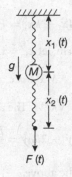

Fig. 1.21

a. Find an integral expression relating the length $x_1(t)$ of the upper spring to the applied force $F(t)$.

b. Using any technique you like, find $x_1(t)$ and $x_2(t)$ for $t > 0$ when $F(t)$ has the particular form

$$F(t) = \begin{cases} 0, & t < 0 \\ \alpha t, & t > 0 \end{cases},$$

where α is a constant.

c. Use a careful sketch of your solutions to show that if α is too small, the upper spring will break. Similarly, shown that if α is too large, then the lower spring could break first.

d. Show that both springs break simultaneously when α is a solution of the equation

$$\sin\left(\frac{kl_c}{\alpha}\sqrt{\frac{k}{M}}\right) = \frac{Mg}{\alpha}\sqrt{\frac{k}{M}}.$$

Sol:

a. The equations of motion for the ball and the lower spring are

$$M\ddot{x}_1 = Mg - kx_1 + kx_2,$$
$$kx_2 = F(t).$$

Eliminating x_2, we obtain

$$M\ddot{x}_1 + kx_1 = F(t) + Mg. \tag{1}$$

To eliminate the constant term, let $x_1 = x + Mg/k$. Equation (1) then becomes

$$M\ddot{x} + kx = F(t).$$

Let $x = e^{i\omega t}y(t)$, where $\omega = \sqrt{k/M}$. The above becomes

$$\ddot{y} + 2i\omega\dot{y} = \frac{F(t)}{M}e^{-i\omega t}. \tag{2}$$

The homogeneous part of the above,

$$\ddot{y} + 2i\omega\dot{y} = 0,$$

can be solved by letting $y = C_1 e^{\alpha t}$, where C_1 and α are constants. Substitution gives $\alpha = -2i\omega$.

A particular solution of (2) is obtained by letting $\dot{y} = e^{-2i\omega t}f(t)$, which gives

$$\frac{df}{dt} = \frac{F(t)}{M}e^{i\omega t},$$

or

$$\dot{y} = e^{-2i\omega t}\int \frac{F(t)}{M}e^{i\omega t}dt.$$

Hence the general solution of (2) is

$$\dot{y} = e^{-2i\omega t}\left[\int \frac{F(t)}{M}e^{i\omega t}dt + c_1\right],$$

giving

$$y = \int e^{-2i\omega t}\left[\int \frac{F(\tau)}{M}e^{i\omega \tau}d\tau + C_1\right]dt + C_2$$

and

$$x_1 = e^{i\omega t}\left\{\int e^{-2i\omega t}\left[\int \frac{F(\tau)}{M}e^{i\omega \tau}d\tau + C_1\right]dt + C_2\right\} + \frac{Mg}{k},\qquad(3)$$

where C_1, C_2 are constants of integration. For application to the problem, either the real or the imaginary part of the last expression is used as the general solution.

b. The equation of motion is

$$M\ddot{x}_1 + kx_1 = Mg\qquad(4)$$

for $t < 0$, and

$$M\ddot{x}_1 + kx_1 = \alpha t + Mg\qquad(5)$$

for $t > 0$. First obtain the solution of (4) by putting $F(t) = 0$ in (3). This gives

$$x_1 = iC_1'e^{-i\omega t} + C_2 e^{i\omega t} + \frac{Mg}{k},$$

where C_1' is a constant of integration in place of C_1. Taking the real part, we have

$$x_1 = C_1'\sin(\omega t) + C_2\cos(\omega t) + \frac{Mg}{k}.$$

The solution of (5) is that of (4) plus a particular solution $\alpha t/k$:

$$x_1 = C_1'\sin(\omega t) + C_2\cos(\omega t) + \frac{\alpha t}{k} + \frac{Mg}{k}.$$

At $t = 0$, $x_1 = Mg/k$, $x_2 = 0$, $\dot{x}_1 = 0$, so that $C_2 = 0$, $C_1' = -\alpha/k\omega$. Hence

$$x_1(t) = \frac{\alpha t}{k} + \frac{Mg}{k} - \frac{\alpha}{k\omega}\sin(\omega t),$$

$$x_2(t) = \frac{\alpha t}{k}.$$

c. In Figs. 1.22 (for large α) and 1.23 (for small α) are plots of the curves for x_1 and x_2. It is seen that the curve for x_1 is given by a line $x = Mg/k + \alpha t/k$, which is parallel to the x_2 line minus an oscillatory term $\alpha \sin(\omega t)/k\omega$ whose amplitude is proportional to α. Hence, if t_1 and t_2 are the instants x_1 and x_2 would reach l_c, the critical length, we have for large α, $t_2 < t_1$, i.e. the lower spring will break first, and for small α, $t_1 < t_2$, i.e. the upper spring will break first.

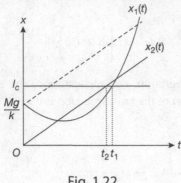
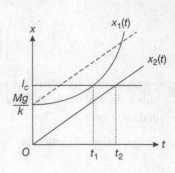

Fig. 1.22   Fig. 1.23

d. For the two springs to break simultaneously, say at time $t = t_0$, we require

$$x_2(t_0) = l_c = \frac{\alpha t_0}{k},$$

or

$$t_0 = \frac{kl_c}{\alpha},$$

and

$$x_1(t_0) = l_c = \frac{Mg}{k} + l_c - \frac{\alpha}{\omega k}\sin\left(\frac{\omega k l_c}{\alpha}\right),$$

or

$$\sin\left(\frac{\omega k l_c}{\alpha}\right) = \frac{Mg\omega}{\alpha},$$

where $\omega = \sqrt{k/M}$.

1036

A pendulum, made up of a ball of mass M suspended from a pivot by a light string of length L, is swinging freely in one vertical plane (see Fig. 1.24). By what factor does the amplitude of oscillations change if the string is very slowly shortened by a factor of 2?

(Chicago)

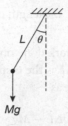

Fig. 1.24

Sol: *Method 1*

For a periodic system with a parameter slowly changing, the action J is an adiabatic invariant. Now

$$J = \oint P_\theta d\theta,$$

where $P_\theta = ML^2 \dot{\theta}$, i.e.

$$J = \oint ML^2 \dot{\theta} \cdot \dot{\theta} dt = ML^2 \langle \dot{\theta}^2 \rangle \frac{2\pi}{\omega}$$

$$= ML^2 \cdot \frac{\omega^2 \theta_0^2}{2} \cdot \frac{2\pi}{\omega} = \pi ML^2 \theta_0^2 \omega$$

$$= \pi Mg^{1/2} \theta_0^2 L^{3/2}.$$

Here we have used $T = 2\pi/\omega$, with $\omega = \sqrt{g/l}$, for the period, and

$$\langle \dot{\theta}^2 \rangle = \langle [-\theta_0 \omega \sin(\omega t + \varphi_0)]^2 \rangle = \frac{\omega^2 \theta_0^2}{2}$$

by taking $\theta = \theta_0 \cos(\omega t + \varphi_0)$. Then, as J is an adiabatic invariant,

$$\theta_0 \propto L^{-3/4}.$$

When

$$L \to L/2, \quad \theta_0 \to 1.68\theta_0,$$

i.e. the amplitude of oscillation is increased by a factor of 1.68.

Method 2

During discussion in a meeting, Einstein used this example to demonstrate what an adiabatic invariant is. His proof is as follows:

$$\text{Tension of string} = Mg\langle \cos\theta \rangle + \left\langle \frac{ML^2 \dot{\theta}^2}{L} \right\rangle$$

$$= Mg\left(1 - \frac{\langle \theta^2 \rangle}{2}\right) + ML\langle \dot{\theta}^2 \rangle$$

$$= Mg\left(1 + \frac{\theta_0^2}{4}\right).$$

It is assumed that over a period, the length of the string is almost unchanged and that θ is a small angle.

When L shortens slowly, the work done on the oscillator is $-\langle N \rangle \Delta L$, where N is the tension of the string, $-\Delta L$ is the displacement of the oscillator. Using the above, we obtain the work done as

$$-Mg\Delta L - Mg \cdot \frac{\theta_0^2}{4} \cdot \Delta L.$$

Under the action of the external force, the change in the oscillator's energy is

$$\Delta(-MgL\cos\theta_0) = \Delta\left[-MgL\left(1 - \frac{\theta_0^2}{2}\right)\right]$$

$$= -Mg\Delta L + \frac{1}{2}Mg\Delta(L\theta_0^2)$$

$$= -Mg\Delta L + \frac{1}{2}Mg\theta_0^2\Delta L + MgL\theta_0\Delta\theta_0.$$

The work done and the increment of energy must balance, giving

$$L\theta_0\Delta\theta_0 + \frac{3\theta_0^2\Delta L}{4} = 0,$$

or

$$L\theta_0^2\Delta\ln(\theta_0 L^{3/4}) = 0.$$

It follows that

$$\theta_0 L^{3/4} = \text{constant},$$

or

$$\theta_0 \propto L^{-3/4}.$$

When

$$L \to \frac{L}{2}, \qquad \theta_0 \to 1.68\theta_0.$$

1037

A perfectly reflecting sphere of radius r and density $\rho = 1$ is attracted to the sun by gravity, and repelled by the sunlight reflecting off its surface. Calculate the value of r for which these effects cancel. The luminosity of the sun is $l_s = 4 \times 10^{33}$ erg/sec and its mass is $M_s = 2 \times 10^{33}$ gm. Give your answer in cm (assume a point-like sun).

(UC, Berkeley).

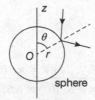

Fig. 1.25

Sol: Let N_v be the number of photons of frequency v passing through a unit area perpendicular to the direction of propagation in unit time, I_v be the energy of sunlight of frequency v radiated by the sun in unit time, and R be the distance from the sun to the sphere. As $R \gg r$, the incident sunlight may be considered parallel and in a direction opposite to the z-axis, as shown in Fig. 1.25. Then

$$I_s = \int I_v dv, \qquad N_v = \frac{I_v}{4\pi R^2 hv}.$$

The photons collide elastically with the perfectly reflecting sphere at its surface. During a time interval Δt, for an elementary surface ΔS at azimuth angle θ, the change of the momentum of photons of frequency v along the z-axis is

$$\Delta P_{vz} = N_v \left[\frac{hv}{c} + \frac{hv}{c} \cos(2\theta) \right] \cos\theta \Delta S \Delta t.$$

This gives rise to a force of magnitude

$$\Delta F_{vz} = \frac{\Delta P_{vz}}{\Delta t} = \frac{2hv}{c} N_v \cos^3\theta \Delta S.$$

Then the total force exerted on the sphere by the sunlight of frequency v is

$$F_{vz} = \int dF_{vz} = \frac{2hv}{c} \cdot \frac{I_v}{4\pi R^2 hv} \int_0^{\pi/2} 2\pi \cos^3\theta \cdot \sin\theta \cdot r^2 d\theta = \frac{I_v r^2}{4R^2 c}.$$

Hence the total repelling force exerted by the sunlight is

$$F_z = \int F_{vz} dv = \frac{I_s r^2}{4R^2 c}.$$

The gravitational force the sun exerts on the sphere is

$$F_g = \frac{GM_s m}{R^2},$$

where $m = \rho \cdot (4/3)\pi r^3 = (4/3)\pi r^3$ is the mass of the sphere. When the two forces balance, we have

$$\frac{I_s r^2}{4R^2 c} = \frac{4GM_s \pi r^3}{3R^2},$$

or

$$r = \frac{3I_s}{16\pi c G M_s}$$

$$= \frac{3 \times 4 \times 10^{33}}{16 \times 3.14 \times 3 \times 10^{10} \times 6.67 \times 10^{-8} \times 2 \times 10^{33}}$$

$$= 5.97 \times 10^{-5} \text{ cm.}$$

1038

A particle of mass m moves along a trajectory given by $x = x_0 \cos \omega_1 t$, $y = y_0 \sin \omega_2 t$.

 a. Find the x and y components of the force. Under what condition is the force a central force?

 b. Find the potential energy as a function of x and y.

 c. Determine the kinetic energy of the particle. Show that the total energy of the particle is conserved.

(Wisconsin)

Sol: **a.**   Differentiating with respect to time, we obtain

$$\dot{x} = -x_0 \omega_1 \sin(\omega_1 t), \quad \ddot{x} = -x_0 \omega_1^2 \cos(\omega_1 t),$$
$$\dot{y} = y_0 \omega_2 \cos(\omega_2 t), \quad \ddot{y} = -y_0 \omega_2^2 \sin(\omega_2 t).$$

Newton's second law gives,

$$\mathbf{F} = m(\ddot{x}\,\mathbf{i} + \ddot{y}\,\mathbf{j}) = -m[x_0 \omega_1^2 \cos(\omega_1 t)\mathbf{i} + y_0 \omega_2^2 \sin(\omega_2 t)\mathbf{j}]$$
$$= -m(\omega_1^2 x \mathbf{i} + \omega_2^2 y \mathbf{j}).$$

The x and y components of the force are therefore

$$F_x = -m\omega_1^2 x,$$
$$F_y = -m\omega_2^2 y.$$

If $\omega_1 = \omega_2$, $\mathbf{F}$ is a central force $\mathbf{F} = -m\omega_1^2 \mathbf{r}$.

 b.   From

$$\mathbf{F} = -\nabla V,$$

i.e.

$$F_x = -\frac{\partial V}{\partial x}, \quad F_y = -\frac{\partial V}{\partial y},$$

we obtain the potential energy

$$V = \frac{1}{2}m(\omega_1^2 x^2 + \omega_2^2 y^2).$$

Note that we take the zero potential level at the origin.

c. The kinetic energy of the particle is

$$T = \frac{1}{2}m(\dot{x}^2 + \dot{y}^2) = \frac{1}{2}m[x_0^2\omega_1^2 \sin^2(\omega_1 t) + y_0^2\omega_2^2 \cos^2(\omega_2 t)].$$

The total energy is then

$$
\begin{aligned}
E &= T + V \\
&= \frac{1}{2}m[x_0^2\omega_1^2 \sin^2(\omega_1 t) + y_0^2\omega_2^2 \cos^2(\omega_2 t) \\
&\quad + \omega_1^2 x_0^2 \cos^2(\omega_1 t) + \omega_2^2 y_0^2 \sin^2(\omega_2 t) \\
&= \frac{1}{2}m(x_0^2\omega_1^2 + y_0^2\omega_2^2) \\
&= \text{constant}.
\end{aligned}
$$

It is therefore conserved.

1039

A particle of mass m is projected with velocity v_0 toward a fixed scattering center which exerts a repulsive force $\mathbf{F} = (mv_1^2/2)\delta(r - a)\hat{\mathbf{r}}$, where $\hat{\mathbf{r}}$ is a unit vector along the radius from the force center, a is a fixed radius at which the force acts, and v_1 is a constant having the dimensions of velocity. The impact parameter is s, as shown in Fig. 1.26.

a. Find the potential energy.

b. Show that if $v_0 < v_1$, the particle does not penetrate the sphere $r = a$, but bounces off, and that the angle of reflection equals the angle of incidence.

c. Sketch carefully the orbit you would expect for $v_0 > v_1$, $s = a/2$.

(*Wisconsin*)

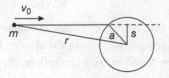

Fig. 1.26

Sol: **a.** The force **F**, being a central force, is conservative. A potential can then be defined:

$$V(\mathbf{r}) = -\int_{\infty}^{r} \mathbf{F}(\mathbf{r}') \cdot d\mathbf{r}' = \frac{1}{2}mv_1^2 \int_{r}^{\infty} \delta(r' - a)dr'$$

$$= \begin{cases} \frac{1}{2}mv_1^2 & \text{for } r < a, \\ 0 & \text{for } r > a \end{cases}.$$

This is the potential energy of the particle in the field of the force.

b. The total energy $T + V$ of the particle is conserved:

$$\frac{1}{2}mv_0^2 = \frac{1}{2}mv'^2 + \frac{1}{2}mv_1^2,$$

i.e. $v_0^2 - v_1^2 = v'^2$, where v' is the speed of the particle inside the sphere $r = a$. For the penetration to take place, v' must be real, i.e. we require that $v_0 > v_1$.

If $v_0 < v_1$, the particle cannot penetrate the sphere $r = a$. Then as the force is radial to the sphere, the radial component of the particle momentum will be reversed in direction but not changed in magnitude, while the component tangential to the sphere will remain the same. Hence, the angles of incidence and reflection, which are determined by the ratio of the magnitude of the tangential component to that of the radial component, are equal. Note that on account of conservation of mechanical energy, the magnitude of the particle momentum will not change on collision.

c. For $v_0 > v_1$ and $s = a/2$, the particle will be incident on the sphere $r = a$ with an incidence angle $\theta_0 = \arcsin[(a/2)/a] = 30°$, and penetrate the sphere. Let the angle it makes with the radial direction be θ. Then conservation of the tangential component of its momentum requires that

$$v' \sin \theta = v_0 \sin 30° = \frac{v_0}{2},$$

so that θ is given by

$$\theta = \arcsin\left(\frac{v_0}{2\sqrt{v_0^2 - v_1^2}}\right).$$

As V is constant (i.e. no force) inside the sphere, the trajectory will be a straight line until the particle leaves the sphere. Deflection of the trajectory again occurs at $r = a$, and outside the sphere, the speed will again be v_0 with

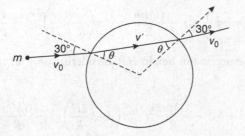

Fig. 1.27

the direction of motion making an angle of 30° with the radial direction at the point of exit, as shown in Fig. 1.27.

1040

A long-range rocket is fired from the surface of the earth (radius R) with velocity $v = (v_r, v_\theta)$ (Fig. 1.28). Neglecting air friction and the rotation of the earth (but using the exact gravitational field), obtain an equation to determine the maximum height H achieved by the trajectory. Solve it to lowest order in (H/R) and verify that it gives a familiar result for the case that **v** is vertical.

(*Wisconsin*)

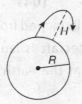

Fig. 1.28

Sol: Both the angular momentum and mechanical energy of the rocket are conserved under the action of gravity, a central force. Considering the initial state and the final state when the rocket achieves maximum height, we have

$$mRv_\theta = m(R + H)v'_\theta,$$

$$\frac{1}{2}m(v_\theta^2 + v_r^2) - \frac{GMm}{R} = \frac{1}{2}mv'^2_\theta - \frac{GMm}{R + H},$$

where the prime refers to the final state at which the radial component of its velocity vanishes, m and M are the masses of the rocket and the earth respectively. Combining the above two equations we obtain

$$\frac{1}{2}m(v_\theta^2 + v_r^2) - \frac{GMm}{R} = \frac{1}{2}m\left(\frac{R}{R+H}\right)^2 v_\theta^2 - \frac{GMm}{R+H},$$

which gives the maximum height H. Considering only terms first order in H/R, we have

$$\frac{1}{2}m(v_r^2 + v_\theta^2) - \frac{GMm}{R} \approx \frac{1}{2}m\left(1 - \frac{2H}{R}\right)v_\theta^2 - \frac{GMm}{R}\left(1 - \frac{H}{R}\right),$$

and hence

$$H \approx \frac{v_r^2 R}{2\left(\frac{GM}{R} - v_\theta^2\right)}.$$

For vertical launching, $v_\theta = 0$, $v_r = v$, and if H/R is small, we can consider g as constant with $g = GM/R^2$. We then obtain the familiar formula

$$H \approx \frac{v^2}{2\left(\frac{GM}{R^2}\right)} = \frac{v^2}{2g}.$$

1041

In a few weeks Mariner 9 will be launched from Cape Kennedy on a mission to Mars. Assume that this spacecraft is launched into an elliptical orbit about the sun with perihelion at the earth's orbit and aphelion at Mar's orbit (Fig. 1.29).

a. Find the values of the parameters λ and ε of the orbit equation $r = \lambda(1 + \varepsilon)/(1 + \varepsilon \cos \theta)$ and sketch the orbit.

b. Use Kepler's third law to calculate the time duration of the mission to Mars on this orbit.

c. In what direction should the launch be made from earth for minimum expenditure of fuel?

Mean distance of Mars from the sun = 1.5 A.U.

Mean distance of the earth from the sun = 1 A.U.

(Wisconsin)

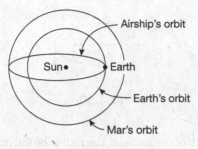

Fig. 1.29

Sol: **a.** Let R_1 be the distance of the earth from the sun and R_2 that of Mars from the sun. Then

$$R_1 = \frac{\lambda(1 + \varepsilon)}{1 + \varepsilon} = \lambda,$$

$$R_2 = \frac{\lambda(1 + \varepsilon)}{1 - \varepsilon}.$$

Solving the equations, we obtain $\lambda = R_1 = 1$ A.U., $\varepsilon = 0.2$.

b. Let T_1 and T be the revolutional periods of the earth and Mariner 9 respectively. According to Kepler's third law, $T^2/a^3 = \text{constant}$,

$$\frac{T^2}{\left(\dfrac{R_1 + R_2}{2}\right)^3} = \frac{T_1^2}{R_1^3},$$

or

$$T = \left(\frac{R_1 + R_2}{2R_1}\right)^{3/2} T_1 = \left(\frac{1}{1 - \varepsilon}\right)^{3/2} T_1 = 1.25^{3/2} T_1 = 1.40 \text{ years}.$$

The mission to Mars on this orbit takes 0.70 year.

c. In order to economize on fuel, the rocket must be launched along the tangent of the earth's orbit and in the same direction as the earth's rotation.

1042

A comet in an orbit about the sun has a velocity 10 km/sec at aphelion and 80 km/sec at perihelion (Fig. 1.30). If the earth's velocity in a circular orbit is 30 km/sec and the radius of its orbit is 1.5×10^8 km, find the aphelion distance R_a for the comet.

(*Wisconsin*)

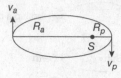

Fig. 1.30

Sol: Let v be the velocity of the earth, R the radius of the earth's orbit, m and m_s the masses of the earth and the sun respectively. Then

$$\frac{mv^2}{R} = \frac{Gmm_s}{R^2},$$

or

$$Gm_s = Rv^2.$$

By the conservation of the mechanical energy and of the angular momentum of the comet, we have

$$\frac{-Gm_cm_s}{R_a} + \frac{m_cv_a^2}{2} = \frac{-Gm_cm_s}{R_p} + \frac{m_cv_p^2}{2},$$
$$m_cR_av_a = m_cR_pv_p,$$

where m_c is the mass of the comet, and v_a and v_p are the velocities of the comet at aphelion and at perihelion respectively. The above equations give

$$R_a = \frac{2Gm_s}{v_a(v_a + v_p)} = \frac{2Rv^2}{v_a(v_a + v_p)} = 3 \times 10^8 \text{ km}.$$

1043

A classical particle with energy E_0 and angular momentum L about point O enters a region in which there is an attractive central potential $V = -G(r)$ centered on point O. The particle is scattered by the potential.

a. Begin by assuming conservation of energy and angular momentum, and find the differential equation for dx/dr in terms of E_0, L, $G(r)$ and r (and the particle mass m).

b. Find an equation for the distance of closest approach, r_{min}, in terms of E, L, $G(r_{min})$, and m.

(Wisconsin)

Sol:

a. $E_0 = \frac{1}{2}m(\dot{r}^2 + r^2\dot{\theta}^2) - G(r),$

$L = mr^2\dot{\theta},$

where θ is shown in Fig. 1.31. Then

$$\dot{r}^2 = \frac{2(E_0 + G)}{m} - \frac{L^2}{m^2r^2}.$$

As

$$\frac{dr}{dt} = \frac{dr}{d\theta}\cdot\frac{d\theta}{dt} = \dot{\theta}\frac{dr}{d\theta}, \qquad \dot{\theta} = \frac{L}{mr^2},$$

the above equation can be written as

$$\left(\frac{dr}{d\theta}\right)^2 = \frac{m^2r^4}{L^2}\left[\frac{2(E_0 + G)}{m} - \frac{L^2}{m^2r^2}\right],$$

giving

$$\frac{dr}{d\theta} = \pm\sqrt{\frac{2m(E_0 + G)r^4}{L^2} - r^2}.$$

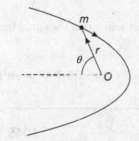

Fig. 1.31

b. At closest approach $r = r_{min}$, $\dot{r} = 0$. Hence

$$E_0 = \frac{1}{2}mr_{min}^2\,\dot{\theta}^2 - G(r_{min})$$

$$= \frac{1}{2}mr_{min}^2\cdot\frac{L^2}{m^2r_{min}^4} - G(r_{min})$$

$$= \frac{L^2}{2mr_{min}^2} - G(r_{min}),$$

or

$$r_{min} = \frac{L}{\sqrt{2m[E_0 + G(r_{min})]}}.$$

The result can also be obtained by putting $dr/d\theta = 0$.

1044

A comet moves toward the sun with initial velocity v_0. The mass of the sun is M and its radius is R. Find the total cross section σ for striking the sun. Take the sun to be at rest and ignore all other bodies.

(Wisconsin)

Sol: Let the impact parameter of the comet be b. At the closest approach to the sun (closest distance r from the sun's center), we have from the conservation of mechanical energy and angular momentum

$$\frac{mV_0^2}{2} = \frac{mV^2}{2} - \frac{GMm}{r},$$
$$mbV_0 = mrV,$$

where m is the mass of the comet and V its velocity at closest approach. From these, we find

$$b = r\sqrt{1 + \frac{2GM}{V_0^2 r}}.$$

If $r < R$, the comet will strike the sun. Hence the total cross section for striking the sun is

$$\sigma = \pi[b(R)]^2 = \pi R^2 \left(1 + \frac{2GM}{V_0^2 R} \right).$$

1045

A particle moves in a circular orbit of radius r under the influence of an attractive central force. Show that this orbit is stable if

$$f(r) > -\left(\frac{r}{3} \right) \frac{\partial f}{\partial r}\bigg|_r,$$

where $f(r)$ is the magnitude of the force as a function of the distance r from the center.

(CUSPEA)

Sol: For the motion of a particle under the influence of a central force, we have

$$mr^2\dot{\theta} = \text{constant} = L, \text{ say,}$$
$$m\ddot{r} = -f + mr\dot{\theta}^2.$$

Consider a particle traveling in a circular orbit of radius r subject to small radial and angular displacements $\delta r, \delta \theta$:

$$r(t) = r + \delta r(t), \qquad \theta = \omega t + \delta \theta(t),$$

where ω is the angular frequency of the particle moving in a circular orbit of radius r given by $m\omega^2 r = f(r)$. As

$$\Delta L \approx mr^2 \delta \dot{\theta} + 2mr\dot{\theta}\delta r,$$
$$m\delta \ddot{r} \approx -\frac{df}{dr}\delta r + m\dot{\theta}^2 \delta r + 2mr\dot{\theta}\delta \dot{\theta},$$
$$\dot{\theta} \approx \omega + \delta \dot{\theta},$$

we have

$$m\delta \ddot{r} \approx -\frac{df}{dr}\delta r + m\omega^2 \delta r + 2\omega \frac{\Delta L - 2mr\omega\delta r}{r}$$
$$= \left(-\frac{df}{dr} - 3\frac{f(r)}{r}\right)\delta r + \frac{2\omega \Delta L}{r}.$$

In the above, we have retained only terms first order in the small quantities.

The circular orbit is stable only if δr varies simple-harmonically. In other words, the stable condition is that the coefficient of δr is negative:

$$-\left[\frac{dt}{dr}\right]_r - \frac{3f(r)}{r} < 0,$$

or

$$f(r) > -\frac{r}{3}\left[\frac{dt}{dr}\right]_r.$$

1046

A body is projected upwards at $\theta_1 = 30°$ from a height 'h' with a velocity 10 m/s. Simultaneously another body is projected horizontally with a velocity from the same height with a velocity $10\sqrt{2}$ m/s. They both meet at the same point on the ground. Determine the horizontal distance travelled by the two bodies, the height h and the time taken for two bodies to reach the ground.

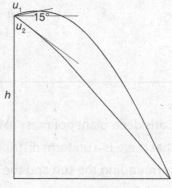

Fig. 1.32

Sol: For the first body: $u_1 = 10$ m/s, $\theta_1 = 30°$, $g = 10$ m/s^2

$$y_1 = h + x_1 \tan \theta_1 - \frac{gx_1^2}{2u_1^2 \cos^2 \theta_1} = h + \frac{1}{\sqrt{3}}x_1 - \frac{20x_1^2}{3 \times 200}$$

$$x_1 = u_1 \cos \theta_1 t_1$$

For the second body: $u_2 = 10\sqrt{2}$ m/s, $\theta_2 = 0$, $g = 10$ m/s^2

$$y_2 = h + x_2 \tan \theta_2 - \frac{gx_2^2}{2u_2^2 \cos^2 \theta_1} = h - \frac{10x_2^2}{400}$$

$$x_2 = u_2 \cos \theta_2 t_2$$

When they meet the two bodies will have $y_1 = y_2$; $x_1 = x_2$

$$\frac{1}{\sqrt{3}}x - \frac{x^2}{30} = -\frac{x^2}{40}$$

$$x\left(\frac{1}{30} - \frac{1}{40}\right) = \frac{1}{\sqrt{3}} \text{ or } x = \frac{120}{\sqrt{3}} = 69.3 \text{ m}$$

The time taken for the first body is

$$t_1 = \frac{x}{u_1 \cos 30} = \frac{120 \times 2}{\sqrt{3} \times 10 \times \sqrt{3}} = 8 \text{ sec}$$

For the second body

$$t_2 = \frac{x}{u_2 \cos 0} = \frac{120}{\sqrt{3} \times 10\sqrt{2}} = \frac{12}{\sqrt{6}} = 5 \text{ sec}$$

'h' will be the same for both bodies and can be determined either from the equation for y_1 or y_2 equating it to '$-h$' when $x = \dfrac{120}{\sqrt{3}}$. Using the equation for y_2

$$-h = h - \frac{10x^2}{400} = h - \frac{10 \times 120 \times 120}{3 \times 400}$$

$$h = 60 \text{ m}$$

1047

Consider a planet of mass m in orbit around a sun of mass M. Assume further that there is a uniform distribution of dust, of density ρ, throughout the space surrounding the sun and the planet.

a. Show that the effect of the dust is to add an additional attractive central force

$$F' = -mkr, \quad \text{where} \quad k = \frac{4\pi\rho G}{3}, \quad G = \text{gravitational constant}.$$

You may neglect any drag force due to collision with the particles.

b. Consider a circular orbit for the planet corresponding to angular momentum L. Give the equation satisfied by the radius of the orbit, r_0, in terms of L, G, M, m and k. You need not solve the equation.

c. Assume F' is small compared with the solar attraction and consider an orbit just slightly deviating from the circular orbit of part (b). By considering the frequencies of the radial and the azimuthal motion, show that the orbit is a precessing ellipse and calculate the angular frequency of precession, ω_p, in terms of r_0, ρ, G and M.

d. Does the axis of the ellipse precess in the same or opposite direction to the orbital angular velocity?

(CUSPEA)

Sol:

a. The mass of the dust in a sphere of radius r centered at the sun is

$$M_{\text{dust}} = \frac{4\pi r^3 \rho}{3}.$$

If r is the distance of the planet from the sun, the gravitational force on the planet due to the attraction of the dust is, on account of the inverse distance square nature of gravitation, as if all the dust were concentrated at the sun. In other words,

$$F' = \frac{-M_{\text{dust}} m G}{r^2} = \frac{-4\pi r^3 \rho}{3} \frac{mG}{r^2} = \frac{-4\pi\rho G m r}{3} = -mkr.$$

b. The planet has acceleration $(\ddot{r} - r\dot{\theta}^2, 2\dot{r}\dot{\theta} + r\ddot{\theta})$ in polar coordinates. Its equations of motion are therefore

$$m\ddot{r} = \frac{-GMm}{r^2} - mkr + mr\dot{\theta}^2, \tag{1}$$

$$mr\ddot{\theta} + 2m\dot{r}\dot{\theta} = 0. \tag{2}$$

Multiplying (2) by r, we have

$$\frac{d(mr^2\dot{\theta})}{dt} = 0,$$

or

$$mr^2\dot{\theta} = L,$$

where L is a constant. Thus the angular momentum L is a constant of the motion. Writing

$$mr\dot{\theta}^2 = \frac{L^2}{mr^3},$$

the radial equation becomes

$$m\ddot{r} = \frac{-GMm}{r^2} - mkr + \frac{L^2}{mr^3}.$$

For a circular orbit, $\ddot{r} = 0$, and we have the equation for the radius r_0 of the orbit:

$$\frac{-GMm}{r_0^2} - mkr_0 + \frac{L^2}{mr_0^3} = 0.$$

c. Let η express a small radial excursion around r_0, i.e. $\eta = r - r_0$, in terms of which (1) becomes

$$
\begin{aligned}
m\ddot{\eta} &= -\frac{GMm}{(\eta + r_0)^2} - mk(\eta + r_0) + \frac{L^2}{m(r_0 + \eta)^3} \\
&= -\frac{GMm}{r_0^2\left(1 + \dfrac{\eta}{r_0}\right)^2} - mkr_0\left(1 + \frac{\eta}{r_0}\right) + \frac{L^2}{mr_0^3\left(1 + \dfrac{\eta}{r_0}\right)^3} \\
&\approx -\frac{GMm}{r_0^2}\left(1 - \frac{2\eta}{r_0}\right) - mkr_0\left(1 + \frac{\eta}{r_0}\right) + \frac{L^2}{mr_0^3}\left(1 - \frac{3\eta}{r_0}\right),
\end{aligned}
$$

as $\eta \ll r_0$. Making use of the equation for circular orbit, we rewrite the above as

$$
\begin{aligned}
m\ddot{\eta} &= \frac{2\eta}{r_0}\left(\frac{GMm}{r_0^2}\right) - \frac{3\eta}{r_0}\frac{L^2}{mr_0^3} - mk\eta \\
&= -\eta\left[\frac{L^2}{mr_0^4} + mk + \frac{2}{r_0}\left(-\frac{GMm}{r_0^2} + \frac{L^2}{mr_0^3}\right)\right] \\
&= -\eta\left[\frac{L^2}{mr_0^4} + 3km\right],
\end{aligned}
$$

or

$$\ddot{\eta} = -\left(\frac{L^2}{m^2 r_0^4} + 3k\right)\eta.$$

This is the equation of a harmonic oscillator with angular frequency

$$\omega_r = \sqrt{\frac{L^2}{m^2 r_0^4} + 3k}.$$

As the radial oscillation frequency is slightly larger than the azimuthal frequency $\frac{L}{mr_0^2}$, the orbit is a precessing ellipse.

To first order in ρ the azimuthal frequency is not affected by the presence of dust:

$$\dot{\theta} = \frac{L}{mr_0^2} = \omega_0.$$

The precession frequency is

$$\omega_p = \omega_r - \omega_0$$

$$= \sqrt{\frac{L^2}{m^2 r_0^4} + 3k} - \frac{L}{mr_0^2}$$

$$= \frac{L}{mr_0^2}\left(\sqrt{1 + \frac{3km^2 r_0^4}{L^2}} - 1\right)$$

$$\approx \frac{L}{mr_0^2} \cdot \frac{3}{2}\frac{km^2 r_0^4}{L^2}$$

$$= \frac{3}{2}\frac{mkr_0^2}{L}.$$

In order to express ω_p in terms of ρ, G, m and r_0, use the expression of L for $k = 0$ (any error is second order in ω_p):

$$L = \sqrt{GMm^2 r_0}.$$

Then

$$\omega_p = \frac{3}{2}m \cdot \frac{4\pi}{3}\rho G \frac{r_0^2}{\sqrt{GMm^2 r_0}} = 2\pi\rho\left(\frac{r_0^3 G}{M}\right)^{1/2}.$$

d. Since the radial oscillation is faster than the orbital revolution, the axis of the ellipse precesses in a direction opposite to the orbital angular velocity as shown in Fig. 1.33.

Fig. 1.33

1048

A meteorite of mass 1.6×10^3 kg moves about the earth in a circular orbit at an altitude of 4.2×10^6 m above the surface. It suddenly makes a head-on collision with another meteorite that is much lighter, and loses 2.0% of its kinetic energy without changing its direction of motion or its total mass.

  a. What physics principles apply to the motion of the heavy meteorite after its collision?

  b. Describe the shape of the meteorite's orbit after the collision.

  c. Find the meteorite's distance of closest approach to the earth after the collision.

(UC, Berkeley)

Sol:

  a. The laws of conservation of mechanical energy and conservation of angular momentum apply to the motion of the heavy meteorite after its collision.

  b. For the initial circular motion, $E < 0$, so after the collision we still have $E < 0$. After it loses 2.0% of its kinetic energy, the heavy meteorite will move in an elliptic orbit.

  c. From

$$\frac{mv^2}{r} = \frac{GmM}{r^2},$$

we obtain the meteorite's kinetic energy before collision:

$$\frac{1}{2}mv^2 = \frac{GmM}{2r} = \frac{mgR^2}{2r}$$

$$= \frac{m \times 9.8 \times 10^3 \times 6400^2}{2(6400 + 4200)} = 1.89 \times 10^7 m \text{ Joules},$$

where m is the mass of the meteorite in kg. The potential energy of the meteorite before collision is

$$-\frac{GmM}{r} = -mv^2 = -3.78 \times 10^7 m \text{ Joules.}$$

During the collision, the heavy meteorite's potential energy remains constant, while its kinetic energy is suddenly reduced to

$$1.89 \times 10^7 m \times 98\% = 1.85 \times 10^7 m \text{ Joules}$$

Hence the total mechanical energy of the meteorite after the collision is

$$E = (1.85 - 3.78) \times 10^7 m = -1.93 \times 10^7 m \text{ Joules.}$$

From

$$E = \frac{-GmM}{2a} = \frac{-mR^2 g}{2a},$$

we obtain the major axis of the ellipse as

$$2a = \frac{R^2 g}{1.93 \times 10^7} = \frac{(6400 \times 10^3)^2 \times 9.8}{1.93 \times 10^7}$$
$$= 2.08 \times 10^7 \text{m} = 2.08 \times 10^4 \text{ km.}$$

As after the collision, the velocity of the heavy meteorite is still perpendicular to the radius vector from the center of the earth, the meteorite is at the apogee of the elliptic orbit. Then the distance of the apogee from the center of the earth is $6400 + 4200 = 10600$ km and the distance of the perigee from the center of the earth is

$$r_{min} = 20800 - 10600 = 10200 \text{ km.}$$

Thus the meteorite's distance of closest approach to the earth after the collision is $10200 - 6400 = 3800$ km.

From the above calculations, we see that it is unnecessary to know the mass of the meteorite. Whatever the mass of the meteorite, the answer is the same as long as the conditions remain unchanged.

1049

Given that an earth satellite near the earth's surface takes about 90 min per revolution and that a moon satellite (of our moon, i.e., a spaceship orbiting our moon) takes also about 90 min per revolution, what interesting statement can you derive about the moon's composition?

(*UC, Berkeley*)

Sol: From the equation $mr\omega^2 = GmM/r^2$ for a body m to orbit around a fixed body M under gravitation, we find

$$r^3\omega^2 = GM.$$

Then if M_e, M_m are the masses and r_e, r_m are the radii of the earth and moon respectively, and the periods of revolution of the earth and moon satellites are the same, we have

$$\frac{r_m^3}{r_e^3} = \frac{M_m}{M_e},$$

or

$$\frac{M_e}{V_e} = \frac{M_m}{V_m},$$

where V_e and V_m are the volumes of the earth and moon respectively. It follows that the earth and moon have the same density.

1050

The interaction between an atom and an ion at distances greater than contact is given by, the potential energy $V(r) = -Cr^{-4}$. ($C = e^2P_a^2/2$, where e is the charge and P_a the polarizability of the atom.)

a. Sketch the effective potential energy as a function of r.

b. If the total energy of the ion exceeds V_0, the maximum value of the effective potential energy, the ion can strike the atom. Find V_0 in terms of the angular momentum L.

c. Find the cross section for an ion to strike an atom (i.e., to penetrate to $r = 0$) in terms of its initial velocity v_0. Assume that the ion is much lighter than the atom.

(UC, Berkeley)

Sol:

a. The effective potential energy as a function of r is

$$V_{\text{eff}}(r) = \frac{-C}{r^4} + \frac{L^2}{2mr^2},$$

where L is the angular momentum of the ion about the force center, and m is the mass of the ion. Its variation with r is shown in Fig. 1.34.

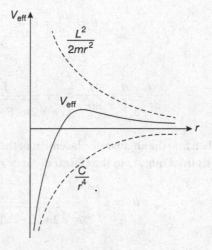

Fig. 1.34

b. To find the maximum of V_{eff}, V_0, we set

$$\frac{dV_{\text{eff}}}{dr} = \frac{4C}{r^5} - \frac{L^2}{mr^3} = \left(\frac{4C}{r^2} - \frac{L^2}{m}\right)\frac{1}{r^3} = 0.$$

The solutions are

$$r_1 = \infty, \qquad r_2 = \frac{2}{L}\sqrt{Cm}.$$

Consider

$$\frac{d^2V_{\text{eff}}}{dr^2} = \frac{-20C}{r^6} + \frac{3L^2}{mr^4}.$$

Substituting r_1 and r_2 in the above we obtain

$$\frac{d^2V_{\text{eff}}}{dr^2}\bigg|_{r=r_1} = 0, \qquad \frac{d^2V_{\text{eff}}}{dr^2}\bigg|_{r=r_2} < 0.$$

Hence at $r = \frac{2}{L}\sqrt{Cm}$, V_{eff} has a maximum value $V_0 = \frac{L^4}{16Cm^2}$.

c. In terms of the total energy

$$E = \frac{1}{2}m\dot{r}^2 + \frac{L^2}{2mr^2} + V,$$

we can write $m\dot{r} = \sqrt{2m(E - V) - \frac{L^2}{r^2}}$. In terms of L we can write $\dot{\theta} = \frac{L}{mr^2}$.
Then as

$$\dot{\theta} = \frac{d\theta}{dt} = \dot{r}\frac{d\theta}{dr},$$

we have

$$\frac{d\theta}{dr} = \frac{\dot{\theta}}{\dot{r}} = \frac{L}{mr^2\dot{r}} = \frac{L}{r^2\sqrt{2m(E-V) - \frac{L^2}{r^2}}}.$$

We can then find the angular displacement of the ion with respect to the atom as it travels from infinity to the closest distance r_{min} from the atom:

$$\theta_1 = L\int_{r_{min}}^{\infty} \frac{dr}{r^2\sqrt{2m(E-V) - \frac{L^2}{r^2}}}.$$

As

$$E = V_0 = \frac{L^4}{16m^2C}, \qquad V = -\frac{C}{r^4},$$

$$\theta_1 = L\int_{r_{min}}^{\infty} \frac{dr}{r^2\left(\frac{L^4}{8mC} - \frac{L^2}{r^2} + \frac{2mC}{r^4}\right)^{1/2}}$$

$$= \sqrt{8mC}\int \frac{d(Lr)}{(L^2r^2 - 4mC)}$$

$$= \frac{1}{\sqrt{2}}\ln\frac{(Lr - 2\sqrt{mC})}{(Lr + 2\sqrt{mC})}\Big|_{r_{min}}^{\infty}.$$

r_{min}, the minimum distance of the ion to the atom, is determined by $\frac{dr}{d\theta} = 0$, i.e.,

$$2m(E-V) - \frac{L^2}{r^2} = 0,$$

or

$$2mEr^4 - L^2r^2 + 2mC = 0.$$

Hence

$$r^2_{min} = \frac{L^2 \pm \sqrt{L^4 - 16m^2EC}}{4mE} = \frac{L^2}{4mE} = \frac{4mC}{L^2},$$

or

$$r_{min} = \frac{2}{L}\sqrt{mC}.$$

Substituting $r_{\min}$ in θ_1 we obtain

$$\theta_1 = \infty.$$

Why cannot we have a finite value for θ_1? It is on account of the fact that, under the condition $E = V_0 = \dfrac{L^4}{16Cm^2}$, while $\dot{r} \to 0$ as $r \to r_{\min}$, the transverse velocity $r\dot{\theta} = \dfrac{L}{mr} \to \dfrac{L}{mr_{\min}}$, a constant, so that with passage of time the trajectory will infinitely approach a circle of radius $r_{\min}$ and no scattering occurs.

If $E > V_0$, $r_{\min}$ as given above is complex, implying that there is no minimum distance from the atom, i.e., the ion will approach the atom infinitely. Physically this can be seen as follows. When the ion reaches the position at which $V_{\text{eff}} = V_0$, $\dot{r} \neq 0$ and the ion continues approaching the atom. As L is conserved, the speed of the ion, $(r^2\dot{\theta}^2 + \dot{r}^2)^{1/2} = \left(\dfrac{L^2}{m^2r^2} + \dot{r}^2 \right)^{1/2}$, will become larger and larger as the atom is approached, if the expression for the potential energy $V(r) = -\dfrac{c}{r^4}$ continues to hold. But this is not so as other ion-atom interactions will come into play when the two bodies are close to each other.

Suppose the ion approaches the atom with impact parameter b and initial velocity v_0. Then to strike the atom we require

$$E = \frac{1}{2}mv_0^2 > V_0 = \frac{L^4}{16Cm^2} = \frac{m^2v_0^4b^4}{16C},$$

or

$$b^4 < \frac{8C}{mv_0^2}.$$

Hence the cross section for the ion to strike the atom is

$$\sigma = \pi b^2 = \frac{2\pi}{v_0}\sqrt{\frac{2C}{m}}.$$

1051

Given a classical model of the tritium atom with a nucleus of charge $+1$ and a single electron in a circular orbit of radius r_0, suddenly the nucleus emits a negatron and changes to charge $+2$. (The emitted negatron escapes rapidly and we can forget about it.) The electron in orbit suddenly has a new situation.

a. Find the ratio of the electron's energy after to before the emission of the negatron (taking the zero of energy, as usual, to be for zero kinetic energy at infinite distance).

b. Describe qualitatively the new orbit.

c. Find the distance of closest and of farthest approach for the new orbit in units of r_0.

d. Find the major and minor axes of the new elliptical orbit in terms of r_0

(*UC, Berkeley*)

Sol:

a. As the negatron leaves the system rapidly, we can assume that its leaving has no effect on the position and kinetic energy of the orbiting electron.

From the force relation for the electron,

$$\frac{mv_0^2}{r_0} = \frac{e^2}{4\pi\varepsilon_0 r_0^2},\tag{1}$$

we find its kinetic energy

$$\frac{mv_0^2}{2} = \frac{e^2}{8\pi\varepsilon_0 r_0}$$

and its total mechanical energy

$$E_1 = \frac{mv_0^2}{2} - \frac{e^2}{4\pi\varepsilon_0 r_0} = \frac{-e^2}{8\pi\varepsilon_0 r_0}$$

before the emission of the negatron. After the emission the kinetic energy of the electron is still $\dfrac{e^2}{8\pi\varepsilon_0 r_0}$, while its potential energy suddenly changes to

$$\frac{-2e^2}{4\pi\varepsilon_0 r_0} = -\frac{e^2}{2\pi\varepsilon_0 r_0}.$$

Thus after the emission the total mechanical energy of the orbiting electron is

$$E_2 = \frac{mv_0^2}{2} - \frac{2e^2}{4\pi\varepsilon_0 r_0} = \frac{-3e^2}{8\pi\varepsilon_0 r_0},$$

giving

$$\frac{E_2}{E_1} = 3.$$

In other words, the total energy of the orbiting electron after the emission is three times as large as that before the emission.

b. As $E_2 = \dfrac{-3e^2}{8\pi\varepsilon_0 r_0}$, the condition Eq. (1) for circular motion is no longer satisfied and the new orbit is an ellipse.

c. Conservation of energy gives

$$\frac{-3e^2}{8\pi\varepsilon_0 r_0} = \frac{-e^2}{2\pi\varepsilon_0 r} + \frac{m(\dot{r}^2 + r^2\dot{\theta}^2)}{2}.$$

At positions where the orbiting electron is at the distance of closest or farthest approach to the atom, we have $\dot{r} = 0$, for which

$$\frac{-3e^2}{8\pi\varepsilon_0 r_0} = \frac{mr^2\dot{\theta}^2}{2} - \frac{e^2}{2\pi\varepsilon_0 r} = \frac{L^2}{2mr^2} - \frac{e^2}{2\pi\varepsilon_0 r}.$$

Then with

$$L^2 = m^2 v_0^2 r_0^2 = \frac{me^2 r_0}{4\pi\varepsilon_0}$$

the above becomes

$$3r^2 - 4r_0 r + r_0^2 = 0,$$

with solutions

$$r = \frac{r_0}{3}, \quad r = r_0.$$

Hence the distances of closest and farthest approach in the new orbit are respectively

$$r_{\min} = \frac{1}{3}, \quad r_{\max} = 1$$

in units of r_0.

d. Let $2a$ and $2b$ be the major and minor axes of the new elliptical orbit respectively, and $2c$ the distance between its two focuses. We have

$$2a = r_{\min} + r_{\max} = \frac{4r_0}{3},$$

$$2c = r_{\max} - r_{\min} = \frac{2r_0}{3},$$

$$2b = 2\sqrt{a^2 - c^2} = \frac{2\sqrt{3}r_0}{3}.$$

1052

A satellite is launched from the earth on a radial trajectory away from the sun with just sufficient velocity to escape from the sun's gravitational field. It is timed so that it will intercept Jupiter's orbit a distance b behind Jupiter, inter-act with Jupiter's gravitational field and be deflected by 90°, i.e., its velocity after the collision is tangential to Jupiter's orbit (Fig. 1.35). How much energy did the satellite gain in the collision? Ignore the sun's gravitational field during the collision and assume that the duration of the collision is small compared with Jupiter's period.

(UC, Berkeley)

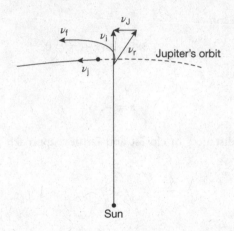

Fig. 1.35

Sol: Let r represent the distance from Jupiter to the sun, v_i the velocity of the satellite with respect to the sun at the time it intercepts Jupiter's orbit a distance b behind it and before any interaction with it, and m and M_s the masses of the satellite and the sun respectively. As the satellite just escapes the sun's gravitational field, we have

$$\frac{mv_i^2}{2} = \frac{GmM_s}{r},$$

giving

$$v_i = \sqrt{\frac{2GM_s}{r}} = \sqrt{\frac{2 \times 4.01 \times 10^{14} \times 3.33 \times 10^5}{7.78 \times 10^{11}}}$$

$$= 1.85 \times 10^4 \text{ m/s} = 18.5 \text{ km/s},$$

where we have used $M_s = 3.33 \times 10^5 M_e$ (M_e is the earth's mass), $GM_e = gR^2$ (R is the radius of the earth) $= 4.01 \times 10^{14} \text{m}^3/\text{s}^2$, $r = 7.78 \times 10^{11}$m.

The velocity v_J of Jupiter with respect to the sun is given by

$$\frac{v_J^2}{r} = \frac{GM_s}{r^2},$$

i.e.

$$v_J = \sqrt{\frac{GM}{r}} = \frac{v_i}{\sqrt{2}} = 13.1 \text{ km/s}.$$

When the satellite just enters the gravitational field of Jupiter, its velocity in the Jupiter frame is

$$\mathbf{v}_r = \mathbf{v}_i - \mathbf{v}_J,$$

or

$$v_r = \sqrt{18.5^2 + 13.1^2} = 22.67 \text{ km/s}.$$

If b does not change during the encounter, conservation of the angular momentum of the satellite in the Jupiter frame shows that this is also the speed of the satellite in the Jupiter frame when it leaves the gravitational field of Jupiter. After the encounter, the satellite leaves the gravitational field of Jupiter with a velocity in the sun's frame tangential to Jupiter's orbit. Thus the speed of the satellite with respect to the sun is

$$v_f = v_r + v_J = 22.67 + 13.1 = 35.77 \text{ km/s}.$$

The energy gained by unit mass of the satellite in the collision is therefore

$$\frac{35.77^2 - 18.5^2}{2} = 468.6 \times 10^6 \text{ J/kg}.$$

1053

By what arguments and using what measurable quantities can one determine the following quantities with good accuracy?

  a. The mass of the earth.

  b. The mass of the moon.

  c. The distance from the earth to the sun.

(Columbia)

Sol:

a. The weight of a body on the earth arises from the gravitational attraction of the earth. We have

$$mg = \frac{Gm_e m}{R^2},$$

whence the mass of the earth is

$$m_e = \frac{gR^2}{G},$$

where the acceleration of gravity g, the radius of the earth R, and the gravitational constant G are measurable quantities.

b. Consider a 2-body system consisting of masses m_1, m_2, separated by r, under gravitational interaction. The force equation is

$$\frac{Gm_1 m_2}{r^2} = \left(\frac{m_1 m_2}{m_1 + m_2}\right) r\omega^2,$$

or

$$G(m_1 + m_2) = \frac{4\pi^2 r^3}{T^2},$$

where $m_1 m_2 / (m_1 + m_2)$ is the reduced mass of the system. Applying this to the moon-earth system, we have

$$G(m_m + m_e) = \frac{4\pi a^3}{T^2},$$

where m_m, a and T are the mass, semimajor axis and period of revolution of the moon respectively. With the knowledge of m_e obtained in (a) and that of a and T determined by astronomical observations, m_m can be obtained.

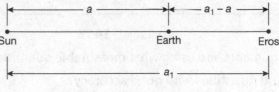

Fig. 1.36

c. Described below is a historical method for determining the earth-sun distance using the asteroid Eros. When the sun, earth and Eros are on a straight line as in Fig. 1.36, two observers A and B at latitudes λ_1 and λ_2 in the meridian plane containing Eros and the sun measure the angles α_1 and α_2 as shown in Fig. 1.37. As

$$\alpha_2 = \lambda_2 - \delta + \beta_2, \quad \alpha_1 = \lambda_1 - \delta + \beta_1,$$

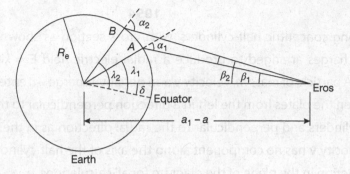

Fig. 1.37

giving

$$\beta_2 - \beta_1 = \alpha_2 - \alpha_1 - \lambda_2 + \lambda_1,$$

and

$$\frac{R_e}{\sin \beta_2} = \frac{a_1 - a}{\sin \alpha_2}, \qquad \frac{R_e}{\sin \beta_1} = \frac{a_1 - a}{\sin \alpha_1},$$

giving

$$\beta_2 - \beta_1 \approx \sin \beta_2 - \sin \beta_1 = \frac{R_e}{a_1 - a}(\sin \alpha_2 - \sin \alpha_1),$$

we have

$$a_1 - a \approx \frac{R_e(\sin \alpha_2 - \sin \alpha_1)}{\alpha_2 - \alpha_1 - \lambda_2 + \lambda_1}.$$

Kepler's third law gives

$$\frac{a^3}{a_1^3} = \frac{T^2}{T_1^2},$$

where T and T_1 are respectively the periods of revolution of the earth and Eros around the sun. The last two equations, used together, determine a.

However, this method is not accurate because the orbits of revolution of the earth and Eros are elliptical, not circular (eccentricity of Eros' orbit = 0.228), and the angle formed by their orbital planes is greater than 10°. More accurate, but non-mechanical, methods are now available for determining the earth-sun distance.

1054

Two long concentric half-cylinders, with cross section as shown in Fig. 1.38, carry charges arranged to produce a radial electric field $\mathbf{E} = ke_r/r$ between them. A particle of mass m, velocity $\mathbf{v}$ and negative charge $-q$ enters the region between the plates from the left in a direction perpendicular to the axis of the half-cylinders and perpendicular to the radial direction as in the figure. Since the velocity $\mathbf{v}$ has no component along the axis of the half-cylinders, consider only motion in the plane of the diagram for all calculations.

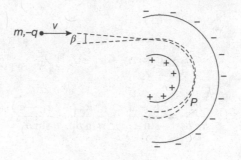

Fig. 1.38

a. If the particle moves on a circular path while between the plates, what must be the radius r of that path?

b. Next consider a trajectory for which the particle enters the region between the plates at the same distance r from the axis and the same speed as in (a), but at a small angle β with the direction of the original path. For small β the point P at which this new trajectory again crosses the trajectory in (a) is independent of β. Find the location of that point P. (Again the particle has no velocity component along the axis of the half-cylinders and remains in the plane of the diagram.)

c. How is the answer to part (a) changed if a uniform magnetic field is introduced parallel to the axis of the half-cylinders?

(Columbia)

Sol:

a. As the particle moves in a circular path, we have

$$\frac{mv^2}{r} = qE = \frac{qk}{r},$$

or

$$v^2 = \frac{qk}{m}.$$

As long as the velocity **v** of the particle satisfies this relation, it will move in a circular path whose radius is equal to the distance between the incident line and the axis of the half-cylinders, while between the plates.

b. The particle enters the region between the half-cyhnders at the same distance r_0 from the axis and with the same speed v as in (a), but at a small angle β with the direction of the original path. Conservation of angular momentum and energy gives

$$mr^2\dot{\theta} = mr_0 v, \tag{1}$$

$$\frac{1}{2}m(\dot{r}^2 + r^2\dot{\theta}^2) + qk\ln\left(\frac{r}{r_0}\right) = \frac{1}{2}mv^2. \tag{2}$$

As the new trajectory deviates from the original one slightly, we set

$$r = r_0 + \delta r,$$

$$\dot{r} = \frac{d}{dt}(\delta r),$$

$$\dot{\theta} = \omega_0 + \delta\dot{\theta},$$

where ω_0 is the angular velocity for the original circular orbit, and δr, $\delta\dot{\theta}$ are small quantities. Substitution in (1) gives

$$\dot{\theta} = \frac{r_0 v}{r^2} = \frac{v}{r_0} \frac{1}{\left(1 + \frac{\delta r}{r_0}\right)^2},$$

or

$$\dot{\theta}^2 = \left(\frac{v}{r_0}\right)^2 \Bigg/ \left(1 + \frac{\delta r}{r_0}\right)^4$$

$$= \left(\frac{v}{r_0}\right)^2 \cdot \left[1 - 4\frac{\delta r}{r_0} + 10\left(\frac{\delta r}{r_0}\right)^2\right].$$

By a similar approximation,

$$\ln\left(\frac{r}{r_0}\right) = \ln\left(1 + \frac{\delta r}{r_0}\right) = \frac{\delta r}{r_0} - \frac{1}{2}\left(\frac{\delta r}{r_0}\right)^2.$$

We can also write

$$r^2 = (r_0 + \delta r)^2 = r_0^2\left[1 + 2\frac{\delta r}{r_0} + \left(\frac{\delta r}{r_0}\right)^2\right].$$

Hence (2) can be written, neglecting small quantities higher than the second order, as

$$\frac{1}{2}m\left(\frac{d\delta r}{dt}\right)^2 + \left(\frac{3}{2}\frac{mv^2}{r_0^2} - \frac{qk}{2r_0^3}\right)(\delta r)^2 + \left(\frac{qk}{r_0} - \frac{mv^2}{r_0}\right)\delta r = 0,$$

or, noting $qk = mv^2$,

$$\frac{1}{2}\left(\frac{d\delta r}{dt}\right)^2 + \left(\frac{v}{r_0}\right)^2(\delta r)^2 = 0.$$

Then taking the time derivative of both sides, we obtain

$$\frac{d^2\delta r}{dt^2} + 2\left(\frac{v}{r_0}\right)^2\delta r = 0,$$

with solution

$$\delta r = A\sin\left(\sqrt{2}\frac{v}{r_0}t + \varphi\right),$$

Where A and φ are constants of integration. The initial conditions

$$r(0) = r_0, \quad \dot{r}(0) = v\sin\beta,$$

or

$$\delta r = 0, \quad \frac{d}{dt}\delta r = v\sin\beta \quad \text{at } t = 0,$$

give

$$\varphi = 0, \quad A = \frac{r_0}{\sqrt{2}}\sin\beta.$$

Hence

$$\delta r = \frac{r_0}{\sqrt{2}}\sin\beta\sin\left(\sqrt{2}\frac{v}{r_0}t\right).$$

At the point where this new trajectory crosses the trajectory in (a), $\delta r = 0$. The second crossing takes place at a time t later given by

$$\sqrt{2}\frac{v}{r_0}t = \pi, \quad \text{or} \quad t = \frac{\pi}{\sqrt{2}}\frac{r_0}{v},$$

and the position of P is given by

$$\theta = \dot{\theta}t \approx \omega_0 t = \frac{v}{r_0}t = \frac{\pi}{\sqrt{2}},$$

which is independent of β.

c. If a uniform magnetic field parallel to the axis of the half-cylinders is used
    instead of the electric field, we will have

$$\frac{mv^2}{r} = qvB.$$

Then the radius r of the circular path will be

$$r = \frac{mv}{qB}.$$

1055

The orbit of a particle moving under the influence of a central force is
$r\theta = $ constant. Determine the potential energy as a function of r.

(*Columbia*)

Sol: Consider a central force $\mathbf{F} = \mathbf{r}F(r)$ acting on a particle of mass m. Newton's second law gives

$$F = m(\ddot{r} - r\dot{\theta}^2), \tag{1}$$

$$0 = m(r\ddot{\theta} + 2\dot{r}\dot{\theta}) \tag{2}$$

in polar coordinates. Equation (2) gives

$$r\ddot{\theta} + 2\dot{r}\dot{\theta} = \frac{1}{r}\frac{d}{dt}(r^2\dot{\theta}) = 0,$$

or

$$r^2\dot{\theta} = \text{constant} = h, \text{ say},$$

or

$$\dot{\theta} = hu^2$$

by putting

$$r = \frac{1}{u}.$$

Then as

$$\dot{r} = \frac{dr}{dt} = \dot{\theta}\frac{dr}{d\theta} = hu^2\frac{dr}{d\theta} = -h\frac{du}{d\theta},$$

$$\ddot{r} = \dot{\theta}\frac{d\dot{r}}{d\theta} = hu^2\frac{d}{d\theta}\left(-h\frac{du}{d\theta}\right) = -h^2u^2\frac{d^2u}{d\theta^2},$$

$$r\dot{\theta}^2 = \frac{1}{u}h^2u^4 = h^2u^3,$$

Eq. (1) becomes

$$F = -mh^2u^2\left(\frac{d^2u}{d\theta^2} + u\right),$$

which is often known as Binet's formula.

In this problem, let $r = \frac{1}{u}$ and write the equation of the trajectory as

$$u = C\theta,$$

where C is a constant. Binet's formula then gives

$$F = -mh^2u^3 = \frac{-mh^2}{r^3}.$$

The potential energy is by definition

$$V = -\int_\infty^r F(r)\,dr = \int_\infty^r \frac{mh^2}{r^3}\,dr = \left[\frac{-mh^2}{2r^2}\right]_\infty^r = \frac{-mh^2}{2r^2},$$

taking an infinity point as the zero potential level.

1056

Mariner 4 was designed to travel from earth to Mars in an elliptical orbit with its perihelion at earth and its aphelion at Mars. Assume that the orbits of earth and Mars are circular with radii R_E and R_M respectively. Neglect the gravitational effects of the planets on Mariner 4.

   a. With what velocity, relative to earth, does Mariner 4 have to leave earth, and in what direction?
   b. How long will it take to reach Mars?
   c. With what velocity, relative to Mars, will it reach the orbit of Mars? (The time at which Mariner 4 leaves earth must be properly chosen if it is to arrive at Mars. Assume this is done.)

(Columbia)

Sol: As the gravitational force on Mariner 4, which is a central force, is conservative, we have

$$E = \frac{m\dot{r}^2}{2} - \frac{GmM}{r} + \frac{mh^2}{2r^2},$$

where m and M are the masses of Mariner 4 and the sun respectively, G is the gravitational constant, and $h = r^2 \dot{\theta}$ is a constant. At the perihelion and aphelion of the elliptical orbit, $\dot{r} = 0$, $r = R_E$ and $r = R_M$ respectively. Then

$$E = \frac{-GmM}{R_M} + \frac{mh^2}{2R_M^2} = \frac{-GmM}{R_E} + \frac{mh^2}{2R_E^2},$$

giving

$$h = \sqrt{\frac{2GMR_M R_E}{R_M + R_E}}.$$

At the perihelion we obtain its velocity relative to the sun as

$$h = \frac{h}{R_E} = \sqrt{\frac{2GMR_M}{R_E(R_M + R_E)}}.$$

Suppose Mariner 4 is launched in a direction parallel to the earth's revolution around the sun. The velocity relative to the earth with which Mariner 4 is to leave the earth is then

$$v_r = v - v_E = \sqrt{\frac{2GMR_M}{R_E(R_M + R_E)}} - \sqrt{\frac{GM}{R_E}},$$

Where v_E is the velocity of revolution of the earth. Similarly at the aphelion the velocity, relative to Mars, which Mariner 4 must have is

$$v'_r = v' - v_M = \sqrt{\frac{2GMR_E}{R_M(R_M + R_E)}} - \sqrt{\frac{GM}{R_E}}.$$

Applying Kepler's third law we have for the period T of revolution of Mariner 4 around the sun

$$T^2 = T_E^2 \left(\frac{R_E + R_M}{2} \right)^3 R_E^{-3},$$

where T_E = period of revolution of the earth = 1 year. Hence the time taken for Mariner 4 to reach Mars in years is

$$t = \frac{T}{2} = \frac{1}{2} \left(\frac{R_E + R_M}{2R_E} \right)^{\frac{3}{2}}.$$

1057

A charged pion (π^+ or π^-) has (non-relativistic) kinetic energy T. A massive nucleus has charge Ze and effective radius b. Considered classical, the pion "hits" the nucleus if its distance of closest approach is b or less. Neglecting nucleus recoil (and atomic-electron effects), show that the collision cross section for these pions is

$$\sigma = \frac{\pi b^2 (T - V)}{T}, \qquad \text{for } \pi^+,$$

and

$$\sigma = \frac{\pi b^2 (T + V)}{T}, \qquad \text{for } \pi^-,$$

where

$$V = \frac{Ze^2}{b}.$$

(Columbia)

Sol: Let d be the impact parameter with which a pion approaches the nucleus. The pion has initial velocity $\sqrt{\dfrac{2T}{m}}$ and angular momentum $\sqrt{2Tm}\,d$, where m is its mass. At the closest approach, the pion has no radial velocity, i.e., $v_r = 0$, $v = b\dot\theta$. Conservation of angular momentum gives

$$\sqrt{2Tm}\,d = mb^2\dot\theta$$

or

$$d = \sqrt{\frac{m}{2T}}b^2\dot\theta.$$

Conservation of energy gives

$$T = V + \frac{1}{2}mb^2\dot\theta^2,$$

as a potential V now comes into play, or

$$b\dot\theta = \sqrt{\frac{2(T - V)}{m}}.$$

The collision cross section is

$$\sigma = \pi d^2 = \frac{m}{2T}\frac{2(T-V)}{m}\pi b^2 = \pi b^2\left(\frac{T-V}{T}\right).$$

Putting $V = \dfrac{Ze^2}{b}$, for π^+ we have

$$\sigma = \pi b^2\left(\frac{T-V}{T}\right),$$

and for π^-, the potential is $-\dfrac{Ze^2}{b}$ and we have

$$\sigma = \pi b^2\left(\frac{T+V}{T}\right).$$

1058

Estimate how big an asteroid you could escape by jumping.

(Columbia)

Sol: Generally speaking, before jumping, one always bends one's knees to lower the center of gravity of the body by about 50 cm and then jumps up. You can usually reach a height 60 cm above your normal height. In the process, the work done is $(0.5 + 0.6)mg$, where m is the mass of your body and g is the acceleration of gravity.

It is reasonable to suppose that when one jumps on an asteroid of mass M and radius R one would consume the same energy as on the earth. Then to escape from the asteroid by jumping we require that

$$1.1mg = \frac{GMm}{R}.$$

If we assume that the density of the asteroid is the same as that of the earth, we obtain

$$\frac{M}{M_E} = \frac{R^3}{R_E^3},$$

where M_E and R_E are respectively the mass and radius of the earth. As $g = GM_E/R_E^2$, we find

$$R = \frac{GM}{1.1g} = \frac{R^3}{1.1R_E},$$

or

$$R = \sqrt{1.1 R_E} = \sqrt{1.1 \times 6400 \times 10^3} = 2.7 \times 10^3 \text{ m}.$$

1059

You know that the acceleration due to gravity on the surface of the earth is 9.8 m/sec², and that the length of a great circle around the earth is 4×10^7m. You are given that the ratios of moon/earth diameters and masses are

$$\frac{D_m}{D_e} = 0.27 \quad \text{and} \quad \frac{M_m}{M_e} = 0.0123$$

respectively.

a. Compute the minimum velocity required to escape from the moon's gravitational field when starting from its surface.

b. Compare this speed with thermal velocities of oxygen molecules at the moon's temperature which reaches 100°C.

(UC, Berkeley)

Sol:

a. Let the velocity required to escape from the moon's gravitational field be v_{min}, then

$$\frac{m v_{min}^2}{2} = \frac{G M_m m}{r_m},$$

giving

$$v_{min} = \sqrt{\frac{2 G M_m}{r_m}} = \sqrt{\left(\frac{0.0123}{0.27}\right)\left(\frac{G M_e}{r_e^2}\right) 2 r_e}$$

$$= \sqrt{\left(\frac{0.0123}{0.27}\right) \cdot g \cdot D_e} = 2.38 \times 10^3 \text{ m/s}$$

using $g = G M_e / r_e^2$, $D_m/D_e = 0.27$ and $M_m/M_e = 0.0123$.

b. The average kinetic energy of the translational motion of oxygen molecules at a temperature of 100°C is $3kT/2$:

$$\frac{1}{2} m v^2 = \frac{3}{2} k T.$$

Hence

$$v = \sqrt{\frac{3kT}{m}} = \sqrt{\frac{3 \times 1.38 \times 10^{-23} \times 373}{32 \times 1.67 \times 10^{-27}}} = 538 \text{ m/s}.$$

v, which is the root-mean-square speed of an oxygen molecule at the highest moon temperature, is smaller than v_{min}, the speed required to escape from the moon.

1060

An object of unit mass orbits in a central potential $U(r)$. Its orbit is $r = ae^{-b\theta}$, where θ is the azimuthal angle measured in the orbital plane. Find $U(r)$ to within a multiplicative constant.

(MIT)

Sol: Let

$$u = \frac{1}{r} = \frac{e^{b\theta}}{a}.$$

Then

$$\frac{d^2u}{d\theta^2} = \frac{b^2 e^{b\theta}}{a} = b^2 u,$$

and Binet's formula (Problem **1055**)

$$F = -mh^2 u^2 \left(\frac{d^2u}{d\theta^2} + u \right)$$

gives for $m = 1$

$$F = -h^2(b^2 + 1)u^3 = -\frac{h^2(b^2 + 1)}{r^3} = -\frac{dU(r)}{dr}.$$

Integrating and taking the reference level for $U(r)$ at $r \to \infty$, we obtain

$$U(r) = \frac{-h^2}{2} \cdot \frac{b^2 + 1}{r^2},$$

where $h = r^2 \dot{\theta}$ is the conserved angular momentum of the object about the force center and is to be determined by the initial condition.

1061

Hard sphere scattering.

Show that the classical cross section for elastic scattering of point particles from an infinitely massive sphere of radius R is isotropic.

(MIT)

Sol: For elastic scattering, the incidence angle equals the angle of reflection. The angle of scattering is then $\varphi = 2\theta$ as shown in Fig. 1.39.

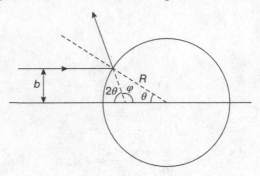

Fig. 1.39

If b is the impact parameter, we have

$$b = R \sin \theta,$$

and

$$db = R \cos \theta \, d\theta.$$

The differential scattering cross section per unit solid angle $\dfrac{d\sigma}{d\Omega}$ is given by

$$2\pi b \, db = 2\pi R^2 \sin \theta \cos \theta \, d\theta = \frac{d\sigma}{d\Omega} 2\pi \sin \varphi \, d\varphi,$$

or

$$\frac{d\sigma}{d\Omega} = \frac{1}{2} \frac{R^2 \sin 2\theta \, d\theta}{\sin \varphi \, d\varphi}$$

$$= \frac{R^2}{4} \frac{\sin \varphi \, d\varphi}{\sin \varphi \, d\varphi} = \frac{R^2}{4}.$$

Thus the classical differential cross section is independent of the angle of scattering. In other words, the scattering is isotropic.

1062

Find the angular distribution and total cross section for the scattering of small marbles of mass m and radius r from a massive billiard ball of mass M and radius R ($m \ll M$). You should treat the scattering as elastic, involving no frictional forces.

(Columbia)

Sol: As $m \ll M$, the massive billiard ball will remain stationary during scattering. As the scattering is elastic (see Fig. 1.40), the scattering angle Θ is related to the angle of incidence by

$$\Theta = \pi - 2\theta,$$

where θ is given by

$$(R + r) \sin \theta = b.$$

The differential scattering cross section is

$$\frac{d\sigma}{d\Omega} = \frac{|2\pi b db|}{d\Omega} = \frac{2\pi |\sin \theta \cos \theta \cdot (R + r)^2 d\theta|}{2\pi d \cos \Theta}$$

$$= \frac{\left| \frac{1}{4}(R + r)^2 \sin \Theta d\Theta \right|}{d \cos \Theta} = \frac{\frac{1}{4}(R + r)^2 d \cos \Theta}{d \cos \Theta}$$

$$= \frac{1}{4}(R + r)^2.$$

As $\dfrac{d\sigma}{d\Omega}$ is isotropic, the total cross section is

$$\sigma_t = 4\pi \frac{d\sigma}{d\Omega} = \pi(R + r)^2.$$

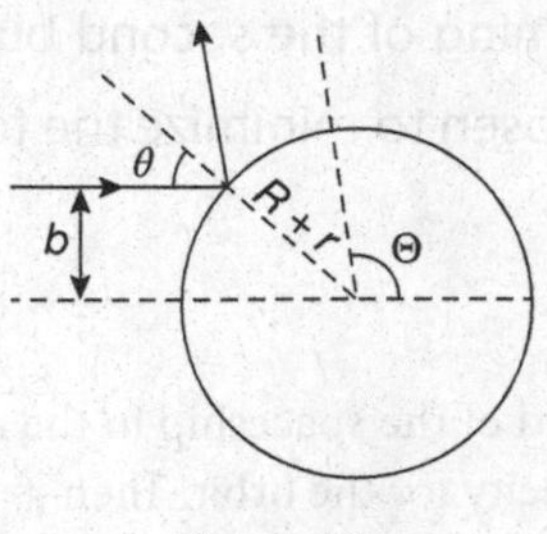

Fig. 1.40

1063

A spaceship is in a circular orbit of radius r_0 around a star of mass M. The spaceship's rocket engine may be fired briefly to alter its velocity (instantaneously) by an amount $\Delta \mathbf{v}$. The direction of firing is specified by the angle θ between the ship's velocity $\mathbf{v}$ and the vector from the tail to the nose of the ship (see Fig. 1.41). To conserve fuel in a sequence of N firings, it is desirable to minimize $\Delta V = \sum_{i=1}^{N} |\Delta \mathbf{v}_i|$. ΔV is known as the specific impulse.

a. Suppose we want to use the ship's engine to escape from the star. What is the minimum specific impulse required if the engine is fired in a single rapid burst? In what direction should the engine be fired?

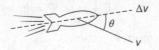

Fig. 1.41

b. Suppose we wish to visit a planet in a circular orbit of radius $r_1 > r_0$. What is the minimum specific impulse required to reach the planet's orbit if the engine is again fired in a single rapid burst?

Suppose we want to use the ship's engine to cause it to crash into the star (assume the radius of the star to be negligible). Calculate the minimum specific impulse for both of the following firing strategies:

c. A single rapid burst at $\theta = 180°$.

d. A single rapid burst at $\theta = 0°$ and then a second burst at $\theta = 180°$ at a later time. The timing of the second burst and the strength of each burst should be chosen to minimize the total specific impulse.

(*MIT*)

Sol:

a. Let v_0 be the speed of the spaceship in the circular orbit of radius r_0, and v_{0e} be the escape velocity for the orbit. Then

$$\frac{mv_0^2}{r_0} = \frac{GMm}{r_0^2}, \qquad \frac{mv_{0e}^2}{2} = \frac{GMm}{r_0},$$

or

$$v_0 = \sqrt{\frac{GM}{r_0}}, \qquad v_{0e} = \sqrt{\frac{2GM}{r_0}}.$$

As

$$v_{0e} = v_0 + |\Delta \mathbf{v}| \cos \theta = v_0 + \Delta V \cos \theta;$$

the specific impulse required for escape is the least for $\theta = 0$, i.e., the initial velocity of the spaceship and the impulse are in the same direction, and is given by

$$\Delta V = v_{0e} - v_0 = \sqrt{\frac{GM}{r_0}}(\sqrt{2} - 1).$$

b. After the first burst, the ship escapes from the circular orbit around the star and moves along a parabolic orbit. When the ship reaches the circular orbit $r = r_1$ of the planet, the engine is again fired in a single rapid burst (see Fig. 1.42). For the ship to move along the circular orbit of radius r_1, its speed must be

$$v_1 = \sqrt{\frac{GM}{r_1}}.$$

Let v_{1e} be the speed of the ship as it arrives at $r = r_1$ and before the burst. Conservation of angular momentum requires

$$v_{0e} r_0 = v_{1e} r_1 \cos \varphi$$

or

$$v_{1e} \cos \varphi = \frac{r_0 v_{0e}}{r_1}.$$

Conservation of energy gives

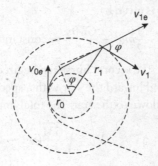

Fig. 1.42

$$\frac{1}{2}mv_{1e}^2 = \frac{GMm}{r_1},$$

or

$$v_{1e} = \sqrt{\frac{2GM}{r_1}}.$$

Then the minimum specific impulse required is given by

$$\Delta V = |\mathbf{v}_1 - \mathbf{v}_{1e}|$$

or

$$(\Delta V)^2 = v_{1e}^2 + v_1^2 - 2v_{1e}v_1 \cos\varphi$$

$$= \frac{2GM}{r_1} + \frac{GM}{r_1} - 2\frac{r_0}{r_1}v_{0e}v_1$$

$$= \frac{3GM}{r_1} - 2\frac{r_0}{r_1}\sqrt{\frac{2GM}{r_0}} \cdot \sqrt{\frac{GM}{r_1}}$$

$$= \frac{GM}{r_1}\left(3 - 2\sqrt{\frac{2r_0}{r_1}}\right).$$

Hence

$$\Delta V = \sqrt{\frac{GM}{r_1}\left(3 - 2\sqrt{\frac{2r_0}{r_1}}\right)}.$$

c. For a single rapid burst at $\theta = 180°$, the minimum specific impulse is that which makes the speed of the spaceship $v' = v_0 - \Delta V = 0$, so that it will fall onto the star. Hence the minimum impulse required is

$$\Delta V = v_0 = \sqrt{\frac{GM}{r_0}}.$$

d. If after the first burst with $\theta = 0°$, the ship acquires the escape velocity $v_{0e} = \sqrt{\frac{2GM}{r_0}}$, i.e., $\Delta V_1 = \sqrt{\frac{GM}{r_0}}(\sqrt{2} - 1)$, it can escape from the orbit. The speed of the ship v is given by

$$\frac{1}{2}mv^2 - \frac{GMm}{r} = \text{constant}.$$

As $r \to \infty$, $v \to 0$. The second burst can be fired when $v \approx 0$ at $\theta = 180°$ to turn the ship around toward the star with a specific impulse $\Delta V_2 \approx 0$; thereafter the ship falls down to the star. The total impulse required is

$$\Delta V \approx \Delta V_1 = \sqrt{\frac{GM}{r_0}}(\sqrt{2} - 1).$$

That this is the minimum impulse can be seen from the following.

Suppose the first burst fired at $\theta = 0°$ is

$$\Delta V_1 \le v_{0e} - v_0,$$

then the ship will move along an elliptic orbit. The speed is then minimum at the aphelion, at which point the second burst at $\theta = 180°$ should be fired to make ΔV_2 small. Suppose at the aphelion the ship is at distance r_2 from the star and has speed v_2, then ΔV_1 is given by the energy equation

$$\frac{1}{2}mv_2^2 - \frac{GMm}{r_2} = \frac{1}{2}m(v_0 + \Delta V_1)^2 - \frac{GMm}{r_0}$$

and the angular momentum equation

$$mr_2v_2 = mr_0(v_0 + \Delta V_1).$$

Eliminating r_2 from the above, we have

$$v_2^2 - \frac{2GMv_2}{r_0(v_0 + \Delta V_1)} + \frac{2GM}{r_0} - (v_0 + \Delta V_1)^2 = 0,$$

giving

$$v_2 = \frac{GM}{r_0(v_0 + \Delta V_1)} \pm \left[\frac{GM}{r_0(v_0 + \Delta V_1)} - (v_0 + \Delta V_1)\right],$$

where the lower sign corresponds to the speed at the perihelion and the upper sign to the speed at the aphelion. At the aphelion,

$$v_2 = \frac{2GM}{r_0(v_0 + \Delta V_1)} - (v_0 + \Delta V_1).$$

The second burst must be such that $\Delta \mathbf{v}_2$ is equal to $\mathbf{v}_2$ in magnitude but opposite in direction in order that the ship can stop and fall down to the star, i.e., $\mathbf{v}_2 + \Delta \mathbf{v}_2 = 0$. Thus

$$\Delta V_2 = |\Delta \mathbf{v}_2| = v_2 = \frac{2GM}{r_0(v_0 + \Delta V_1)} - v_0 - \Delta V_1,$$

or

$$\Delta V_1 + \Delta V_2 = \frac{2GM}{r_0(v_0 + \Delta V_1)} - v_0.$$

From the above it can be seen that the larger the value of ΔV_1, the smaller is the specific impulse $\Delta V = \Delta V_1 + \Delta V_2$, under the condition

$$\Delta V_1 \le v_{1e} - v_0.$$

Hence in order to minimize the total specific impulse, the first burst should carry impulse $\Delta V_1 = \sqrt{GM/r_0}(\sqrt{2} - 1)$ and after an infinitely long period of time, a second burst is fired with an infinitesimal impulse ΔV_2.

1064

"Interstellar bullets" are thought to be dense clumps of gas which move like ballistic particles through lower-density interstellar gas clouds. Consider a uniform spherical cloud of radius R, mass M, and a "bullet" of radius $\ll R$ and mass $m \ll M$. Ignore all non-gravitational interactions.

a. Obtain expressions for the force $\mathbf{F}(r)$, $0 < r < \infty$, suffered by the bullet in terms of the distance r from the cloud center, and for the potential energy $V(r)$, $0 < r < \infty$. Sketch $V(r)$.

b. The bullet has angular momentum $L = m(GMR/32)^{1/2}$ about $r = 0$ and total energy $E = -5GMm/4R$. Find the orbit turning point(s). Is the bullet always in the cloud, outside the cloud, or sometimes inside and sometimes outside?

c. For L and E as in (b), obtain an expression for the differential orbit angle $d\theta$ in terms of dr, r and R.

d. Obtain an orbit equation $r(\theta, R)$ by integrating your answer to (c), you may wish to use

$$\int \frac{dx}{\sqrt{a + bx + cx^2}} = \frac{1}{\sqrt{-c}} \arcsin \frac{-2cx - b}{\sqrt{b^2 - 4ac}}, \quad c < 0.$$

Find the turning points r_1 and sketch the orbit.

(*MIT*)

Sol:

a. The force $\mathbf{F}$ acting on the bullet is

$$\mathbf{F} = \begin{cases} -\dfrac{GMm}{R^3}\mathbf{r} & (0 < r \le R), \\ -\dfrac{GMm}{r^3}\mathbf{r} & (R \le r < \infty). \end{cases}$$

From the definition of potential energy $V(r)$, $\mathbf{F} = -\nabla V(r)$, we have

$$V(r) = -\int_{\infty}^{R} \mathbf{F} \cdot d\mathbf{r} - \int_{R}^{r} \mathbf{F} \cdot d\mathbf{r} \quad (0 < r \le R),$$

$$V(r) = -\int_{\infty}^{r} \mathbf{F} \cdot d\mathbf{r} \quad (R \le r < \infty).$$

Substituting in the appropriate expressions for **F** and integrating, we find

$$V(r) = \begin{cases} -\dfrac{GMm}{2R^3}(3R^3 - r^2) & (0 < r \le R), \\[2mm] -\dfrac{GMm}{r^2} & (R \le r < \infty). \end{cases}$$

A sketch of $V(r)$ is shown in Fig. 1.43.

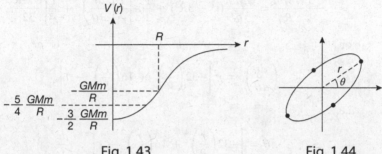

Fig. 1.43　　　　　　　　Fig. 1.44

b. As shown in Fig. 1.43, with total energy $E = -5GMm/4R$, the bullet can only move inside the gas cloud in a region bounded by the turning points. At the turning points, $\dot{r} = 0$, $v = v_\theta$. Hence

$$rmv_\theta = m\sqrt{\frac{GMR}{32}},$$

$$\frac{GMm}{2R^3}(r^2 - 3R^2) + \frac{mv_\theta^2}{2} = \frac{-5GMm}{4R}.$$

Eliminating v_θ, we have for the turning distance r

$$32\left(\frac{r}{R}\right)^4 - 16\left(\frac{r}{R}\right)^2 + 1 = 0,$$

which has solutions

$$\left(\frac{r}{R}\right)^2 = \frac{2 \pm \sqrt{2}}{8},$$

giving

$$r\pm = \sqrt{\frac{2 \pm \sqrt{2}}{8}}R.$$

c. Conservation of energy and of angular momentum give

$$E = V(r) + \frac{m(\dot{r}^2 + r^2\dot{\theta}^2)}{2},$$

$$L = mr^2\dot{\theta}.$$

Substituting in the above

$$\dot{\theta} = \frac{L}{mr^2},$$

$$\dot{r} = \frac{dr}{d\theta}\frac{d\theta}{dt} = \dot{\theta}\frac{dr}{d\theta} = \frac{L}{mr^2}\frac{dr}{d\theta},$$

we have

$$-\frac{5}{4}\frac{GMm}{R} = \frac{GMm}{2R^3}(r^2 - 3R^2) + \frac{1}{2}m\left[\frac{1}{r^4}\left(\frac{dr}{d\theta}\right)^2 + \frac{1}{r^2}\right]\frac{GMR}{32},$$

or

$$\left(\frac{dr}{d\theta}\right)^2 = r^2\left[-32\left(\frac{r}{R}\right)^4 + 16\left(\frac{r}{R}\right)^2 - 1\right],$$

i.e.

$$d\theta = \left[-32\left(\frac{r}{R}\right)^4 + 16\left(\frac{r}{R}\right)^2 - 1\right]^{-1/2}\frac{dr}{r}.$$

d. To integrate the last equation, let $x = (r/R)^{-2}$ and rewrite the equation as

$$-2d\theta = \frac{dx}{\sqrt{-32 + 16x - x^2}}.$$

Integrating, we obtain

$$\alpha - 2\theta = \arcsin\frac{2x - 16}{8\sqrt{2}}$$

or

$$x = 8 + 4\sqrt{2}\sin(\alpha - 2\theta) = 8 + 4\sqrt{2}\cos(2\theta + \beta),$$

i.e.

$$\left(\frac{r}{R}\right)^2 = \frac{1}{4[2 + \sqrt{2}\cos(2\theta + \beta)]},$$

where β is a constant of integration. By a suitable choice of the coordinate axes, we can make $\beta = 0$. At a turning point, r is either maximum or minimum, i.e. $\cos\theta = \pm 1$. Hence the turning points are given by

$$\frac{r_\pm}{R} = \pm\sqrt{\frac{1}{4(2 \pm \sqrt{2})}} = \pm\sqrt{\frac{2 \mp \sqrt{2}}{8}}.$$

Thus there are a total of 4 turning points as shown in Fig. 1.44.

1065

A very broad parallel beam of small particles of mass m and charge zero is fired from space towards the moon, with initial velocity V_0 relative to the moon.

- **a.** What is the collision cross section for the particles to hit the moon? Express the cross section σ in terms of the moon's radius R, the escape velocity V_{esc} from the surface of the moon, and V_0. Neglect the existence of the earth and of the sun.

- **b.** If you are unable to derive the formula, partial credit will be given for a good formula guessed on the basis of dimensional analysis, and an argument as to what should be the answer in the two limits that V_0 goes to zero and V_0 goes to infinity.

(UC, Berkeley)

Solution:

- **a.** Let the maximum impact parameter be b_{max}. The particles will hit the moon if their distances of closest approach are b_{max} or less. Conservation of energy and angular momentum give

$$\frac{mV_0^2}{2} = \frac{mV^2}{2} - \frac{GMm}{R}, \tag{1}$$

$$mV_0 b_{max} = mVR. \tag{2}$$

From Eq. (1), we obtain

$$V^2 = V_0^2 + \frac{2GM}{R} = V_0^2 + V_{esc}^2.$$

Equation (2) then gives

$$b_{max}^2 = \frac{V^2R^2}{V_0^2} = R^2 + \frac{V_{esc}^2 R^2}{V_0^2}.$$

Hence the collision cross section for the particles to hit the moon is

$$\sigma = \pi b_{max}^2 = \pi R^2 \left(1 + \frac{V_{esc}^2}{V_0^2} \right).$$

- **b.** For the two limiting cases we have

$$\sigma \to \infty \quad \text{for } V_0 \to 0,$$
$$\sigma \to \pi R^2 \quad \text{for } V_0 \to \infty.$$

These results can be understood as follows. For very small V_0, all the particles will be attracted to the moon as we have neglected the effects of the earth and the sun. For very large velocities, only those aimed at the moon will arrive there as the potential energy due to the moon's attraction is negligible compared with the kinetic energy.

To apply the method of dimensional analysis, we make a guess that the cross section will be the geometrical cross section of the moon with some dimensionless correction factor involving V_0 and V_{esc}:

$$\sigma = \pi R^2\left[1 + b\left(\frac{V_{esc}}{V_0}\right)^a\right],$$

Where a and b are unknown constants which cannot be determined by this method alone. a however must be positive to satisfy our expectations for the two limiting cases.

1066

Consider the potential energy curve $U = U_0(-4 + e^{x/a} + e^{-x/a})$, where a has dimensions of distance and $U_0 > 0$, identify the regions where the particles in these regions can be treated as simple harmonic oscillators.

Sol: First we find the positions of the minima, $\dfrac{dU}{dx} = 0 = \dfrac{U_0}{a}(e^{x/a} - e^{-x/a})$,   (1)

i.e. $x = 0$.

Taylor expanding the potential U about $x = 0$,

$$U \approx -4U_0 + \left(1 + \frac{x}{a} + \frac{x^2}{2!a^2} + \dots\right)U_0 + \left(1 - \frac{x}{a} + \frac{x^2}{2!a^2} + \dots\right)U_0 \quad (2)$$

$$= U_0\left(-2 + \frac{x^2}{a^2}\right)$$

$$F = -dU/dx = -\frac{2x}{a^2} \propto -x$$

Therefore, the motion is simple harmonic about $x = 0$.

1067

A particle of mass m is bound by a linear potential $U = kr$.

a. For what energy and angular momentum will the orbit be a circle of radius r about the origin?

b. What is the frequency of this circular motion?

c. If the particle is slightly disturbed from this circular motion, what will be the frequency of small oscillations?

(UC, Berkeley)

Sol: The force acting on the particle is

$$\mathbf{F} = -\frac{dU}{dr}\hat{\mathbf{r}} = -k\hat{\mathbf{r}}.$$

a. If the particle moves in a circle of radius r, we have

$$m\omega^2 r = k,$$

i.e.

$$\omega^2 = \frac{k}{mr}.$$

The energy of the particle is then

$$E = kr + \frac{mv^2}{2} = kr + \frac{m\omega^2 r^2}{2} = \frac{3kr}{2}$$

and its angular momentum about the origin is

$$L = m\omega r^2 = mr^2\sqrt{\frac{k}{mr}} = \sqrt{mkr^3}.$$

b. The angular frequency of circular motion is $\omega = \sqrt{\dfrac{k}{mr}}$.

c. The effective potential is

$$U_{\text{eff}} = kr + \frac{L^2}{2mr^2}.$$

The radius r_0 of the stationary circular motion is given by

$$\left(\frac{dU_{\text{eff}}}{dr}\right)_{r=r_0} = k - \frac{L^2}{mr_0^3} = 0,$$

i.e.
$$r_0 = \left(\frac{L^2}{mk}\right)^{1/3}.$$

As
$$\left(\frac{d^2U_{\text{eff}}}{dr^2}\right)_{r=r_0} = \frac{3L^2}{mr^4}\Bigg|_{r=r_0} = \frac{3L^2}{m}\left(\frac{mk}{L^2}\right)^{4/3} = 3k\left(\frac{mk}{L^2}\right)^{1/3},$$

the angular frequency of small radial oscillations about r_0, if it is slightly disturbed from the stationary circular motion, is (Problem **1066**)

$$\omega_r = \sqrt{\frac{1}{m}\left(\frac{d^2U_{\text{eff}}}{dr^2}\right)_{r=r_0}} = \sqrt{\frac{3k}{m}\left(\frac{mk}{L^2}\right)^{1/3}}$$

$$= \sqrt{\frac{3k}{mr_0}} = \sqrt{3}\omega_0,$$

where ω_0 is the angular frequency of the stationary circular motion.

1068

A planet has a circular orbit around a star of mass M. The star explodes, ejecting its outer envelope at a velocity much greater than the orbital motion of the planet, so that the mass loss may be considered instantaneous. The remnant of the star has a mass M' which is much greater than the mass of the planet. What is the eccentricity of the planetary orbit after the explosion? (Neglect the hydrodynamic force exerted on the planet by the expanding shell. Recall that the eccentricity is given in terms of the energy E and angular momentum L by

$$e^2 = 1 + \frac{2EL^2}{M_pK^2},$$

where M_p is the mass of the planet and where the magnitude of the gravitational force between the star and planet is K/r_0^2.)

(UC, Berkeley)

Sol: Before the explosion the planet moves in a circle of radius R around the star. As the eccentricity e of the orbit is zero, we have from the given equation for e

$$E = \frac{-M_pK^2}{2L^2}.$$

As

$$\frac{M_p v^2}{R} = \frac{K}{R^2}, \qquad L = M_p R v,$$

we have

$$R = \frac{L^2}{M_p K}.$$

Let L' and E' be respectively the angular momentum and total energy of the planet after the explosion. Then

$$L' = L,$$
$$E' = E + \frac{G(M - M')M_p}{R}.$$

With $K = GMM_p$ and $K' = GM'M_p$ we have for the eccentricity e of the orbit after the explosion

$$e^2 = 1 + \frac{2E'L'^2}{M_p K'^2}$$

$$= 1 + \frac{2\left[-\dfrac{M_p K^2}{2L^2} + \dfrac{M_p K}{L^2}G(M - M')M_p\right]L^2}{M_p K'^2}$$

$$= 1 + \left(\frac{M}{M'}\right)^2\left(1 - \frac{2M'}{M}\right),$$

giving

$$e = \sqrt{1 + \left(\frac{M}{M'}\right)^2\left(1 - \frac{2M'}{M}\right)}.$$

1069

A satellite traveling in an elliptical orbit about the earth is acted on by two perturbations:

a. a non-central component to the earth's gravitational field arising from polar flattening,

    b. an atmospheric drag which, because of the rapid decrease in pressure with altitude, is concentrated near the perigee.

Give qualitative arguments showing how these perturbations will alter the shape and orientation of a Keplerian orbit.

(UC, Berkeley)

Sol: **a.** Owing to polar flattening of the earth (shaded area in Fig. 1.45), the equipotential surface in the neighboring space is a flattened sphere (dashed ellipsoid).

Suppose the orbital plane N of the satellite makes an angle θ with the equatorial plane M of the earth.

As the equipotential surface deviates from the spherical shape, the earth's gravitational force acting on the satellite, which is normal to the equipotential surface, does not direct toward the center of the earth (e.g, the forces on the satellite at A and B in Fig. 1.45). As the effect is quite small, the orbit of the satellite can still be considered, to first approximation, as circular. The effect of the non-radial component of the force cancels out over one period, but its torque with respect to the center of the earth does not. This "equivalent" torque is directed into the plane of the paper and is perpendicular to the orbiting angular momentum $\mathbf{L}$ of the satellite, which is perpendicular to the orbit plane N and is in the plane of the paper. It will cause the total angular momentum vector to precess about $\mathbf{L}$.

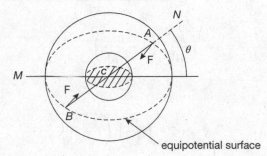

Fig. 1.45

   b. Because the atmospheric drag is concentrated near the perigee, it makes the satellite slow down at the perigee and reduces the energy and angular momentum of the satellite every time it crosses the perigee. This will make the apogee of the satellite's orbit come closer and closer to the earth and finally the orbit will become a circle with a radius equal to the distance of the perigee to the center of the earth. Further action by the drag will further reduce its distance to the earth until it falls to the earth.

1070

A particle of mass m moves under the influence of an attractive central force $f(r)$.

a. Show that by a proper choice of initial conditions a circular orbit can result.

The circular orbit is now subjected to a small radial perturbation.

b. Determine the relation that must hold among $f(r)$, r and $\partial f/\partial r$ for this orbit to be stable.

Now assume that the force law is of the form $f(r) = -K/r^n$.

c. Determine the maximum value of n for which the circular orbit can be stable.

(*Princeton*)

Sol: a.   The effective potential of the particle is

$$V^* = \frac{J^2}{2mr^2} + V(r),$$

where J is a constant and V is related to f by $f = \frac{-dV}{dr}$, in terms of which the total energy is

$$E = \frac{m\dot{r}^2}{2} + V^*.$$

The motion can then be treated as one dimensional, along the radial direction. The circular motion of the particle in the V field corresponds to the particle being at rest in the equilibrium position in the V^* field.

At the equilibrium position $r = r_0$,

$$\frac{\partial V^*}{\partial r} = 0,$$

or

$$\frac{\partial V}{\partial r} = \frac{J^2}{mr^3}. \tag{1}$$

If the initial condition satisfies the above equality and $E = V^*(r_0)$, the orbit is a circular one.

b.   For the orbit to be stable, V^* must be minimum at $r = r_0$. This requires that

$$\left(\frac{\partial^2 V^*}{\partial r^2}\right)_{r=r_0} > 0,$$

i.e.

$$\frac{3J^2}{mr^4} + \frac{\partial^2 V}{\partial r^2} > 0, \quad \text{or} \quad \frac{3J^2}{mr^4} - \frac{\partial f}{\partial r} > 0,$$

at $r = r_0$.

c.   If $f = -K/r^n$, then $\partial f/\partial r = nK/r^{n+1}$ and (1) gives

$$J^2 = mK/r_0^{n-3}.$$

Hence the condition

$$\frac{3J^2}{mr^4} - \frac{\partial f}{\partial r} > 0,$$

i.e.

$$\frac{3K}{r^{n+1}} - \frac{nK}{r^{n+1}} > 0,$$

requires that $n < 3$ for the circular orbit to be stable.

1071

Consider a planet of mass m moving in a nearly circular orbit of radius R around a star of mass M. There is, in addition to gravitation, a repulsive force on the planet proportional to the distance r from the star, $\mathbf{F} = A\mathbf{r}$. Compute the angular velocity of precession of the periastron (point of closest approach to the star).

(Princeton)

Sol: The force on the planet is

$$f = \frac{-GMm}{r^2} + Ar.$$

With $u = \frac{1}{r}$, Binet's formula (Problem **1055**) gives the equation for the orbit:

$$-mh^2u^2\left(\frac{d^2u}{d\theta^2} + u\right) = -GMmu^2 + \frac{A}{u}.$$

For nearly circular orbit we set $u = u_0 + \delta u$, where δu is a small quantity. The above equation then gives, retaining only the lowest-order small quantities,

$$mh^2 \left[\frac{d^2(\delta u)}{d\theta^2} + u_0 + \delta u \right] = GMm - \frac{A}{u_0^3 \left(1 - \dfrac{\delta u}{u_0} \right)^3}$$

$$\approx GMm - \left(\frac{A}{u_0^3} \right) \left(1 - \frac{3\delta u}{u_0} \right). \tag{1}$$

If the orbit is exactly circular, $u = u_0$, $\delta u = 0$, the above becomes

$$mh^2 u_0 = GMm - \frac{A}{u_0^3}. \tag{2}$$

Using this equation in (1) we obtain

$$mh^2 \left[\frac{d^2(\delta u)}{d\theta^2} + \delta u \right] = \frac{3A}{u_0^4} \delta u,$$

or

$$\frac{d^2(\delta u)}{d\theta^2} + \left(1 - \frac{3A}{mh^2 u_0^4} \right) \delta u = 0. \tag{3}$$

Choosing suitable coordinate axes we can write its solution as

$$\delta u = B \sin(a\theta),$$

where

$$a = \sqrt{1 - \frac{3A}{mh^2 u_0^4}} = \sqrt{1 - \frac{3AR^3}{GMm - AR^3}}$$

as $h^2 u_0 = GM - A/mu_0^3$, $u_0 = 1/R$. Then if θ_1 and θ_2 are the angles for two successive periastrons, we have

$$a\theta_2 - a\theta_1 = 2\pi.$$

or

$$\Delta\theta = \frac{2\pi}{a}.$$

As $a < 1$, the angle of precession is

$$\Delta\theta_p = \Delta\theta - 2\pi = \frac{2\pi(1-a)}{a}.$$

The time required for the line joining the periastron and the star to rotate through an angle $\Delta\theta_p$ is

$$\Delta t = \frac{\Delta\theta}{\dot\theta} = \frac{2\pi}{a\dot\theta}.$$

Hence the angular velocity of precession is

$$\omega_p = \frac{\Delta\theta_p}{\Delta t} = (1-a)\,\dot\theta.$$

As the angular velocity of revolution of the planet is by the definition of h

$$\dot\theta = \frac{h}{R^2} = \sqrt{\frac{GM}{R^3} - \frac{A}{m}},$$

we have

$$\omega_p = \sqrt{\frac{GM}{R^3} - \frac{A}{m}} - \sqrt{\frac{GM}{R^3} - \frac{4A}{m}}.$$

1072

a. A planet of mass m is orbiting a star of mass M. The planet experiences a small drag force $\mathbf{F} = -\alpha\mathbf{v}$ due to motion through the star's dense atmosphere. Assuming an essentially circular orbit with radius $r = r_0$ at $t = 0$, calculate the time dependence of the radius.

b. Now ignore the drag force. Assume that in addition to the Newtonian gravitational potential, the planet experiences a small additional potential so that its potential energy is actually

$$V(r) = -\frac{GMm}{r} + \frac{\varepsilon}{r^2},$$

where ε is a small constant.

Calculate the rate of precession of the planetary perihelion, to lowest order in ε. You may assume the orbit is almost circular. In other words, you are to calculate the angle φ sketched in Fig. 1.46.

(Princeton)

Fig. 1.46

Sol: **a.**   As the drag force **F** is small, it can be considered as a small perturbation on the circular motion of the planet under the gravitational force of the star. The unperturbed energy equation is

$$E = \frac{1}{2}m(\dot{r}^2 + r^2\dot{\theta}^2) - \frac{GMm}{r}.$$

If the orbit is circular with radius r, we have

$$\dot{r} = 0, \qquad mr\dot{\theta}^2 = \frac{mv^2}{r} = \frac{GMm}{r^2},$$

and thus

$$E = -\frac{GMm}{2r}.$$

The drag force causes energy loss at the rate

$$-\mathbf{F} \cdot \mathbf{v} = \alpha\mathbf{v} \cdot \mathbf{v} = \alpha v^2 = \frac{\alpha GM}{r}.$$

This must be equal to

$$-\frac{dE}{dt} = -\frac{GMm\dot{r}}{2r^2},$$

giving rise to

$$\dot{r} = -\frac{2\alpha}{m}r,$$

whose integration gives

$$r = r_0 e^{-\frac{2\alpha t}{m}},$$

where we have used $r = r_0$ at $t = 0$.

b. The planet is now moving in a central potential $V(r) = -\dfrac{GMm}{r} + \dfrac{\varepsilon}{r^2}$ and its total energy is

$$E = \frac{1}{2}m\dot{r}^2 + \frac{J^2}{2mr^2} + V(r),$$

where J is the conserved angular momentum, $J = mr^2\dot{\varphi}$.

As

$$\dot{r} = \frac{dr}{d\varphi}\frac{d\varphi}{dt} = \dot{\varphi}\frac{dr}{d\varphi},$$

we have

$$dr = \sqrt{2m(E - V) - \frac{J^2}{r^2}} \cdot \left(\frac{r^2}{J}\right)d\varphi,$$

or

$$\varphi = \int \frac{J\,dr}{r^2\sqrt{2m(E - V) - \dfrac{J^2}{r^2}}} + \text{const.}$$

In the unperturbed field $V_0 = -GMm/r$, the orbit is in general an ellipse. However, in the perturbed field V, the orbit is not closed. During the time in which r varies from $r_{\min}$ to $r_{\max}$ and back to $r_{\min}$ again, the radius vector has turned through an angle $\Delta\varphi$ given by

$$\Delta\varphi = 2\int_{r_{\min}}^{r_{\max}} \frac{J\,dr}{r^2\sqrt{2m(E - V) - \dfrac{J^2}{r^2}}}$$

$$= -2\frac{\partial}{\partial J}\int_{r_{\min}}^{r_{\max}} \sqrt{2m(E - V) - \frac{J^2}{r^2}}\,dr.$$

Writing $V = \dfrac{-GMm}{r} + \dfrac{\varepsilon}{r^2} = V_0 + \delta V$, we expand the integrand as a Taylor series in powers of δV:

$$f(V) = f(V_0) + \frac{1}{1!}\left(\frac{\partial f}{\partial V}\right)_{V=V_0}\delta V + \frac{1}{2!}\left(\frac{\partial^2 f}{2V^2}\right)_{V=V_0}(\delta V)^2 + \cdots .$$

The zero-order term gives 2π as the corresponding orbit is an ellipse. The first-order term gives

$$\varphi = \frac{\partial}{\partial J}\int_{r_{\min}}^{r_{\max}} \frac{2m\delta V\,dr}{\sqrt{2m(E - V_0) - \dfrac{J^2}{r^2}}},$$

the angle shown in Fig. 1.46.

The variable over which the integration is to be carried out can be changed in the following way. We have

$$\dot{r} = \dot{\varphi}\frac{dr}{d\varphi} = \frac{J}{mr^2}\frac{dr}{d\varphi},$$

i.e.

$$\frac{1}{m}\sqrt{2m(E - V_0) - \frac{J^2}{r^2}} = \frac{J}{mr^2}\frac{dr}{d\varphi}.$$

So the last integral can be written as

$$\varphi = \frac{\partial}{\partial J}\left[\frac{2m}{J}\int_0^{\pi} r^2 \delta V d\varphi\right].$$

With $\delta V = \varepsilon/r^2$ we obtain

$$\varphi = \frac{\partial}{\partial J}\left[\frac{2m}{J}\int_0^{\pi} \varepsilon d\varphi\right] = -\frac{2\pi\varepsilon m}{J^2}.$$

1073

a. Find the central force which results in the following orbit for a particle:

$$r = a(1 + \cos\theta).$$

b. A particle of mass m is acted on by attractive force whose potential is given by $U \propto r^{-4}$. Find the total cross section for capture for the particle coming from infinity with an initial velocity V_∞.

Note: Parts (a) and (b) may refer to different forces.

(Princeton)

Sol: **a.** In the central force field, the equations of motion for the particle are

$$m(\ddot{r} - r\dot{\theta}^2) = F(r), \tag{1}$$

$$r^2\dot{\theta} = \text{const.} = h, \text{ say.}$$

Then

$$\dot{\theta} = \frac{h}{r^2}, \quad \ddot{\theta} = \frac{d}{dt}\left(\frac{h}{r^2}\right) = \frac{-2h\dot{r}}{r^3}.$$

With $r = a(1 + \cos\theta)$, we also have

$$\dot{r} = -a\dot{\theta}\sin\theta,$$

$$\ddot{r} = -a(\ddot{\theta}\sin\theta + \dot{\theta}^2\cos\theta) = -\frac{ah^2}{r^4}\left(\frac{2\sin^2\theta}{1 + \cos\theta} + \cos\theta\right)$$

$$= \frac{-ah^4}{r^4}(2 - \cos\theta) = h^2\left(\frac{r - 3a}{r^4}\right).$$

Using the above we can write (1) as

$$F(r) = m \left[\frac{h^2(r - 3a)}{r^4} - r\left(\frac{h}{r^2}\right)^2 \right] = -\frac{3mh^2a}{r^4},$$

which is the central force required.

b.   As $U = -\dfrac{\alpha}{r^4}$, the effective potential is

$$U_{\text{eff}} = \frac{L^2}{2mr^2} - \frac{\alpha}{r^4},$$

where $L = mbV_\infty$ is the angular momentum, which is conserved in a central force field, b being the impact parameter. To find the maximum of U_{eff}, consider

$$\frac{dU_{\text{eff}}}{dr} = \frac{-L^2}{mr^3} + \frac{4\alpha}{r^5} = 0,$$

which gives

$$r_0 = \sqrt{\frac{4m\alpha}{L^2}}$$

as the distance where U_{eff} is maximum. Then

$$(U_{\text{eff}})_{\text{max}} \equiv U_0 = \frac{L^4}{16m^2\alpha}.$$

The form of U_{eff} is shown in Fig. 1.47. It is seen that only particles with total energy $E > U_0$ will "fall" to the force center. Thus the maximum impact parameter for capture is given by $E = U_0$, or

$$\frac{m^4 b^4 V_\infty^4}{16m^2\alpha} = \frac{1}{2}mV_\infty^2,$$

giving

$$b_{\text{max}} = \left(\frac{8\alpha}{mV_\infty^2}\right)^{1/4}.$$

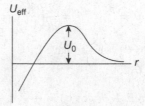

Fig. 1.47

Hence the total capture cross section is

$$\sigma = \pi b^2_{\text{max}} = 2\pi\sqrt{\frac{2\alpha}{mV^2_\infty}}.$$

1074

a. A particle of mass m moves in a potential $V(r) = k/r^2$, $k > 0$. Consider motion in the X–Y plane, letting r and ϕ be the polar coordinates in that plane, and solve for r as a function of ϕ, angular momentum l and energy E (Fig. 1.48).

b. Use the result of part (a) to discuss (classical) scattering in this potential. Let θ be the scattering angle. Relate the impact parameter to θ and E and thereby compute the differential cross section as a function of θ and E.

(Princeton)

Sol:

a. The force on the particle is

$$F = -\frac{\partial V}{\partial r} = \frac{2k}{r^3}.$$

Binet's formula (Problem **1055**) then becomes

$$h^2 u^2 \left(\frac{\partial^2 u}{\partial \phi^2} + u\right) = -\frac{F}{m} = -\frac{2k}{m}u^3,$$

or

$$\frac{\partial^2 u}{\partial \phi^2} + \left(1 + \frac{2k}{mh^2}\right)u = 0,$$

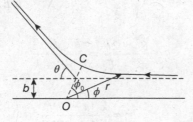

Fig. 1.48

where $h = r^2\dot\phi$, $u = 1/r$. Its solution is

$$u = A\sin(\omega\phi + \psi),$$

where $\omega^2 = 1 + 2k/mh^2$, and A and ψ are constants of integration to be determined from the initial conditions.

It can be seen from Fig. 1.48 that for $r \to \infty$, i.e. $u \to 0$, $\phi \to 0$. Hence $\psi = 0$. Also for $r \to \infty$, $\dot r \to \dot r_\infty$ given by $E = \frac{1}{2}m\dot r_\infty^2$, i.e. $\dot r_\infty = -\sqrt{\dfrac{2E}{m}}$, where the minus sign is chosen because for incidence r decreases with increasing t. Then as

$$\dot r = \frac{dr}{d\phi}\dot\phi = \frac{dr}{d\phi}\frac{h}{r^2} = -h\frac{du}{d\phi} = -Ah\omega\cos(\omega\phi),$$

we have, with $l = hm$,

$$A = \frac{1}{l\omega}\sqrt{2mE}$$

and hence

$$\frac{1}{r} = \frac{\sqrt{2mE}}{l\omega}\sin(\omega\phi),$$

where ω is given by

$$\omega^2 = 1 + \frac{2mk}{l^2}.$$

b. From the above result it can be seen that r is at a minimum when $\omega\phi = \dfrac{\pi}{2}$, i.e. at $\phi = \phi_0 = \dfrac{\pi}{2\omega}$. This is the distance of closest approach OC shown in Fig. 1.48. Due to the symmetry of the scattering, the scattering angle is

$$\theta = \pi - 2\phi_0 = \pi\left(1 - \frac{1}{\omega}\right).$$

Then as $l^2 = m^2 b^2 \dot r_\infty^2 = 2b^2 mE$, we have

$$1 - \frac{\theta}{\pi} = \left(1 + \frac{2mk}{l^2}\right)^{-\frac{1}{2}} = \left(1 + \frac{k}{b^2 E}\right)^{-\frac{1}{2}},$$

i.e.

$$\frac{\theta^2}{\pi^2} - \frac{2\theta}{\pi} = -\frac{k}{b^2 E + k},$$

giving

$$b^2 = \frac{k\,(\pi - \theta)^2}{E(2\pi - \theta)\theta}$$

as the relation between θ and b.

Particles with impact parameters between b and $b + db$ will be scattered into angles between θ and $\theta + d\theta$, i.e. into a solid angle $d\Omega = 2\pi \sin \theta d\theta$. Hence the differential cross section at scattering angle θ per unit solid angle is

$$\frac{d\sigma}{d\Omega} = \left| \frac{2\pi b db}{2\pi \sin \theta d\theta} \right| = \left| \frac{b}{\sin \theta} \frac{db}{d\theta} \right|$$

$$= \frac{k}{E \sin \theta} \frac{\pi^2(\pi - \theta)}{(2\pi - \theta)^2 \theta^2}.$$

1075

Derive formulas and calculate the values of (a) the gravitational acceleration at the surface of the moon, and (b) the escape velocity from the moon.

(SUNY, Buffalo)

Sol: **a.** Let M and R be the mass and radius of the moon respectively. Then by the law of universal gravitation and the definition of gravitational acceleration at the surface of the moon we have

$$\frac{GMm}{R^2} = mg,$$

where m is the mass of a body on the surface of the moon. The relation gives the gravitational acceleration at the surface of the moon as

$$g = \frac{GM}{R^2} = \frac{6.67 \times 10^{-11} \times 7.35 \times 10^{22}}{(1.74 \times 10^6)^2} = 1.62 \text{ m/s}^2.$$

b. The potential energy of a projectile of mass m at infinite distance from the moon $\rho \to \infty$ is

$$-\frac{GmM}{\rho} = -\frac{mgR^2}{\rho} \to 0.$$

Its kinetic energy, a positive quantity, is at least zero. Hence for the projectile to reach infinity from the surface of the moon, its total mechanical energy must be at least zero, by the conservation of energy.

At the surface of the moon, the projectile has total energy

$$E = \frac{1}{2}mv_0^2 - mgR.$$

If v_0 is the escape velocity, we require $E = 0$, or

$$v_0 = \sqrt{2gR} = \sqrt{2 \times 162 \times 1.74 \times 10^6} = 2.37 \times 10^3 \text{m/s}.$$

1076

Consider the motion of a particle of mass m under the influence of a force $\mathbf{F} = -K\mathbf{r}$, where K is a positive constant and $\mathbf{r}$ is the position vector of the particle.

  a. Prove that the motion of the particle lies in a plane.

  b. Find the position of the particle as a function of time, assuming that at $t = 0$, $x = a$, $y = 0$, $V_x = 0$, $V_y = V$.

  c. Show that the orbit is an ellipse.

  d. Find the period.

  e. Does the motion of the particle obey Kepler's laws of planetary motion?

(SUNY, Buffalo)

Sol: **a.**   For a central force field $\mathbf{F} = -K\mathbf{r}$,

$$\mathbf{r} \times \mathbf{F} = K\mathbf{r} \times \mathbf{r} = 0.$$

Then as $\mathbf{F} = m d\mathbf{V}/dt$ we have

$$\mathbf{r} \times \frac{d\mathbf{V}}{dt} = 0,$$

or

$$\frac{d(\mathbf{r} \times \mathbf{V})}{dt} = \mathbf{V} \times \mathbf{V} + \mathbf{r} \times \frac{d\mathbf{V}}{dt} = 0.$$

Integrating we obtain

$$\mathbf{r} \times \mathbf{V} = \mathbf{h},$$

a constant vector.

It follows that

$$\mathbf{r} \cdot \mathbf{h} = \mathbf{r} \cdot \mathbf{r} \times \mathbf{V} = \mathbf{r} \times \mathbf{r} \cdot \mathbf{V} = 0,$$

which shows that $\mathbf{r}$ is perpendicular to the constant vector $\mathbf{h}$, i.e. $\mathbf{r}$ lies in a plane perpendicular to $\mathbf{h}$. This proves that the motion of the particle is confined to a plane. We shall choose the plane as the xy plane with the origin at the center of the force.

  b.   The equation of the motion of the particle is

$$m\ddot{\mathbf{r}} = -K\mathbf{r},$$

or, in Cartesian coordinates,

$$\ddot{x} + \omega^2 x = 0,$$
$$\ddot{y} + \omega^2 y = 0,$$

where $\omega^2 = K/m$. The general solution of the above set of equations is

$$x = A_1 \sin(\omega t + \phi_1),$$
$$y = A_2 \sin(\omega t + \phi_2).$$

With the initial conditions given, i.e.

$$x = A_1 \sin \phi_1 = a,$$
$$y = A_2 \sin \phi_2 = 0,$$
$$\dot{x} = A_1 \omega \cos \phi_1 = 0,$$
$$\dot{y} = A_2 \omega \cos \phi_2 = V_0,$$

we find $\phi_1 = \pi/2$, $\phi_2 = 0$, $A_1 = a$, $A_2 = V_0/\omega = \sqrt{m}/KV_0$. Hence

$$x = a \sin\left(\sqrt{\frac{K}{m}}t + \frac{\pi}{2}\right),$$

$$y = \sqrt{\frac{m}{K}}V_0 \sin\left(\sqrt{\frac{K}{m}}t\right).$$

c. The last set of equations describes an ellipse. Eliminating the parameter t we obtain the standard equation for an ellipse:

$$\frac{x^2}{a^2} + \frac{y^2}{b^2} = 1,$$

with

$$b = \sqrt{\frac{m}{K}}V_0.$$

d. (x, y) return to the same values when t increases by T such that

$$\sqrt{\frac{K}{m}}T = 2\pi.$$

Hence the period is

$$T = 2\pi\sqrt{\frac{m}{K}}.$$

e. Kepler's third law states that ratio of the square of the period of revolution of a planet to the cube of the length of the semimajor axis of its orbit is a constant. Hence we have

$$\frac{(\text{period})^2}{(\text{length of semimajor axis})^3} = \begin{cases} \dfrac{4\pi^2 m}{Ka^3} & \text{if } a > b, \\[3mm] \dfrac{4\pi^2}{V_0^3}\sqrt{\dfrac{K}{m}} & \text{if } a < b, \end{cases}$$

As this ratio depends on m and a or m and V_0, Kepler's third law is not obeyed.

1077

a. A particle of mass m moves in a central field of potential energy $U(r)$. From the constants of the motion obtain the equation of the trajectory. Express the polar angle φ in terms of r.

b. If the particle moves in from infinitely far away with initial speed V_0, impact parameter b, and is scattered to a particular direction θ, define the differential cross section in terms of b.

c. Calculate the differential and total cross section for the scattering from a hard sphere.

(SUNY, Buffalo)

Sol:

a.   If a particle of mass m moves in a central force field of potential energy $U(r)$, its mechanical energy E and angular momentum with respect to the center of the force mh are conserved quantities. Thus

$$\frac{1}{2}m(\dot{r}^2 + r^2\dot{\varphi}^2) + U(r) = E,$$

$$r^2\dot{\varphi} = h, \quad \text{or} \quad \dot{\varphi} = \frac{h}{r^2}.$$

As we also have

$$\dot{r} = \frac{dr}{dt} = \frac{dr}{d\varphi}\frac{d\varphi}{dt} = \dot{\varphi}\frac{dr}{d\varphi} = \frac{h}{r^2}\frac{dr}{d\varphi},$$

the energy equation becomes

$$\frac{1}{2}m\left[\frac{h^2}{r^4}\left(\frac{dr}{d\phi}\right)^2 + r^2\frac{h^2}{r^4}\right] + U(r) = E,$$

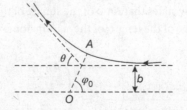

Fig. 1.49

i.e.

$$d\phi = \frac{h\,dr}{r\sqrt{\dfrac{2r^2}{m}[E - U(r)] - h^2}},$$

or

$$\phi = \int \frac{h\,dr}{r\sqrt{\dfrac{2r^2}{m}[E - U(r)] - h^2}},$$

which express ϕ in terms of r.

b. The orbit of the particle in the central force field is symmetrical with respect to the line joining the center of the force to the point of closest approach (OA in Fig. 1.49). The angle of scattering of the particle is then

$$\theta = \pi - 2\varphi_0$$

with φ_0 given by

$$\varphi_0 = \int_{r_{\min}}^{\infty} \frac{h\,dr}{r\sqrt{\dfrac{2r^2}{m}[E - U(r)] - h^2}},$$

where $r_{\min}$ is given by $\dot{r} = 0$ in the energy equation, or $E = U(r) + \dfrac{mh^2}{2r^2}$. The conservation laws give

$$E = \frac{mV_0^2}{2}, \qquad mh = mbV_0,$$

so that

$$\varphi_0 = \int_{r_{\min}}^{\infty} \frac{b\,dr}{r^2\sqrt{1 - \dfrac{b^2}{r^2} - \dfrac{2U(r)}{mV_0^2}}}.$$

The scattering angle θ can then be determined.

Let dN denote the number of particles scatterred per unit time into the solid angle corresponding to scattering angles θ and $\theta + d\theta$, and n denote the number of particles passing through unit cross sectional area of the beam per unit time. The differential cross section is defined as

$$d\sigma = \frac{dN}{n}.$$

As the scattering angle θ corresponds to a unique impact parameter b, we have

$$dN = 2\pi nb\,db,$$

i.e.

$$d\sigma = 2\pi b\, db.$$

We can write the above as

$$d\sigma = 2\pi b\left|\frac{db}{d\theta}\right|d\theta = \frac{b}{\sin\theta}\left|\frac{db}{d\theta}\right|d\Omega,$$

where $d\Omega$ is the solid angle between two right circular cones of opening angles θ and $\theta + d\theta$:

$$d\Omega = 2\pi\sin\theta\, d\theta.$$

Note that $\dfrac{d\sigma}{d\Omega}$ is known as the differential cross section per unit solid angle.

c. A particle moves freely before it hits the hard sphere. Because it cannot enter into the interior of the sphere, momentum conservation requires that the incidence and reflected angles are equal as shown in Fig. 1.50.

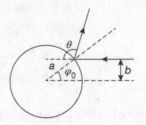

Fig. 1.50

Then

$$b = a\sin\varphi_0 = a\sin\left(\frac{\pi-\theta}{2}\right) = a\cos\left(\frac{\theta}{2}\right).$$

Hence

$$\frac{d\sigma}{d\Omega} = \frac{ba}{2\sin\theta}\sin\left(\frac{\theta}{2}\right) = \frac{a^2}{4}.$$

As this is independent of the scattering angle, the total scattering cross section is $\sigma = 4\pi\dfrac{d\sigma}{d\Omega} = \pi a^2$, which is equal to the geometrical cross section of the hard sphere.

1078

When displaced and released, the 2 kg mass in Fig. 1.51 oscillates on the frictionless horizontal surface with period $\pi/6$ seconds.

a. How large a force is necessary to displace the mas 2 cm from equilibrium?

b. If a small mass is placed on the 2 kg block and the coefficient of static friction between the small mass and the 2 kg block is 0.1, what is the maximum amplitude of oscillation before the small mass slips? (Assume the period is unaffected by adding the small mass.)

(Wisconsin)

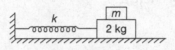

Fig. 1.51

Sol: Let k be the spring constant. The equation of the motion of the mass is

$$2\ddot{x} + kx = 0,$$

or

$$\ddot{x} + \omega^2 x = 0,$$

where x is the displacement of the block from its equilibrium postition, and $\omega^2 = \dfrac{k}{2}$. The general solution is

$$x = A\cos(\omega t + \phi).$$

The period of oscillation is

$$T = \frac{2\pi}{\omega} = 2\pi\sqrt{\frac{2}{k}} = \frac{\pi}{6},$$

giving $\omega = 12\ \text{s}^{-1}$, $k = 288\ \text{Nm}^{-1}$. If $x = x_0$ at $t = 0$, then $\phi = 0$, $A = x_0$ and the solution is $x = x_0 \cos(12t)$.

a. The force needed is

$$f = kx = 288 \times 2 \times 10^{-2} = 5.76\ \text{N}.$$

b. If the small mass moves together with the 2 kg block, it has the same acceleration as the latter, i.e. $\ddot{x} = -144x_0 \cos(12t)$. Let its mass be m. When it starts to slip, the maximum horizontal force on it just exceeds the static friction:

$$0.1 \times mg = 144mx_0,$$

giving

$$x_0 = \frac{0.98}{144} = 6.8 \times 10^{-3}\ \text{m}.$$

If x_0 exceeds this value m will slip. Hence it gives the maximum amplitude for no slipping.

1079

Two synchronous tuning forks of identical frequency and loudness produce zero net intensity at some point A. However if either one is sounded alone, a loudness *l* is heard at A. Explain in detail, as to a sophomore, what became of the law of conservation of energy.

(Wisconsin)

Sol: Let s_1 and s_2 be the distances between a point in space and the two tuning forks. Each of the forks alone produces oscillations at this point represented by

$$y_1 = I_1 \sin\left[\omega\left(t - \frac{s_1}{c}\right)\right],$$

and

$$y_2 = I_2 \sin\left[\omega\left(t - \frac{s_2}{c}\right)\right],$$

where c is the speed of sound.

If s_1 and s_2 are both much larger than the distance between the two forks, we can regard I_1 and I_2 as approximately the same, i.e. $I_1 \approx I_2 \approx I_0$. Then the resultant oscillation is

$$y = y_1 + y_2 = I_0 \left\{ \sin\left[\omega\left(t - \frac{s_1}{c}\right)\right] + \sin\left[\omega\left(t - \frac{s_2}{c}\right)\right] \right\}$$

$$= 2I_0 \sin\left[\omega\left(t - \frac{s_1 + s_2}{2c}\right)\right] \cos\left[\omega\left(\frac{s_2 - s_1}{2c}\right)\right].$$

Hence $y = 0$ if

$$\frac{\omega(s_2 - s_1)}{2c} = (2n + 1)\frac{\pi}{2}, \qquad n = 0, 1, 2, \ldots.$$

Thus the resultant oscillation is zero at points where $s_2 - s_1$ is some odd multiple of $\lambda/2$. This does not violate the law of conservation of energy as is evident when we consider the energy stored in the whole wave field. Although the amplitude and energy of oscillation are zero at the nodes, at the antinodes, the amplitude of oscillation is twice and the energy is four times that of the individual value. Detailed calculations will demonstrate that the energy of the resultant oscillation is equal to the sum of that of the individual oscillations.

1080

A mass m moves in a plane in uniform circular motion with angular frequency ω. The centripetal force is provided by a spring whose force constant is K (ignore gravity). A very small radial impulse is given to the mass. Find the frequency of the resulting radial oscillation.

(Wisconsin)

Sol: In polar coordinates the equations of motion for the mass are

$$m(\ddot{r} - r\dot{\theta}^2) = -K(r - r_0),$$
$$m(r\ddot{\theta} + 2\dot{r}\dot{\theta}) = 0.$$

The second equation gives

$$r^2\dot{\theta} = \text{const.}$$

Let R be the radius of the uniform circular motion of the mass. We have

$$mR\omega^2 = K(R - r_0), \qquad r^2\dot{\theta} = R^2\omega.$$

Let $r' = r - R$ for departure from uniform circular motion. The radial equation can be written as

$$\ddot{r} - \frac{r^4\dot{\theta}^2}{r^3} = \ddot{r}' - \frac{R^4\omega^2}{(R + r)^3} = -\frac{K}{m}(r' + R - r_0).$$

If the radial impulse is very small, $r' \ll R$ and the above becomes

$$\ddot{r}' - R\omega^2\left(1 - \frac{3r'}{R}\right) = \frac{-Kr'}{m} - \frac{K(R - r_0)}{m},$$

or

$$\ddot{r}' + \left(3\omega^2 + \frac{K}{m}\right)r' = 0.$$

It follows that the frequency of radial oscillation is $\omega' = \sqrt{3\omega^2 + \dfrac{K}{m}}.$

1081

A particle of mass m moves under the action of a restoring force $-Kx$ and a resisting force $-Rv$, where x is the displacement from equilibrium and v is the particle's velocity. For fixed K and arbitrary initial conditions, find the value of $R = R_c$ giving the most rapid approach to equilibrium. Is it possible to pick initial conditions (other than $x = v = 0$) so that the approach is more rapid for $R > R_c$ and $R < R_c$? Explain.

(Wisconsin)

Sol: The equation of motion is $\ddot{m} + R\dot{x} + Kx = 0$. Assume $x = Ae^{\alpha t}$, we obtain the indicial equation $m\alpha^2 + R\alpha + K = 0$, giving

$$\alpha = \frac{-R \pm \sqrt{R^2 - 4Km}}{2m}.$$

In general, if $R = R_c = 2\sqrt{Km}$ (critical damping), the mass approaches equilibrium most rapidly. However, if $R > R_c$, the mass may approach equilibrium even more rapidly under certain particular conditions. For now the general solution is

$$x = A \exp\left(\frac{-R + \sqrt{R^2 - 4Km}}{2m}\right)t + B \exp\left(\frac{-R - \sqrt{R^2 - 4Km}}{2m}\right)t$$

We can choose initial conditions so that $A = 0$. Then the remaining term has a damping coefficient

$$-\alpha = \frac{R + \sqrt{R^2 - 4Km}}{2m} = \frac{R + \sqrt{R^2 - R_c^2}}{2m} > \frac{R_c}{2m},$$

so that approach to equilibrium is even faster than for critical damping. If $R < R_c$, we have

$$\alpha = \frac{-R \pm i\sqrt{R_c^2 - R^2}}{2m},$$

so that the general solution is

$$x = A \exp\left(-\frac{Rt}{2m}\right)\exp\left(\frac{i\sqrt{R_c^2 - R^2}t}{2m}\right)$$

$$+ B \exp\left(-\frac{Rt}{2m}\right)\exp\left(\frac{-i\sqrt{R_c^2 - R^2}t}{2m}\right).$$

Then the approach to equilibrium is oscillatory with a damping coefficient

$$\frac{R}{2m} < \frac{R_c}{2m}.$$

The approach is always slower than for critical damping.

1082

A freely running motor rests on a thick rubber pad to reduce vibration (Fig. 1.52). The motor sinks 10 cm into the pad. Estimate the rotational speed (revolutions per minute, i.e. RPM) at which the motor will exhibit the largest vertical vibration.

(UC, Berkeley)

10 cm

Fig. 1.52

Sol: Let the elastic coefficient of the rubber pad be k. Then $kx = mg$, where m is the mass of the motor. As $x = 0.1$ m, $\frac{k}{m} = \frac{g}{x} = 98$ s^{-2}. Then the natural frequency of the system is

$$\omega = \sqrt{\frac{k}{m}} = 9.9\,s^{-1}.$$

Hence when the motor is rotating at a rate

$$\frac{\omega}{2\pi} = \frac{60 \times 9.9}{2\pi} = 94.5 \text{ RPM},$$

resonance will take place and the motor will exhibit the largest vertical vibration.

1083

A car is traveling in the x-direction and maintains constant horizontal speed v. The car goes over a bump whose shape is described by $y_0 = A[1 - \cos(\pi x/l)]$ for $0 \le x \le 2l$; $y_0 = 0$ otherwise (Fig. 1.53). Determine the motion of the center of mass of the car while passing over the bump. Represent the car as a mass m attached to a massless spring of relaxed length l_0 and spring constant k. Ignore friction and assume that the spring is vertical at all times.

(MIT)

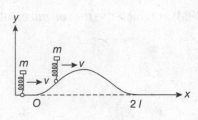

Fig. 1.53

Sol: Let the location of the mass at time t be (x, y). Choose the origin so that $x(0) = 0$. Then $x(t) = vt$. The equation of the motion of the mass in the y-direction is

$$m\ddot{y} = -k(y - y_0 - l_0) - mg$$
$$= -k\left(y - A - l_0 + \frac{mg}{k}\right) - kA\cos\left(\frac{\pi vt}{l}\right).$$

Putting $Y = y - A - l_0 + mg/k$, we can write the above equation as

$$m\ddot{Y} + kY = -kA\cos\left(\frac{\pi vt}{l}\right).$$

This equation describes the motion of a driven harmonic oscillator. Trying a particular solution of the form $Y = B\cos\left(\frac{\pi vt}{l}\right)$, we find

$$-mB\left(\frac{\pi v}{l}\right)^2 + kB = -kA,$$

i.e.

$$B = \frac{kl^2A}{m\pi^2v^2 - kl^2}.$$

Hence, the general solution of the equation of motion for the mass is

$$y = C_1\cos(\omega t) + C_2\sin(\omega t) + B\cos\left(\frac{\pi vt}{l}\right) + A + l_0 - \frac{mg}{k},$$

with $\omega = \sqrt{\frac{k}{m}}$.

The initial conditions are $y(0) = l_0 - mg/k$, $\dot{y}(0) = 0$, giving $C_2 = 0$, $C_1 = -(B + A) = m\pi^2 v^2 A/(kl^2 - m\pi^2 v^2)$. Therefore the motion of the center of mass of the car is described by

$$y(t) = C_1 \cos(\omega t) + B \cos\left(\frac{\pi v t}{l}\right) + A + l_0 - \frac{mg}{k}$$

with $\omega = \sqrt{\frac{k}{m}}$, $C_1 = -\frac{m\pi^2 v^2 A}{m\pi^2 v^2 - kl^2}$, $B = \frac{kl^2 A}{m\pi^2 v^2 - kl^2}$.

1084

A thin ring of mass M and radius r lies flat on a frictionless table. It is constrained by two extended identical springs with relaxed length $l_0 (l_0 \gg r)$ and spring constant k as shown in Fig. 1.54.

a. What are the normal modes of small oscillations and their frequencies?

b. What qualitative changes in the motion would occur if the relaxed lengths of the springs were $2l_0$?

(*MIT*)

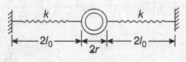

Fig. 1.54

Sol: **a.** As $l_0 \gg r$, any rotation of the ring will cause a negligible change of length in the springs, any elastic force so arising is also negligible. Newton's second law then gives

$$M\ddot{x} = -k[\sqrt{(2l_0 + x)^2 + y^2} - l_0]\frac{2l_0 + x}{\sqrt{(2l_0 + x)^2 + y^2}}$$

$$+ k[\sqrt{(2l_0 - x)^2 + y^2} - l_0]\frac{2l_0 - x}{\sqrt{(2l_0 - x)^2 + y^2}},$$

$$M\ddot{y} = -k[\sqrt{(2l_0 + x)^2 + y^2} - l_0]\frac{y}{\sqrt{(2l_0 + x)^2 + y^2}}$$

$$- k[\sqrt{(2l_0 - x)^2 + y^2} - l_0]\frac{y}{\sqrt{(2l_0 - x)^2 + y^2}},$$

Where x, y give the displacement of the center of the ring from the equilibrium position. Neglecting terms higher than the first order in the small quantities x, y, we have

$$\sqrt{(2l_0 \pm x)^2 + y^2} \approx \sqrt{4l_0^2 \pm 4l_0 x} \approx 2l_0\left(1 \pm \frac{x}{2l_0}\right) \approx 2l_0 \pm x.$$

The above equations then become

$$m\ddot{x} = -2kx,$$
$$m\ddot{y} = -ky,$$

with solutions

$$x = A_x \cos(\omega_x t + \varphi_x),$$
$$y = A_y \cos(\omega_y t + \varphi_y),$$

where $\omega_x = \sqrt{\dfrac{2k}{M}}$, $\omega_y = \sqrt{\dfrac{k}{M}}$, and the constants A_x, A_y, φ_x, φ_y are determined by the initial conditions. These are the two normal modes of small oscillations.

b. With the relaxed length increased to $2l_0$, during the motion, one spring is extended while the other compressed. The latter will exert an elastic force on the ring opposite to that when extended. Assuming that the spring constant is the same for compression as for extension, the equations of motion are now

$$
\begin{aligned}
M\ddot{x} &= -k[\sqrt{(2l_0 + x)^2 + y^2} - 2l_0]\frac{2l_0 + x}{\sqrt{(2l_0 + x)^2 + y^2}} \\
&\quad - k[\sqrt{(2l_0 - x)^2 + y^2} - 2l_0]\frac{2l_0 - x}{\sqrt{(2l_0 - x)^2 + y^2}} \\
&\approx -2kx,
\end{aligned}
$$

$$
\begin{aligned}
M\ddot{y} &= -k[\sqrt{(2l_0 + x)^2 + y^2} - 2l_0]\frac{y}{\sqrt{(2l_0 + x)^2 + y^2}} \\
&\quad + k[\sqrt{(2l_0 - x)^2 + y^2} - 2l_0]\frac{y}{\sqrt{(2l_0 - x)^2 + y^2}} \\
&\approx -\frac{kxy}{l_0},
\end{aligned}
$$

retaining only the lowest order terms in the small qualities x, y. It can be seen that the motion of the ring in the x-direction is similar to that in part (a) while the motion in the y-direction, though quite complicated, is of a higher order.

1085

Two particles are connected by a spring of spring constant K and zero equilibrium length. Each particle has mass m and positive charge q. A constant horizontal electric field $\mathbf{E} = E_0\mathbf{i}$ is applied. Take into account the particles' Coulomb interaction but neglect magnetic effects, radiation, relativistic effects, etc. Assume the particles do not collide.

a. If the particles slide along a frictionless straight wire in the x direction and the distance d between them is constant, find d.

b. Find the acceleration of the center of mass in (a).

c. In (a), suppose the distance $d(t)$ undergoes small oscillations around the equilibrium value you found. What is the frequency?

d. Suppose the particles slide along a horizontal frictionless table instead of the wire. Find the general solution of the equations of motion. You may leave your answer in terms of integrals.

(*MIT*)

Sol: a. Considering the forces on the two particles as shown in Fig. 1.55, we obtain the equations of motion

$$qE + k(x_2 - x_1) - F_c = m\ddot{x}_1, \tag{1}$$
$$qE + k(x_1 - x_2) + F_c = m\ddot{x}_2, \tag{2}$$

where F_c is the mutual Coulomb force between the particles

$$F_c = \frac{1}{4\pi\varepsilon_0}\frac{q^2}{(x_2 - x_1)^2}.$$

As $x_2 - x_1 = d$, a constant, $\ddot{x}_2 = \ddot{x}_1$. Subtracting (2) from (1), we obtain

$$2kd = 2F_c = \frac{2}{4\pi\varepsilon_0}\frac{q^2}{d^2},$$

or

$$d = \left(\frac{1}{4\pi\varepsilon_0}\frac{q^2}{k}\right)^{1/3}.$$

b. Adding (1) and (2) we have

$$2qE = m(\ddot{x}_1 + \ddot{x}_2),$$

or

$$\ddot{x}_0 = \frac{qE}{m},$$

where $x_0 = \frac{1}{2}(x_1 + x_2)$ is the center of mass of the system.

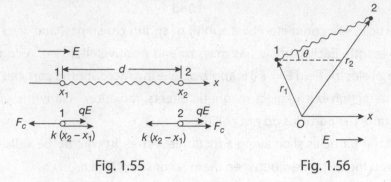

Fig. 1.55   Fig. 1.56

c. Subtracting (1) from (2) we obtain

$$m(\ddot{x}_2 - \ddot{x}_1) + 2k(x_2 - x_1) = \frac{q^2}{2\pi\varepsilon_0(x_2 - x_1)^2}.$$

Putting $x_2 - x_1 = d + \Delta d$, where $\Delta d \ll d$, the above becomes

$$m(\Delta\ddot{d}) + 2k(d + \Delta d) = \frac{q^2}{2\pi\varepsilon_0(d + \Delta d)^2},$$

where $\Delta\ddot{d} \equiv \dfrac{d^2\Delta d}{dt^2}$. As $d^3 = \dfrac{1}{4\pi\varepsilon_0}\dfrac{q^2}{k}$ and $\Delta d \ll d$, the above can be written as

$$m\Delta\ddot{d} + 6k\Delta d = 0$$

by retaining only the first order terms in $\dfrac{\Delta d}{d}$. It follows that the angular frequency of small oscillations is

$$\omega = \sqrt{\frac{6k}{m}}.$$

d. With $\mathbf{r}_1$, $\mathbf{r}_2$ and θ as defined in Fig. 1.56, we can write the equations of the two-dimensional motion as

$$m\ddot{\mathbf{r}}_1 = qE\mathbf{i} + k(\mathbf{r}_2 - \mathbf{r}_1) - \mathbf{F}_c, \tag{3}$$

$$m\ddot{\mathbf{r}}_2 = qE\mathbf{i} + k(\mathbf{r}_1 - \mathbf{r}_2) + \mathbf{F}_c, \tag{4}$$

with

$$\mathbf{F}_c = \frac{q^2}{4\pi\varepsilon_0}\frac{(\mathbf{r}_2 - \mathbf{r}_1)}{|\mathbf{r}_2 - \mathbf{r}_1|^3}.$$

Adding (3) and (4) we obtain

$$\ddot{\mathbf{r}}_1 + \ddot{\mathbf{r}}_2 = \left(\frac{2qE}{m}\right)\mathbf{i},$$

which is equivalent to two scalar equations

$$\ddot{x}_1 + \ddot{x}_2 = \frac{2qE}{m},$$

$$\ddot{y}_1 + \ddot{y}_2 = 0.$$

Integration gives

$$x_1 + x_2 = \frac{qEt^2}{m} + C_1 t + C_2, \tag{5}$$

$$y_1 + y_2 = D_1 t + D_2, \tag{6}$$

where C_1, C_2, D_1, D_2 are constants of integration. Subtracting (3) from (4) we obtain

$$m(\ddot{\mathbf{r}}_2 - \ddot{\mathbf{r}}_1) = 2\mathbf{F}_c - 2k(\mathbf{r}_2 - \mathbf{r}_1).$$

Put $\mathbf{r}_2 - \mathbf{r}_1 = \mathbf{r}$ and rewrite the above as

$$\ddot{\mathbf{r}} = \frac{1}{m}\left(\frac{q^2}{2\pi\varepsilon_0 r^2} - 2kr\right)\mathbf{e}_r.$$

In polar coordinates we have

$$\ddot{\mathbf{r}} = (\ddot{r} - r\dot{\theta}^2)\mathbf{e}_r + (r\ddot{\theta} + 2\dot{r}\dot{\theta})\mathbf{e}_\theta,$$

so that

$$r\ddot{\theta} + 2\dot{r}\dot{\theta} = \frac{1}{r}\frac{d}{dt}(r^2\dot{\theta}) = 0,$$

giving

$$r^2\dot{\theta} = \text{constant} = H, \text{ say.}$$

We also have

$$\ddot{r} - r\dot{\theta}^2 = \frac{q}{2\pi\varepsilon_0 mr^2} - \frac{2kr}{m}.$$

As

$$\ddot{r} = \frac{d\dot{r}}{dt} = \dot{r}\frac{d\dot{r}}{dr} = \frac{1}{2}\frac{d}{dr}(\dot{r}^2), \qquad r\dot{\theta}^2 = \frac{H^2}{r^3},$$

it can be written as

$$\frac{d}{dr}(\dot{r}^2) = \frac{q^2}{\pi\varepsilon_0 mr^2} + \frac{2H^2}{r^3} - \frac{4kr}{m}.$$

Integrating we obtain

$$\dot{r}^2 = F - \frac{q^2}{\pi\varepsilon_0 mr} - \frac{2kr^2}{m} - \frac{H^2}{r^2},$$

where F is a constant. Integrating again we obtain

$$\int^r \frac{dr}{\sqrt{F - \dfrac{q^2}{\pi\varepsilon_0 mr} - \dfrac{2kr^2}{m} - \dfrac{H^2}{r^2}}} = t + W, \tag{7}$$

where W is a constant. Also, as

$$r = \sqrt{(x_2 - x_1)^2 + (y_2 - y_1)^2}$$

we have

$$H = r^2\dot{\theta} = [(x_2 - x_1)^2 + (y_2 - y_1)^2]\frac{d}{dt}\left[\arctan\left(\frac{y_2 - y_1}{x_2 - x_1}\right)\right],$$

or

$$Ht + V = \int [(x_2 - x_1)^2 + (y_2 - y_1)^2]d\left[\arctan\left(\frac{y_2 - y_1}{x_2 - x_1}\right)\right], \tag{8}$$

where V is a constant.

The four equations (5)–(8) allow us to find x_1, x_2, y_1 and y_2 as functions of t. Note that the constants of integrations C_1, C_2, D_1, D_2, H, V, F and W are to be determined from the initial conditions.

1086

A clockwork governor employs a vibrating weight on the end of a horizontal flywheel-driven (i.e. uniformly rotating) shaft, as shown in Fig. 1.57. The flat spring has a spring constant K and can neither twist nor bend except in a direction perpendicular to its (relaxed) flat side. The angular velocity ω of the shaft, externally driven, gradually increases until a "resonance" occurs ("resonance" here means that the weight swings in a circular orbit). Air friction (proportional to the velocity of the weight) dissipates the input energy and this limits the resonance to a finite amplitude. You may assume the spring deviation to be so small that the spring is always in its linear regime. For this problem, you need not explicitly include the air friction.

   a. Show that there are two different angular frequencies at which a "resonance" can occur. What are the frequencies?

   b. Describe the orbit of the weight for each of the two resonant frequencies (i.e. draw a picture of what the problem looks like).

c. At the lower frequency resonance, write down an equation for the steady-state shaft torque as a function of ω and time.

d. Show that there is an upper bound on the shaft torque at the lower resonance. What happens if the driving clock spring yields a torque greater than this upper bound?

(UC, Berkeley)

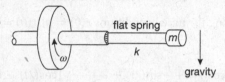

Fig. 1.57

Sol:

a. When the flywheel rotates with angular velocity ω, the mass m undergoes three-dimensional motion. However, as the longitudinal oscillation of the spring is small, we can consider the mass as not moving in the direction of the axis. As the spring can only bend in one direction, let r be the displacement of m in that direction, as shown in Fig. 1.58. The angular velocity is constant when "resonance" occurs and we shall consider the equation of the motion at resonance.

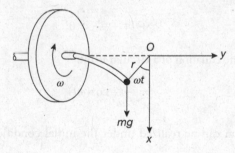

Fig. 1.58

As the elastic force is $-Kr$ and the component of the gravitational force in the direction of r is $mg\cos(\omega t)$, we can, neglecting the air friction, write the equation of the motion of the vibrating weight as

$$mg\cos(\omega t) - Kr = m(\ddot{r} - r\omega^2),$$

i.e.

$$\ddot{r} + \lambda^2 r = g\cos(\omega t),$$

where

$$\lambda^2 = \frac{K}{m} - \omega^2.$$

Trying a particular solution $r = A\cos(\omega t)$, we find $A = \dfrac{g}{\lambda^2 - \omega^2}$. The homogeneous equation $\ddot{r} + \lambda^2 r = 0$ has general solution

$$r = B\cos(\lambda t) + C\sin(\lambda t).$$

Then assuming the initial conditions $r(0) = a$, $\dot{r}(0) = b$, we obtain the general solution

$$r = a\cos(\lambda t) + \frac{b}{\lambda}\sin(\lambda t) + \frac{g}{\omega^2 - \lambda^2}[\cos(\lambda t) - \cos(\omega t)]. \tag{1}$$

A circle of radius R can be described by an equation in polar coordinates of the form

$$r = 2R\cos\theta.$$

Equation (1) can be written in this form under certain particular conditions as follows. If we let λ in (1) to go to zero, we shall obtain

$$r = a + bt + \frac{g}{\omega^2}(1 - \cos\omega t).$$

If we then put

$$a + bt + \frac{g}{\omega^2} = 0,$$

we obtain the equation of a circular orbit:

$$r = -\frac{g}{\omega^2}\cos(\omega t).$$

This solution can be realized under the initial conditions $a = -\dfrac{g}{\omega^2}$, $b = 0$, and with the angular velocity ω satisfying $\lambda = 0$, or

$$\omega = \omega_1 = \sqrt{\frac{K}{m}},$$

which is one of the resonant frequencies.

Another resonance is obtained if we put $\lambda = \omega$ in (1), which then becomes

$$r = a\cos(\omega t) + \frac{b}{\omega}\sin(\omega t) + \frac{gt}{2\omega}\sin(\omega t).$$

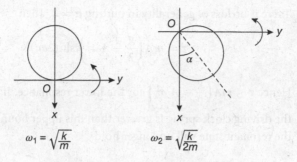

$$\omega_1 = \sqrt{\frac{k}{m}} \qquad\qquad \omega_2 = \sqrt{\frac{k}{2m}}$$

Fig. 1.59

The last term on the right-hand side has an amplitude which diverges as time goes on. However, the air friction will dissipate the input energy and limit the resonance to a finite amplitude. Thus this term can be set to zero (which can be seen by inserting a damping term $-\beta \dot{r}$ in the equation). We therefore neglect the last term and obtain

$$r = a \cos(\omega t) + \frac{b}{\omega} \sin(\omega t) = A \cos(\omega t - \alpha),$$

which again describes a circular orbit. The corresponding resonant frequency is given by $\lambda = \omega$, or

$$\omega_2 = \sqrt{\frac{K}{2m}}.$$

b. The orbits corresponding to the resonances are shown in Fig. 1.59. For the resonance at ω_1, the initial conditions must be chosen properly. On the other hand, the resonance at ω_2 can occur under any initial conditions which, however, determine the amplitude A and the angle α. ω_2 is therefore the practical resonance frequency.

c. Consider the equation of the transverse motion of the mass

$$F - mg \sin(\omega t) = m(r\dot{\omega} + 2\omega \dot{r}).$$

At the lower resonance, $r = A \cos(\omega t - \alpha)$, we have $\dot{\omega} = 0$, $\dot{r} = -\omega A \sin(\omega t - \alpha)$, so that

$$F = m[-2A\omega^2 \sin(\omega t - \alpha) + g \sin(\omega t)],$$

giving the torque as

$$\tau = Fr = mA \cos(\omega t - \alpha)[-2A\omega^2 \sin(\omega t - \alpha) + g \sin(\omega t)].$$

d. There is no loss of generality in putting $\alpha = 0$. Then

$$\tau = mA\left(\frac{g}{2} - A\omega^2\right)\sin(2\omega t).$$

Hence $\tau \leq mA\left(\frac{g}{2} - A\omega^2\right)$ for the lower resonance. If the torque yielded by the driving clock spring is greater than this upper bound, ω will increase and the resonant state will no longer hold.

1087

A mass m_1 moves around a hole on a frictionless horizontal table. The mass is tied to a string which passes through the hole. A mass m_2 is tied to the other end of the string (Fig. 1.60).

a. Given the initial position $\mathbf{R}_0$ and velocity $\mathbf{V}_0$ in the plane of the table and the masses m_1 and m_2, find the equation that determines the maximum and minimum radial distances of the orbit. (Do not bother to solve it!)

b. Find the frequency of oscillation of the radius of the orbit when the orbit is only sightly different from circular.

(*Princeton*)

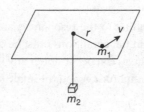

Fig. 1.60

Sol: **a.** The equations of motion of m_1 and m_2 are

$$m_1(\ddot{r} - r\dot{\theta}^2) = -T, \tag{1}$$

$$m_1 r^2 \dot{\theta} = m_1 h, \tag{2}$$

$$T - m_2 g = m_2 \ddot{r}, \tag{3}$$

Where $m_1 h$ is the angular momentum, a constant. Eliminating T from (1) and (3), we obtain

$$(m_1 + m_2)\ddot{r} - m_1 r\dot{\theta}^2 + m_2 g = 0. \tag{4}$$

Equations (2) and (4) give

$$(m_1 + m_2)\ddot{r} - \frac{m_1 h^2}{r^3} = -m_2 g. \tag{5}$$

As $\dot{r} = \dot{r}\dfrac{d\dot{r}}{dr} = \dfrac{1}{2}\dfrac{d\dot{r}^2}{dr}$, the above can be readily integrated to give

$$\frac{1}{2}(m_1 + m_2)\dot{r}^2 + \frac{m_1 h^2}{2r^2} = -m_2 gr + C. \tag{6}$$

At $t = 0$, $r = R_0$, $\dot{r} = V_0 \cos\phi$, $r\dot{\theta} = V_0 \sin\phi$, so that $h = R_0 V_0 \sin\phi$, where ϕ is the angle between $\mathbf{R_0}$ and $\mathbf{V_0}$. Then the constant of integration C can be evaluated as

$$C = \frac{1}{2}[(m_1 + m_2) V_0^2 \cos^2\phi + m_1 V_0^2 \sin^2\phi] + m_2 g R_0.$$

For r to be an extremum, $\dot{r} = 0$, with which (6) becomes

$$2m_2 gr^3 - 2Cr^2 + m_1 h^2 = 0,$$

whose solution gives the maximum and minimum radial distances of r.

b. When the orbit of m_1 is circular, $\ddot{r} = 0$, and (5) gives

$$h^2 = \frac{m_2 g r_0^3}{m_1}, \tag{7}$$

where r_0 is the radius of the circular orbit. When the orbit is slightly different from circular, let $r = r_0 + x$, where $x \ll r_0$. Equation (5) then becomes

$$(m_1 + m_2)\ddot{x} - m_1 h^2/(r_0 + x)^3 = -m_2 g.$$

As

$$(r_0 + x)^{-3} = r_0^{-3}\left(1 + \frac{x}{r_0}\right)^{-3} \approx r_0^{-3}\left(1 - \frac{3x}{r_0}\right),$$

the above becomes

$$(m_1 + m_2)\ddot{x} - m_1 h^2 (r_0^{-3} - 3x r_0^{-4}) = -m_2 g.$$

Then using (7) we have

$$(m_1 + m_2)\ddot{x} + \frac{3m_2 gx}{r_0} = 0.$$

This shows that x oscillates simple-harmonically with frequency

$$\frac{\omega}{2\pi} = \frac{1}{2\pi}\sqrt{\frac{3m_2 g}{(m_1 + m_2)r_0}}.$$

1088

a. Consider a damped, driven harmonic oscillator (in one dimension) with equation of motion

$$m\ddot{x} = -m\omega_0^2 x - \gamma\dot{x} + A\cos(\omega t).$$

What is the time-averaged rate of energy dissipation?

b. Consider an anharmonic oscillator with equation of motion

$$m\ddot{x} = -m\omega_0^2 x + \alpha x^2 + A\cos b\,(\omega t),$$

where α is a small constant.

At time $t = 0$, $x = 0$ and $\dot{x} = 0$. Solve for the subsequent motion, including terms of first order in α.

(Princeton)

Sol:

a. The equation of motion is the real part of

$$m\ddot{z} + m\omega_0^2 z + \gamma\dot{z} = Ae^{i\omega t}.$$

For the steady-state solution we try $z = z_0 e^{i\omega t}$. Substitution gives

$$z_0 = \frac{A}{m(\omega_0^2 - \omega^2) + i\gamma\omega} = Be^{-i\phi},$$

with

$$B = \frac{A}{\sqrt{m^2(\omega_0^2 - \omega^2)^2 + \gamma^2\omega^2}}, \qquad \phi = \arctan\frac{\gamma\omega}{m(\omega_0^2 - \omega^2)}.$$

The rate of work done by the force $F = Re(Ae^{i\omega t})$ is

$$P = ReF \cdot Re\dot{z} = \frac{1}{4}(F + F^*)(\dot{z} + \dot{z}^*)$$

$$= \frac{1}{4}(F\dot{z} + F^*\dot{z}^* + F^*\dot{z} + F\dot{z}^*) = \frac{1}{4}(F^*\dot{z} + F\dot{z}^*),$$

when averaged over one period as $F\dot{z}$ and $F^*\dot{z}^*$ each carries a time factor $e^{\pm 2i\omega t}$ which vanishes on integration over one period. Thus the average work done is

$$\langle P \rangle = \frac{i\omega AB}{4}(e^{-i\varphi} - e^{i\varphi}) = \frac{\omega AB}{2}\sin\varphi$$

$$= \frac{\omega AB}{2}\frac{B}{A}\gamma\omega = \frac{\gamma\omega^2 B^2}{2} = \frac{\gamma\omega^2 A^2}{2[m^2(\omega_0^2 - \omega^2) + \gamma^2\omega^2]}.$$

In steaWdy state, this is equal to the rate of energy dissipation of the oscillator, which is given by the work done against the dissipative term, i.e.

$$\left\langle P' \right\rangle = \gamma (Re\dot{z})^2 = \gamma \frac{\dot{z}\dot{z}^*}{2} = \gamma \frac{\omega^2 B^2}{2}.$$

As noted, the two approaches give the same result.

b. The equation of motion is now

$$m\ddot{x} + m\omega_0^2 x - A\cos(\omega t) = \alpha x^2. \tag{1}$$

As α is a small number, we can write the solution as

$$x = x_0 + \alpha x_1 + \alpha^2 x_2 + \cdots \approx x_0 + \alpha x_1$$

in first approximation. x_0 is the solution for $\alpha = 0$, i.e. of

$$m\ddot{x}_0 + m\omega_0^2 x_0 = A\cos(\omega t).$$

A particular solution is obtained by putting $x_0 = B'\cos(\omega t)$. Substitution gives $B' = \dfrac{A}{m(\omega_0^2 - \omega^2)}$. The general solution of the homogeneous part of the equation is harmonic. Hence the complete general solution is

$$x_0 = C\cos(\omega_0 t + \psi) + B'\cos(\omega t).$$

The initial condition $x_0 = \dot{x}_0 = 0$ at $t = 0$ then gives $\psi = 0, C = -B'$, or

$$x_0 = B'[\cos(\omega t) - \cos(\omega_0 t)].$$

Substituting $x = x_0 + \alpha x_1$ in (1) and neglecting powers of α higher than one, we have

$$\ddot{x}_1 + \omega_0^2 x_1 \approx \frac{x_0^2}{m}$$

$$= \frac{B'^2}{m}[\cos(\omega t) - \cos(\omega_0 t)]^2$$

$$= \frac{B'^2}{m}\left\{ 1 + \frac{1}{2}\cos(2\omega t) + \frac{1}{2}\cos(2\omega_0 t) - \cos[(\omega_0 - \omega)t] \right.$$

$$\left. -\cos[(\omega_0 + \omega)t], \right.$$

or, in complex form,

$$\ddot{z}_1 + \omega_0^2 z_1 \approx \frac{B'^2}{m}\left[1 + \frac{1}{2}e^{i2\omega t} + \frac{1}{2}e^{i2\omega_0 t} - e^{i(\omega_0 - \omega)t} - e^{i(\omega_0 + \omega)t} \right]. \tag{2}$$

For a particular solution try

$$z_1 = a + be^{i2\omega t} + ce^{i2\omega_0 t} + de^{i(\omega_0 - \omega)t} + fe^{i(\omega_0 + \omega)t}.$$

Substitution gives

$$a = \frac{B'^2}{m\omega_0^2}, \qquad b = \frac{B'^2}{2m(\omega_0^2 - 4\omega^2)}, \qquad c = -\frac{B'^2}{6m\omega_0^2},$$

$$d = \frac{B'^2}{m(\omega^2 - 2\omega\omega_0)}, \qquad f = \frac{B'^2}{m(\omega^2 + 2\omega\omega_0)}.$$

The general solution of (1), to first order in α, is then

$$x = x_0 + \alpha x_1$$
$$= B'[\cos(\omega t) - \cos(\omega_0 t)]$$
$$\quad + \alpha\{D\cos(\omega_0 t + \theta) + a + b\cos(2\omega t) + c\cos(2\omega_0 t)$$
$$\quad + d\cos[(\omega_0 - \omega)t] + f\cos[(\omega_0 + \omega)t]\}.$$

The initial conditions $x = \dot{x} = 0$ at $t = 0$ then give $\theta = 0$ and

$$D = -(a + b + c + d + f) = \frac{10B'^2(\omega_0^2 - \omega^2)^2}{3m\omega_0^2(\omega^2 - 4\omega_0^2)(\omega_0^2 - 4\omega^2)}.$$

Hence the motion of the anharmonic oscillator is described approximately by

$$x \approx \frac{A[\cos(\omega t) - \cos(\omega_0 t)]}{m(\omega_0^2 - \omega^2)} + \frac{10\alpha A^2 \cos(\omega_0 t)}{3m^3\omega_0^2(\omega^2 - 4\omega_0^2)(\omega_0^2 - 4\omega^2)} + \frac{\alpha A^2}{m^3(\omega_0^2 - \omega^2)^2}$$

$$\times \left\{ \frac{1}{\omega_0^2} + \frac{\cos(2\omega t)}{2(\omega_0^2 - 4\omega^2)} - \frac{\cos(2\omega_0 t)}{6\omega_0^2} + \frac{\cos[(\omega_0 - \omega)t]}{\omega^2 - 2\omega\omega_0} + \frac{\cos[(\omega_0 + \omega)t]}{\omega^2 + 2\omega\omega_0} \right\}.$$

1089

It is well known that if you drill a small tunnel through the solid, non-rotating earth of uniform density from Buffalo through the earth's center and to Olaffub on the other side, and drop a small stone into the hole, it will be seen at Olaffub after a time $T_1 = \dfrac{\pi}{\omega_0}$, where ω_0 is a constant. Now, instead of just dropping the stone, you throw it into the hole with an initial velocity v_0. How big should v_0 be, so that it now appears at Olaffub after a time $T_2 = T_1/2$? Your answer should be given in terms of ω_0 and R, the radius of the earth.

(Princeton)

Sol: Let r be the distance of the stone, of mass m, from the center of the earth. The gravitational force on it is $F = -\dfrac{Gm4\pi r^3\rho}{3r^2} = -\omega_0^2 mr$, where $\omega_0 = \sqrt{\dfrac{4\pi G\rho}{3}}$,

ρ being the density of the uniform earth. The equation of the motion of the stone is then

$$\ddot{r} = -\omega_0^2 r.$$

Thus the stone executes simple harmonic motion with a period $T = \dfrac{2\pi}{\omega_0}$. Then if the stone starts from rest at Buffalo, it will reach Olaffub after a time $T_1 = \dfrac{T}{2} = \dfrac{\pi}{\omega_0}$.

The solution of the equation of motion is

$$r = A\cos(\omega t + \varphi).$$

Suppose now the stone starts at $r = R$ with initial velocity $\dot{r} = -v_0$. We have

$$R = A\cos\varphi, \qquad -v_0 = -A\omega_0\sin\varphi,$$

giving

$$\varphi = \arctan\left(\frac{v_0}{R\omega_0}\right), \qquad A = \sqrt{R^2 + \left(\frac{v_0}{\omega_0}\right)^2}.$$

To reach Olaffub at $t = \dfrac{T_1}{2} = \dfrac{\pi}{2\omega_0}$, we require

$$-R = \sqrt{R^2 + \left(\frac{v_0}{\omega_0}\right)^2}\cos\left(\frac{\pi}{2} + \varphi\right) = -\sqrt{R^2 + \left(\frac{v_0}{\omega_0}\right)^2}\sin\varphi.$$

As $\sin^2\varphi + \cos^2\varphi = 1$, we have

$$\frac{R^2}{R^2 + \left(\dfrac{v_0}{\omega_0}\right)^2} + \frac{R^2}{R^2 + \left(\dfrac{v_0}{\omega}\right)^2} = 1,$$

giving

$$v_0 = R\omega_0.$$

1090

A body of mass m moves along the x direction under the influence of a potential $U(x) = \dfrac{U_0}{a^3}\left(\dfrac{x^4}{a} - x^3\right)$. (a) Find the points where the particle is in equilibrium and identify the nature of the equilibrium. (b) For what region of space is

Problems and Solutions on Mechanics

the particle likely to undergo simple harmonic motion (SHM)? What is the minimum velocity that the particle must have to escape this region?

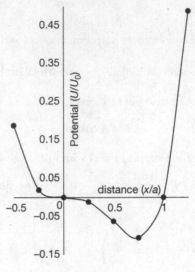

Fig. 1.61

Sol:

a.

$$F(x) = -\frac{dU}{dx} = -\frac{U_0}{a^3}\left(\frac{4x^3}{a} - 3x^2\right)$$

For equilibrium

$$F(x) = -\frac{dU}{dx} = -\frac{U_0}{a^3}\left(\frac{4x^3}{a} - 3x^2\right) = 0$$

i.e., either $x = 0$ or $\frac{3a}{4}$.

Slope of the tangent to the curve along the points slightly greater than 0 and slightly less than 0 is 0, then the point $x = 0$ is a neutral equilibrium. Slope of the tangent to the curve along the points slightly less than $3a/4$ is negative and that slightly greater than $3a/4$ is positive, then the point $x = 3a/4$ is a stable equilibrium.

Alternate method to identify the nature of the equilibrium:

When $x = 0$

$$\frac{d^2U}{dx^2} = \frac{U_0}{a^3}\left(\frac{12x^2}{a} - 6x\right) = 0,$$

indicating neutral equilibrium.

When $x = \dfrac{3a}{4}$

$$\frac{d^2U}{dx^2} = \frac{9U_0}{4a^2} > 0,$$

indicating stable equilibrium.

b. The particle will undergo SHM around $x = 3a/4$.

The total energy (E) in this region $T + U =$ constant at all times oscillating between totally potential at $x = 0$ and $x = 1$ and totally kinetic at $x = 3a/4$.

The particle will escape this region if its total energy is positive. The deepest point of the potential at $x = 3a/4$ has $U = -\dfrac{27}{256}U_0$, so the particle needs

$$T = \frac{1}{2}mv^2 > \frac{27U_0}{256} \text{ or } v > \frac{3}{16}\sqrt{\frac{6U_0}{m}}.$$

1091

A spring mass system is set up as shown in the figure. If the mass is released from rest, what is the initial displacement of the spring when the mass comes to rest momentarily? Is the subsequent motion of the mass simple harmonic? Assume that there is no friction between the block and the surface.

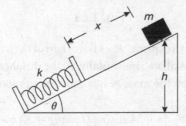

Fig. 1.62

Sol: Let "x" represents the compression of the spring.

Once the mass reaches the top of the spring, it moves against the restoring force till it momentarily comes to rest after displacing the spring by an amount x'. At this time,

$$mgh = \frac{kx'^2}{2}$$

$$x' = \sqrt{\frac{2mgh}{k}}.$$

From the force equation,

$$mg \sin \theta - kx = m\ddot{x}.$$

The subsequent motion is simple harmonic because the restoring force is proportional to the displacement.

1092

A mass m hangs in equilibrium by a spring which exerts a force $F = -K(x - l)$, where x is the length of the spring and l is its length when relaxed. At $t = 0$ the point of support to which the upper end of the spring is attached begins to oscillate sinusoidally up and down with amplitude A, angular frequency ω as shown in Fig. 1.63. Set up and solve the equation of motion for $x(t)$.

(SUNY, Buffalo)

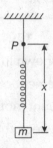

Fig. 1.63

Sol: Take the upper end of the spring, P, as the origin of the x coordinate of the mass m. At $t = 0$, P starts to oscillate sinusoidally, so the distance of P from the fixed support is $A \sin(\omega t)$. Then the mass m has equation of motion

$$m \frac{d^2}{dt^2}[x + A \sin(\omega t)] = mg - K(x - l).$$

Let $y = x - l - \dfrac{mg}{K}$, $\omega_0^2 = \dfrac{K}{m}$ The above can be written as

$$\ddot{y} + \omega_0^2 y = \omega^2 A \sin(\omega t).$$

Try a particular solution $y = B \sin(\omega t)$. Substitution gives

$$B = \frac{\omega^2 A}{\omega_0^2 - \omega^2}.$$

Hence the general solution is

$$y = C\cos(\omega_0 t) + D\sin(\omega_0 t) + \frac{\omega^2 A \sin(\omega t)}{\omega_0^2 - \omega^2}.$$

Using the initial condition

$$mg = K(x - l), \text{ i.e. } x = \frac{mg}{K} + l, \text{ or } y = 0$$

and $\dot{y} = 0$, we have

$$C = 0, \quad D = \frac{\omega^3 A}{\omega_0(\omega_0^2 - \omega^2)}$$

and hence

$$x(t) = \frac{\omega^2 A}{\omega_0^2 - \omega^2}\left[\sin(\omega t) - \frac{\omega}{\omega_0}\sin(\omega_0 t)\right] + \frac{mg}{K} + l.$$

1093

A body of mass m is given a velocity v up a smooth incline of angle θ. Find expressions for the time when the body would be at a distance of half its maximum distance along the incline.

Sol: When moving up the incline, $ma = -mg\sin\theta$ or $a = -g\sin\theta$

At the topmost point

$$0 = v^2 - 2gs\sin\theta \text{ or } s = \frac{v^2}{2g\sin\theta}$$

Let t be the time taken by the body to reach half the maximum distance.

Use the equation $x = ut + \frac{1}{2}at^2$ and set $x = \frac{s}{2} = \frac{v^2}{4g\sin\theta}$, we get

$$\frac{v^2}{4g\sin\theta} = vt - \frac{1}{2}g(\sin\theta)t^2$$

Solve for t.

$$t \doteq (2 - \sqrt{2})\frac{v}{2g\sin} \text{ or } t = (2 + \sqrt{2})\frac{v}{2g\sin}$$

1094

A merry-go-round (carousel) starts from rest and accelerates with a constant angular acceleration of 0.02 revolution per second per second. A woman sitting on a chair 6 meters from the axis of revolution holds a 2 kg ball (see Fig. 1.64). Calculate the magnitude and direction of the force she must exert to hold the ball 5 seconds after the merry-go-round begins to rotate. Specify the direction with respect to the radius of the chair on which she is sitting.

(Wisconsin)

2 kg

6 m

Fig. 1.64

Sol: Consider two coordinate frames L, R with the same origin. L is fixed to the laboratory and R rotates with angular velocity ω. The time derivatives of a vector **A** in the two frames are related by

$$\left(\frac{d\mathbf{A}}{dt}\right)_L = \left(\frac{d\mathbf{A}}{dt}\right)_R + \omega \times \mathbf{A}.$$

Then for a point of radius vector **r** from the origin we have

$$\left(\frac{d\mathbf{r}}{dt}\right)_L = \left(\frac{d\mathbf{r}}{dt}\right)_R + \omega \times \mathbf{r},$$

and

$$\left(\frac{d^2\mathbf{r}}{dt^2}\right)_L = \left(\frac{d}{dt}\right)_L \left(\frac{d\mathbf{r}}{dt}\right)_R + \omega \times \left(\frac{d\mathbf{r}}{dt}\right)_L + \left(\frac{d\omega}{dt}\right)_L \times \mathbf{r}.$$

As

$$\left(\frac{d}{dt}\right)_L \left(\frac{d\mathbf{r}}{dt}\right)_R = \left(\frac{d^2\mathbf{r}}{dt^2}\right)_R + \omega \times \left(\frac{d\mathbf{r}}{dt}\right)_R,$$

$$\omega \times \left(\frac{d\mathbf{r}}{dt}\right)_L = \omega \times \left(\frac{d\mathbf{r}}{dt}\right)_R + \omega \times (\omega \times \mathbf{r}).$$

Putting

$$\left(\frac{d^2\mathbf{r}}{dt^2}\right)_R = \mathbf{a}', \quad \left(\frac{d\mathbf{r}}{dt}\right)_R = \mathbf{v}', \quad \left(\frac{d\omega}{dt}\right)_L = \dot{\omega},$$

we have

$$\left(\frac{d^2\mathbf{r}}{dt^2}\right)_L = \mathbf{a}' + 2\omega \times \mathbf{v}' + \omega \times (\omega \times \mathbf{r}) + \dot{\omega} \times \mathbf{r}.$$

In the rotating frame attached to the carousel, the equation of the motion of the ball $\mathbf{F} = m\left(\dfrac{d^2\theta}{dt^2}\right)_L$ then gives

$$ma' = \mathbf{F} + m\omega^2\mathbf{r} - m\dot{\omega} \times \mathbf{r} - 2m\omega \times \mathbf{v}',$$

as $\omega \perp \mathbf{r}$ so that $\mathbf{r} \times (\omega \times \mathbf{r}) = -\omega^2\mathbf{r}$. As the ball is held stationary with respect to the carousel, $\mathbf{a}' = 0, \mathbf{v}' = 0$, and

$$\mathbf{F} = -m\omega^2\mathbf{r} + m\dot{\omega} \times \mathbf{r}.$$

For the rotating frame R, take the z-axis along the axis of rotation and the x-axis from the center toward the chair. Then

$$\dot{\omega} = \dot{\omega}\mathbf{k}, \qquad \mathbf{r} = r\mathbf{i}.$$

The force $\mathbf{F}$ acting on the ball is the resultant of the holding force $\mathbf{f}$ exerted by the woman and the gravity of the earth:

$$\mathbf{F} = \mathbf{f} - mg\mathbf{k}.$$

Hence

$$\mathbf{f} = -m\omega^2 r\mathbf{i} + m\dot{\omega}r\mathbf{j} + mg\mathbf{k}.$$

With $\dot{\omega} = 0.02 \times 2\pi$ rad/s^2, $\omega = 5\dot{\omega}$, $m = 2$ kg, $r = 6$ m, we have

$$\mathbf{f} = -4.74\mathbf{i} + 1.5l\mathbf{j} + 19.6\mathbf{k} \text{ N},$$

of magnitude $= 20.2$ N.

1095

A planet of uniform density spins about a fixed axis with angular velocity ω. Due to the spin the planet's equatorial radius R_E is slightly larger than its polar radius R_P as described by the parameter $\varepsilon = (R_E - E_P)/R_E$. The contribution to the gravitational potential resulting from this distortion is

$$\Phi(R, \theta) = \frac{2GM_e\varepsilon R_E^2 P_2(\cos\theta)}{5R^3},$$

where θ is the polar angle and $P_2(\cos\theta) = \frac{3\cos^2\theta}{2} - \frac{1}{2}$. State a reasonable condition for equilibrium of the planet's surface and compute the value of ε in terms of the parameter $\lambda = \dfrac{\omega^2 R_E}{g}$, where g is the gravitational acceleration. Make a numerical estimate of ε for the earth.

(Wisconsin)

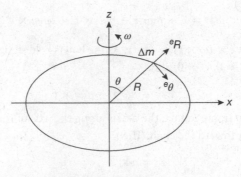

Fig. 1.65

Sol: The forces acting on a mass element Δm on the surface of the planet are gravity, centrifugal force, and the constraint exerted by the rest of the planet. The condition for equilibrium of the surface is that the resultant of gravity and centrifugal force is perpendicular to the surface, i.e. it has no tangential component.

Suppose the surface of the planet is an ellipsoid of revolution with the z-axis as its axis of symmetry as shown in Fig. 1.65. The line of intersection of the ellipsoid with the xz-plane is an ellipse:

$$z = R_P\cos\alpha, \qquad x = R_E\sin\alpha,$$

where α is a parameter. The polar angle θ of a point on the ellipse is given by

$$\tan\theta = \frac{x}{z} = \frac{R_E}{R_P}\tan\alpha.$$

The unit tangent τ to the ellipse at this point is

$$\tau \propto \mathbf{i}dx + \mathbf{k}dz = \left(\mathbf{i}\frac{dx}{d\alpha} + \mathbf{k}\frac{dz}{d\alpha}\right)d\alpha$$

$$= (\mathbf{i}R_E \cos\alpha - \mathbf{k}R_P \sin\alpha)\,d\alpha = \frac{\cos\alpha}{R_E}(\mathbf{i}R_E^2 - \mathbf{k}R_P^2 \tan\theta)\,d\alpha.$$

The centrifugal force **f1** on Δm is

$$\mathbf{f}_1 = \mathbf{i}\Delta m R\omega^2 \sin\theta$$

and the gravitational force on it is

$$\mathbf{f}_2 = -\nabla V = \nabla\left[\frac{GM_e \Delta m}{R} + \frac{2GM_e \varepsilon R_E^2 \Delta m}{5R^3} P_2(\cos\theta)\right]$$

$$= GM_e \Delta m\left(-\frac{1}{R^2} - \frac{6\varepsilon R_E^2}{5R^4}P_2(\cos\theta)\right)\mathbf{e}_r - \frac{6GM_e \varepsilon R_E^2 \Delta m}{5R^4}\sin\theta\cos\theta\,\mathbf{e}_\theta.$$

As

$$\mathbf{e}_r = \mathbf{i}\sin\theta + \mathbf{k}\cos\theta,$$

$$\mathbf{e}_\theta = \mathbf{i}\cos\theta - \mathbf{k}\sin\theta,$$

$$\mathbf{f}_2 = GM_e \Delta m\left[-\frac{\sin\theta}{R^2} - \frac{6\varepsilon R_E^2}{5R^4}\sin\theta P_2(\cos\theta) - \frac{6\varepsilon R_E^2}{5R^4}\sin\theta\cos^2\theta\right]\mathbf{i}$$

$$+ GM_e \Delta m\left[-\frac{\cos\theta}{R^2} - \frac{6\varepsilon R_e^2}{5R^4}\cos\theta P_2(\cos\theta) + \frac{6\varepsilon R_E^2}{5R^4}\sin^2\theta\cos\theta\right]\mathbf{k}$$

$$= GM_e \Delta m\left\{\mathbf{i}\left[-\frac{1}{R^2} - b\left(\frac{3}{2}\cos^2\theta - \frac{1}{2} + \cos^2\theta\right)\right]\sin\theta\right.$$

$$\left. + \mathbf{k}\left[-\frac{1}{R^2} - b\left(\frac{3}{2}\cos^2\theta - \frac{1}{2} - \sin^2\theta\right)\right]\cos\theta,\right.$$

with $b = 6\varepsilon R_E^2/5R^4$.

The condition for equilibrium of the surface is

$$(\mathbf{f}_1 + \mathbf{f}_2)\cdot\tau = 0,$$

which gives for $R \approx R_P \approx R_E$

$$R_E^3\omega^2 \sin\theta - R_E^2 bGM_e \sin\theta \approx 0.$$

Hence

$$\varepsilon = \frac{5R_E^2 b}{6} \approx \frac{5R_E^3\omega^2}{6GM_e} = \frac{5R_E\omega^2}{6g} = \frac{5\lambda}{6}$$

as $g = \dfrac{GM_e}{R\,\dfrac{2}{E}}$. For earth, $R_E = 6378 \times 10^3$ m, $\omega = \dfrac{2\pi}{24 \times 3600}$ rad/s, $g = 9.8$ m/s^2,

we have

$$\varepsilon \approx 2.9 \times 10^{-3}.$$

1096

A satellite moves in a circular orbit around the earth. Inside, an astronaut takes a small object and lowers it a distance Δr from the center of mass of the satellite towards the earth. If the object is released from rest (as seen by the astronaut), describe the subsequent motion as seen by the astronaut in the satellite's frame of reference.

(Wisconsin)

Sol: The satellite revolves around the earth with an angular velocity ω. We assume that one side of the satellite always faces the earth, i.e. its spin angular velocity is also ω. Choose a coordinate frame attached to the satellite such that the origin is at the center of mass of the satellite and the center of the earth is on the y-axis as shown in Fig. 1.66, where R is the distance of the satellite from the center of the earth.

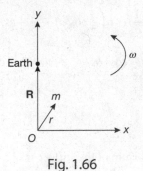

Fig. 1.66

The equation of the motion of the small object of mass m in the satellite frame is given by (Problem **1094**)

$$\mathbf{F} = m\ddot{\mathbf{r}} + m\omega \times (\omega \times \mathbf{r}) + 2m\omega \times \dot{\mathbf{r}} + m\dot{\omega} \times \mathbf{r}$$
$$= m\ddot{\mathbf{r}} - m\omega^2\mathbf{r} + 2m\omega \times \dot{\mathbf{r}},$$

since $\dot{\omega} = 0$, $\omega \cdot \mathbf{r} = 0$. Thus

$$m\ddot{\mathbf{r}} = \mathbf{F} + m\omega^2\mathbf{r} - 2m\omega \times \dot{\mathbf{r}}.$$

In the above, $\mathbf{F}$ is the gravitational force exerted by the earth:

$$\mathbf{F} = \frac{GMm}{|\mathbf{R} - \mathbf{r}|^3}(\mathbf{R} - \mathbf{r}) \approx \frac{GMm}{R^3}\frac{(\mathbf{R} - \mathbf{r})}{\left(1 - \dfrac{\mathbf{r} \cdot \mathbf{R}}{R^2}\right)^3}$$

$$\approx \frac{GMm}{R^3}\left(1 + \frac{3y}{R}\right)(\mathbf{R} - \mathbf{r})$$

$$\approx \frac{GMm}{R^3}\mathbf{R} - \frac{GMm}{R^3}\mathbf{r} + \frac{3GMmy}{R^3}\mathbf{e}_y,$$

where $m\omega^2\mathbf{r}$ is the centrifugal force and

$$-2m\omega \times \dot{\mathbf{r}} = -2m\omega\mathbf{e}_z \times (\dot{x}\mathbf{e}_x + \dot{y}\mathbf{e}_y + \dot{z}\mathbf{e}_z)$$
$$= -2m\omega(\dot{x}\mathbf{e}_y - \dot{y}\mathbf{e}_x)$$

is the Coriolis force.

As initially, $\mathbf{r} = \Delta r\mathbf{e}_y$, and all the forces are in the xy-plane, the object always moves in this plane. Then $\mathbf{r} = x\mathbf{e}_x + y\mathbf{e}_y$. If the satellite has mass m', we have

$$\frac{GMm'}{R^2} = m'R\omega^2,$$

or $\omega^2 = \dfrac{GM}{R^3}$. Then the second term of $\mathbf{F}$ cancels out the centrifugal force. The first term of $\mathbf{F}$ acts on the satellite as a whole and is of no interest to us. Hence the equation of motion becomes

$$\ddot{x}\mathbf{e}_x + \ddot{y}\mathbf{e}_y = 3\omega^2y\mathbf{e}_y - 2\omega(\dot{x}\mathbf{e}_y - \dot{y}\mathbf{e}_x)$$

or, in component form,

$$\ddot{y} = 3\omega^2y - 2\omega\dot{x}, \tag{1}$$

$$\ddot{x} = 2\omega\dot{y}. \tag{2}$$

Integrating (2) and making use of the initial conditions $\dot{x} = 0, y = \Delta r$ at $t = 0$, we find

$$\dot{x} = 2\omega(y - \Delta r). \tag{3}$$

Substitution in (1) gives

$$\ddot{y} = -\omega^2y + 4\omega^2\Delta r,$$

whose general solution is $y = A\cos(\omega t) + B\sin(\omega t) + 4\Delta r$, A, B being constants. With the initial conditions $y = \Delta r, \dot{y} = 0$ at $t = 0$, we find

$$y = -3\Delta r\cos(\omega t) + 4\Delta r.$$

Equation (3) now becomes

$$\dot{x} = 6\omega\Delta r[1 - \cos(\omega t)].$$

Integrating and applying the initial condition $x = 0$ at $t = 0$, we obtain

$$x = 6\Delta r[\omega t - \sin(\omega t)],$$

Hence, the subsequent motion as seen by the astronaut in the satellite's frame of reference is described by

$$x = 6\Delta r[\omega t - \sin(\omega t)],$$
$$y = \Delta r[4 - 3\cos(\omega t)].$$

1097

Consider a hoop of radius a in a vertical plane rotating with angular velocity ω about a vertical diameter. Consider a bead of mass m which slides without friction on the hoop as indicated in Fig. 1.67.

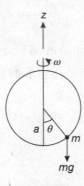

Fig. 1.67

a. Under what specific condition will the equilibrium of the bead at $\theta = 0$ be stable?

b. Find another value of θ for which, in certain circumstances, the bead will be in stable equilibrium. Indicate the values of ω for which this stable equilibrium takes place.

c. Explain your answer with the aid of appropriate graphs of the potential energy versus θ as measured in the rotating frame.

(Wisconsin)

Sol: Consider a coordinate frame (r, θ) attached to the hoop and use the derivation in Problem **1094**. As ω is constant, we have in the rotating frame

$$\mathbf{g} = \ddot{\mathbf{r}} + 2\boldsymbol{\omega} \times \dot{\mathbf{r}} + \boldsymbol{\omega} \times (\boldsymbol{\omega} \times \mathbf{r}).$$

As

$$\begin{aligned}
\mathbf{g} &= -g\mathbf{e}_z = g \cos \theta \mathbf{e}_r - g \sin \theta \mathbf{e}_\theta, \\
\ddot{\mathbf{r}} &= -a\dot{\theta}^2 \mathbf{e}_r + a\ddot{\theta}\mathbf{e}_\theta \\
\boldsymbol{\omega} &= \omega \mathbf{e}_z = -\omega \cos \theta \mathbf{e}_r + \omega \sin \theta \mathbf{e}_\theta, \\
\mathbf{r} &= a\mathbf{e}_r,
\end{aligned}$$

we have the equation of the motion of the bead in the $\mathbf{e}_\theta$ direction in the rotating frame as

$$a\ddot{\theta} = -g \sin \theta + a\omega^2 \sin \theta \cos \theta. \tag{1}$$

To find the equilibrium positions, let $\ddot{\theta} = 0$. The above then gives, for equilibrium, $\theta = 0$ and $\cos \theta = \dfrac{g}{a\omega^2}$.

a. When θ is in the neighborhood of zero,

$$\sin \theta \approx \theta, \qquad \cos \theta \approx 1 - \frac{\theta^2}{2}.$$

We can approximate (1) to

$$\begin{aligned}
\ddot{\theta}\left(\frac{g}{a} - \omega^2\right)\theta = 0 \qquad &\text{if } \frac{g}{a} - \omega^2 \neq 0, \\
\ddot{\theta} + \frac{\omega^2}{2}\theta^3 = 0 \qquad &\text{if } \frac{g}{a} - \omega^2 = 0.
\end{aligned}$$

It is evident that if and only if $\omega^2 \leq g/a$, in which case the resultant force acting on the bead is always directed toward the equilibrium position, will the equilibrium of the bead at $\theta = 0$ be stable.

b. The other value of θ for which the bead will be in equilibrium is

$$\theta_0 = \arccos\left(\frac{g}{a\omega^2}\right).$$

Let $\theta = \theta_0 + \delta\theta$, where $\delta\theta \ll \theta_0$. Then

$$\begin{aligned}
\sin \theta &= \sin(\theta_0 + \delta\theta) \approx \sin \theta_0 + \cos \theta_0 \delta\theta, \\
\cos \theta &= \cos(\theta_0 + \delta\theta) \approx \cos \theta_0 - \sin \theta_0 \delta\theta.
\end{aligned}$$

Substitution in (1) gives

$$\frac{d^2\delta\theta}{dt^2} + \left(1 - \frac{g^2}{a^2\omega^4}\right)\omega^2 \delta\theta = 0.$$

Hence the condition of stable equilibrium is

$$1 - \frac{g^2}{a^2\omega^4} > 0, \quad \text{or} \quad \omega > \sqrt{\frac{g}{a}}.$$

c. The potential energy of the bead in the rotating frame consists of two parts, i.e. gravitational potential energy V_1 and centrifugal potential energy V_2, given by

$$-\frac{\partial V_1}{\partial z} = -mg,$$

i.e.

$$V_1 = mgz = mga(1 - \cos\theta),$$

$$-\frac{\partial V_2}{\partial r} = mr\omega^2$$

i.e.

$$V_2 = -\frac{1}{2}m\omega^2 r^2 = -\frac{1}{2}ma^2\omega^2 \sin^2\theta.$$

Thus

$$V = V_1 + V_2 = mga(1 - \cos\theta) - \frac{1}{2}ma^2\omega^2 \sin^2\theta.$$

The two equilibrium positions are given by $\dfrac{\partial V}{\partial \theta} = 0$:

$$\sin\theta = 0, \quad \text{or} \quad \theta = 0,$$

$$\cos\theta = \frac{g}{a\omega^2}, \quad \text{or} \quad \theta = \arccos\left(\frac{g}{a\omega^2}\right).$$

Figures 1.68 (a), (b), and (c) are the graphs of the potential energy versus θ as measured in the rotating frame for $\omega < \sqrt{g/a}, \omega = \sqrt{g/a}$ and $\omega > \sqrt{g/a}$ respectively.

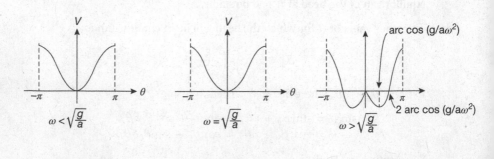

(a)   (b)   (c)

Fig. 1.68

The potential energy V must be a minimum for the equilibrium to be stable. This is the case for $\theta = 0$ in Figs. (a) and (b) and for $\theta = \arccos\left(\dfrac{g}{a\omega^2}\right)$ in Fig. (c). The point $\theta = 0$ in Fig. (c) is an equilibrium position but the equilibrium is unstable as V is a maximum there.

1098

A perfectly smooth horizontal disk is rotating with an angular velocity ω about a vertical axis passing through its center. A person on the disk at a distance R from the origin gives a perfectly smooth coin (negligible size) of mass m a push toward the origin. This push gives it an initial velocity V relative to the disk. Show that the motion for a time t, which is such that $(\omega t)^2$ is negligible, appears to the person on the disk to be a parabola, and give the equation of the parabola.

(Wisconsin)

Sol: Use a Cartesian coordinate frame attached to the disk such that the z-axis is along the axis of rotation and the x-axis is opposite to the initial velocity V of the coin, both x, y-axis being on the plane of the disk. In this rotating frame, we have (Problem **1094**),

$$\frac{md\mathbf{v}}{dt} = \mathbf{F} - \frac{md\boldsymbol{\omega}}{dt} \times \mathbf{r} - m\boldsymbol{\omega} \times (\boldsymbol{\omega} \times \mathbf{r}) - 2m\boldsymbol{\omega} \times \mathbf{v}.$$

As there is no horizontal force on the coin after the initial push and $\boldsymbol{\omega} = \omega\mathbf{k}$, $\dot{\boldsymbol{\omega}} = 0$, the above gives

$$\ddot{x} = \omega^2 x + 2\omega\dot{y}, \tag{1}$$
$$\ddot{y} = \omega^2 y - 2\omega\dot{x} \tag{2}$$

Let $z = x + iy$. Then (1) + (2) × i gives

$$\ddot{z} + 2i\omega\dot{z} - \omega^2 z = 0. \tag{3}$$

With $z = e^{\gamma t}$, we have the characteristic equation

$$\gamma^2 + 2i\omega\gamma - \omega^2 = (\gamma + i\omega)^2 = 0.$$

This has a double root $\gamma = -i\omega$, so that the general solution of (3) is

$$z = (A + iB)e^{-i\omega t} + (C + iD)te^{-i\omega t}.$$

The initial conditions are $x = R, y = 0, \dot{x} = -V, \dot{y} = 0$, or $z = R, \dot{z} = -V$, at $t = 0$, which give

$$R = A + iB, \quad -V = \omega B + C + i(D - \omega A),$$

or

$$A = R, \ B = 0, \quad C = -V, \quad D = \omega R.$$

Hence

$$z = [(R - Vt) + iR\omega t]e^{-i\omega t},$$

or

$$x = (R - Vt)\cos(\omega t) + R\omega t \sin(\omega t),$$
$$y = -(R - Vt)\sin(\omega t) + R\omega t \cos(\omega t).$$

Neglecting the $(\omega t)^2$ terms, the above become

$$x \approx R - Vt,$$
$$y \approx -(R - Vt)\omega t + R\omega t = V\omega t^2.$$

Hence the trajectory is approximately a parabola $y = \dfrac{\omega}{V}(R - x)^2$.

1099

A body is dropped from rest at a height h above the surface of the earth and at a latitude 40°N. For $h = 100$ m, calculate the lateral displacement of the point of impact due to the Coriolis force.

(Columbia)

Sol: If the body has mass m, in the rotating frame of the earth, a Coriolis force $-2m\boldsymbol{\omega} \times \dot{\mathbf{r}}$ is seen to act on the body. We choose a frame with origin at the point on the earth's surface below the starting point of the body, with x-axis pointing south, y-axis pointing east and z-axis pointing vertically up (Fig. 1.69). Then the equation of the motion of the body in the earth frame is

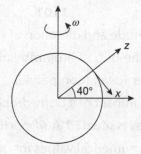

$$\text{Fig. 1.69}$$

$$m\ddot{r} = -mg\mathbf{k} - 2m\omega \times \dot{\mathbf{r}}.$$

$$= -mg\mathbf{k} - 2m \begin{vmatrix} \mathbf{i} & \mathbf{j} & \mathbf{k} \\ -\omega\cos 40° & 0 & \omega\sin 40° \\ \dot{x} & \dot{y} & \dot{z} \end{vmatrix}.$$

From the above, expressions for $\ddot{x}$, $\ddot{y}$ and $\ddot{z}$, can be obtained, which are readily integrated to give $\dot{x}$, $\dot{y}$ and $\dot{z}$. These results are then used in the expressions for $\ddot{x}$, $\ddot{y}$ and $\ddot{z}$ and. As the time of the drop of the body is short compared with the period of rotation of the earth, we can ignore terms of order ω^2 and write the following:

$$\ddot{x} = 0,$$
$$\ddot{y} = 2gt\omega\cos 40°,$$
$$\ddot{z} = -g.$$

Integrating the above twice and using the initial conditions, we obtain

$$x = 0,$$
$$y = \frac{1}{3}gt^2\omega\cos 40°,$$
$$z - h = -\frac{gt^2}{2}.$$

The last equation gives the time of arrival of the body at the earth's surface $z = 0$:

$$t = \sqrt{\frac{2h}{g}}.$$

Then the lateral displacement of the body at impact is

$$y = \frac{1}{3}\sqrt{\frac{8h^3}{g}}\omega\cos 40° = 0.017 \text{ m}.$$

1101

a. What are the magnitude and direction of the deflection caused by the earth's rotation to the bob of a plumb-line hung from the top to the bottom of the Sather Tower (Companile).

b. What is the point of impact of a body dropped from the top?

Assume that Berkeley is situated at $\theta°$ north latitude and that the tower is L meters tall. Give numerical values for (a) and (b) based on your estimates of L and θ.

(Columbia)

Sol:

a. In Fig. 1.70, $\mathbf{F}_e$ is the fictitious centrifugal force, α is the angle that $m\mathbf{g}$, the apparent gravity, makes with the direction pointing to the center of the earth. The gravity $m\mathbf{g}_0$ for a non-rotating earth is related to the above quantities by

$$m\mathbf{g} = m\mathbf{g}_0 + \mathbf{F}_e.$$

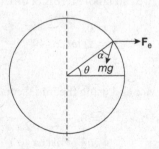

Fig. 1.70

Then by the force triangle, we have

$$\frac{F_e}{\sin \alpha} = \frac{mg}{\sin \theta},$$

or

$$\sin \alpha = \frac{F_e \sin \theta}{mg} = \frac{mR\omega^2 \cos \theta \sin \theta}{mg} = \frac{R\omega^2 \sin 2\theta}{2g}.$$

Hence the magnitude of the deflection of the bob is

$$L\alpha = L \arcsin\left(\frac{R\omega^2 \sin 2\theta}{2g}\right).$$

b.   The lateral displacement of a body falling from rest at height L in the northern hemisphere due to the Coriolis force is to the east and has magnitude (Problem **1099**)

$$\delta = \frac{1}{3}\sqrt{\frac{8L^3}{g}}\,\omega\cos\theta.$$

1101

Under especially favorable conditions, an ocean current circulating counter-clockwise when viewed from directly overhead was discovered in a well-isolated layer beneath the surface. The period of rotation was 14 hours. At what latitude and in which hemisphere was the current detected?

(Columbia)

Sol: We choose a coordinate frame attached to the earth with origin at the point on the earth's surface where the ocean current is, x-axis pointing south, y-axis pointing east and z-axis pointing vertically upward. The circulation in the ocean is due to the Coriolis force which causes an additional acceleration (Problem **1094**)

$$\mathbf{a} = -2\boldsymbol{\omega}\times\mathbf{v},$$

where $\boldsymbol{\omega} = \omega\cos\theta\,\mathbf{i} + \omega\sin\theta\,\mathbf{k}$ is the earth's rotational angular velocity, θ is the latitude, and $\mathbf{v}$ is the velocity of ocean current. Thus

$$\mathbf{a} = -2\omega\begin{vmatrix} \mathbf{i} & \mathbf{j} & \mathbf{k} \\ -\cos\theta & 0 & \sin\theta \\ v_x & v_y & 0 \end{vmatrix}.$$

The horizontal component of the acceleration which affects the circulation of the ocean current is

$$\mathbf{a}_H = -2\omega\sin\theta\,(-v_y\mathbf{i} + v_x\mathbf{j}) = -2\omega_z\mathbf{k}\times\mathbf{v}.$$

As $\mathbf{a}_H$ is always normal to $\mathbf{v}$, it does not change the magnitude of the latter but only its direction. It causes the current to circulate in a circular path. Let Ω be the angular velocity of the circular motion. Then

$$|\mathbf{a}_H| = 2\omega v\sin\theta = \frac{v^2}{r} = v\Omega,$$

where r is the radius of the circular path. Hence

$$\sin\theta = \frac{\Omega}{2\omega} = \frac{2\pi}{14}\cdot\frac{24}{4\pi} = \frac{6}{7},$$

or

$$\theta = 59°.$$

If the ocean current is on the northern hemisphere, $\omega_z\mathbf{k}$ points toward the north pole and $\mathbf{a}_H$ always points to the right of the velocity $\mathbf{v}$. This makes $\mathbf{v}$ turn right and gives rise to clockwise circulation. In a similar way, in the southern hemisphere, the Coriolis force causes counter-clockwise circulation. Hence the circulating ocean current was detected at a latitude of 59°S.

1102

A small celestial object, held together only by its self-gravitation, can be disrupted by the tidal forces produced by another massive body, if it comes near enough to that body. For an object of diameter 1 km and density 2 g/cm³, find the critical distance from the earth (Roche limit).

(UC, Berkeley)

Sol: Suppose the earth is fixed in space and the small celestial object orbits around it at a distance l away as shown in Fig. 1.71. Let M be the mass of the earth, m the mass and ρ the density of the small celestial object. Consider a unit mass of the object on the line OC at distance x from C. We have from the balance of forces on it

$$(l - x)\omega^2 = \frac{GM}{(l - x)^2} - \frac{G\left(\frac{4}{3}\right)\pi x^3\rho}{x^2}.$$

We also have for the celestial body

$$ml\omega^2 = \frac{GMm}{l^2},$$

which gives ω^2 to be used in the above. Then as $\dfrac{x}{l} \ll 1$, retaining only the lowest order in $\dfrac{x}{l}$, we have

$$l = \left(\frac{9M}{4\pi\rho}\right)^{\frac{1}{3}}.$$

With $M = 6 \times 10^{27}$g, $\rho = 2$g/cm³, we find

$$l = 1.29 \times 10^9 \text{ cm} = 1.29 \times 10^4 \text{ km}.$$

If l is less than this value, the earth's attraction becomes too large for the unit mass to be held by the celestial body and disruption of the latter occurs.

If the unit mass is located to the right of C on the extended line of OC, x is negative but the above conclusion still holds true. We may also consider a unit mass located off the line OC such as the point P in Fig. 1.72. We now have

$$\sqrt{(l - x)^2 + y^2}\omega^2 \cos\theta = \frac{GM}{(l - x)^2 + y^2}\cos\theta - \frac{4}{3}\pi\rho G \sqrt{x^2 + y^2}\cos\varphi,$$

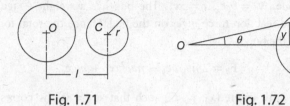

Fig. 1.71   Fig. 1.72

with

$$\cos\theta = \frac{l - x}{\sqrt{(l - x)^2 + y^2}}, \quad \cos\varphi = \frac{x}{\sqrt{x^2 + y^2}}.$$

As $x/l \ll 1$, $y/l \ll 1$, and retaining only the first-order terms we would obtain the same result.

1103

A merry-go-round (MGR) has two orthogonal axes (x, y) painted on it, and is rotating on the earth (assume to be an inertial frame x_0, y_0, z_0) with constant angular velocity ω about a vertical axis. A bug of mass m is crawling outward without slipping along the x-axis with constant velocity v_0 (Fig. 1.73). What is the total force $\mathbf{F}_b$ exerted by the MGR on the bug? Give all components of $\mathbf{F}_b$ in the earth-frame coordinates x_0, y_0, z_0 of the bug.

(UC, Berkeley)

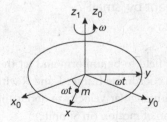

Fig. 1.73

Sol:

In the rotating coordinate system(x, y, z), the bug, which crawls with constant velocity v_0 along the x-axis, has no acceleration, so that the horizontal force acting on it by the MGR is (Problem **1094**)

$$\mathbf{F} = 2m\omega \times \mathbf{v}' + m\omega \times (\omega \times \mathbf{r}),$$

where $\omega = \omega \mathbf{e}_z$, $\mathbf{v}' = v_0 \mathbf{e}_x$, $\mathbf{r} = x\mathbf{e}_x$. The bug has a weight $-mg\mathbf{e}_z$, so that the MGR exerts a reaction force $mg\mathbf{e}_z$ on the bug. Hence the total force exerted by the MGR on the bug is

$$\mathbf{F}_b = 2mv_0\omega \mathbf{e}_y - m\omega^2 x\mathbf{e}_x + mg\mathbf{e}_z.$$

Choose the earth frame (x_0, y_0, z_0) such that at $t = 0$, the corresponding axes coincide with those of the rotating frame. Then, denoting the unit vectors along the x_0-, y_0-, z_0-axes by $\mathbf{i}, \mathbf{j}, \mathbf{k}$ respectively, we have

$$\mathbf{e}_x = \cos(\omega t)\mathbf{i} + \sin(\omega t)\mathbf{j},$$
$$\mathbf{e}_y = -\sin(\omega t)\mathbf{i} + \cos(\omega t)\mathbf{j},$$
$$\mathbf{e}_z = \mathbf{k}.$$

For simplicity, assume that the bug was at the origin at $t = 0$, then $x = v_0 t$. In the earth frame, $\mathbf{F}_b$ can thus be written as

$$\mathbf{F}_b = -mv_0\omega[2\sin(\omega t) + \omega t\cos(\omega t)]\mathbf{i}$$
$$+ mv_0\omega[2\cos(\omega t) - \omega t\sin(\omega t)]\mathbf{j} + mg\mathbf{k}.$$

1104

Consider a collection of charged particles, all with the same charge/mass ratio (e/m), interacting via conservative central forces. Prove that the motion of the particles in a small magnetic field **B** is identical with that in the absence of the field, when viewed in a coordinate system rotating with an appropriately chosen angular velocity ω (Larmor's theorem). What is the appropriate value of ω and what is meant by "small"?

(Chicago)

Sol: Assume the magnetic field to be uniform and let the central force on a particle be $\mathbf{F}(\mathbf{r})$. Consider two coordinate frames L and R with origins at the force center such that R rotates with angular velocity ω about the common origin. Problem **1094** gives the equation of motion (in SI units)

$$\mathbf{F}(\mathbf{r}) + e\mathbf{v} \times \mathbf{B} = m\mathbf{a} \tag{1}$$

in L and

$$\mathbf{F}(\mathbf{r}) + e\mathbf{v} \times \mathbf{B} = m\mathbf{a}' + 2m\boldsymbol{\omega} \times \mathbf{v}' + m\boldsymbol{\omega} \times (\boldsymbol{\omega} \times \mathbf{r}) \tag{2}$$

in R. As $\mathbf{v} = \mathbf{v}' + \boldsymbol{\omega} \times \mathbf{r}$, (2) can be written as

$$
\begin{aligned}
m\mathbf{a}' &= \mathbf{F}(\mathbf{r}) + e\mathbf{v} \times \mathbf{B} - 2m\boldsymbol{\omega} \times (\mathbf{v} - \boldsymbol{\omega} \times \mathbf{r}) - m\boldsymbol{\omega} \times (\boldsymbol{\omega} \times \mathbf{r}) \\
&= \mathbf{F}(\mathbf{r}) + \mathbf{v} \times (e\mathbf{B} + 2m\boldsymbol{\omega}) + m\boldsymbol{\omega} \times (\boldsymbol{\omega} \times \mathbf{r}).
\end{aligned}
$$

If R is chosen with

$$\boldsymbol{\omega} = -\frac{e\mathbf{B}}{2m}$$

and if the centrifugal term $m\boldsymbol{\omega} \times (\boldsymbol{\omega} \times \mathbf{r})$ can be neglected, the above becomes

$$\mathbf{F}(\mathbf{r}) = m\mathbf{a}',$$

i.e. the motion of the particle when viewed in the rotating frame is the same as that in the absence of the magnetic field.

This conclusion applies to a system of particles with the same e/m and subject to central forces with the same center. The particles will move as if the magnetic field were absent but the system as a whole precesses in the laboratory frame with angular velocity $\boldsymbol{\omega}$.

We have assumed that for every particle in the system,

$$m|\boldsymbol{\omega} \times (\boldsymbol{\omega} \times \mathbf{r})| \ll 2m|\boldsymbol{\omega} \times \mathbf{v}|,$$

i.e.

$$\omega \ll \frac{2v}{r},$$

or

$$B \ll \frac{4mv}{er},$$

which limits the strength of the field.

1105

The pivot point of a rigid pendulum is in forced vertical oscillation, given by $\eta(t) = \eta_0\cos(\omega t)$. The pendulum consists of a massless rod of length L with a mass m attached at the end.

a. Derive an equation of motion for θ, where θ is the pendulum angle indicated in Fig. 1.74. Assume $\theta \ll 1$ and $\eta_0 \ll L$.

b. Solve the equation to first order in η_0 for the initial conditions

   i. $\theta = a, \quad \dot{\theta} = 0,$ and

   ii. $\theta = 0, \quad \dot{\theta} = a\sqrt{\dfrac{g}{L}}.$

c. Evaluate the solutions for (i) and (ii) at resonance and describe the difference in the two motions.

                                                         (MIT)

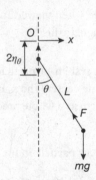

Fig. 1.74

Sol:

a. Use Cartesian coordinates with origin at point O in Fig. 1.76, x-axis horizontal and y-axis vertically downward. We have

$$x = L\sin\theta, \qquad y = L\cos\theta + \eta_0[1 - \cos(\omega t)],$$
$$m\ddot{x} = -F\sin\theta, \qquad m\ddot{y} = mg - F\cos\theta.$$

As $\theta \ll 1$ rad, we can omit terms involving θ^2 and take the approximation $\cos\theta \approx 1$, $\sin\theta \approx \theta$. Then

$$\ddot{x} \approx -L\theta\dot{\theta}^2 + L\ddot{\theta}, \quad \ddot{y} \approx -L\theta\ddot{\theta} + \eta_0\omega^2\cos(\omega t),$$

and the equation of motion for θ is

$$\ddot{\theta} + \frac{1}{L}[g - \eta_0\omega^2\cos(\omega t)]\theta = 0. \tag{1}$$

b. For an approximate solution to first order in $\alpha = \dfrac{\eta_0}{L}$, let

$$\theta = \varphi + \alpha\xi(t),$$

where φ satisfies the equation $\ddot{\varphi} + \omega_0^2\varphi = 0$, with $\omega_0^2 = \dfrac{g}{L}$, as well as the initial conditions for θ.

i. For initial conditions $\theta = a$, $\dot{\theta} = 0$, we have

$$\varphi = a\cos(\omega_0 t),$$
$$\theta = a\cos(\omega_0 t) + \alpha\xi(t).$$

Substitution of the above in (1) gives, retaining only first order terms of α,

$$\ddot{\xi} + \omega_0^2\xi = a\omega^2\cos(\omega_0 t)\cos(\omega t) = \frac{a\omega^2}{2}\{\cos[(\omega_0 + \omega)t]$$

$$+ \cos[(\omega_0 - \omega)t]\}.$$

This has a particular solution

$$\xi = \frac{a\omega}{2}\left\{-\frac{\cos[(\omega_0 + \omega)t]}{2\omega_0 + \omega} + \frac{\cos[(\omega_0 - \omega)t]}{2\omega_0 - \omega}\right\},$$

so that the general solution is

$$\xi = C_1\cos(\omega_0 t) + C_2\sin(\omega_0 t) - \frac{a\omega\cos[(\omega_0 + \omega)t]}{2(2\omega_0 + \omega)}$$

$$+ \frac{a\omega\cos[(\omega_0 - \omega)t]}{2(2\omega_0 - \omega)}.$$

The initial conditions $\xi = 0$, $\dot{\xi} = 0$ at $t = 0$ then give

$$C_1 = \frac{-a\omega^2}{(2\omega_0 + \omega)(2\omega_0 - \omega)} \qquad C_2 = 0,$$

and

$$\theta = a\cos(\omega_0 t)$$

$$+ \frac{\eta_0}{L}\left\{-\frac{a\omega^2\cos(\omega_0 t)}{(2\omega_0 + \omega)(2\omega_0 - \omega)} - \frac{a\omega\cos[(\omega_0 + \omega)t]}{2(2\omega_0 + \omega)}\right.$$

$$\left. + \frac{a\omega\cos[(\omega_0 - \omega)t]}{2(2\omega_0 - \omega)}\right\}.$$

ii. For initial conditions $\theta = 0$, $\dot{\theta} = a\sqrt{\dfrac{g}{L}} = a\omega_0$, let

$$\varphi = a\sin(\omega_0 t),$$
$$\theta = a\sin(\omega_0 t) + \alpha\xi(t).$$

Substitution in (1) gives

$$\ddot{\xi} + \omega_0^2 \xi = a\omega^2 \cos(\omega t)\sin(\omega_0 t)$$
$$= \frac{a\omega^2}{2}\{\sin[(\omega_0 + \omega)t] + \sin[(\omega_0 - \omega)t]\},$$

which has general solution

$$\xi = D_1 \cos(\omega_0 t) + D_2 \sin(\omega_0 t) - \frac{a\omega \sin[(\omega_0 + \omega)t]}{2(2\omega_0 + \omega)}$$

$$+ \frac{a\omega \sin[(\omega_0 - \omega)t]}{2(2\omega_0 - \omega)}.$$

The initial conditions $\xi = 0$, $\dot{\xi} = 0$ at $t = 0$ then give

$$D_1 = 0, \qquad D_2 = \frac{a\omega^2}{(2\omega_0 + \omega)(2\omega_0 - \omega)}$$

and hence

$$\theta = a \sin(\omega_0 t)$$

$$+ \frac{\eta_0}{L}\left\{\frac{a\omega^2 \sin(\omega_0 t)}{(2\omega_0 + \omega)(2\omega_0 - \omega)} - \frac{a\omega \sin[(\omega_0 + \omega)t]}{2(2\omega_0 + \omega)} + \frac{a\omega \sin[(\omega_0 - \omega)t]}{2(2\omega_0 - \omega)}\right\}.$$

c.   Resonance occurs at $\omega = 2\omega_0$. As $\omega \approx 2\omega_0$, we have for case (i)

$$\theta = a \cos(\omega_0 t) - \frac{\eta_0 a}{4L}\cos(3\omega_0 t) = a\left[1 + \frac{3\eta_0}{4L} - \frac{\eta_0}{L}\cos^2(\omega_0 t)\right]\cos(\omega_0 t),$$

and for case (ii)

$$\theta = a \sin(\omega_0 t) - \frac{\eta_0 a}{4L}\sin(3\omega_0 t) = a\left[1 - \frac{3\eta_0}{4L} + \frac{\eta_0}{L}\sin^2(\omega_0 t)\right]\sin(\omega_0 t).$$

It is seen that the amplitude at resonance is limited to $\approx a$ in both cases. However, the two resonances occur at phases differing by $\frac{\pi}{2}$.

1106

A hemispherical bowl of radius R rotates around a vertical axis with constant angular speed Ω. A particle of mass M moves on the interior surface of the bowl under the influence of gravity (Fig. 1.75). In addition, this particle is subjected to a frictional force $\mathbf{F} = -k\mathbf{V}_{rel}$, where k is a constant and $\mathbf{V}_{rel}$ is the velocity of the particle relative to the bowl.

a. If the particle is at the bottom of the bowl ($\theta = 0$), it is clearly in equilibrium. Show that if $\Omega > \sqrt{\dfrac{g}{R}}$, there is a second equilibrium value of θ and determine its value.

b. Suppose the particle is in equilibrium at the bottom of the bowl. To describe the motion of the particle in the vicinity of the equilibrium point, we construct a local inertial Cartesian coordinate system (x, y, z) and neglect the curvature of the bowl except in calculating the gravitational restoring force. Show that for $|x| \ll R$, $|y| \ll R$, the particle position satisfies $x = Re(x_0 e^{\lambda t})$, $y = Re(y_0 e^{\lambda t})$, where

$$\left(\lambda^2 + \frac{k\lambda}{M} + \frac{g}{R}\right)^2 + \left(\frac{k}{M}\right)^2 \Omega^2 = 0.$$

c. Find the angular speed of the bowl, Ω_0, for which the particle's motion is periodic.

d. There is a transition from stable to unstable at $\Omega = \Omega_0$. By considering behavior of frequencies in the vicinity of Ω_0, prove that the motion is stable for $\Omega < \Omega_0$ and unstable for $\Omega > \Omega_0$.

(MIT)

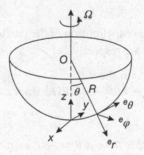

Fig. 1.75

Sol: In a frame rotating with angular velocity Ω, the equation of motion of a particle of mass M is by Problem **1094**

$$\mathbf{F} = M\mathbf{a}' + 2M\mathbf{\Omega} \times \mathbf{v}' + M\mathbf{\Omega} \times (\mathbf{\Omega} \times \mathbf{r}) + M\dot{\mathbf{\Omega}} \times \mathbf{r},$$

where $\mathbf{a}'$, $\mathbf{v}'$ are the acceleration and velocity in the rotating frame.

For the rotating frame, choose a spherical frame (r, θ, φ) attached to the bowl with origin O at the center of the bowl, then $\dot{\Omega} = 0$. In the spherical coordinate frame, we have

$$\dot{\mathbf{e}}_r = \dot{\theta}\mathbf{e}_\theta + \dot{\varphi}\sin\theta\boldsymbol{\theta}_\varphi,$$
$$\dot{\mathbf{e}}_\theta = -\dot{\theta}\mathbf{e}_r + \dot{\varphi}\cos\theta\mathbf{e}_\varphi,$$
$$\dot{\mathbf{e}}_\varphi = -\dot{\varphi}\sin\theta\mathbf{e}_r - \dot{\varphi}\cos\theta\mathbf{e}_\theta.$$

Then for a particle at $\mathbf{r} = r\mathbf{e}_r$, the velocity is

$$\mathbf{v} = \dot{r}\mathbf{e}_r + r\dot{\theta}\mathbf{e}_\theta + r\dot{\varphi}\sin\theta\mathbf{e}_\varphi,$$

and the acceleration is

$$\begin{aligned}
\mathbf{a} = {} & (\ddot{r} - r\dot{\theta}^2 - r\dot{\varphi}^2\sin^2\theta)\mathbf{e}_r \\
& + (r\ddot{\theta} + 2\dot{r}\dot{\theta} - r\dot{\varphi}^2\sin\theta\cos\theta)\mathbf{e}_\theta \\
& + (r\ddot{\varphi}\sin\theta + 2\dot{r}\dot{\varphi}\sin\theta + 2r\dot{\theta}\dot{\varphi}\cos\theta)\mathbf{e}_\varphi.
\end{aligned}$$

a. For the particle in the rotating bowl, we have

$$\boldsymbol{\Omega} = -\Omega\cos\theta\mathbf{e}_r + \Omega\sin\theta\mathbf{e}_\theta,$$
$$\mathbf{r} = R\mathbf{e}_r, \quad \dot{r} = \ddot{r} = 0, \quad \mathbf{V}_{\text{rel}} = R\dot{\theta}\mathbf{e}_\theta + R\dot{\varphi}\sin\theta\mathbf{e}_\varphi,$$
$$\mathbf{F} = M\mathbf{g} + \mathbf{N} - k\mathbf{V}_{\text{rel}}$$

where

$$M\mathbf{g} = Mg\cos\theta\mathbf{e}_r - Mg\sin\theta\mathbf{e}_\theta,$$
$$\mathbf{N} = -N\mathbf{e}_r.$$

Hence the equations of the motion of the particle in the rotating frame in the $\mathbf{e}_\theta$ and $\mathbf{e}_\varphi$ directions are respectively

$$\begin{aligned}
MR&\ddot{\theta} - MR\dot{\varphi}^2\sin\theta\cos\theta \\
&= -Mg\sin\theta - kR\dot{\theta} - 2MR\Omega\dot{\varphi}\sin\theta\cos\theta + MR\Omega^2\sin\theta\cos\theta,
\end{aligned}$$

and

$$MR\ddot{\varphi}\sin\theta + 2MR\dot{\theta}\dot{\varphi}\cos\theta = -kR\dot{\varphi}\sin\theta + 2MR\Omega\dot{\theta}\cos\theta.$$

At equilibrium, $\dot{\theta} = 0$, $\dot{\varphi} = 0$, $\ddot{\theta} = 0$, $\ddot{\varphi} = 0$, and we obtain

$$-Mg\sin\theta + MR\Omega^2\sin\theta\cos\theta = 0,$$

which gives

$$\sin\theta = 0, \quad \text{or} \quad \cos\theta = \frac{g}{R\Omega^2}.$$

Hence $\theta = 0$ is an equilibrium position. If $\Omega > \sqrt{\dfrac{g}{R}}$ there is another equilibrium position

$$\theta = \arccos\left(\frac{g}{R\Omega^2}\right).$$

b. Use Cartesian coordinates (x, y, z) for the local inertial frame with origin at $\theta = 0$ at the bottom of the bowl and the z-axis along the axis of rotation. In this frame, the position vector of the particle, which is near the bottom of the bowl, is

$$\mathbf{r}' = x\mathbf{i} + y\mathbf{j} + z\mathbf{k} \approx x\mathbf{i} + y\mathbf{j},$$

neglecting the curvature of gthe bowl, and its equation of motion is

$$M\ddot{\mathbf{r}}' = M\mathbf{g} - k\mathbf{V}_{\text{rel}} + \mathbf{N}.$$

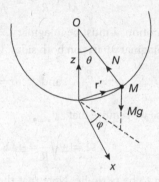

Fig. 1.76

As shown in Fig. 1.76, the component of the force $\mathbf{N}$ along $\mathbf{r}$ is approximately zero and the component of $M\mathbf{g}$ along $\mathbf{r}$ is

$$-Mg \sin\theta \cos\varphi\,\mathbf{i} - Mg \sin\theta \sin\varphi\,\mathbf{j} \approx -\frac{Mgx}{R}\mathbf{i} - \frac{Mgy}{R}\mathbf{j}$$

as $\sin\theta \approx \dfrac{r'}{R}$, $\cos\varphi \approx \dfrac{x}{r'}$, $\sin\varphi \approx \dfrac{y}{r'}$. Also as $\dot{\mathbf{r}}' = \mathbf{V}_{\text{rel}} + \boldsymbol{\Omega} \times \mathbf{r}'$,

$$-k\mathbf{V}_{\text{rel}} = -k\dot{\mathbf{r}}' + k\boldsymbol{\Omega} \times \mathbf{r}'$$
$$= -k(\dot{x} + y\Omega)\mathbf{i} - k(\dot{y} - x\Omega)\mathbf{j}.$$

Let $x = x_0 e^{\lambda t}$, $y = y_0 e^{\lambda t}$ and the above becomes

$$\begin{cases} \left(\lambda^2 + \dfrac{k}{M}\lambda + \dfrac{g}{R}\right)x_0 + \dfrac{k\Omega}{M}y_0 = 0, \\[3mm] -\dfrac{k\Omega}{M}x_0 + \left(\lambda^2 + \dfrac{k}{M}\lambda + \dfrac{g}{R}\right)y_0 = 0 \end{cases}$$

For a non-zero solution, we require

$$
\begin{vmatrix} \lambda^2 + \dfrac{k}{M}\lambda + \dfrac{g}{R} & \dfrac{k\Omega}{M} \\[2ex] -\dfrac{k\Omega}{M} & \lambda^2 + \dfrac{k\lambda}{M} + \dfrac{g}{R} \end{vmatrix} = \left(\lambda^2 + \dfrac{k\lambda}{M} + \dfrac{g}{R}\right)^2 + \left(\dfrac{k\Omega}{M}\right)^2 = 0. \qquad (1)
$$

Hence if this condition holds, we can describe the particle's position by

$$
x = Re(x_0 e^{\lambda t}), \qquad y = Re(y_0 e^{\lambda t}).
$$

This conclusion is valid only for $|x| \ll R$, $|y| \ll R$ since we have neglected the curvature and considered the particle as moving in a horizontal plane.

c.   The left-hand side of (1) can be factorized and shown to have solutions

$$
\lambda^2 + \dfrac{k\lambda}{M} + \dfrac{g}{R} = \pm i\,\dfrac{k\Omega}{M}.
$$

For periodic motion, λ must be imaginary, $\lambda = i\omega$, where ω is real. Equating the real and imaginary parts on both sides, we have

$$
\omega^2 = \dfrac{g}{R} \quad \text{and} \quad \omega = \pm\Omega.
$$

To satisfy, these we require that

$$
\Omega = \pm\sqrt{\dfrac{g}{R}} = \pm\Omega_0
$$

for the motion to be periodic. Note that the '+' and '−', signs correspond to two opposite directions of rotation.

d.   As has been shown in (a), if $\Omega < \Omega_0$, there is only one equilibrium position $\theta = 0$. The equilibrium at this point is stable. For $\Omega > \Omega_0$, there are two equilibrium positions $\theta = 0$ and $\theta = \arccos\left(\dfrac{\Omega_0^2}{\Omega^2}\right)$. However the equilibrium at the former position is now unstable, so that the stable equilibrium is shifted to the latter position if $\Omega > \Omega_0$. Hence for $\theta = 0$, there is a transition from stable to unstable at $\Omega = \Omega_0$.

1107

A particle of mass m can slide without friction on the inside of a small tube bent in the form of a circle of radius a. The tube rotates about a vertical diameter at a constant rate of ω rad/sec as shown in Fig. 1.77. Write the differential

equation of motion. If the particle is disturbed slightly from its unstable equilibrium position at $\theta = 0$, find the position of maximum kinetic energy.

(SUNY, Buffalo)

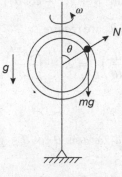

Fig. 1.77

Sol: In a rotating coordinate frame (r, θ, φ) attached to the circular tube, we have
(Problem **1094**)

$$\mathbf{F} = m\mathbf{a}' + 2m\omega \times \mathbf{v}' + m\omega \times (\omega \times \mathbf{r}),$$

with

$$\mathbf{F} = m\mathbf{g} + \mathbf{N}$$
$$= -mg\cos\theta\,\mathbf{e}_r + mg\sin\theta\,\mathbf{e}_\theta + N\mathbf{e}_r,$$
$$\omega = \omega\cos\theta\,\mathbf{e}_r - \omega\sin\theta\,\mathbf{e}_\theta,$$
$$\mathbf{a}' = -a\dot{\theta}^2\mathbf{e}_r + a\ddot{\theta}\mathbf{e}_\theta,$$
$$\mathbf{v}' = a\dot{\theta}\mathbf{e}_\theta, \quad \mathbf{r} = a\mathbf{e}_r.$$

The equation for the motion in the $\mathbf{e}_\theta$ direction is then

$$ma\ddot{\theta} = mg\sin\theta + ma\omega^2\sin\theta\cos\theta.$$

As $\ddot{\theta} = \dfrac{d\dot{\theta}}{d\theta}\dfrac{d\theta}{dt} = \dfrac{1}{2}\dfrac{d\dot{\theta}^2}{d\theta}$, the above with the initial condition $\theta = \dot{\theta} = 0$ at $t = 0$
gives

$$a\dot{\theta}^2 = a\omega^2\sin^2\theta + 2g(1 - \cos\theta).$$

In an inertial frame that instantaneously coincides with the rotating frame, the
velocity of the particle is

$$\mathbf{v} = \mathbf{v}' + \omega \times \mathbf{r} = a\dot{\theta}\mathbf{e}_\theta + a\omega\sin\theta\,\mathbf{e}_\phi,$$

and its kinetic energy is

$$E = \frac{1}{2} m(a^2 \dot{\theta}^2 + a^2 \omega^2 \sin^2 \theta)$$

$$= \frac{1}{2} m[\omega^2 a^2 \sin^2 \theta + 2ga(1 - \cos \theta) + \omega^2 a^2 \sin^2 \theta]$$

$$= ma[\omega^2 a \sin^2 \theta + g(1 - \cos \theta)].$$

For E to be a maximum at θ_0, we require

$$\left(\frac{dE}{d\theta} \right)_{\theta_0} = 0, \qquad \left(\frac{d^2 E}{d\theta^2} \right)_{\theta_0} < 0.$$

As

$$\frac{dE}{d\theta} = ma[2\omega^2 a \sin \theta \cos \theta + g \sin \theta] = 0,$$

$$\frac{d^2 E}{d\theta^2} = ma[4\omega^2 a \cos^2 \theta + g \cos \theta - 2\omega^2 a],$$

we have for the position of maximum kinetic energy

$$\theta_0 = \pi \qquad\qquad \text{if} \quad \omega^2 < \frac{g}{2a},$$

$$\theta_0 = \arccos \left(\frac{-g}{2\omega^2 a} \right) \qquad \text{if} \quad \omega^2 > \frac{g}{2a}.$$

1108

Let S be a set of axes centered at the earth's center, with the z-axis pointing north, forming an inertial frame. Let S' be similarly placed, but rotating with the earth.

a. Write down the non-relativistic equation giving the transformation of the time derivative of any vector from S' to S. Use this to derive an expression for the Coriolis force experienced by a body moving in S'. Define all symbols.

b. In the northern hemisphere, find the direction of the Coriolis force on a body moving eastward and on one moving vertically upward.

c. Consider a body dropped from a height of 10 feet at a latitude of 30°N. Find, approximately, the horizontal deflection due to the Coriolis effect when it reaches the ground. Neglect air resistance.

(SUNY, Buffalo)

Sol:

a. Let XYZ be the inertial reference frame S and $X'Y'Z'$ the rotating frame S' fixed to the earth which rotates with angular velocity ω. In S', an arbitrary vector **A** can be written as

$$\mathbf{A} = A_x\mathbf{i} + A_y\mathbf{j} + A_z\mathbf{k}.$$

In S, the time derivative of **A** is

$$\frac{d\mathbf{A}}{dt} = \left(\frac{dA_x}{dt}\mathbf{i} + \frac{dA_y}{dt}\mathbf{j} + \frac{dA_z}{dt}\mathbf{k} \right) + \left(A_x\frac{d\mathbf{i}}{dt} + A_y\frac{d\mathbf{j}}{dt} + A_z\frac{d\mathbf{k}}{dt} \right).$$

Let d^*/dt denote time derivative in S', then

$$\frac{d^*\mathbf{A}}{dt} = \frac{dA_x}{dt}\mathbf{i} + \frac{dA_y}{dt}\mathbf{j} + \frac{dA_z}{dt}\mathbf{k}.$$

The kinematics of a rigid body gives

$$\frac{d\mathbf{i}}{dt} = \omega \times \mathbf{i}, \quad \frac{d\mathbf{j}}{dt} = \omega \times \mathbf{j}, \quad \frac{d\mathbf{k}}{dt} = \omega \times \mathbf{k}.$$

Hence

$$\frac{d\mathbf{A}}{dt} = \frac{d^*\mathbf{A}}{dt} + \omega \times (A_x\mathbf{i} + A_y\mathbf{j} + A_z\mathbf{k}) = \frac{d^*\mathbf{A}}{dt} + \omega \times \mathbf{A}.$$

Thus for the radius vector **r** to a point p, we have

$$\frac{d\mathbf{r}}{dt} = \frac{d^*\mathbf{r}}{dt} + \omega \times \mathbf{r},$$

$$\frac{d^2r}{dt^2} = \frac{d^*}{dt}\left(\frac{d^*\mathbf{r}}{dt} + \omega \times \mathbf{r} \right) + \omega \times \left(\frac{d^*\mathbf{r}}{dt} + \omega \times \mathbf{r} \right)$$

$$= \frac{d^{*2}\mathbf{r}}{dt^2} + 2\omega \times \frac{d^*\mathbf{r}}{dt} + \omega \times (\omega \times \mathbf{r}) + \frac{d^*\omega}{dt} \times \mathbf{r}.$$

Note that in the above

$$\frac{d^*\omega}{dt} = \frac{d\omega}{dt} - \omega \times \omega = \frac{d\omega}{dt}.$$

Newton's second law applies to the inertial frame, so for a particle of mass m at P acted on by a force **F**, we have

$$\mathbf{F} = m\frac{d^2\mathbf{r}}{dt^2} = m\frac{d^{*2}\mathbf{r}}{dt^2} + 2m\omega \times \frac{d^*\mathbf{r}}{dt} + m\omega \times (\omega \times \mathbf{r}) + m\frac{d^*\omega}{dt} \times \mathbf{r}.$$

Denoting $\dfrac{d^*}{dt}$ by a dot and noting that for the earth $\dot{\omega} = 0$, we write the above as

$$m\ddot{\mathbf{r}} = \mathbf{F} - 2m\omega \times \dot{\mathbf{r}} - m\omega \times (\omega \times \mathbf{r})$$

for the rotating frame. This shows that Newton's second law can still be considered valid if, in addition to **F**, we introduce two fictitious forces: $-2m\omega \times \dot{\mathbf{r}}$, the Coriolis force, and $-m\omega \times (\omega \times \mathbf{r})$, the centrifugal force. Thus a body of mass m moving on earth with a velocity $\mathbf{v}'$ is seen by an observer on the earth to suffer a Coriolis force $-2m\omega \times \mathbf{v}'$.

b. Choose for S' a frame fixed at a point on the surface of the earth at latitude λ and let its orthogonal unit vectors **i**, **j**, **k** be directed toward the south, the east and vertically upward respectively. Then

$$\omega = -\omega \cos \lambda \mathbf{i} + \omega \sin \lambda \mathbf{k}.$$

1. When the body moves eastward, $\mathbf{v}' = \dot{y}\mathbf{j}$, the Coriolis force is

$$\mathbf{F}_\varepsilon = -2m\omega \times \mathbf{v}' = 2m\omega\dot{y} \sin \lambda \mathbf{i} + 2m\omega\dot{y} \cos \lambda \mathbf{k},$$

which has magnitude

$$|\mathbf{F}_c| = \sqrt{(2m\omega\dot{y} \sin \lambda)^2 + (2m\omega\dot{y} \cos \lambda)^2} = 2m\omega\dot{y}$$

and is pointing south at an inclination angle ϕ given by

$$\tan \phi = \frac{\cos \lambda}{\sin \lambda} = \tan\left(\frac{\pi}{2} - \lambda\right).$$

2. When the body moves upward, $\mathbf{v}' = \dot{z}\mathbf{k}$, the Coriolis force is

$$\mathbf{F}_c = -2m\omega \times \mathbf{v}' = -2m\omega\dot{z} \cos \lambda \mathbf{j},$$

which has magnitude $2m\omega\dot{z} \cos \lambda$ and direction toward the west.

c. The equations of motion for the free-fall body in S' are

$$\begin{cases} m\ddot{x} = 2m\omega\dot{y} \sin \lambda, \\ m\ddot{y} = -2m\omega(\dot{x} \sin \lambda + \dot{z} \cos \lambda), \\ m\ddot{z} = -mg + 2m\omega\dot{y} \cos \lambda, \end{cases}$$

with initial conditions $x = y = 0$, $z = h = 10 =$ ft, $\dot{x} = \dot{y} = \dot{z} = 0$ at $t = 0$. Integrating and using the initial conditions, we obtain

$$\begin{cases} \dot{x} = 2\omega y \sin \lambda, \\ \dot{y} = -2\omega[x \sin \lambda + (z - h) \cos \lambda], \\ \dot{z} = -gt + 2\omega y \cos \lambda. \end{cases}$$

Substituting these into the original set of equations, we obtain

$$\begin{cases} \ddot{x} = -4\omega^2[x \sin \lambda + (z - h) \cos \lambda] \sin \lambda, \\ \ddot{y} = 2gt\omega \cos \lambda - 4\omega^2 y, \\ \ddot{z} = -g - 4\omega^2[x \sin \lambda + (z - h) \cos \lambda] \cos \lambda. \end{cases}$$

Neglecting the terms involving ω^2, we have approximately

$$\begin{cases} \ddot{x} = 0, \\ \ddot{y} = 2gt\omega \cos \lambda y, \\ \ddot{z} = -g. \end{cases}$$

Integrating, applying the initial conditions and eliminating t, we obtain

$$y^2 = \left(\frac{8\omega^2 \cos^2 \lambda}{9g} \right) (h - z)^3.$$

When the body reaches the ground, $z = 0$,

$$y = \left(\frac{2\omega \cos \lambda}{3} \right) \sqrt{\frac{2h^3}{g}}.$$

With $h = 10 \text{ ft} = 3.05 \text{ m}$, $\lambda = 30°$, we find $y = 1.01 \times 10^{-4} \text{ m}$. Hence the deflection due to the Coriolis effect is toward the east and has a magnitude 0.01 cm.

DYNAMICS OF A SYSTEM OF POINT MASSES (1109–1144)

1109

A cart of mass M has a pole on it from which a ball of mass μ hangs from a thin string attached at point P. The cart and ball have initial velocity V. The cart crashes onto another cart of mass m and sticks to it (Fig. 1.78). If the length of the string is R, show that the smallest initial velocity for which the ball can go in circles around point P is $V = [(m + M)/m]\sqrt{5gR}$. Neglect friction and assume $M, m \gg \mu$.

(Wisconsin)

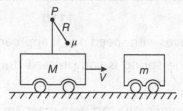

Fig. 1.78

Sol: As $\mu \ll m, M$, momentum conservation

$$MV = (M + m)V'$$

173

gives for the velocity of the two carts after collision,

$$V' = \frac{MV}{M + m}.$$

Consider the circular motion of the ball atop the cart M if it were stationary. If a the lowest and highest points the ball has speeds V_1 and V_2 respectively, we have

$$\frac{1}{2}\mu V_1^2 = \frac{1}{2}\mu V_2^2 + 2\mu gR,$$

$$\frac{\mu V_2^2}{R} = T + \mu g,$$

where T is the tension in the string when the ball is at the highest point. Th smallest V_2 is given by $T = 0$. Hence the smallest V_1 is given by

$$\frac{1}{2}\mu V_1^2 = \frac{1}{2}\mu gR + 2\mu gR,$$

i.e.

$$V_1 = \sqrt{5gR}.$$

With the cart moving, V_1 is the velocity of the ball relative to the cart. As the bal has initial velocity V and the cart has velocity V' after the collision, the velocity o the ball relative to the cart after the collision is $V - V'$. Hence the smallest V fo the ball to go round in a circle after the collision is given by

$$V - V' = V - \frac{MV}{M + m} = \sqrt{5gR},$$

i.e.

$$V = \frac{M + m}{m}\sqrt{5gR}.$$

1110

A cart of mass m moves with speed v as it approaches a cart of mass $3m$ tha is initially at rest. The spring is compressed during the head-on collisior (Fig. 1.79).

  a. What is the speed of the cart with mass $3m$ at the instant of maximum spring compression assuming conservation of energy?

  b. How would your answer differ if energy is not conserved?

c. What is the final velocity of the heavier cart after a long time has passed, if energy is conserved?

d. Give the final velocity of the heavier cart in a completely inelastic collision.

(Wisconsin)

Sol:

a. When the spring is at maximum compression, the two carts are nearest each other and at that instant move with a velocity v', say. Conservation of momentum gives

$$mv = (m + 3m)v',$$

i.e.

$$v' = \frac{v}{4}.$$

Thus the heavier cart has speed $\frac{v}{4}$ at that instant.

b. Even if mechanical energy is not conserved, the above result still holds since it has been derived from conservation of momentum which holds as long as no external force is acting.

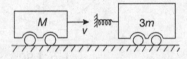

Fig. 1.79

c. Conservation of energy and of momentum give

$$\frac{mv^2}{2} = \frac{mv_1'^2}{2} + \frac{3mv_2'^2}{2},$$
$$mv = mv_1' + 3mv_2',$$

where v_1', v_2' are respectively the velocities after collision of the lighter and heavier carts. Hence the heavier cart has final velocity

$$v_2' = \frac{2mv}{m + 3m} = \frac{v}{2}.$$

d. If the collision is completely inelastic, the two carts will move together after collision. Their velocity is then $\frac{v}{4}$ as given in (a).

1111

Determine the ratio of the centripetal to the gravitational force experience by a body at (i) the equator, (ii) a latitude of 42°, and (iii) the poles. Consider the Earth to be a sphere of radius $R = 6.4 \times 10^6$ m, mass $M = 6.0 \times 10^{24}$ kg and gravitational constant $G = 6.67 \times 10^{-11}$ Nm²kg⁻²

Sol:

i.   At the equator

$$\frac{F_c}{F_g} = \frac{\omega^2 R^3}{GM} = 0.0035$$

ii. At a latitude of 42°

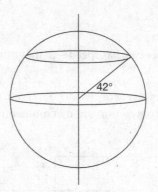

Fig. 1.80

$$\frac{F_c}{F_g} = \frac{\omega^2 R^3 (\cos 42)^3}{GM} = 0.0014$$

iii. At the poles, cos 90° = 0

$$\frac{F_c}{F_g} = 0$$

1112

A small block slides down a loop from a height that is twice that required to just keep the block on the track at its highest point. What is its maximum and minimum acceleration as it traverses the loop? After exiting the loop, the block travels up another circular loop. What should the radius of the loop be so that the block remains on the track at all times? Assume the radius of the loop to be R and the tracks to be frictionless.

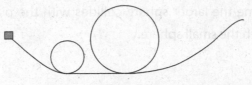

Fig. 1.81

Sol: For the block to not leave the track, at the highest point in the circle, the normal reaction on the track should be equal to the centripetal force

$$mg = \frac{mv^2}{R} \Rightarrow v = \sqrt{Rg}$$

Writing the energy equation when the block is on top of the loop

$$\frac{1}{2}mv^2 + mg2R = mgh \text{ or } h = \frac{5}{2}R$$

If the initial height is $5R$, the energy equation at the top of the loop would be

$$\frac{mv^2}{2} + mg2R = mg5R \Rightarrow v^2 = 6Rg$$

The minimum acceleration at the top of the loop

$$a = \frac{v^2}{R} = 6g \text{ and } a_{min} = 6g - g = 5g$$

The energy equation at the bottom of the loop

$$\frac{mv^2}{2} + 0 = mg5R \Rightarrow v^2 = 10Rg$$

The maximum acceleration at the bottom of the loop

$$a = \frac{v^2}{R} = 10g \text{ and } a_{max} = 10g + g = 11g$$

The block now enters the second loop. Let the radius of the second loop be r. Using the energy equation,

$$\frac{1}{2}mv^2 + mgr = mg\frac{5}{2}R$$

Solving we get,

$$r = 2R$$

1113

Two steel spheres, the lower of radius $2a$ and the upper of radius a, are dropped from a height h (measured from the center of the large sphere) above a steel plate as shown in Fig. 1.82. Assuming the centers of the spheres always lie on a vertical line and all collisions are elastic, what is the maximum height the upper sphere will reach?

Hint: Assume the larger sphere collides with the plate and recoils before it collides with the small sphere.

(Wisconsin)

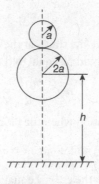

Fig. 1.82

Sol: Let the mass of the smaller sphere be m_1 and that of the larger one m_2. Then $m_2 = 8m_1$. The landing velocity of the larger sphere is $v_2 = \sqrt{2g(h - 2a)}$ and its velocity immediately after bouncing back from the steel plate is still in magnitude. At this point, the descending velocity of the smaller sphere is $v_1 = \sqrt{2g(h - 2a)} = v_2$. Let the velocities of the larger and smaller spheres after elastic collision be v_2' and v_1' respectively and take the upward direction as positive. Conservation of momentum and of mechanical energy give

$$m_2 v_2 - m_1 v_1 = m_2 v_2' + m_1 v_1',$$

$$\frac{m_1 v_1^2}{2} + \frac{m_2 v_2^2}{2} = \frac{m_2 v_2'^2}{2} + \frac{m_1 v_1'^2}{2},$$

whose solution is

$$v_2' = \frac{5v_2}{9} = \frac{5}{9}\sqrt{2g(h - 2a)},$$

$$v_1' = \frac{23v_2}{9} = \frac{23}{9}\sqrt{2g(h - 2a)}.$$

Conservation of the mechanical energy of the smaller sphere thus gives the maximum height (measured from the steel plate) of the smaller sphere as

$$H = 3a + \frac{v_1'^2}{2g} = 3a + \frac{529}{81}(h - 2a).$$

1114

A railroad flatcar of mass M can roll without friction along a straight horizontal track as shown in Fig. 1.83. N men, each of mass m, are initially standing on the car which is at rest.

a. The N men run to one end of the car in unison; their speed relative to the car is V_r just before they jump off (all at the same time). Calculate the velocity of the car after the men have jumped off.

b. The N men run off the car, one after the other (only one man running at a time), each reaching a speed V_r relative to the car just before jumping off. Find an expression for the final velocity of the car.

c. In which case, (a) or (b), does the car attain the greater velocity?

(CUSPEA)

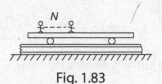

Fig. 1.83

Sol:

a. As there is no horizontal external force acting, the center of mass of the system consisting of the flatcar and N men remains stationary. Taking the x-axis along the track, we have for the center of mass,

$$x_{cm} = \frac{Mx_{car} + Nmx_{man}}{M + Nm},$$

$$\dot{x}_{cm} = 0 = M\dot{x}_{car} + Nm\dot{x}_{man},$$

where $\dot{x}_{car}$ and $\dot{x}_{man}$ are respectively the velocities of the car and each man after the men have jumped off. Writing $\dot{x}_{car} = V_{car}$ and noting that $\dot{x}_{man} = V_{car} - V_r$, we have

$$MV_{car} + Nm(V_{car} - V_r) = 0,$$

giving

$$V_{car} = \frac{NmV_r}{M + Nm}.$$

b. Consider the transition from n men to $(n - 1)$ men on the car. Let V_n be th velocity of the car when n men are left on it. The total momentum of the car wit the n men is

$$P_n = MV_n + nmV_n.$$

When the nth man jumps off the car with a speed V_r relative to the car, th momentum of the system consisting of the car and n men is

$$P_{n-1} = MV_{n-1} + (n - 1)mV_{n-1} + m(V_{n-1} - V_r).$$

Momentum conservation $P_{n-1} = P_n$ gives

$$(M + nm)V_n = (M + nm)V_{n-1} - mV_r,$$

or

$$V_{n-1} = V_n + \frac{mV_r}{M + nm}.$$

Hence

$$V_{n-s} = V_n + \sum_{i=1}^{s} \frac{mV_r}{M + (n - i + 1)m}.$$

As $n = N$, $V_N = 0$ initially, we have for $s = N$,

$$V_0 = \sum_{i=1}^{N} \frac{mV_r}{M + (N - i + 1)m} = \sum_{n=1}^{N} \frac{mV_r}{M + nm}.$$

c. As

$$\sum_{n=1}^{N} \frac{1}{M + nm} > \frac{N}{M + Nm},$$

the car in case (b) attains a greater final velocity.

1115

Consider the same system as the previous question, but with friction. Let μ be the coefficient of friction between the block and the surface. What is the initial displacement of the spring when the mass comes to rest momentar ily? Obtain the extent to which the spring is compressed and the acceleration during its motion.

Sol: Along the plane,

$$ma = mg \sin \theta - \mu mg \cos \theta.$$

Velocity when it reaches the spring,

$$u = \sqrt{2xg(\sin\theta - \mu\cos\theta)}.$$

Subsequently, when the block comes to rest,

$$F = mg\sin\theta - \mu mg\cos\theta - kx' = 0,$$

where x' is the extent to which the spring compresses as the mass come to rest

$$x' = \frac{mg}{k}(\sin\theta - \mu\cos\theta).$$

During its transit, its acceleration is $a' = \dfrac{mg\sin\theta - \mu mg\cos\theta - kx'}{m}$.

1116

Determine the gravitational potential for points at 2R, R, and R/2 from the center of the Earth assuming it to be a sphere of radius R. Assume that the density of the Earth varies as $\rho = \rho_0\left(1 - \dfrac{r}{R}\right)$

Sol:

i.   For $r = 2R$

$$M = \int_0^R dm = \int_0^R 4\pi r^2\, dr \rho_0\left(1 - \frac{r}{R}\right) = \pi\rho_0\frac{R^3}{3}$$

$$V = -\frac{GM}{2R} = -G\pi\rho_0\frac{R^2}{6}$$

ii.   For $r = R$

$$V = -\frac{GM}{R} = -G\pi\rho_0\frac{R^2}{3}$$

iii.   For $r = R/2$

$$V = -G\left(\frac{2}{R}\int_0^{R/2} 4\pi r^2\, dr\rho_0\left(1 - \frac{r}{R}\right) + \frac{2}{R}\int_{R/2}^R 4\pi r^2\, dr\rho_0\left(1 - \frac{r}{R}\right)\right)$$

$$= -5G\pi\rho_0\frac{R^2}{24} - 11G\pi\rho_0\frac{R^2}{24} = -G\pi\rho_0\frac{2R^2}{3}$$

1117

Two bodies of mass m and $2m$ initially at rest at a very large distance from each other are moving toward each other under the influence of the gravitational force. Determine the distance between the two masses when the smaller mass is travelling at a velocity $2v$ and the larger mass with a velocity toward each other.

Sol:

$$a_m = -\frac{2Gm}{r^2}$$

$$a_{2m} = -\frac{Gm}{r^2}$$

The relative acceleration

$$a = a_m + a_{2m} = -\frac{3Gm}{r^2}$$

But

$$a = \frac{dv_{rel}}{dt} = \frac{dv_{rel}}{dr}\frac{dr}{dt} = v_{rel}\frac{dv_{rel}}{dr}$$

$$\int_{\infty}^{s} -\frac{3Gm}{r^2}\, dr = \int_{0}^{3v} v_{rel}\, dv_{rel}$$

$$s = \frac{2}{3}\frac{Gm}{v^2}$$

1118

Given a system of N point-masses with pairwise additive central forces use Newton's second and third laws to demonstrate that the total angular momentum of this system is a constant. Does this calculation depend upon what point is chosen as the origin of coordinates?

(UC, Berkeley)

Sol: The angular momentum of a system of N point-masses about a fixed origin is by definition

$$\mathbf{L} = \sum_i \mathbf{r}_i \times \mathbf{p}_i.$$

Newton's second law $\mathbf{F}_i = \dfrac{d\mathbf{p}i}{dt}$ then gives

$$\frac{d\mathbf{L}}{dt} = \sum_i \mathbf{r}_i \times \mathbf{F}_i = \sum_i \mathbf{r}_i \times \sum_{j\neq i} \mathbf{f}_{ij} = \sum_i \sum_{j\neq i} \mathbf{r}_i \times \mathbf{f}_{ij},$$

where $\mathbf{f}_{ij}$ is the force the jth mass exerts on the ith mass. For two masses i and j Newton's third law gives

$$\mathbf{f}_{ij} = -\mathbf{f}_{ji},$$

and so

$$\mathbf{r}_i \times \mathbf{f}_{ij} + \mathbf{r}_j \times \mathbf{f}_{ji} = (\mathbf{r}_i - \mathbf{r}_j) \times \mathbf{f}_{ij} = 0,$$

since $\mathbf{r}_i - \mathbf{r}_j$ is parallel to $\mathbf{f}_{ij}$. As the sum $\sum_i \sum_{j \neq i} (\mathbf{r}_i \times \mathbf{f}_{ij})$ is due to pairs of forces like $\mathbf{f}_{ij}$ and $\mathbf{f}_{ji}$, we have

$$\frac{d\mathbf{L}}{dt} = 0,$$

i.e.

$$\mathbf{L} = \text{constant}.$$

As the origin in this proof is arbitrary, the conclusion is independent of the choice of the origin of coordinates.

1119

Two masses are connected through a pulley as indicated in the figure. The inclined and vertical surfaces have μ_1 and μ_2 as coefficients of static friction, respectively. The angle θ is such that it is greater than the minimum angle that causes the body m_1 to start moving in the absence of the rope. Obtain an inequality condition on the normal reaction N on the vertical wall that will keep the entire configuration at rest.

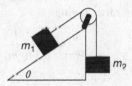

Fig. 1.84

Sol: According to the given condition $\theta > \tan^{-1} \mu_1$ or $\sin \theta > \mu_1 \cos \theta$

For the body m_1 to be stationary

$$m_1 g \sin \theta - \mu_1 m_1 g \cos \theta - T = 0$$

For the body m_2 to be stationary

$$m_2 g - \mu_2 N - T = 0$$

$$m_2 g - \mu_2 N = m_1 g (\sin \theta - \mu_1 \cos \theta)$$

Since $\sin \theta > \mu_1 \cos \theta$,

$$N > \frac{g}{\mu_2} (m_2 - m_1 (\sin \theta - \mu_1 \cos \theta))$$

1120

The captain of a small boat becalmed in the equatorial doldrums decides to resort to the expedience of raising the anchor ($m = 200$ kg) to the top of the mast ($s = 20$ m). The rest of the boat has a mass of $M = 1000$ kg.

a. Why will the boat begin to move?

b. In which direction will it move?

c. How fast will it move?

(Chicago)

Sol:

a. The vertical motion of the anchor causes a Coriolis force $-2m\omega \times v$, where v is the velocity of the anchor and ω the angular velocity of the earth, and so the boat moves.

b. As ω points to the north and v is vertically upward, the Coriolis force points toward the west. Hence the boat will move westward.

c. As the total angular momentum of the boat and anchor with respect to the center of mass of the earth in an inertial frame is conserved, we have

$$(M + m)r^2\omega_0 = [Mr^2 + m(r + s)^2]\omega,$$

where ω_0 and ω are the angular velocities of the earth and the boat respectively, is the radius of the earth, giving

$$\frac{\omega}{\omega_0} \approx \frac{(M + m)r^2}{(M + m)r^2 + 2mrs},$$

or

$$\frac{\omega - \omega_0}{\omega_0} \approx \frac{-2ms}{(M + m)r + 2ms} \approx \frac{-2ms}{(M + m)r}.$$

Hence the relative velocity of the boat with respect to the earth is

$$u = r(\omega - \omega_0) = \frac{-2ms\omega_0}{M + m} = -4.9 \times 10^{-4} \, \text{m/s}.$$

The negative sign indicates that the boat moves westward.

1121

A simple classical model of the CO_2 molecule would be a linear structure of three masses with the electrical forces between the ions represented by two identical springs of equilibrium length l and force constant k, as shown in Fig. 1.85. Assume that only motion along the original equilibrium line is possible, i.e. ignore rotations. Let m be the mass of O^- and M be the mass of C^{++}.

a. How many vibrational degrees of freedom does this system have?

b. Define suitable coordinates and determine the equation of motion of the masses.

c. Seek a solution to the equations of motion in which all particles oscillate with a common frequency (normal modes) and calculate the possible frequencies.

d. Calculate the relative amplitudes of the displacements of the particles for each of these modes and describe the nature of the motion for each mode. You may use a sketch as part of your description.

e. Which modes would you expect to radiate electromagnetically and what is the multipole order of each?

(*MIT*)

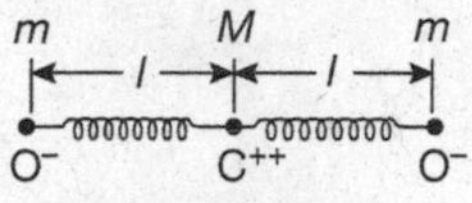

Fig. 1.85 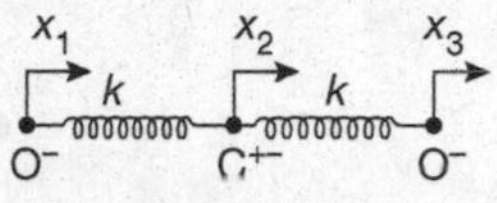   Fig. 1.86

Sol:

a. The system has two vibrational degrees of freedom.

b. Let x_1, x_2 and x_3 be the displacements of O^-, C^{++}, and O^- from their equilibrium positions respectively, as shown in Fig. 1.86. The equations of motion are

$$m\ddot{x}_1 = k(x_2 - x_1),$$
$$M\ddot{x}_2 = k(x_3 - x_2) - k(x_2 - x_1) = k(x_1 - 2x_2 + x_3),$$
$$m\ddot{x}_3 = -k(x_3 - x_2).$$

c. Let $x_1 = A_1 \cos \omega t$, $x_2 = A_2 \cos \omega t$ and $x_3 = A_3 \cos \omega t$ in the equations above. We have

$$(k - m\omega^2)A_1 - kA_2 = 0,$$
$$-kA_1 + (2k - M\omega^2)A_2 - kA_3 = 0,$$
$$-kA_1 + (k - m\omega^2)A_3 = 0.$$

For A_1, A_2, A_3 not to be identically zero, we require

$$\begin{vmatrix} k - m\omega^2 & -k & 0 \\ -k & 2k - M\omega^2 & -k \\ 0 & -k & k - m\omega^2 \end{vmatrix} = 0,$$

which has solutions

$$\omega_1 = \sqrt{\frac{k}{m}}, \qquad \omega_2 = \sqrt{\frac{(2m + M)k}{mM}}, \qquad \text{and} \qquad \omega_3 = 0.$$

Angular frequencies ω_1 and ω_2 correspond to possible vibrations, while ω_3 corresponds to the translational oscillations of the molecule as a whole.

d. Substituting ω_1 and ω_2 into the equations for A_1, A_2 and A_3, we find that relative amplitudes are $\begin{pmatrix} 1 \\ 0 \\ -1 \end{pmatrix}$ for ω_1 and $\begin{pmatrix} 1 \\ -\frac{2m}{M} \\ 1 \end{pmatrix}$ for ω_2, as depicted in Fig. 1.87.

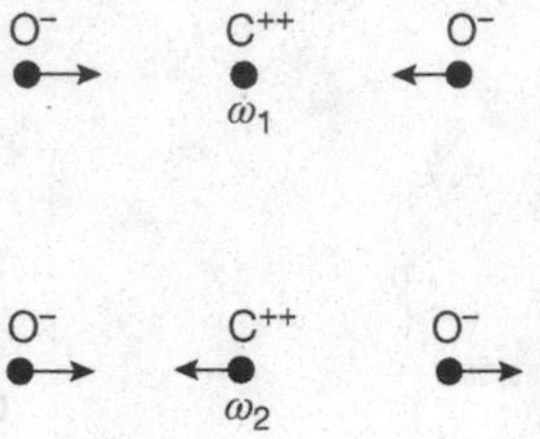

Fig. 1.87

e. The ω_1 mode will not give rise to radiation because the center of the charge remains stationary in the oscillations. The ω_2 mode can give rise to dipole radiation, while quadruple and higher multipole radiations are possible for both ω and ω_3 modes.

1122

Take a very long chain of beads connected by identical springs of spring constant K and equilibrium length a, as shown in Fig. 1.88. Each bead is hee to oscillate along the x direction. All beads have mass m except for one which has mass $m_0 < m$. The mass of the spring is negligibly small.

a. Far from the "special" bead, what is the relation between the wave vector and the frequency of the resulting oscillation?

b. For a wave of wave vector k, what is the reflection probability when the wave hits the special bead?

Hint for part (b): Try a solution of the form

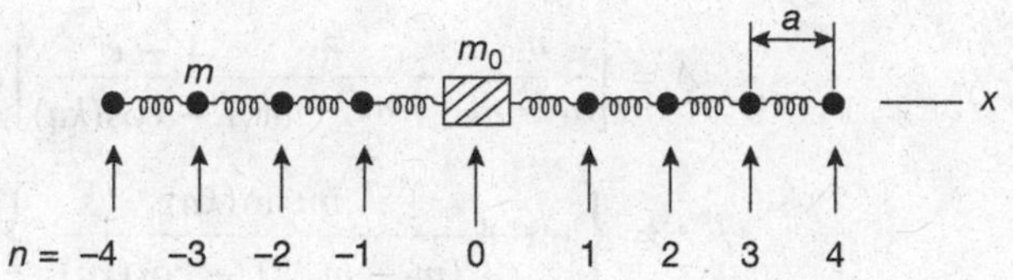

Fig. 1.88

$$x_n = Ae^{ikan} + Be^{-ikan} \qquad \text{for} \qquad n < 0,$$
$$x_n = Ce^{ikan} \qquad \text{for} \qquad n > 0,$$

where A, B, and C are functions of time.

(*Chicago*)

Sol:

a. For $n \neq 0$,

$$\ddot{x}_n = -\frac{K}{m}[(x_n - x_{n-1}) + (x_n - x_{n+1})] = -\frac{K}{m}(2x_n - x_{n+1} - x_{n-1}).$$

Setting $x_n = Ae^{i(kan-\omega t)}$ in the above, we obtain

$$-\omega^2 x = -\frac{K}{m}(2 - e^{ika} - e^{-ika})x_n = -\frac{2K}{m}[1 - \cos(ka)]x_n,$$

or

$$\omega^2 = \frac{2K}{m}[1 - \cos(ka)].$$

b. Try a solution of the form

$$x_n = (Ae^{ikan} + Be^{-ikan})e^{-i\omega t} \qquad \text{for} \qquad n \leq 0,$$
$$x_n = Ce^{i(kan-\omega t)} \qquad \text{for} \qquad n \geq 0.$$

For $n = 0$, the above implies $C = A + B$. Substituting the solution into the equation of the motion of the $n = 0$ bead,

$$\ddot{x}_0 = -\frac{K}{m_0}(2x_0 - x_1 - x_{-1}),$$

we find

$$\frac{\omega^2 m_0}{K}(A + B) = 2(A + B) - (A + B)e^{ika} - Ae^{-ika} - Be^{ika},$$

or

$$A = \left[\frac{m_0}{m - m_0} - \frac{m}{m - m_0}\frac{1 - e^{ika}}{1 - \cos(ka)}\right]B$$

$$= \left\{-1 + \frac{im\sin(ka)}{(m - m_0)[1 - \cos(ka)]}\right\}B.$$

Hence the reflection probability is

$$\left|\frac{B}{A}\right|^2 = \left\{1 + \left(\frac{m}{m - m_0}\right)^2\left[\frac{\sin(ka)}{1 - \cos(ka)}\right]^2\right\}^{-1}.$$

1123

A body of mass m slides down a frictionless ramp at an angle θ and lands on a spring at the bottom of the ramp. Obtain an equation that describes the subsequent motion of the body. Assume that when the body is placed on the spring without sliding, it produces a displacement x. How would your result change if the coefficient of friction between the block and the ramp is μ?

Sol: For the spring

$$k = \frac{mg\sin\theta}{x}$$

The spring mass system will undergo SHM along the ramp. Measuring displacements along the ramp

$$m\frac{d^2x}{dt^2} + \frac{mg\sin\theta}{x}x = 0$$

The system undergoes SHM with frequency of

$$\frac{1}{2\pi}\sqrt{\frac{g\sin\theta}{x}}.$$

In the presence of friction,

Assuming that this force is proportional to velocity

$$F_f = \mu mg\cos\theta = c\frac{dx}{dt}$$

$$m\frac{d^2x}{dt^2} + c\frac{dx}{dt} + \frac{mg\sin\theta}{x}x = 0$$

The system will undergo damped SHM and eventually come to rest.

1124

Three identical objects, each of mass m, are connected by springs of spring constant K, as shown in Fig. 1.89. The motion is confined to one dimension.

At $t = 0$, the masses are at rest at their equilibrium positions. Mass A is then subjected to an external time-dependent driving force $F(t) = f\cos(\omega t)$, $t > 0$. Calculate the motion of mass C.

(Princeton)

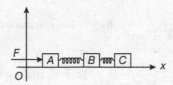

Fig. 1.89

Sol: Let x_A, x_B, x_C be the coordinates of the three masses and a the relaxed length of each spring. The equations of motion are

$$f\cos(\omega t) + K(x_B - x_A - a) = m\ddot{x}_A,$$
$$K(x_C - x_B - a) - K(x_B - x_A - a) = m\ddot{x}_B,$$
$$-K(x_C - x_B - a) = m\ddot{x}_C.$$

The above set of equations can be written as

$$f\cos(\omega t) = m(\ddot{x}_A + \ddot{x}_B + \ddot{x}_C),$$
$$f\cos(\omega t) - 2aK = m(\ddot{x}_A - \ddot{x}_C) + K(x_A - x_C),$$
$$f\cos(\omega t) = m(\ddot{x}_A - 2\ddot{x}_B + \ddot{x}_C) + 3K(x_A - 2x_B + x_C),$$

or

$$f\cos(\omega t) = m\ddot{y}_1, \tag{1}$$
$$f\cos(\omega t) - 2aK = m\ddot{y}_2 + Ky_2, \tag{2}$$
$$f\cos(\omega t) = m\ddot{y}_3 + 3Ky_3, \tag{3}$$

with $y_1 = x_A + x_B + x_C$, $y_2 = x_A - x_C$, $y_3 = x_A - 2x_B + x_C$. It can be seen tha y_1, y_2 and y_3 are the three normal coordinates of this vibrational system. Th initial conditions are that at $t = 0$,

$$x_A = 0, \quad x_B = a, \quad x_C = 2a, \quad \dot{x}_A = \dot{x}_B = \dot{x}_C = 0,$$

or

$$y_1 = 3a, \quad y_2 = -2a, \quad y_3 = 0, \quad \dot{y}_1 = \dot{y}_2 = \dot{y}_3 = 0.$$

Equation (1) can be integrated, with the use of initial conditions, to give

$$y_1 = \frac{f}{m\omega^2}[1 - \cos(\omega t)] + 3a.$$

To solve (2), we try a particular solution

$$y_2 = A_2 \cos(\omega t) + B_2$$

and obtain $A_2 = \dfrac{f}{K - m\omega^2}$, $B_2 = -2a$. The general solution is

$$y_2 = \frac{f\cos(\omega t)}{K - m\omega^2} - 2a + C_2 \cos(\omega_2 t) + D_2 \sin(\omega_2 t),$$

where $\omega_2 = \sqrt{\dfrac{k}{m}}$. Initial conditions then give

$$C_2 = -\frac{f}{K - m\omega^2}, \quad D_2 = 0.$$

To solve (3), we try a particular solution

$$y_3 = A_3 \cos(\omega t)$$

and obtain $A_3 = f/(3K - m\omega^2)$. The general solution is

$$y_3 = \frac{f\cos(\omega t)}{3K - m\omega^2} + C_3 \cos(\omega_3 t) + D_3 \sin(\omega_3 t),$$

where $\omega_3 = \sqrt{\dfrac{3k}{m}}$. Initial conditions then give

$$C_3 = -\frac{f}{3K - m\omega^2}, \quad D_3 = 0.$$

Therefore the solutions are as follows:

$$\begin{cases} y_1 = \dfrac{f}{m\omega^2}[1 - \cos(\omega t)] + 3a, \\[3mm] y_2 = \dfrac{f}{K - m\omega^2}[\cos(\omega t) - \cos(\omega_2 t)] - 2a, \\[3mm] y_3 = \dfrac{f}{3K - m\omega^2}[\cos(\omega t) - \cos(\omega_3 t)]. \end{cases}$$

The motion of C is a linear combination of y_1, y_2 and y_3:

$$x_C = \frac{y_1}{3} - \frac{y_2}{2} + \frac{y_3}{6} = 2a + \frac{f}{3m\omega^2}[1 - \cos(\omega t)]$$

$$+ \frac{f}{2m(\omega_2^2 - \omega^2)}[\cos(\omega t) - \cos(\omega_2 t)]$$

$$+ \frac{f}{6m(\omega_3^2 - \omega^2)}[\cos(\omega t) - \cos(\omega_3 t)].$$

Note that ω_2 and ω_3 are the normal frequencies of the system.

1125

A model of benzene ring useful for some purposes is a wire ring strung with 6 frictionless beads, with springs taut between the beads, as shown in Fig. 1.90. The beads each has mass m and the springs all have spring constant K. The masses have been numbered for the grader's convenience. The ring is fixed in space.

a. Calculate, or write down by intuition, the eigenfrequencies of the normal modes, indicating any degeneracies. In Fig.1.91, picture each mode by drawing an arrow near each mass indicating the direction of motion and shading those masses at rest.

b. With what frequencies can the center of mass oscillate?

c. Which modes could be related to the modes of the real benzene molecule?

Hint: Much algebra can be eliminated by considering the symmetries of the problem.

(Princeton)

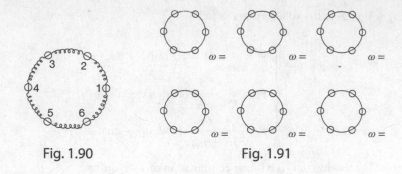

Fig. 1.90   Fig. 1.91

Sol:

a. Let ψ_n be the displacement of the nth bead. Its equation of motion is

$$m\ddot{\psi}_n = K(\psi_{n+1} - \psi_n) - K(\psi_n - \psi_{n-1})$$
$$= K(\psi_{n-1} + \psi_{n+1} - 2\psi_n).$$

Setting $\psi_n = A_n e^{i\omega t}$, we obtain

$$-m\omega^2 A_n = K(A_{n-1} + A_{n+1} - 2A_n),$$

or

$$A_{n-1} + \varepsilon A_n + A_{n+1} = 0, \qquad n = 1, 2, \ldots, 6,$$

where

$$\varepsilon = \frac{m\omega^2}{K} - 2.$$

For the set of linear homogeneous equations to have a non-zero solution, the following determinant must vanish, i.e.

$$\begin{vmatrix} \varepsilon & 1 & 0 & 0 & 0 & 1 \\ 1 & \varepsilon & 1 & 0 & 0 & 0 \\ 0 & 1 & \varepsilon & 1 & 0 & 0 \\ 0 & 0 & 1 & \varepsilon & 1 & 0 \\ 0 & 0 & 0 & 1 & \varepsilon & 1 \\ 1 & 0 & 0 & 0 & 1 & \varepsilon \end{vmatrix} = 0,$$

or

$$\varepsilon^6 - 6\varepsilon^4 + 9\varepsilon^2 - 4 = (\varepsilon + 1)^2(\varepsilon - 1)^2(\varepsilon + 2)(\varepsilon - 2) = 0.$$

Thus the solutions are

$$\varepsilon_1 = 2, \qquad \varepsilon_2 = -2, \qquad \varepsilon_3 = 1,$$
$$\varepsilon_4 = 1, \qquad \varepsilon_5 = -1, \qquad \varepsilon_6 = -1.$$

The corresponding eigenfrequencies are given by

$$\omega_1^2 = \frac{4K}{m}, \qquad \omega_2^2 = 0, \qquad \omega_3^2 = \frac{3K}{m},$$

$$\omega_4^2 = \frac{3K}{m}, \qquad \omega_5^2 = \frac{K}{m}, \qquad \omega_6^2 = \frac{K}{m}.$$

It can be seen that modes 3 and 4 as well as modes 5 and 6 are degenerate. Substituting $\varepsilon_1, \varepsilon_2, \dots, \varepsilon_6$ one by one into the set of equations $A_{n-1} + \varepsilon A_n + A_{n+1} = 0$, we can find the ratios of amplitudes for each normal mode. The results are depicted in Fig. 1.92. The displacements of the six beads have the same magnitude in modes 1, 2, 3, and 5 except in modes 3 and 5 the first and fourth beads are stationary. Their directions are shown in the figure. Mode 2 corresponds to rotation of the system as a whole. In mode 4, the displacements of the second bead and the fifth bead are twice as large as those of the others, and in mode 6, the displacements of the third bead and the sixth bead are also twice as large as those of the others. These larger displacements are indicated by two arrows in the same direction in the figures.

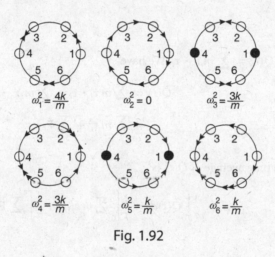

Fig. 1.92

b. It is seen from Fig. 1.92 that only in modes 5 and 6 can the center of mass oscillate with a frequency $\sqrt{k/m}$.

c. As the center of mass of a real benzene molecule cannot oscillate, only modes 1, 2, 3, and 4 can be related to the real benzene molecule.

1126

Consider a classical system of point masses m_i with position vectors $\mathbf{r}_i$, each experiencing a net applied force $\mathbf{F}_i$.

a. Consider the quantity $\sum_i m_i \dot{\mathbf{r}}_i \cdot \mathbf{r}_i$, assumed to remain finite at all times and prove the virial theorem

$$T = -\frac{1}{2}\overline{\sum_i \mathbf{F}_i \cdot \mathbf{r}_i},$$

where T is the total kinetic energy of the system and the bar denotes time average.

b. In the case of a single particle acted on by a central inverse-square law force, show that

$$T = -\frac{\overline{V}}{2},$$

where V is the potential energy.

(SUNY, Buffalo)

Sol:

a. Let $Q(t) = \sum_i m_i \dot{\mathbf{r}}_i \cdot \mathbf{r}_i$. We have

$$\dot{Q}(t) = \sum_i m_i \dot{\mathbf{r}}_i \cdot \dot{\mathbf{r}}_i + \sum_i m_i \ddot{\mathbf{r}}_i \cdot \mathbf{r}_i$$

$$= \sum_i m_i \dot{\mathbf{r}}_i^2 + \sum_i \mathbf{F}_i \cdot \mathbf{r}_i.$$

The time average of $\dot{Q}(t)$ is

$$\frac{1}{\tau}\int_0^\tau \dot{Q}(t)\,dt = \frac{1}{\tau}\int_0^\tau \sum_i m_i \dot{\mathbf{r}}_i^2\,dt + \frac{1}{\tau}\int_0^\tau \sum_i \mathbf{F}_i \cdot \mathbf{r}_i\,dt,$$

i.e.

$$\frac{1}{\tau}[Q(\tau) - Q(0)] = 2\overline{T} + \sum_i \overline{\mathbf{F}_i \cdot \mathbf{r}_i},$$

where τ is the period if the motions are periodic with the same period, or $\tau \to \infty$ otherwise. In both cases, the left-hand side of the equation is equal to zero and we have

$$2\overline{T} + \sum_i \overline{\mathbf{F}_i \cdot \mathbf{r}_i} = 0$$

as stated.

b. For a single particle acted on by a central inverse-square law force,

$$F = \frac{C}{r^2}, \quad \text{or} \quad V = \frac{C}{r},$$

where C is a constant. The virial theorem then gives

$$\overline{T} = -\frac{1}{2}\overline{\frac{C}{r^3}\mathbf{r}\cdot\mathbf{r}} = -\frac{1}{2}\overline{\left(\frac{C}{r}\right)} = -\frac{\overline{V}}{2}.$$

1127

At the vertices A, B, and C of the isosceles triangle of sides l, $2l$, $2l$ shown in the figure are placed masses of 1, 2, and 3 kg, respectively. Determine the center of mass of the system. If another mass of 4 kg is placed above the triangle at a height $3l$ above the center of the perpendicular bisector AD, what would be the center of mass of the system change?

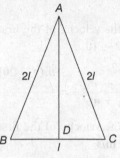

Fig. 1.93

Sol: Let B be the origin, then the coordinates of the vertices are $A:\left(\frac{l}{2}, \sqrt{3}l\right)$, $B:(0, 0)$, $C:(l, 0)$. The center of mass is given as

$$x_{cm} = \frac{m_1 x_1 + m_2 x_2 + m_3 x_3}{m_1 + m_2 + m_3} = \frac{\frac{l}{2} + 0 + 3l}{6} = \frac{7}{12}l.$$

$$y_{cm} = \frac{m_1 y_1 + m_2 y_2 + m_3 y_3}{m_1 + m_2 + m_3} = \frac{\sqrt{3}l + 0 + 0}{6} = \frac{\sqrt{3}}{6}l.$$

In the second case, the coordinates of the points are $A:\left(\frac{l}{2}, \sqrt{3}l, 0\right)$, $B:(0, 0, 0)$,

$C:(l, 0, 0)$ and $\left(\frac{l}{2}, \frac{\sqrt{3}}{2}l, 3l\right)$. The center of mass is now given as

$$x_{cm} = \frac{m_1x_1 + m_2x_2 + m_3x_3 + m_4x_4}{m_1 + m_2 + m_3 + m_4} = \frac{\dfrac{l}{2} + 0 + 3l + \dfrac{4l}{2}}{10} = \frac{11}{20}l.$$

$$y_{cm} = \frac{m_1y_1 + m_2y_2 + m_3y_3 + m_4y_4}{m_1 + m_2 + m_3 + m_4} = \frac{\sqrt{3}l + 0 + 0\dfrac{4\sqrt{3}}{2}l}{10} = \frac{3\sqrt{3}}{10}l.$$

$$z_{cm} = \frac{m_1z_1 + m_2z_2 + m_3z_3 + m_4z_4}{m_1 + m_2 + m_3 + m_4} = \frac{0 + 0 + 0 + 12l}{10} = \frac{6}{5}l.$$

1128

A body of mass "*m*" is thrown up with a velocity of "*v*." One second later another body of mass 2*m* is thrown up along the same path with a velocity "*v*/2." What is the velocity of the center of mass of the system after 2 seconds? Assume *g* = 10 m/s. If the second body was dropped after 1 second, how would the result change?

Sol: After 2 seconds, the velocity of the first body is $v - 20$ and the velocity of the second body is $v/2 - 10$.

$$v_{cm} = \frac{m_1v_1 + m_2v_2}{m_1 + m_2} = \frac{m(v - 20) + 2m\left(\dfrac{v}{2} - 10\right)}{3m} = -\left(\frac{40}{3}\right)\text{m/s}.$$

After 2 seconds, the velocity of the first body is $v - 20$ and the velocity of the second body is 10 m/s.

$$v_{cm} = \frac{m_1v_1 + m_2v_2}{m_1 + m_2} = \frac{m(v - 20) + 10m}{3m} = \frac{v - 10}{3}\text{m/s}.$$

1129

A rocket is projected straight up and explodes into three equally massive fragments just as it reaches the top of its flight (Fig. 1.94). One of the fragments is observed to come straight down in a time t_1, while the other two land at a time t_2, after the burst. Find the height $h(t_1, t_2)$ at which the fragmentation occurred.

(Wisconsin

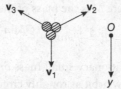

Fig. 1.94

Sol: The velocity and momentum of the rocket are zero when it reaches the top of its flight. Conservation of momentum gives, after the burst,

$$m_1\mathbf{v}_1 + m_2\mathbf{v}_2 + m_3\mathbf{v}_3 = 0.$$

As

$$m_1 = m_2 = m_3, \quad v_1 + v_2 + v_3 = 0.$$

As the second and third fragments land at the same time, the vertical components of $\mathbf{v}_2$ and $\mathbf{v}_3$ are the same. As $\mathbf{v}_1$ is vertically downward the vertical components of $\mathbf{v}_2$ and $\mathbf{v}_3$ are each $-v_1/2$. Hence for the first and second fragments we have

$$h = v_1 t_1 + \frac{g t_1^2}{2},$$

$$h = \frac{-v_1 t_2}{2} + \frac{g t_2^2}{2},$$

giving

$$v_1 = \frac{g(t_2^2 - t_1^2)}{2t_1 + t_2},$$

$$h = \frac{g t_1 t_2}{2} \cdot \frac{t_1 + 2t_2}{2t_1 + t_2}.$$

1130

A satellite of mass m moves in a circular orbit of radius R with speed v about the earth. It abruptly absorbs a small mass δm which was stationary prior to the collision. Find the change in the total energy of the satellite and, assuming the new orbit is roughly circular, find the radius of the new orbit.

(Wisconsin)

Sol: Before picking up the small mass, the satellite moves in a circular orbit so that

$$\frac{mv^2}{R} = \frac{GMm}{R^2},$$

giving $Rv^2 = GM$, where M is the mass of the earth. Hence its total energy is

$$E = \frac{1}{2}mv^2 - \frac{GMm}{R} = -\frac{1}{2}mv^2.$$

After absorbing the stationary small mass δm, the speed of the satellite change to (considering the new orbit as roughly circular, although it is actually elliptic)

$$v' = \frac{mv}{m + \delta m},$$

and its total energy becomes

$$E' = -\frac{1}{2}(m + \delta m)v'^2 = -\frac{1}{2}\frac{m^2 v^2}{m + \delta m}.$$

Hence the energy loss due to the collision is

$$E' - E = \frac{1}{2}mv^2\left(1 - \frac{m}{m + \delta m}\right) = \frac{1}{2}mv^2\frac{\delta m}{m + \delta m}$$

$$\approx \frac{1}{2}v^2\delta m.$$

If the new radius is R' we also have

$$R'v'^2 = GM = Rv^2,$$

giving

$$R' = \left(\frac{v}{v'}\right)^2 R = \left(\frac{m + \delta m}{m}\right)^2 R \approx \left(1 + \frac{2\delta m}{m}\right)R.$$

1131

A spring mass system is arranged as given below. Considering that the masses are on a frictionless surface, determine the expression for the possible angular frequencies with which the system oscillates. Assume that the oscillation i initially started by moving both bodies to the right.

Sol: Let the displacement of m_1 be x_1 and that of m_2 is x_2. Then if F_1 and F_2 are the forces on the two masses,

$$F_1 = -k_1 x_1 + k_2(x_2 - x_1) = m_1\ddot{x}_1.$$

$$F_2 = -k_1 x_2 + k_2(x_1 - x_2) = m_2\ddot{x}_2.$$

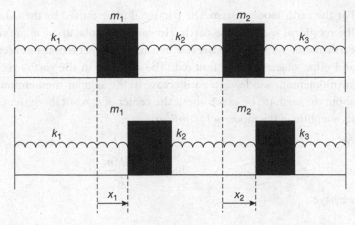

Fig. 1.95

The solutions to the two equations are of the form

$$x_1 = Ae^{i\omega t} \text{ and } x_2 = Be^{i\omega t},$$

resulting in

$$-(k_1 + k_2 - m\omega^2)A + k_2B = 0 \text{ and}$$

$$k_2A - (k_1 + k_2 - m\omega^2)B = 0.$$

Solving the two equations,

$$(k_1 + k_2 - m\omega^2)^2 AB - k_2^2 AB = 0.$$

A and B represent the amplitudes of oscillations; they cannot be 0. Hence,

$$k_1 + k_2 - m\omega^2 = \pm k_2.$$

Two possible frequencies exist with

$$\omega = \sqrt{\frac{k_1}{m}} \quad \text{and} \quad \omega = \sqrt{\frac{k_1 + 2k_2}{m}}.$$

1132

Consider the earth-moon system and for simplicity assume that any interaction with other objects can be ignored. The moon, which moves around the earth more slowly than the earth rotates, creates tides on the earth. A similar situation exists on Mars, but with the difference that one of its moons revolves about Mars faster than the planet rotates. Show that one consequence of tidal friction is that in one system the moon-planet distance is increasing, and in the other it is decreasing. In which one is it decreasing?

(Wisconsin)

Sol: For the earth-moon system, the frictional force caused by the tides slows dow
the rotational speed of the earth. However, the total angular momentum of th
earth-moon system is conserved because the interaction between this syster
and other objects can be ignored. The decrease in the earth's rotational angu
lar momentum will lead to an increase in the angular momentum of the moo
about the earth (to be exact, about the center of mass of the system). The angula
momentum of the moon is $J = mR^2\omega$.

As

$$mR\omega^2 = \frac{GMm}{R^2},$$

we have

$$J = mR^2\sqrt{\frac{GM}{R^3}} = m\sqrt{GMR}.$$

Here we consider the center of the earth to be approximately fixed, so that R is th
earth-moon distance. Then as J increases, R will increase also. Thus for the earth
moon system, the effect of tides is to increase the distance between the moon an
the earth.

For the Mars-moon system, the moon revolves about Mars faster than the latte
rotates, so the frictional force caused by tides will speed up the rotation of Mars
whose rotational angular momentum consequently increases. As the total angu
lar momentum is conserved, the angular momentum of the moon will decrease
The argument above then shows that the distance between Mars and its moor
will decrease.

1133

Two mass points, each of mass m, are at rest on a frictionless horizontal sur
face. They are connected by a spring of equilibrium length l and constant K
An impulse I is given at time $t = 0$ to one of the mass points in a directior
perpendicular to the spring. Assume that the spring always lines up along the
connecting length l, i.e. there is no bending.

   a.   After a time t, what will be the total energy and total momentum of the
       two mass points?

   b.   What will be the velocity of the center of mass (including direction) and
       the total angular momentum about the center of mass?

c. What will be the maximum separation between the two mass points during the motion that follows the impulse?

d. What will be the maximum instantaneous speed achieved by either particle? Explain your answer.

(UC, Berkeley)

Sol:

a. On account of the conservation of momentum and of mechanical energy, the total momentum and total energy of the two mass points at time t are the same as those at time $t = 0$ just after the impulse is applied:

$$\mathbf{P} = \mathbf{I}, \quad E = \frac{P^2}{2m} = \frac{I^2}{2m}.$$

b. The system has total mass $2m$, total momentum I, so that the center of mass has velocity

$$v_c = \frac{I}{2m}.$$

Just after the impulse is given, the angular momentum of the system about the center of mass is $L = \frac{Il}{2}$. By the conservation of angular momentum in the center of mass frame, L is the angular momentum about the center of mass at all later times.

c. Let l_M denote the maximum separation required. Conservation of angular momentum and of mechanical energy

$$2m\left(\frac{l_M}{2}\right)^2\dot{\varphi} = \frac{Il}{2},$$

$$m\left(\frac{l_M}{2}\dot{\varphi}\right)^2 + m\left(\frac{I}{2m}\right)^2 + \frac{1}{2}K(l_M - l)^2 = \frac{I^2}{2m}$$

give

$$2mKl_M^4 - 4mKll_M^3 + (2mKl^2 - I^2)l_M^2 + I^2l^2 = 0,$$

whose positive real root is the maximum distance between the two mass points during the motion that follows the impulse.

d. Let x denote the distance between the two masses. Conservation of mechanical energy gives

$$\frac{1}{2}\cdot 2m\left[\left(\frac{x}{2}\dot{\varphi}\right)^2 + \left(\frac{\dot{x}}{2}\right)^2\right] + \frac{1}{2}\cdot 2m\left(\frac{I}{2m}\right)^2 + \frac{1}{2}K(x - l)^2 = \frac{I^2}{2m},$$

or

$$\frac{m}{4}(x^2\dot\varphi^2 + \dot x^2) + \frac{1}{2}K(x - l)^2 = \text{constant},$$

shows that when $x = l$, the kinetic energy of the two mass points, given by the first term on the left-hand side, is maximum. This is the case at $t = 0$. Also, at $t = 0$, only the mass that had been given the impulse has a velocity while the other mass is still instantaneously at rest. Thus the first mass achieves a maximum speed

$$v_M = \frac{I}{m}$$

at time $t = 0$.

The condition that $x = l$ can be satisfied again from time to time. However, as the speed of the first mass will not be zero, the second mass cannot achieve this maximum speed. Therefore the maximum speed that can be achieved by the second mass is less than v_m.

1134

A heavy chain of length L and linear density $\rho = k(1 + x)$ lies on a table with a section l of the chain hanging over the edge of the table. The coefficient of friction between the chain and the table is μ. What is the maximum length of the chain that can hang out and the chain still remain stationary on the table? Compare your answer when the heavier part is over the edge of the table with that when the lighter part is over the edge of the table.

Sol: Consider the lighter part to be on the table.

Let the section that is hanging over the edge be l.

At equilibrium, frictional force of the chain on the table = weight of chain that is hanging.

$$\mu kg\int_0^{L-l}(1 + x)\,dx = kg\int_{L-l}^L(1 + x)\,dx$$

$$\mu kg\left(\frac{L^2 + l^2}{2} + L - Ll - l\right) = kg\left(l + lL - \frac{l^2}{2}\right)$$

$$(\mu + 1)\frac{l^2}{2} - (L + 1)(\mu + 1)l + \mu L + \frac{\mu}{2}L^2 = 0$$

$$l = \frac{(L+1)(\mu+1) - \sqrt{[(L+1)(\mu+1)]^2 - \mu(\mu+1)(2L+L^2)}}{\mu+1}$$

The positive sign is disregarded as l must be $<L$.

Now consider the heavier part to be on the table. Then

$$\mu kg \int_L^{L-1} (1+x)dx = kg \int_{L-l}^0 (1+x)dx$$

$$\frac{(\mu+1)}{2}l^2 - (L+1)(\mu+1)l + L + \frac{L^2}{2} = 0$$

$$l = \frac{(L+1)(\mu+1) - \sqrt{[(L+1)(\mu+1)]^2 - (\mu+1)(2L+L^2)}}{(\mu+1)}$$

1135

Use the rocket equation to find the rocket residual mass m (in terms of the initial mass) at which the momentum of the rocket is a maximum, for a rocket of mass m starting at rest in free space. The exhaust velocity is a constant v_0.

(*Wisconsin*)

Sol: The equation of motion for a rocket, velocity v, in hee space is

$$\frac{mdv}{dt} = \frac{-v_0 dm}{dt},$$

i.e.

$$dv = \frac{-v_0 dm}{m}.$$

Integrating, we obtain

$$v = -v_0 \ln\left(\frac{m}{m_0}\right),$$

where m_0 is the rocket mass at firing. The momentum of the rocket is

$$P = mv = -mv_0 \ln\left(\frac{m}{m_0}\right).$$

For it to be a maximum, we require

$$\frac{dP}{dm} = -v_0 \ln\left(\frac{m}{m_0}\right) - v_0 = 0.$$

Hence the rocket has maximum momentum when its residual mass is

$$m = \frac{m_0}{e}.$$

1136

Assume that the Earth is a sphere of radius R and a density given as

$$\rho = \rho_0(1 - kr)$$

Determine the ratio of the distances above and below the Earth where the acceleration due to gravity is the same.

Sol: At the Earth's surface

$$g_0 = \frac{GM}{R^2} = \frac{G}{R^2}\int_0^R 4\pi\rho_0 r^2(1 - kr)\,dr = G4\pi\rho_0\left(\frac{R}{3} - \frac{kR^2}{4}\right)$$

At a height h above the Earth's surface

$$g = \frac{GM}{(R + h)^2}$$

where

$$g = \frac{G}{(R + h)^2}\int_0^R 4\pi\rho_0 r^2(1 - kr)\,dr = \frac{G4\pi\rho_0\left(\dfrac{R^3}{3} - \dfrac{kR^4}{4}\right)}{(R + h)^2}$$

$$g = g_0\left(\frac{R}{R + h}\right)^2 \approx g_0\left(1 - \frac{2h}{R}\right)$$

since $\dfrac{h}{R} \ll 1$.

Below the surface of the Earth,

$$g' = \frac{G}{(R - h')^2}\int_0^{R-h'} 4\pi\rho_0 r^2(1 - kr)\,dr = G4\pi\rho_0(R - h')\left(\frac{1}{3} - \frac{k(R - h')}{4}\right)$$

$$g' = g_0\left(\frac{(R - h')(4 - 3kR + 3kh')}{R(4 - 3kR)}\right) = g_0\left(1 - \frac{h'}{R}\right)\left(1 + \frac{3kh'}{4 - 3kR}\right)$$

Equating g and g', we get

$$1 - \frac{2h}{R} \cong 1 - \frac{h'}{R} + \frac{3kh'}{4 - 3kR}$$

since h'/R is very small.

$$\frac{h}{h'} = \frac{1}{2}\left(1 + \frac{1}{\frac{4}{3kR} - 1}\right)$$

1137

A bucket of mass M (when empty) initially at rest and containing a mass of water is being pulled up a well by a rope exerting a steady force P. The water is leaking out of the bucket at a steady rate such that the bucket is empty after a time T. Find the velocity of the bucket at the instant it becomes empty.

(Wisconsin)

Sol: Let the total mass of the bucket and water be M'. Then

$$M' = M + m - \frac{mt}{T},$$

where m is the initial mass of the water. As the leaking water has zero velocity relative to the bucket, the equation of motion is

$$M'\frac{dv}{dt} = P - M'g,$$

or

$$dv = \frac{P - M'g}{M'}dt = \left(\frac{P}{M + m - \frac{m}{T}t} - g\right)dt.$$

The velocity of the bucket at the instant it becomes empty is

$$v = \int_0^T \frac{P\,dt}{M + m - \frac{m}{T}t} - gT = \frac{PT}{m}\ln\left(\frac{M + m}{M}\right) - gT.$$

1138

A rocket ship with mass M_0 and loaded with fuel of mass m_0 takes off vertically in a uniform gravitational field as shown in Fig. 1.96. It ejects fuel with velocity U_0 with respect to the rocket ship. The fuel is completely ejected during a time T_0.

a. Find the equation of motion of the rocket in terms of dM/dt, U_0, g, an M, where M is the mass of the rocket at time t.

b. What is the velocity of the vehicle at the instant t_0 when all the fuel ha been ejected, in terms of M_0, m_0, g and t_0?

(MIT

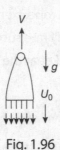

Fig. 1.96

Sol:

a. Consider a rocket, having mass M and velocity V, in time interval Δt ejecting mass ΔM at a velocity U_0 relative to the rocket and gaining an additional velocit ΔV. Taking the vertical upward direction as positive, we have by the momentum theorem

$$(M - \Delta M)(V + \Delta V) + (V + U_0)\Delta M - MV = -Mg\Delta t,$$

i.e.

$$M\frac{\Delta V}{\Delta t} = -U_0\frac{\Delta M}{\Delta t} - Mg,$$

or, in the limit $\Delta t \to 0$,

$$M\frac{dV}{dt} - U_0\frac{dM}{dt} - Mg.$$

b. The equation can be rewritten as

$$dV = -U_0\frac{dM}{M} - gdt.$$

Integrating we obtain

$$V = -U_0 \ln M - gt + K.$$

As $M = M_0 + m_0$, $V = 0$ at $t = 0$,

$$K = U_0 \ln(M_0 + m_0).$$

Hence when $M = M_0$ at $t = t_0$, we have

$$V = U_0 \ln\left(\frac{M_0 + m_0}{M_0}\right) - gt_0.$$

1139

A droplet nucleates in uniform quiescent fog. It then falls, sweeping up the fog which lies in its path. Assume that it retains all the fog which it collects, that it remains spherical and experiences no viscous drag. Asymptotically, it falls with a uniform acceleration a:

$$V(t) \rightarrow at, \qquad \text{for large} \quad t.$$

Find a.

(MIT)

Sol: Let ρ_1, ρ_2 be the densities of the droplet and fog respectively, $R(t)$ be the radius and $V(t)$ the velocity of the droplet, and assume that the buoyancy of the air can be neglected. Making use of the "rocket equation" in Problem **1138**, with

$$M = \frac{4}{3}\pi R^3 \rho_1, \quad U_0 = 0 - (-V) = V,$$

and the replacement

$$V \rightarrow -V,$$
$$\frac{dM}{dt} \rightarrow -\frac{dM}{dt},$$

we have

$$\frac{4}{3}\pi R^3 \rho_1 \frac{dV}{dt} + V\frac{d}{dt}\left(\frac{4}{3}\pi R^3 \rho_1\right) = \frac{4}{3}\pi R^3 \rho_1 g,$$

or

$$R\frac{dV}{dt} + 3V\frac{dR}{dt} = Rg.$$

The droplet sweeps out a cylinder $\pi R^2 V$ in unit time so the rate of change of its mass m is

$$\frac{dm}{dt} = \frac{d}{dt}\left(\frac{4}{3}\pi R^3 \rho_1\right) = 4\pi\rho_1 R^2 \frac{dR}{dt} = \pi R^2 V \rho_2,$$

giving

$$V = 4\eta \dot{R},$$

where $\eta = \rho_1/\rho_2$. We thus have

$$4\eta R\ddot{R} + 12\eta \dot{R}^2 = Rg.$$

As for large t, $V = at$ or $\dot{R} = \dfrac{at}{4\eta}$, we set $R = bt^2 + c$, where b, c are constants and substitute it in the differential equation. Equating the coefficients of t^2 and separately on the two sides of the equation, we have

$$56\eta b - g = 0, \quad (8\eta b - g)c = 0.$$

For a consistent solution, we take

$$b = \frac{g}{56\eta}, \quad c = 0.$$

Hence $V = 4\eta\dot{R} = 8\eta bt = \dfrac{g}{7}t$, i.e. the asymptotic acceleration is

$$a = \frac{g}{7}.$$

1140

An hour glass sits on a scale. Initially all the sand (mass m) in the glass (mass M) is held in the upper reservoir. At $t = 0$, the sand is released. If it exits the upper reservoir at a constant rate $dm/dt = \lambda$, draw (and label quantitatively) a graph showing the reading of the scale at all times $t > 0$.

(MIT)

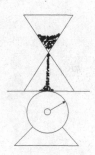

Fig. 1.97

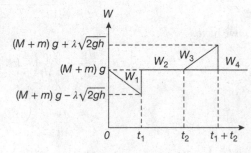

Fig. 1.98

Sol: Suppose all the sand falls to the bottom of the lower reservoir so that for all grains the falling height is h. A grain falling through this distance will acquire a velocity $V = \sqrt{2gh}$ when it reaches the bottom and the whole trip takes a time

$$t_1 = \sqrt{\frac{2h}{g}}.$$

For the reading of the scale, consider the following four periods of time:

Period 1: The time $t = 0$ when the sand is released, to the time t_1 when the sand begins to arrive at the bottom of the lower reservoir. The reading of the scale in this period is

$$W_1 = (M + m)g - \lambda tg, \quad 0 < t < t_1,$$

where $t_1 = \sqrt{2gh}$.

Period 2: The time t_1 when the sand begins to arrive at the bottom, to the time t_2 when all the sand has left the upper reservoir. In this period, the force on the scale consists of two parts: weight of the sand as given by the above equation with $t = t_1$ and a part due to the impulse of the sand on the bottom of the reservoir with magnitude

$$V \frac{dm}{dt} = \lambda \sqrt{2gh}.$$

Hence the scale reads

$$W_2 = [(M + m)g - \lambda t_1 g] + \lambda \sqrt{2gh} = (M + m)g, \quad t_1 < t < t_2,$$

where $t_2 = m/\lambda$.

Period 3: The time t_2 when all the sand has left the upper reservoir to the time t_3 when all sand has reached the bottom. The scale reads

$$W_3 = W_2 + \lambda(t - t_2)g, \quad t_2 < t < t_3,$$

where $t_3 = t_2 + t_1$.

Period 4: The time after all the sand has reached the bottom. The reading of the scale is constant at

$$W_4 = (M + m)g, \quad t > t_3.$$

The reading of the scale is depicted in Fig. 1.98.

1141

A rocket of instantaneous mass m achieves a constant thrust F by emitting propellant at a low rate with high relative speed. The rocket directs its thrust always along the direction of its instantaneous velocity $\mathbf{u}$. By so doing it moves from an initial radius $\mathbf{r}_1$ (measured from the center of the earth) to a larger radius r_2, remaining in the same plane and following a path roughly hke a spiral. The starting radius r_1 is close to the earth's radius r_0, where the gravitational acceleration is g, while $r_2 \gg r_0$. The angular coordinate from the earth's center is ϕ.

   a. Is the angular momentum of the rocket per unit mass a constant of th
motion? Discuss.

   b. In terms of r, r_0, $\dot{r}$, and $\dot{\phi}$, derive expressions for the instantaneous veloc
ity u and the gravitational acceleration g.

   c. Derive expressions for $\ddot{r}$ and $\ddot{\phi}$ in terms of the quantities listed above

(*Princeton*

Sol:

   a. Use polar coordinates (r, ϕ) as given. The angular momentum of the rocket pe
unit mass is $j = r^2 \dot{\phi}$. Although gravity is a central force, the rocket thrust is no
Hence the angular momentum is not a conserved quantity.

   b. The instantaneous velocity of the rocket is

$$\mathbf{u} = u_r \mathbf{e}_r + u_\varphi \mathbf{e}_r = \dot{r} \mathbf{e}_r + r\dot{\varphi} \mathbf{e}_\varphi$$

with magnitude

$$u = \sqrt{\dot{r}^2 + r^2 \varphi^2}.$$

The gravitational acceleration g is

$$g = \frac{GM}{r^2}.$$

As $g_0 = \dfrac{GM}{r_0^2}$, it can be written as

$$g = g_0 \frac{r_0^2}{r^2}.$$

   c. The equation of the motion of the rocket is

$$\mathbf{f} = m\frac{d\mathbf{u}}{dt},$$

where $\mathbf{f} = \mathbf{F} + m\mathbf{g}$, $m = m(t)$, terms involving $\dfrac{dm}{dt}$ having been neglected
As the thrust $\mathbf{F}$ is always parallel to $\mathbf{u}$, its components are $F_r = F\dfrac{\dot{r}}{u}$, $F_\varphi = F\dfrac{r\dot{\varphi}}{u}$
The gravitational acceleration is $\mathbf{g} = -g\mathbf{e}_r$. Hence the equation of motion ha
component equations

$$m(\ddot{r} - r\dot{\phi}^2) = f_r = \frac{F\dot{r}}{\sqrt{\dot{r}^2 + r^2\dot{\phi}^2}} - \frac{mg_0 r_0^2}{r^2},$$

$$m(r\ddot{\phi} + 2\dot{r}\dot{\phi}) = f_\phi = \frac{F\dot{r}\phi}{\sqrt{\dot{r}^2 + r^2\dot{\phi}^2}},$$

from which the expressions for $\ddot{r}$ and $\ddot{\phi}$ can be obtained.

1142

A rocket ship far from any gravitational field has a source of energy E on board. The ship has initial mass m_1 and final mass m_2.

a. Find the maximum velocity v that the ship can achieve starting from rest. E, m_1, and m_2 are fixed, but the exhaust velocity ω (relative to the ship) may vary as a function of the instantaneous mass m of the ship.

b. What is the maximum velocity v that can be obtained if the exhaust velocity w is constrained to be constant?

(Princeton)

Sol:

a. Integrating the equation of motion for the rocket ship (Problem **1138**)

$$m\frac{dv}{dt} = -\omega(m)\frac{dm}{dt},$$

and taking account of the initial conditions $v = 0$, $m = m_1$ at $t = 0$, we have

$$v = -\int_{m_1}^{m} \frac{\omega(m)}{m} dm.$$

Hence the maximum velocity is

$$v_{max} = -\int_{m_1}^{m_2} \frac{\omega(m)}{m} dm.$$

b. If ω is constant, the maximum velocity is

$$v_{max} = \omega \ln \frac{m_1}{m_2}.$$

1143

A spherical dust particle falls through a water mist cloud of uniform density such that the rate of accretion onto the droplet is proportional to the volume of the mist cloud swept out by the droplet per unit time. If the droplet starts from rest in the cloud, find the value of the acceleration of the drop for large times.

(Princeton)

Sol: Suppose the spherical dust particle initially has mass M_0 and radius R_0. Take the initial position of the dust particle as the origin and the x-axis along the downward vertical. Let $M(t)$ and $R(t)$ be the mass and radius of the droplet at time respectively. Then

$$M(t) = M_0 + \frac{4}{3}\pi(R^3 - R_0^3)\rho,$$

where ρ is the density of the water mist, giving

$$\frac{dM}{dt} = \rho 4\pi R^2 \frac{dR}{dt}.$$

The droplet has a cross section πR^2 and sweeps out a cylinder of volume $\pi R^2 \dot{x}$ in unit time, where $\dot{x}$ is its velocity. As the rate of accretion is proportional to this volume, we have

$$\frac{dM}{dt} = \alpha \pi R^2 \dot{x},$$

α being a positive constant. Hence

$$\dot{x} = \frac{4\rho}{\alpha}\dot{R}.$$

The momentum theorem gives

$$M(t + dt)\dot{x}(t + dt) - M(t)\dot{x}(t) = Mgdt.$$

Using Taylor's theorem to expand $M(t + dt)$ and $\dot{x}(t + dt)$ and retaining only the lowest-order terms, we obtain

$$\dot{x}\frac{dM}{dt} + M\ddot{x} = Mg.$$

For large t, $M(t) \approx \frac{4}{3}\pi R^3\rho$, $dM/dt \approx 3M\dot{R}/R$, and the above becomes

$$\ddot{R} + \frac{3\dot{R}^2}{R} = \frac{\alpha g}{4\rho}.$$

For a particular solution valid for large t, setting

$$R(t) = at^2,$$

where a is a constant, in the above we obtain

$$a = \frac{\alpha g}{56\rho}.$$

Thus for large t,

$$\dot{x} = \frac{4\rho}{\alpha} \cdot 2at = \frac{gt}{7}.$$

Hence the acceleration for large times is $g/7$.

1144

Suppose a spacecraft of mass m_0 and cross-sectional area A is coasting with velocity v_0 when it encounters a stationary dust cloud of density ρ as shown in Fig. 1.99. If the dust sticks to the spacecraft, solve for the subsequent motion of the spacecraft. Assume A is constant over time.

(Princeton)

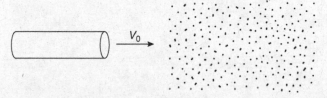

Fig. 1.99

Sol: Suppose the dust offers no resistance to the spacecraft. Newton's second law

$$\frac{d(mv)}{dt} = 0,$$

or

$$m\frac{dv}{dt} + v\frac{dm}{dt} = 0,$$

implies that $mv = m_0 v_0$. Then as

$$dm = \rho v A dt, \qquad \frac{dm}{dt} = \rho A v,$$

we have

$$\frac{dv}{v^3} + \frac{\rho A dt}{m_0 v_0} = 0.$$

Integrating, we obtain

$$\frac{1}{v^2} = \frac{2\rho A t}{m_0 v_0} + C,$$

where C is a constant. If we measure time from the instant the spacecraft first encounters the dust, then $v = v_0$ at $t = 0$, giving $C = v_0^{-2}$. Hence the motion of the spacecraft can be described by

$$\frac{1}{v^2} = \frac{1}{v_0^2} + \frac{2\rho A t}{m_0 v_0}.$$

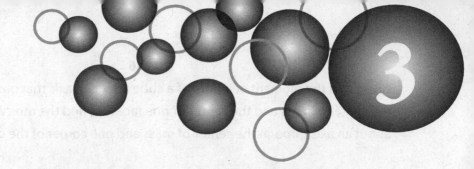

DYNAMICS OF RIGID BODIES (1145–1223)

Two circular metal disks have the same mass M and the same thickness t. Disk 1 has a uniform density ρ_1 which is less than ρ_2, the uniform density of disk 2. Which disk, if either, has the larger moment of inertia?

(Wisconsin)

Sol: Let the radii of the disks be R_1 and R_2 respectively.

Since the disks have the same mass and thickness, we have $\rho_1 R_1^2 = \rho_2 R_2^2$,

or

$$\frac{R_1^2}{R_2^2} = \frac{\rho_2}{\rho_1}.$$

The moments of inertia of the disks are

$$I_1 = \frac{MR_1^2}{2}, \qquad I_2 = \frac{MR_2^2}{2},$$

so

$$\frac{I_1}{I_2} = \frac{R_1^2}{R_2^2} = \frac{\rho_2}{\rho_1}.$$

As $\rho_1 < \rho_2$, $I_1 > I_2$. Hence disk 1 has the larger moment of inertia.

1146

Given that the moment of inertia of a cube about an axis that passes throug[h] the center of mass and the center of one face is I_0, find the moment of inerti[a] about an axis through the center of mass and one corner of the cube.

(UC, Berkeley)

Sol: Use Cartesian coordinates with origin at the center of mass and the axes throug[h] the centers of the three pairs of faces of the cube. We have

$$I_{xx} = I_{yy} = I_{zz} = I_0,$$
$$I_{xy} = I_{yz} = I_{zx} = 0,$$

The moment of inertia about an axis having direction cosines λ, μ, ν is

$$I = \lambda^2 I_{xx} + \mu^2 I_{yy} + \nu^2 I_{zz} - 2\mu\nu I_{yz} - 2\nu\lambda I_{zx} - 2\lambda\mu I_{xy}$$
$$= (\lambda^2 + \mu^2 + \nu^2) I_0.$$

To find the direction cosines of a radius vector $\mathbf{r}$ from the origin to one corner o[f] the cube, without loss of generality, we can just consider the corner with its $x, y,$ coordinates all positive. Then

$$\mathbf{r} = a\mathbf{i} + a\mathbf{j} + a\mathbf{k},$$

where we have taken $2a$ as the length of a side of the cube. As $|\mathbf{r}| = \sqrt{3}a$ we have

$$\lambda = \mu = \nu = \frac{1}{\sqrt{3}},$$

so that

$$I = I_0.$$

1147

A thin disk of radius R and mass M lying in the xy-plane has a point mass $m = 5M/4$ attached on its edge (as shown in Fig. 1.100). The moment of inerti[a] of the disk about its center of mass is (the z-axis is out of the paper)

$$I = \frac{MR^2}{4} \begin{bmatrix} 1 & 0 & 0 \\ 0 & 1 & 0 \\ 0 & 0 & 2 \end{bmatrix}$$

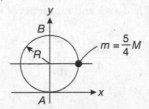

Fig. 1.100

a. Find the moment of inertia tensor of the combination of disk and point mass about point A in the coordinate system shown.

b. Find the principal moments and the principal axes about point A.

c. The disk is constrained to rotate about the y-axis with angular velocity ω by pivots at A and B. Describe the angular momentum about A as a function of time and find the vector force applied at B (ignore gravity).

(*UC, Berkeley*)

Sol:

a. The contribution of a mass element Δm at radius vector $\mathbf{r} = (x_1, x_2, x_3)$ to the moments and products of inertia about the origin is

$$I_{ij} = \Delta m(r^2\delta_{ij} - x_i x_j),$$

where $\delta_{ij} = 1$ if $i = j$, $\delta_{ij} = 0$ if $i \neq j$. Thus the moment of inertia tensor of the point mass about A is

$$\frac{5MR^2}{4}\begin{pmatrix} 1 & -1 & 0 \\ -1 & 1 & 0 \\ 0 & 0 & 2 \end{pmatrix}.$$

The moment of inertia tensor of the disk about A, according to the theorem of parallel axes, is

$$\frac{MR^2}{4}\begin{pmatrix} 5 & 0 & 0 \\ 0 & 1 & 0 \\ 0 & 0 & 6 \end{pmatrix}.$$

Hence the moment of inertia tensor of the disk and point mass about A is

$$\frac{MR^2}{4}\begin{pmatrix} 10 & -5 & 0 \\ -5 & 6 & 0 \\ 0 & 0 & 16 \end{pmatrix}.$$

b.   To find the principal axes and moments of inertia, solve the secular equation

$$\frac{MR^2}{4}\begin{vmatrix} 10-\gamma & -5 & 0 \\ -5 & 6-\gamma & 0 \\ 0 & 0 & 16-\gamma \end{vmatrix} = 0,$$

or

$$(16-\gamma)(\gamma^2 - 16\gamma + 35) = 0.$$

The solutions are

$$\gamma_1 = 16, \qquad \gamma_2 = 8 - \sqrt{29}, \qquad \gamma_3 = 8 + \sqrt{29}.$$

Hence the three principal moments of inertia about A are

$$I_1 = 4MR^2, \qquad I_2 = \left(2 - \frac{\sqrt{29}}{4}\right)MR^2, \qquad I_3 = \left(2 + \frac{\sqrt{29}}{4}\right)MR^2.$$

The direction cosines (λ, μ, ν) of the principal axes corresponding to I_1 are given by

$$\begin{pmatrix} -6 & -5 & 0 \\ -5 & -10 & 0 \\ 0 & 0 & 0 \end{pmatrix}\begin{pmatrix} \lambda \\ \mu \\ \nu \end{pmatrix} = 0,$$

i.e.

$$-6\lambda - 5\mu = 0,$$
$$-5\lambda - 10\mu = 0,$$
$$0\nu = 0.$$

The solution is $\lambda = \mu = 0, \nu =$ arbitrary. As

$$\lambda^2 + \mu^2 + \nu^2 = 1,$$

by definition, the direction cosines are

$$\lambda = 0, \qquad \mu = 0, \qquad \nu = 1.$$

The principal axes for I_2 and I_3 have direction cosines given by

$$\begin{pmatrix} 2 \pm \sqrt{29} & -5 & 0 \\ -5 & -2 \pm \sqrt{29} & 0 \\ 0 & 0 & -2 \pm \sqrt{29} \end{pmatrix}\begin{pmatrix} \lambda \\ \mu \\ \nu \end{pmatrix} = 0,$$

i.e.

$$(2 \pm \sqrt{29}))\lambda - 5\mu = 0,$$
$$-5\lambda + (-2 \pm \sqrt{29})\mu = 0,$$
$$(8 \pm \sqrt{29})\nu = 0,$$

where the top sign is for I_2 and the bottom sign, I_3.

The solutions are

$$\frac{\lambda}{\mu} = \frac{-2 \pm \sqrt{29}}{5} = \begin{cases} 0.677 \\ -1.477 \end{cases}, \quad \nu = 0.$$

Then as $\lambda^2 + \mu^2 + \nu^2 = \lambda^2 + \mu^2 = 1$, we have

$$|\mu| = \left(\frac{\lambda^2}{\mu^2} + 1\right)^{-\frac{1}{2}} = \begin{cases} 0.828 \\ 0.561 \end{cases},$$

$$|\nu| = (1 - \mu^2)^{\frac{1}{2}} = \begin{cases} 0.561 \\ 0.828 \end{cases}.$$

We also require that the principal axes for I_2 and I_3 be orthogonal:

$$\lambda_2\lambda_3 + \mu_2\mu_3 + \nu_2\nu_3 = 0.$$

We therefore take the principal axes as

$$(0, 0, 1),$$
$$(0.561, 0.828, 0),$$
$$(-0.828, 0.561, 0).$$

c. The moment of inertia tensor I of the system of disk and mass point about the origin A found in (a) refers to a coordinate frame (x, y, z) attached to the disk. In this frame the angular momentum of the system rotating with angular velocity ω is

$$L = I\omega,$$

or

$$\begin{pmatrix} L_x \\ L_y \\ L_z \end{pmatrix} = \frac{MR^2}{4} \begin{pmatrix} 10 & -5 & 0 \\ -5 & 6 & 0 \\ 0 & 0 & 16 \end{pmatrix} \begin{pmatrix} 0 \\ \omega \\ 0 \end{pmatrix} = \frac{MR^2\omega}{4} \begin{pmatrix} -5 \\ 6 \\ 0 \end{pmatrix}.$$

Consider a laboratory frame (x', y', z') having the same y-axis as the rotating frame (x, y, z) such that the respective axes coincide at $t = 0$, as shown in Fig. 1.101.

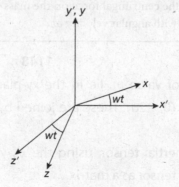

Fig. 1.101

As

$$x' = x\cos(\omega t) + z\sin(\omega t), \quad y' = y,$$
$$z' = -x\sin(\omega t) + z\cos(\omega t),$$

we can define a transformation tensor

$$S = \begin{pmatrix} \cos(\omega t) & 0 & \sin(\omega t) \\ 0 & 1 & 0 \\ -\sin(\omega t) & 0 & \cos(\omega t) \end{pmatrix}$$

so that a vector **V** is transformed according to

$$V' = SV.$$

Applying the above to the angular momentum vector, we find the angular momentum about A in the laboratory frame:

$$\begin{pmatrix} L'_x \\ L'_y \\ L'_z \end{pmatrix} = SL = \frac{MR^2\omega}{4} \begin{pmatrix} -5\cos(\omega t) \\ 6 \\ 5\sin(\omega t) \end{pmatrix},$$

i.e.

$$L'_x = \frac{-5MR^2\omega}{4}\cos(\omega t), \quad L'_y = \frac{3MR^2\omega}{2}, \quad L'_z = \frac{5MR^2\omega}{4}\sin(\omega t)$$

considering the disk alone. The y-axis is a principal axis of inertia and so rotation about it will not cause any force to be exerted on the pivots. Hence the forces on the pivots are due entirely to the rotating mass point. In the rotating frame the mass point suffers a centrifugal force of magnitude $\frac{5MR\omega^2}{4}$, which is balanced by forces exerted on the disk by the pivots. The forces on the pivots are reaction to these forces. Hence pivot B suffers a force of magnitude $\frac{5MR\omega^2}{8}$ in the same direction as the centrifugal force on the mass point. In the laboratory frame this force rotates with angular velocity ω.

1148

Four masses, all of value m, lie in the xy-plane at positions $(x, y) = (a, 0)$ $(-a, 0)$, $(0, +2a)$, $(0, -2a)$. These are joined by massless rods to form a rigid body.

a. Find the inertial tensor, using the x-, y-, z-axes as reference system. Exhibit the tensor as a matrix.

b. Consider a direction given by unit vector $\hat{\mathbf{n}}$ that lies "equally between" the positive x-, y-, z-axes, i.e. it makes equal angles with these three directions. Find the moment of inertia for rotation about this axis.

c. Given that at a certain time t the angular velocity vector lies along the above direction $\hat{\mathbf{n}}$, find, for that instant, the angle between the angular momentum vector and $\hat{\mathbf{n}}$.

(UC, Berkeley)

Sol:

a. The elements I_{ij} of the inertial tensor are given by

$$I_{ij} = \sum_n m_n (r_n^2 \delta_{ij} - x_{n_i} n_{n_j})$$

where

$$r_n^2 = x_{n_1}^2 + x_{n_2}^2 + x_{n_3}^2.$$

As at least one of the coordinates of each mass is zero, $x_i x_j = 0$ so that $I_{ij} = 0$ for all $i \neq j$. For $i = j$, because of symmetry we have

$$I_{11} = 2m(a^2 - a^2) + 2m(4a^2 - 0) = 8ma^2,$$
$$I_{22} = 2m(a^2 - 0) + 2m(4a^2 - 4a^2) = 2ma^2,$$
$$I_{33} = 2m(a^2 - 0) + 2m(4a^2 - 0) = 10ma^2.$$

Hence the inertial tensor is given by the matrix

$$\begin{pmatrix} 8ma^2 & 0 & 0 \\ 0 & 2ma^2 & 0 \\ 0 & 0 & 10ma^2 \end{pmatrix}.$$

b. As the given direction makes the same angle with the axes, its direction cosines λ, μ, ν are equal. The moment of inertia about this direction is then

$$I = \lambda^2 I_{11} + \mu^2 I_{22} + \nu^2 I_{33} - 2\mu\nu I_{23} - 2\nu\lambda I_{31} - 2\lambda\mu I_{12}$$
$$= (8ma^2 + 2m\omega^2 + 10ma^2)\lambda^2$$
$$= 20ma^2\lambda^2.$$

The direction cosines are subject to the condition

$$\lambda^2 + \mu^2 + \nu^2 = 3\lambda^2 = 1,$$

giving $\lambda^2 = \dfrac{1}{3}$. Hence

$$I = \frac{20}{3}ma^2.$$

c. The direction $\hat{\mathbf{n}}$ is given by

$$\hat{\mathbf{n}} = \begin{pmatrix} \lambda \\ \mu \\ \nu \end{pmatrix} = \lambda \begin{pmatrix} 1 \\ 1 \\ 1 \end{pmatrix}.$$

At time T, ω is parallel to $\hat{\mathbf{n}}$:

$$\omega = \omega\hat{\mathbf{n}} = \lambda\,\omega \begin{pmatrix} 1 \\ 1 \\ 1 \end{pmatrix}.$$

The angular momentum at this instant is given by

$$L = I\omega,$$

or

$$\mathbf{L} = \lambda ma^2\omega \begin{pmatrix} 8 \\ 2 \\ 10 \end{pmatrix},$$

with magnitude

$$L = \lambda ma^2\omega\sqrt{8^2 + 2^2 + 10^2} = \sqrt{168}\lambda ma^2\omega.$$

The angle ϕ between $\mathbf{L}$ and $\hat{\mathbf{n}}$ is then given by

$$\cos\phi = \frac{\mathbf{L}\cdot\hat{\mathbf{n}}}{L} = \frac{\lambda^2 ma^2\omega(8 + 2 + 10)}{\lambda ma^2\omega\sqrt{168}} = \frac{1}{\sqrt{3}}\frac{20}{\sqrt{168}} = 0.891,$$

i.e.

$$\phi = 27°.$$

1149

Due to polar flattening, the earth has a slightly larger moment of inertia about its polar axis than about its equatorial axis. Assume axial symmetry about the polar axis.

a. Show that the dominant terms of the gravitational potential above the surface of the earth can be expressed as

$$U = -\frac{GM}{r}\left[1 - \frac{C - A}{Ma^2}\left(\frac{a}{r}\right)^2\left(\frac{3\cos^2\theta - 1}{2}\right)\right],$$

where C and A are the moments of inertia about the polar and equatorial axes respectively, M is the earth's mass, a is the mean earth radius and r is the distance to the center of mass of the earth. The coefficient $(C - A)/Ma^2$ is about 10^{-3}.

b. What secular effect will the second term have upon a satellite traveling in a circular orbit around the earth?

c. If the normal to the plane of the satellite is inclined at an angle α to the polar axis of the earth, derive an expression for the magnitude of this effect by taking a time average over the circular orblt.

(UC, Berkeley)

Sol:

a. Choose the polar axis as the z-axis and the equatorial plane as the xy plane. Let a mass element dM of the earth have position vector $\mathbf{r}' = (x', y', z')$ and let a satellite above the surface of the earth have position vector $\mathbf{r} = (x, y, z)$. Then the gravitational potential energy per unit mass of the satellite is

$$U = -\int \frac{GdM}{|\mathbf{r} - \mathbf{r}'|} = -\int \frac{GdM}{\left[r^2 - 2\mathbf{r} \cdot \mathbf{r}' + r'^2\right]^{\frac{1}{2}}}$$

$$= -\int \frac{GdM}{r}\left[1 - \frac{2\mathbf{r} \cdot \mathbf{r}'}{r^2} + \frac{r'^2}{r^2}\right]^{-\frac{1}{2}}$$

integrating over the entire earth. Taylor expansion gives, neglecting terms of order higher than $\left(\frac{r'}{r}\right)^2$,

$$U = -\int \frac{GdM}{r}\left[1 + \frac{\mathbf{r} \cdot \mathbf{r}'}{r^2} - \frac{r'^2}{2r^2} + \frac{3}{2}\frac{(\mathbf{r} \cdot \mathbf{r}')^2}{r^4}\right].$$

As $\mathbf{r}$ is a constant vector and the earth is assumed to be a symmetrical ellipsoid,

$$\int \mathbf{r} \cdot \mathbf{r}'dM = \mathbf{r} \cdot \int \mathbf{r}'dM = 0.$$

Hence

$$U = -\frac{GM}{r} - \frac{G}{r^3}\int\left[\frac{3}{2}\frac{(r \cdot r')^2}{r^2} - \frac{r'^2}{2}\right]dM$$

$$= -\frac{GM}{r}$$

$$= -\frac{G}{r^3}\int\left[\frac{3(xx' + yy' + zz')^2 - (x^2 + y^2 + z^2)(x'^2 + y'^2 + z'^2)}{2r^2}\right]dM.$$

Due to the symmetry of the earth, the integrals of $x'y'$, $y'z'$ and $z'x'$ are all zero and we have

$$U = -\frac{GM}{r}$$

$$-\frac{G}{r^3}\int[2(x^2x'^2 + y^2y'^2 + z^2z'^2)$$

$$-(x^2y'^2 + x^2z'^2 + y^2x'^2 + y^2z'^2 + z^2x'^2 + z^2y'^2)]\frac{dM}{2r^2}.$$

Now, the choice of the x- and y-axes is arbitrary (as long as they are in the equatorial plane), so that the integral of x' is equal to that of y'. Thus

$$U = -\frac{GM}{r} - \frac{G}{r^3}\int\left[\frac{x^2x'^2 + y^2y'^2 + 2z^2x'^2 - x^2z'^2 - y^2z'^2 - 2z^2x'^2}{2r^2}\right]dM$$

$$= -\frac{GM}{r} - \frac{G}{r^3}\int\left[\frac{(x^2 + y^2)x'^2 - 2z^2x'^2 + (2z^2 - x^2 - y^2)z'^2}{2r^2}\right]dM$$

$$= -\frac{GM}{r} - \frac{G}{r^3}\int\frac{(3z^2 - r^2)(z'^2 - x'^2)}{2r^2}dM$$

$$= -\frac{GM}{r} - \frac{G}{r^3}\left(\frac{3z^2}{2r^2} - \frac{1}{2}\right)\int[(z'^2 + y'^2) - (x'^2 + y'^2)]dM$$

$$= -\frac{GM}{r} - \frac{G}{r^3}\left(\frac{3z^2}{2r^2} - \frac{1}{2}\right)(I_x - I_y). \tag{1}$$

As $I_x = I_y = A$, $I_z = C$, $z = r\cos\theta$, where θ is the angle between $\mathbf{r}$ and the polar axis, the above can be written as

$$U = -\frac{GM}{r}\left[1 - \frac{C - A}{Ma^2}\left(\frac{a}{r}\right)^2\left(\frac{3\cos^2\theta - 1}{2}\right)\right].$$

b. Equation (1) can be written as $U = U_1 + U_2$. $U_1 = -\frac{GM}{r}$ is the potential energy per unit mass the satellite would have if the earth were a perfect sphere.

U_2 arises from polar flattening. It gives rise to an additional force per unit mass of the satellite of $\mathbf{F} = -\nabla U_2$. As $\nabla r = \dfrac{\mathbf{r}}{r}$, $\nabla r^{-5} = \dfrac{-5\mathbf{r}}{r^7}$, $\nabla r^{-3} = \dfrac{-3\mathbf{r}}{r^5}$, $\nabla z^2 = 2z\mathbf{k}$,

$$\mathbf{F} = \frac{3G(A-C)}{2r^5}\left[\left(1-\frac{5z^2}{r^2}\right)\mathbf{r} + 2z\mathbf{k}\right],$$

Note that the first part in the square brackets is still a central force, albeit not of the inverse-square type. It does not change the magnitude and direction of the angular momentum about the center of the earth; hence it has no effect on the plane of orbit, but only makes the satellite deviate from circular orbit slightly. The second part,

$$\mathbf{F}_2 = \frac{3G(A-C)}{r^5}z\mathbf{k},$$

is not a central force; it makes the orbit plane precess about the z-axis.

c. As the motion of the satellite is very nearly a uniform circular motion with center at the origin, because of symmetry the integral of $\mathbf{F}_2 dt$ over a period of the circular motion is equal to zero, so that its average effect on the motion is zero. The torque caused by $\mathbf{F}_2$ with respect to the center of the earth is

$$\mathbf{M} = \mathbf{r} \times \mathbf{F}_2 = \frac{3G(C-A)}{r^5}(-yz\mathbf{i} + xz\mathbf{j}).$$

Let the intersection of the orbital plane and the equatorial plane of the earth be the x-axis (Fig. 1.102). In the course of rotation, yz is always positive while the average value of zx is zero. So over one period, the average torque is directed in the $-x$ direction. As the angular momentum vector lies in the yz-plane and is thus perpendicular to the average torque, the latter does not change the magnitude of the angular momentum.

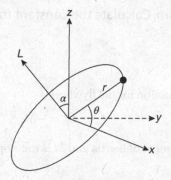

Fig. 1.102

The angular momentum vector **L** has two components Ly and Lz. As the average torque, which is in the $-x$ direction, is perpendicular to Lz, it does not affect the latter. Hence it does not change the angle α between **L** and the z-axis. The result is that **L** will precess about the z-axis, describing a cone of semivertex angle α in a frame fixed to a distant star. As the (x, y, z) frame is fixed with respect to the orbit, the x-axis will rotate around the center of the earth in the equatorial plane.

Let θ be the angle between the position vector of the satellite and the x-axis. Setting $\theta = 0$ at time $t = 0$, we have $\theta = \omega t$, ω being the angular velocity of the satellite. As $y = r \sin \theta \sin \alpha$, $z = r \sin \theta \cos \alpha$, the average of **M** over one period $T = \dfrac{2\pi}{\omega}$ is

$$\langle \mathbf{M} \rangle = \mathbf{i} \frac{3G(C - A)}{r^5 T} \int_0^T - yz\,dt$$

$$= -\frac{\mathbf{i}3G(C - A)\sin(2\alpha)}{2r^3 T} \int_\theta^T \sin^2(\omega t)\,dt$$

$$= -\mathbf{i} \frac{3G(C - A)\sin(2\alpha)}{4r^3}.$$

As $\langle \mathbf{M} \rangle$ is perpendicular to the angular momentum **L**, this will cause the angular momentum vector to precess about the z-axis with angular velocity

$$\dot{\phi} = \frac{|\langle \mathbf{M} \rangle|}{r^2 \omega} = \frac{3G(C - A)\sin(2\alpha)}{4r^5 \omega}.$$

1150

A flywheel in the form of a uniformly thick disk 4 ft in diameter weighs 600 lb and rotates at 1200 rpm. Calculate the constant torque necessary to stop it in 2.0 min.

(*Wisconsin*)

Sol: The equation of motion for the flywheel is

$$I\ddot{\theta} = -M,$$

where I is the moment of inertia and M is the stopping torque. Hence

$$\dot{\theta} = \omega_0 - \frac{Mt}{I}.$$

When the flywheel stops at time t, $\dot{\theta} = 0$ and

$$M = \frac{I\omega_0}{t}.$$

With $I = \frac{MR^2}{2} = 1200$ lb ft^2, $\omega_0 = 40\pi$ rad/s, $t = 120$ s,

$$M = 400\pi \text{ pdl ft} = 39 \text{ Ib ft}.$$

1151

A structure is made of equal-length beams, 1 to 11, as shown in Fig. 1.103, hinged at the joints $A, B, \ldots, G$. Point A is supported rigidly while G is only supported vertically. Neglect the beam weights. A weight ω is placed at E. Each member is under pure tension T or compression C. Solve for the vertical support forces at A and G and find the tension T or compression C in each member.

(Columbia)

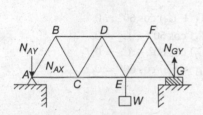

Fig. 1.103

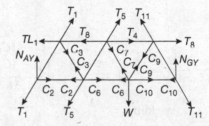

Fig. 1.104

Sol: Consider the structure as a whole. The equilibrium conditions for forces at A and G and for torques about A give

$$N_{AX} = 0,$$
$$N_{AY} + N_{GY} - W = 0,$$
$$\overline{AE} \cdot W - \overline{AG} \cdot N_{GY} = 0,$$

whence

$$N_{AX} = 0, \quad N_{AY} = \frac{W}{3}, \quad N_{GY} = \frac{2W}{3}.$$

Then let the tensions and compressions in the rods be as shown in Fig. 1.104. Considering the equilibrium conditions for joint A we have

$$N_{AY} - T_1 \sin 60° = 0,$$
$$C_2 - T_1 \cos 60° = 0,$$

yielding

$$T_1 = \frac{2\sqrt{3}}{9}W, \quad C_2 = \frac{\sqrt{3}}{9}W.$$

Consider the balance of vertical forces at B, C, D, G, F. We obtain by inspection of Fig. 1.104

$$C_3 = T_1 = \frac{2\sqrt{3}}{9}W, \quad T_5 = C_3 = \frac{2\sqrt{3}}{9}W, \quad C_7 = T_5 = \frac{2\sqrt{3}}{9}W,$$

$$T_{11} = \frac{N_{GY}}{\sin 60°} = \frac{4\sqrt{3}}{9}W, \quad C_9 = T_{11} = \frac{4\sqrt{3}}{9}W.$$

Then considering the balance of horizontal forces at B, C, E, F we have

$$T_4 - (T_1 + C_3) \cos 60° = 0,$$
$$C_6 - (C_3 + T_5) \cos 60° - C_2 = 0,$$
$$C_{10} - (C_9 - C_7) \cos 60° - C_6 = 0,$$
$$T_8 - (T_{11} + C_9) \cos 60° = 0,$$

yielding

$$T_4 = \frac{2\sqrt{3}}{9}W, \quad C_6 = \frac{\sqrt{3}}{3}W, \quad C_{10} = \frac{2\sqrt{3}}{9}W, \quad T_8 = \frac{4\sqrt{3}}{9}W.$$

1152

A uniform thin rigid rod of mass M is supported by two rapidly rotating rollers whose axes are separated by a fixed distance a. The rod is initially placed at rest asymmetrically, as shown in Fig. 1.105.

   a. Assume that the rollers rotate in opposite directions as shown in the figure. The coefficient of kinetic friction between the bar and the rollers is μ. Write down the equation of motion of the bar and solve for the displacement

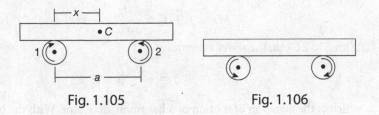

Fig. 1.105   Fig. 1.106

$x(t)$ of the center C of the bar from roller 1 assuming $x(0) = x_0$ and $\dot{x}(0) = 0$.

b. Now consider the case in which the directions of rotation of the rollers are reversed, as shown in Fig. 1.106. Calculate the displacement $x(t)$, again assuming $x(0) = x_0$ and $\dot{x}(0) = 0$.

(*Princeton*)

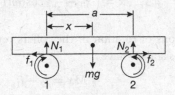

Fig. 1.107

Sol:

a. The forces exerted by the rollers on the rod are as shown in Fig. 1.107. For equilibrium along the vertical direction we require

$$N_1 + N_2 = Mg, \qquad aN_2 = xM_g,$$

giving

$$N_1 = \left(1 - \frac{x}{a}\right)Mg, \qquad N_2 = \frac{x}{a}Mg.$$

The kinetic friction forces are

$$f_1 = \mu N_1, \qquad f_2 = \mu N_2,$$

with directions as shown in the figure. Note that as the rollers rotate rapidly, a change in the direction of motion of the rod will not affect the directions of these forces. Newton's second law then gives

$$M\ddot{x} = f_1 - f_2 = \frac{\mu M g}{a}(a - 2x).$$

With $\xi = 2x - a$, the above becomes

$$\ddot{\xi} = -\frac{2\mu g}{a}\xi,$$

which is the equation of motion of a harmonic oscillator. With the initial conditions $\xi = 2x_0 - a$, $\dot{\xi} = 0$ at $t = 0$, the solution is

$$\xi = (2x_0 - a)\cos(\omega t),$$

where

$$\omega = \sqrt{\frac{2\mu g}{a}}.$$

Hence

$$x = \left(x_0 - \frac{a}{2}\right)\cos(\omega t) + \frac{a}{2}.$$

b. With the directions of rotation of the rollers reversed, the friction forces also reverse directions and we have

$$M\ddot{x} = f_2 - f_1,$$

or

$$\ddot{\xi} = \frac{2\mu g}{a}\xi,$$

where $\xi = 2x - a$ as before. The motion is no longer simple harmonic. With the same initial conditions, the solution is

$$\xi = \left(x_0 - \frac{a}{2}\right)(e^{-\omega t} + e^{\omega t}) = (2x_0 - a)\cosh(\omega t),$$

i.e.

$$x = \left(x_0 - \frac{a}{2}\right)\cosh(\omega t) + \frac{a}{2},$$

where $\omega = \sqrt{\frac{2\mu g}{a}}$. Note that if $x_0 \neq \frac{a}{2}$, the rod will move in one direction until it loses contact with one roller, at which time the equation ceases to apply.

1153

A torsion pendulum consists of a vertical wire attached to a mass which may rotate about the vertical. Consider three torsion pendulums which consist of identical wires from which identical homogeneous solid cubes are hung. One cube is hung from a corner, one from midway along an edge, and one from the middle of a face, as shown in Fig. 1.108. What are the ratios of the periods of the three pendulums?

(MIT)

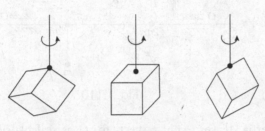

Fig. 1.108

Sol: In all the three cases, the vertical wire passes through the center of mass of the solid cube. As the ellipsoid of inertia of a homogeneous solid cube is a sphere, the rotational inertia about any direction passing through the center of mass is the same. Hence the periods of the three torsion pendulums are equal.

1154

Figure 1.109 shows a simple-minded abstraction of a camshaft with point masses m and $2m$ fixed on massless rods, all in a plane. It rotates with constant angular velocity ω around the axis OO' through the long shaft, held by frictionless bearings at O and O'.

a. What is the torque with respect to the mid-point of the long shaft exerted by the bearings? (Give magnitude and direction.)

b. Locate an axis, fixed in the plane of the masses, around which the thing could rotate with zero torque when the angular velocity is constant.

(UC, Berkeley)

Sol: Choose a coordinate system attached to the shaft with origin at the mid-point C of the long shaft, the z-axis along the axis OO' and the x-axis in the plane of the point masses, as shown in Fig. 1.109.

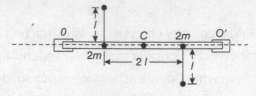

Fig. 1.109

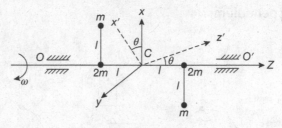

Fig. 1.110

The inertia tensor with respect to C is calculated using the formula $I_{ij} = \sum_n m_n(r_n^2 \delta_{ij} - x_{ni}x_{nj})$, where $r_n^2 = x_{n1}^2 + x_{n2}^2 + x_{n3}^2$. As the masses have coordinates

$$2m : (0, 0, l), \qquad m : (l, 0 - l),$$
$$2m : (0, 0, -l), \qquad m : (-l, 0, l),$$

we have

$$\mathbf{I} = \begin{pmatrix} 6ml^2 & 0 & 2ml^2 \\ 0 & 8ml^2 & 0 \\ 2ml^2 & 0 & 2ml^2 \end{pmatrix}.$$

Considering the angular momentum $\mathbf{J}$ and torque $\mathbf{M}$ about C, we have

$$\mathbf{M} = \frac{d\mathbf{J}}{dt} = \frac{d^*\mathbf{J}}{dt} + \boldsymbol{\omega} \times \mathbf{J} = \boldsymbol{\omega} \times \mathbf{J},$$

where the star denotes differentiation with respect to the rotating coordinate system (x, y, z), as the angular velocity is constant. As

$$\mathbf{J} = \mathbf{I}\boldsymbol{\omega} = \mathbf{I}\begin{pmatrix} 0 \\ 0 \\ \omega \end{pmatrix} = 2ml^2\omega\begin{pmatrix} 1 \\ 0 \\ 1 \end{pmatrix},$$

we find

$$M = \omega \times J = \begin{vmatrix} \mathbf{i} & \mathbf{j} & \mathbf{k} \\ 0 & 0 & \omega \\ 2ml^2\omega & 0 & 2ml^2\omega \end{vmatrix} = 2ml^2\omega\mathbf{j}.$$

The torque with respect to the mid-point of the shaft exerted by the bearings has magnitude $2ml^2\omega^2$ and is in the y direction.

(b) Denote the axis in the xz-plane about which the torque is zero as the z'-axis and suppose it makes an angle θ with the z-axis. As shown in Fig. 1.110, the x'-, y'- and z'-axes form a Cartesian frame, where the x'-axis is also in the xz-plane. In this frame, the angular velocity ω is

$$\omega = \begin{pmatrix} \omega \sin \theta \\ 0 \\ \omega \cos \theta \end{pmatrix}$$

and

$$J = I\omega = \begin{pmatrix} 6ml^2\omega \sin \theta + 2ml^2\omega \cos \theta \\ 0 \\ 2ml^2\omega \sin \theta + 2ml^2\omega \cos \theta \end{pmatrix}.$$

Hence

$$M = \omega \times J$$

$$= \begin{vmatrix} \mathbf{i} & \mathbf{j} & \mathbf{k} \\ \omega \sin \theta & 0 & \omega \cos \theta \\ ml^2\omega(6 \sin \theta + 2 \cos \theta) & 0 & 2ml^2\omega(\cos \theta + \sin \theta) \end{vmatrix}$$

$$= 2ml^2\omega^2(\sin 2\theta + \cos 2\theta)\mathbf{j}$$

For $M = 0$, we require that

$$\tan 2\theta = -1,$$

i.e. $\theta = -22.5°$ or $67.5°$. Note that the z'-axis, about which the torque vanishes, is a principal axis of inertia. As such it can also be found by the method of Problem **1147**.

1155

A coin with its plane vertical and spinning with angular velocity ω in its plane as shown in Fig. 1.111 is set down on a flat surface. What is the final angular velocity of the coin? (Assume the coin stays vertical; neglect rolling friction.)

(Wisconsin)

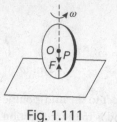

Fig. 1.111

Sol: The spinning coin is on a horizontal plane. As the forces acting on the coin namely, the supporting force **F** and gravity **P**, both pass through the center of mass, the angular momentum of the coin about its center of mass is conserved. Hence the angular velocity is still ω after it is set down on the surface.

1156

Human legs are such that a person of normal size finds it comfortable to walk at a natural, swinging pace of about one step per second, but uncomfortable to force a pace substantially faster or slower. Neglecting the effect of the knee joint, use the simplest model you can to estimate the frequency which determines this pace, and to find what characteristic of the leg it depends on.

(Wisconsin)

Sol: Consider the human leg to be a uniform pole of length l. In the simplest model, the swinging frequency of the leg should be equal to the characteristic frequency of the pole when it swings about its end as a fixed point. The motion is that of a compound pendulum described by

$$\frac{1}{3}ml^2\ddot{\theta} = -\frac{1}{2}mgl\sin\theta,$$

or

$$\ddot{\theta} + \frac{3g}{2l}\theta = 0,$$

for θ small. Then the frequency of swing is $\nu = \frac{1}{2\pi}\sqrt{\frac{3g}{2l}}$. If we take $l \approx 0.4$ m, $\nu \approx 1$ s^{-1}.

1157

A body of mass m is pulled up an incline by another mass m connected to it through a single smooth pulley. The acceleration of the system is $g/4$. The two masses are now connected through a pulley system as shown in the figure. What is the acceleration of the system now? Neglect any friction in the system.

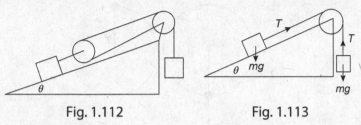

Fig. 1.112   Fig. 1.113

Sol:

In the first case $T - mg \sin \theta = ma$
$$mg - T = ma$$
Hence $g(1 - \sin \theta) = 2a$
$$g(1 - \sin\theta) = 2\frac{g}{4}$$
$$\sin\theta = \frac{1}{2}$$
$$\theta = 30°$$

In the second case,
$$2T - mg \sin \theta = ma'$$
$$mg - T = ma'$$

giving
$$2g - g \sin 30° = 3a'$$

Or $a' = \frac{1}{2}g$

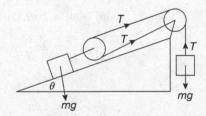

Fig. 1.114

1158

A body is tied to a string and rotated in a vertical circle of radius R. Obtain an expression for its angular acceleration as it travels from the lowest to the highest positions of its path if that the string is just taut at the highest point. Also obtain an expression for the average angular velocity.

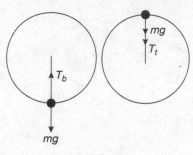

Fig. 1.115

Sol: Writing the force equations in each case

$$T_b - mg = \frac{mv_b^2}{R}$$

$$T_t + mg = \frac{mv_t^2}{R}$$

For the string is to be just taut at the top, $T_t = 0$ giving

$$v_t = \sqrt{Rg}$$

Equating the energies of the body at the top and the bottom, the law of conservation of energy gives

$$\frac{1}{2}mv_t^2 + 2Rmg = \frac{1}{2}mv_b^2$$

yielding

$$v_b = \sqrt{5Rg}$$

To determine the acceleration

$$\omega_t^2 - \omega_b^2 = 2\alpha\theta$$

$$\frac{v_t^2}{R^2} - \frac{v_b^2}{R^2} = 2\alpha\frac{s}{R}$$

$$s = \pi R$$

$$\frac{v_t^2}{R^2} - \frac{v_b^2}{R^2} = 2\pi\alpha$$

$$\frac{Rg - 5Rg}{R^2} = 2\pi\alpha$$

$$\alpha = \frac{g - 5g}{2\pi R}$$

$$\alpha = \frac{-2g}{\pi R}$$

The time it takes to reach to the top is obtained from the equation

$$\omega_t = \omega_b + \alpha t$$

$$\frac{v_t}{R} = \frac{v_b}{R} + \alpha t$$

$$\sqrt{Rg} = \sqrt{5Rg} - \frac{2gt}{\pi}$$

$$t = (\sqrt{5Rg} - \sqrt{Rg})\frac{\pi}{2g}$$

The average angular velocity is given as

$$\omega = \frac{\pi R}{t} = \frac{2\sqrt{gR}}{(\sqrt{5} - 1)} = \frac{\sqrt{gR}(\sqrt{5} + 1)}{2}$$

1159

An ultra-high speed rotor consists of a homogeneous disc of mass M, radius R and width $2l$. It is mounted on a shaft supported on bearings separated by a distance $2d$ as shown in Fig. 1.116. The two additional masses, of equal mass m, are arranged symmetrically so that the rotor remains in "static" balance. Find the time-varying force on the bearings if the rotor turns at angular velocity ω.

(Wisconsin)

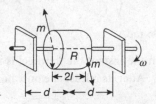

Fig. 1.116

Sol: In the rotating frame attached to the disk, the additional masses each suffers a centrifugal force $mR\omega^2$, resulting in a torque $T = 2mR\omega^2 l$. This torque is balanced by a torque of the same magnitude but opposite in direction, supphed by the bearings which are separated by a distance $2d$. Hence the bearings each suffers a force

$\dfrac{T}{2d} = \dfrac{mR\omega^2 l}{d}$ in the same direction as that of the centrifugal force on the neare

mass. In the fixed frame these rotate with angular velocity ω.

1160

A 100 m^2 solar panel is coupled to a flywheel such that it converts inciden
sunlight into mechanical energy of rotation with 1% efficiency.

a. With what angular velocity would a solid cylindrical flywheel of mass
500 kg and radius 50 cm be rotating (if it started from rest) at the end of
8 hours of exposure of the solar panel?
Take the solar constant to be 2 cal/cm^2/min, for the full time interval
(1 cal = 4.2 Joules)

b. Suppose the flywheel, whose axle is horizontal, were suddenly released
from its stationary bearings and allowed to start rolhng along a horizontal surface with kinetic coefficient of friction $\mu = 0.1$. How far will it
roll before it stops slipping?

c. With what speed is the center of mass moving at that moment?

d. How much energy was dissipated in heat?

(UC, Berkeley)

Sol:

a. The kinetic energy of rotation of the flywheel is $E = \dfrac{1}{2} I \omega_0^2$, where $I = \dfrac{1}{2} m R^2$,
giving

$$\omega_0 = \sqrt{\dfrac{2E}{I}} = \sqrt{\dfrac{2 \times 0.01 \times 100 \times 10^4 \times 8 \times 60 \times 2 \times 4.22E}{\dfrac{1}{2} \times 500 \times 0.5^2}}$$

$$= 1136 \text{ rad/s}.$$

b. Measure time from the instant the flywheel is released, when it is rotating with
angular velocity ω_0. After its release the only horizontal force

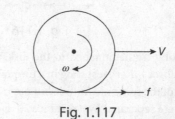

Fig. 1.117

on the flywheel is the frictional force as shown in Fig. 1.117. The equations of motion are

$$I\dot{\omega} = -fR, \quad m\dot{v} = f.$$

At time t_1 when the flywheel stops slipping, let its angular velocity be ω_1. The boundary conditions are $\omega = \omega_0, v = 0$ at $t = 0, \omega = \omega_1, v = v_1 = R\omega_1$ at $t = t_1$. The above equations integrate to give

$$I(\omega_1 - \omega_0) = -fRt_1,$$
$$mv = mR\omega_1 = ft_1.$$

Note that these equations can also be obtained directly by an impulse consideration. Solving these we have

$$\omega_1 = \frac{\omega_0}{3}, \quad t_1 = \frac{\omega_0 R}{3\mu g},$$

as $I = \frac{1}{2}mR^2, f = \mu mg$. The distance covered by the flywheel before it stops slipping is

$$S = \frac{1}{2}\left(\frac{f}{m}\right)t_1^2 = \frac{1}{2}\mu g\left(\frac{\omega_0 R}{3\mu g}\right)^2 = \frac{(\omega_0 R)^2}{18\mu g} = 18290 \text{ m}.$$

c. At t_1 the speed of the center of mass is

$$v_1 = R\omega_1 = \frac{\omega_0 R}{3} = 189.3 \text{ ms}^{-1}.$$

d. At time $0 < t < t_1$, the equation of motion integrates to

$$I(\omega - \omega_0) = -fRt,$$
$$mv = ft.$$

At $0 < t < t_1$, the flywheel both slips and rolls. Only the slipping part of the motion causes dissipation of energy into heat. The slipping velocity is

$$v - R\omega = \frac{3ft}{m} - R\omega_0$$

and the total dissipation of energy into heat is

$$Q = -\int_0^{t_1} (v - R\omega)f dt$$

$$= -\frac{3f^2 t_1^2}{2m} + R\omega_0 ft_1$$

$$= \frac{mR^2\omega_0^2}{6} = 2.688 \times 10^7 \text{ J}.$$

This can also be obtained by considering the change in the kinetic energy of the flywheel:

$$Q = \frac{1}{2}I\omega_0^2 - \left(\frac{1}{2}I\omega_1^2 + \frac{1}{2}mv_1^2\right)$$
$$= \frac{mR^2}{4} \cdot \frac{8\omega_0^2}{9} - \frac{m(R\omega_0)^2}{2} \cdot \frac{1}{9} = \frac{mR^2\omega_0^2}{6},$$

same as the above.

1161

A man wishes to break a long rod by hitting it on a rock. The end of the rod which is in his hand rotates without displacement as shown in Fig. 1.118. The man wishes to avoid having a large force act on his hand at the time the rod hits the rock. Which point on the rod should hit the rock? (Ignore gravity).

(CUSPEA)

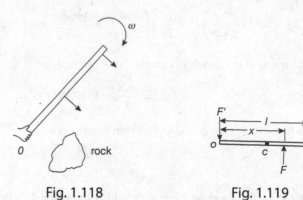

Fig. 1.118   Fig. 1.119

Sol: Let the point of impact be at distance x from the end O held by the hand and the reaction to the force acting on the hand as a result of the application of F be F', as shown in Fig. 1.119. Considering the motion of the center of mass C, we have

$$\int (F - F')\,dt = mv,$$

$$\int \left[F\left(x - \frac{l}{2}\right) - F'\frac{l}{2}\right]dt = I\omega,$$

where v is the velocity of C, ω the angular velocity about C immediately after the application of F, and $I = \dfrac{ml^2}{12}$, m being the mass of the rod. As O is to remain stationary, we require

$$v - \frac{\omega l}{2} = 0,$$

or

$$v = \frac{\omega l}{2}.$$

We also require $F' \approx 0$, so that

$$\int F dt = mv, \qquad \left(x - \frac{l}{2}\right)\int F dt = I\omega,$$

which give

$$x = \frac{l}{6} + \frac{l}{2} = \frac{2l}{3}.$$

1162

The two flywheels in Fig. 1.120 are on parallel frictionless shafts but initially do not touch. The larger wheel has $f = 2000$ rev/min while the smaller is at rest. If the two parallel shafts are moved until contact occurs, find the angular velocity of the second wheel after equilibrium occurs (i.e. no further sliding at the point of contact), given that $R_1 = 2R_2$, $I_1 = 16I_2$.

(Wisconsin)

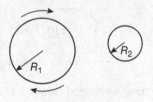

Fig. 1.120

Sol: Suppose the impulse of the interacting force between the two wheels from contact to equilibrium is J. Then the torque of the impulse acting on the larger wheel is $J R_1$ and that on the smaller wheel is $J R_2$.

We have $I_1(\omega_1 - \omega_1') = JR_1$, $I_2\omega_2' = JR_2$, where ω_1 and ω_1' are the angular veloci ties of the larger wheel before contact and after equilibrium respectively, and ω_2' i the angular velocity of the smaller wheel after equilibrium. As there is no slidin between the wheels when equilibrium is reached,

$$\omega_1'R_1 = \omega_2'R_2.$$

The above equations give

$$\omega_2' = \frac{I_1 R_1 R_2 \omega_1'}{I_1 R_2^2 + I_2 R_1^2} = 1.6\omega_1 = 3200 \text{ rev/ min} .$$

1163

Two uniform cylinders are spinning independently about their axes, which are parallel. One has radius R_1 and mass M_1, the other R_2 and M_2. Initially they rotate in the same sense with angular speeds Ω_1 and Ω_2 respectively as shown in Fig. 1.121. They are then displaced until they touch along a common tangent. After a steady state is reached, what is the final angular velocity of each cylinder?

(CUSPEA)

Sol: Let ω_1, ω_2 be the final angular velocities of the two cylinders respectively after steady state is reached. Then

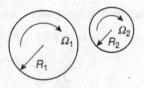

Fig. 1.121

$$\omega_1 R_1 = -\omega_2 R_2.$$

Let J_1 and J_2 be the time-integrated torque 2 exerts on 1 and 1 on 2, then

$$\frac{J_1}{R_1} = \frac{J_2}{R_2},$$

$$J_1 = I_1(\omega_1 - \Omega_1), \qquad J_2 = I_2(\omega_2 - \Omega_2);$$

or

$$\frac{I_1(\omega_1 - \Omega_1)}{R_1} = \frac{I_2(\omega_2 - \Omega_2)}{R_2}.$$

As $I \propto MR^2$, the last equation becomes

$$M_1 R_1(\omega_1 - \Omega_1) = M_2 R_2(\omega_2 - \Omega_2),$$

i.e.

$$M_1 R_1 \omega_1 - M_2 R_2 \omega_2 = M_1 R_1 \Omega_1 - M_2 R_2 \Omega_2.$$

Hence

$$\omega_1 = \frac{M_1 R_1 \Omega_1 - M_2 R_2 \Omega_2}{R_1(M_1 + M_2)},$$

$$\omega_2 = \frac{M_2 R_2 \Omega_2 - M_1 R_1 \Omega_1}{R_2(M_1 + M_2)}.$$

1164

Three identical cylinders rotate with the same angular velocity Ω about parallel central axes. They are brought together until they touch, keeping the axes parallel. A new steady state is achieved when, at each contact line, a cylinder does not slip with respect to its neighbor as shown in Fig. 1.122. How much of the original spin kinetic energy is now left?
(The precise order in which the first and second touch, and the second and third touch, is irrelevant.)

(CUSPEA)

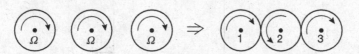

Fig. 1.122

Sol: As there is no slipping, if Ω' is the final angular velocity of cylinder 1, then cylinders 2 and 3 have final angular velocities $-\Omega'$ and Ω' respectively. Let I be the moment of inertia of each cylinder about its axis of rotation, M_{ij} be the angular impulse imparted to the ith cylinder with respect to its axis of rotation by the jth cylinder. Newton's third law requires that, as the cylinders have the same radius,

$$M_{ij} = M_{ji}. \quad (i, j = 1, 2, 3; \quad i \neq j)$$

Dynamical considerations give

$$I(\Omega' - \Omega) = M_{12}, \tag{1}$$

$$I(-\Omega' - \Omega) = M_{21} + M_{23}, \tag{2}$$

$$I(\Omega' - \Omega) = M_{32}. \tag{3}$$

(1) + (3) − (2) gives

$$I(3\Omega' - \Omega) = 0,$$

or

$$\Omega' = \frac{\Omega}{3}.$$

The ratio of the spin kinetic energies after and before touching is

$$\frac{T'}{T} = \frac{\dfrac{1}{2}(3I\Omega'^2)}{\dfrac{1}{2}(3I\Omega^2)} = \left(\frac{\Omega'}{\Omega}\right)^2 = \frac{1}{9}.$$

1165

Find the ratio of the periods of the two torsion pendula shown in Fig. 1.123. The two differ only by the addition of cylindrical masses as shown in the figure. The radius of each additional mass is 1/4 the radius of the disc. Each cylinder and disc have equal mass.

(Wisconsin)

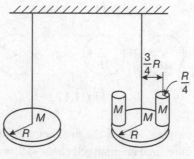

Fig. 1.123

Sol: Let I_1 and I_2 be the moments of inertia of the two torsion pendula respectively. If A is the restoring coefficient of each wire, then the equations of motion are $I_1\ddot{\theta} + A\theta = 0$, $I_2\ddot{\theta} + A\theta = 0$. Hence the angular frequencies of oscillation of the torsion pendula are $\omega_1 = \sqrt{A/I_1}$ and $\omega_2 = \sqrt{A/I_2}$. For the first pendulum, $I_1 = MR^2/2$, and for the second,

$$I_2 = \frac{MR^2}{2} + 2\left[\frac{M}{2}\left(\frac{R}{4}\right)^2 + M\left(\frac{3R}{4}\right)^2\right] = \frac{27}{16}MR^2.$$

Hence the ratio of the periods is

$$\frac{T_1}{T_2} = \frac{\omega_2}{\omega_1} = \sqrt{\frac{I_1}{I_2}} = \left(\frac{2}{3}\right)^{\frac{3}{2}}.$$

1166

A long thin uniform bar of mass M and length L is hung from a fixed (assumed frictionless) axis at A as shown in Fig. 1.124. The moment of inertia about A is $ML^2/3$.

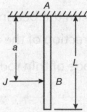

Fig. 1.124

a. An instantaneous horizontal impulse J is delivered at B, a distance a below A. What is the initial angular velocity of the bar?

b. In general, as a result of J, there will be an impulse J' on the bar from the axis at A. What is J'?

c. Where should the impulse J be delivered in order that J' be zero?

(Wisconsin)

Sol:

a. $Ja = I(\omega - \omega_0)$, where ω_0 is the angular velocity of the bar before the impulse is delivered. As $\omega_0 = 0$, the initial angular velocity is

$$\omega = \frac{Ja}{I} = \frac{3Ja}{ML^2}.$$

b. The initial velocity of the center of mass of the bar is $v = \omega L/2$. So the change in the momentum of the bar is $Mv = M\omega L/2$. As this is equal to the total impulse on the bar, we have

$$J + J' = \frac{M\omega L}{2}.$$

Hence

$$J' = \frac{M\omega L}{2} - J = J\left(\frac{3a}{2L} - 1\right).$$

c.
$$J' = 0, \quad \text{if} \quad a = \frac{2L}{3}.$$

Hence there will be no impulsive from the axis if J is delivered at a point $2L/3$ below A.

1167

The crankshaft shown in Fig. 1.125 rotates with constant angular velocity ω Calculate the resultant forces on the bearings. In a sketch show the directions of these reactions and the direction of the angular momentum.

(Assume the crankshaft is made of thin rods with uniform density).

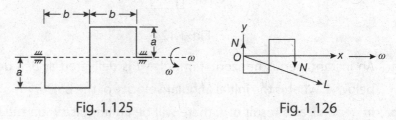

Fig. 1.125   Fig. 1.126

Sol: Consider the motion in a frame attached to the crankshaft as shown in Fig. 1.126. As the rods are either parallel or perpendicular to the axis of rotation, the centrifugal force on each rod can be considered as that on a point of the same mass located at its center of mass. Let N denote the constraint force exerted by the bearing on each shaft. As there is no rotation about the z-axis, we require that the moments of the forces about O should balance:

$$2b \cdot N + \frac{b}{2} \cdot \rho b \cdot a\omega^2 = \frac{3b}{2} \cdot \rho b \cdot a\omega^2 + 2b \cdot \rho a \cdot \frac{a}{2}\omega^2,$$

giving

$$N = \frac{\rho a \omega^2}{2}(a + b),$$

where ρ is the mass per unit length of the rods. The reactions on the bearings are equal and opposite to N as shown in Fig. 1.125. In a fixed frame these forces are rotating, together with the crankshaft, with angular velocity ω about the axle. The angular momentum of the crankshaft is given by

$$\begin{pmatrix} L_x \\ L_y \\ L_z \end{pmatrix} = I \begin{pmatrix} \omega \\ 0 \\ 0 \end{pmatrix} = \begin{pmatrix} I_{xx}\omega \\ I_{yx}\omega \\ I_{zx}\omega \end{pmatrix},$$

where I is the moment of inertia tensor about O with elements

$$I_{ij} = \sum_n \Delta m_n (r^2 \delta_{ij} - x_i x_j).$$

As all $z = 0$, $I_{zx} = 0$. Furthermore it can be seen that $I_{xx} > 0$, $I_{yx} < 0$. Hence the angular momentum **L** has direction in the rotating coordinate frame as shown in Fig. 1.126. Note that gravity has been neglected in the calculation, otherwise there is an additional constant force acting on each bearing, $(2a + b)\rho g$ in magnitude and vertically downward in direction in the fixed frame.

1168

Two equal point masses M are connected by a massless rigid rod of length $2A$ (a dumbbell) which is constrained to rotate about an axle fixed to the center of the rod at an angle θ (Fig. 1.127). The center of the rod is at the origin of coordinates, the axle along the z-axis, and the dumbbell lies in the xz-plane at $t = 0$. The angular velocity ω is a constant in time and is directed along the z-axis.

a. Calculate all elements of the inertia tensor. (Be sure to specify the coordinate system you use.)

b. Using the elements just calculated, find the angular momentum of the dumbbell in the laboratory frame as a function of time.

c. Using the equation **L** = **r** × **p**, calculate the angular momentum and show that it is equal to the answer for part (b).

d. Calculate the torque on the axle as a function of time.

e. Calculate the kinetic energy of the dumbbell.

(UC, Berkeley)

Sol:

a. Use a coordinate frame *xyz* attached to the dumbbell such that the two point masses are in the *xz*-plane. The elements of the inertia tensor about O, given by $I_{ij} = \sum_n m_n(r^2\delta_{ij} - x_i x_j)$, are

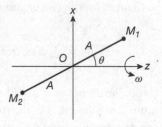

Fig. 1.127

$$I_{xx} = 2MA^2\cos^2\theta, \qquad I_{yy} = 2MA^2, \qquad I_{zz} = 2MA^2\sin^2\theta,$$
$$I_{xy} = I_{yz} = 0, \qquad I_{zx} = -2MA^2\cos\theta \qquad \sin\theta = -MA^2\sin^2\theta.$$

Thus

$$I = \begin{pmatrix} 2MA^2\cos^2\theta & 0 & -MA^2\sin 2\theta \\ 0 & 2MA^2 & 0 \\ -MA^2\sin 2\theta & 0 & 2MA^2\sin^2\theta \end{pmatrix}.$$

b. Use a laboratory frame $x'y'z'$ such that the z'-axis coincides with the z-axis of the rotating frame in (a) and that all the respective axes of the two frames coincide at $t = 0$. The unit vectors along the axes of the two frames are related by

$$\begin{pmatrix} e_1 \\ e_2 \\ e_3 \end{pmatrix} = \begin{pmatrix} \cos\omega t & \sin\omega t & 0 \\ -\sin\omega t & \cos\omega t & 0 \\ 0 & 0 & 1 \end{pmatrix} \begin{pmatrix} e_1' \\ e_2' \\ e_3' \end{pmatrix}.$$

Then the inertia tensor in the laboratory frame is

$$I' = S'IS \begin{pmatrix} \cos \omega t & -\sin \omega t & 0 \\ \sin \omega t & \cos \omega t & 0 \\ 0 & 0 & 1 \end{pmatrix}$$

$$\times \begin{pmatrix} 2MA^2 \cos^2 \theta & 0 & -MA^2 \sin 2\theta \\ 0 & 2MA^2 & 0 \\ -MA^2 \sin 2\theta & 0 & 2MA^2 \sin^2 \theta \end{pmatrix} \begin{pmatrix} \cos \omega t & \sin \omega t & 0 \\ -\sin \omega t & os \omega t & 0 \\ 0 & 0 & 1 \end{pmatrix}.$$

Hence the angular momentum of the dumbbell in the laboratory frame is

$$\mathbf{L} = I' \begin{pmatrix} 0 \\ 0 \\ \omega \end{pmatrix} = MA^2\omega \begin{pmatrix} -\sin 2\theta \cos \omega t \\ -\sin 2\theta \sin \omega t \\ 2 \sin^2 \theta \end{pmatrix}.$$

c. The radius vectors of M_1 and M_2 from O are respectively

$$\mathbf{r}_1 = A(\sin \theta, 0, \cos \theta),$$
$$\mathbf{r}_2 = A(-\sin \theta, 0, -\cos \theta)$$

in the rotating frame. Using the transformation for the unit vectors we have

$$\mathbf{r}_1 = A[\sin \theta(\mathbf{e}_1' \cos \omega t + \mathbf{e}_2' \sin \omega t) + \mathbf{e}_3' \cos \theta]$$
$$= A(\sin \theta \cos \omega t, \sin \theta \sin \omega t, \cos \theta),$$
$$\mathbf{r}_2 = A(-\sin \theta \cos \omega t, -\sin \theta \sin \omega t, -\cos \theta)$$

in the laboratory frame. The angular momentum of the system in the laboratory frame is

$$\mathbf{L} = \sum \mathbf{r} \times \mathbf{p} = \sum M\mathbf{r} \times (\omega \times \mathbf{r}) = \sum M[r^2\omega - (\mathbf{r} \cdot \omega)\mathbf{r}]$$
$$= 2MA^2\omega\mathbf{e}_3'$$
$$\quad - MA^2\omega \cos \theta(\mathbf{e}_1' \sin \theta \cos \omega t + \mathbf{e}_2' \sin \theta \sin \omega t + \mathbf{e}_3 \cos \theta)$$
$$\quad + MA^2\omega \cos \theta(-\mathbf{e}_1' \sin \theta \cos \omega t - \mathbf{e}_2' \sin \theta \sin \omega t - \mathbf{e}_3' \cos \theta)$$
$$= MA^2\omega(-\mathbf{e}_1' \sin 2\theta \cos \omega t - \mathbf{e}_2' \sin 2\theta \sin \omega t + \mathbf{e}_3' 2 \sin^2 \theta),$$

same as that obtained in (b).

d. The torque on the axle is

$$\tau = \frac{d\mathbf{L}}{dt} = MA^2\omega^2[\sin 2\theta \sin \omega t\mathbf{e}_1' - \sin 2\theta \cos \omega t\mathbf{e}_2'].$$

e. As $\omega = (0, 0, \omega)$ the rotational kinetic energy of the dumbbell is

$$T = \frac{I_{zz}\omega^2}{2} = MA^2\omega^2 \sin^2 \theta.$$

1169

A squirrel of mass m runs at a constant speed V_0 relative to a cylindrical exer cise cage of radius R and moment of inertia I as shown in Fig. 1.128. The cage has a damping torque proportional to its angular velocity. Neglect the dimen sions of the squirrel compared with R. If initially the cage is at rest and the squirrel is at the bottom and running, find the motion of the squirrel relative to a fixed coordinate system in the small oscillation, underdamped case. Find the squirrel's angular velocity in terms of its angle relative to the vertical for arbitrary angular displacements for the undamped case. Discuss any design criteria for the cage in this case.

(*Wisconsin*)

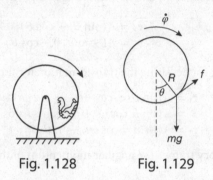

Fig. 1.128   Fig. 1.129

Sol: In a fixed coordinate frame, define θ as shown in Fig. 1.129. For the squirrel, the equation of motion is

$$mR\ddot{\theta} = f - mg \sin\theta,$$

and for the cage the equation of motion is

$$I\ddot{\varphi} = -fR - k\dot{\varphi},$$

where f is the friction between the squirrel and the cage and k is a constant. In addition, as the squirrel has a constant speed V_0 relative to the cage, we have

$$R(\dot{\theta} - \dot{\varphi}) = V_0,$$

which means $\ddot{\varphi} = \ddot{\theta}$, $\dot{\varphi} = \dot{\theta} - \dfrac{V_0}{R}$. Making use of these and eliminating f from the equations of motion give

$$(I + mR^2)\ddot{\theta} + k\dot{\theta} + mgR \sin\theta = \frac{kV_0}{R}.$$

For small oscillations, $\theta \ll 1$ and the above reduces to

$$(I + mR^2)\,\ddot{\theta} + k\dot{\theta} + mgR\theta = \frac{kV_0}{R}.$$

A particular solution of this equation is

$$\theta = \frac{kV_0}{mgR^2},$$

while for underdamping the general solution for the homogeneous equation is

$$\theta = e^{-bt}(A\,\sin \omega t + B\,\cos \omega t),$$

where

$$b = \frac{k}{2(I + mR^2)}, \qquad \omega = \sqrt{\frac{mgR}{I + mR^2} - b^2}.$$

Hence the general solution of the above equation is

$$\theta = \frac{kV_0}{mgR^2} + e^{-bt}(A\,\sin \omega t + B\,\cos \omega t).$$

Using the initial condition that at $t = 0$, $\theta = 0$, $\varphi = 0$, $\dot{\varphi} = 0$, $\dot{\theta} = \dfrac{V_0}{R}$, we find

$$\theta = \frac{kV_0}{mgR^2} - \frac{kV_0}{mgR^2}\left[\cos \omega t + \left(\frac{b}{\omega} - \frac{mgR}{\omega K}\right)\sin \omega t\right]e^{-bt}$$

For the undamped case ($k = 0$), the differential equation is

$$(I + mR^2)\,\ddot{\theta} + mgR\theta = 0$$

or, as $\ddot{\theta} = \dfrac{1}{2}\dfrac{d\dot{\theta}^2}{d\theta}$,

$$(I + mR^2)d\dot{\theta}^2 = -2mgR\theta\,d\theta,$$

which integrates to

$$\dot{\theta}^2 = \left(\frac{V_0}{R}\right)^2 - \frac{mgR}{I + mR^2}\theta^2$$

using the initial condition for $\dot{\theta}$. Hence

$$\dot{\theta}^2 = \sqrt{\left(\frac{V_0}{R}\right)^2 - \frac{mgR}{I + mR^2}\theta^2}.$$

We require $I + mR^2 \gg k$ for the undamped case to hold. Hence the cage should be designed with a large moment of inertia.

1170

A thin square plate with side length a rotates at a constant angular frequency ω about an axis through the center tilted by an angle θ with respect to the normal to the plate.

a. Find the principal moments of inertia.

b. Find the angular momentum **J** in the laboratory system.

c. Calculate the torque on the axis.

(UC, Berkeley)

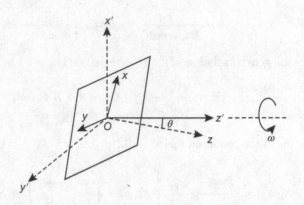

Fig. 1.130

Sol:

a. Take origin at the center O of the square. For a coordinate frame attached to the square, take the plane of the square as the xy-plane with the x- and y-axis parallel to the sides. The z-axis, which is along the normal, makes an angle θ with the z'-axis of the laboratory frame about which the square rotates, as shown in Fig. 1.130. We also assume that the x-, z- and z'-axes are coplanar.

Then by symmetry the x-, y- and z-axes are the principal axes of inertia about O, with corresponding moments of inertia

$$I_{xx} = I_{yy} = \frac{ma^2}{12}, \quad I_{zz} = \frac{ma^2}{6},$$

where m is the mass of the square.

b. The angular momentum **J** resolved along the rotating frame coordinate axes is

$$
\begin{pmatrix} J_x \\ J_y \\ J_z \end{pmatrix} = \begin{pmatrix} \dfrac{ma^2}{12} & 0 & 0 \\ 0 & \dfrac{ma^2}{12} & 0 \\ 0 & 0 & \dfrac{ma^2}{6} \end{pmatrix} \begin{pmatrix} \omega \sin\theta \\ 0 \\ \omega \cos\theta \end{pmatrix} = \begin{pmatrix} \dfrac{ma^2}{12}\omega \sin\theta \\ 0 \\ \dfrac{ma^2}{6}\omega \cos\theta \end{pmatrix}.
$$

We can choose the laboratory frame so that its y'-axis coincides with the y-axis at $t = 0$. Then the unit vectors of the two frames are related by

$$
\begin{cases} \mathbf{e}_x = \cos\theta \cos\omega t\, \mathbf{e}_{x'} + \cos\theta \sin\omega t\, \mathbf{e}_{y'} + \sin\theta\, \mathbf{e}_{z'}, \\ \mathbf{e}_y = -\sin\omega t\, \mathbf{e}_{x'} + \cos\omega t\, \mathbf{e}_{y'}, \\ \mathbf{e}_z = -\sin\theta \cos\omega t\, \mathbf{e}_{x'} - \sin\theta \sin\omega t\, \mathbf{e}_{y'} + \cos\theta\, \mathbf{e}_{z'}. \end{cases}
$$

Hence the angular momentum resolved along the laboratory frame coordinate axes is

$$
\begin{pmatrix} J_{x'} \\ J_{y'} \\ J_{z'} \end{pmatrix} = \begin{pmatrix} \cos\theta \cos\omega t & -\sin\omega t & -\sin\theta \cos\omega t \\ \cos\theta \sin\omega t & \cos\omega t & -\sin\theta \sin\omega t \\ \sin\theta & 0 & \cos\theta \end{pmatrix} \begin{pmatrix} \dfrac{ma^2}{12}\omega \sin\theta \\ 0 \\ \dfrac{ma^2}{6}\omega \cos\theta \end{pmatrix}
$$

$$
= \begin{pmatrix} -\dfrac{ma^2}{12}\omega \sin\theta \cos\theta \cos\omega t \\ -\dfrac{ma^2}{12}\omega \sin\theta \cos\theta \sin\omega t \\ \dfrac{ma^2}{12}\omega(1 + \cos^2\theta) \end{pmatrix}.
$$

c. The torque on the axis is given by

$$
\mathbf{M} = \left(\frac{d\mathbf{J}}{dt}\right)_{\text{lab}} = \left(\frac{d\mathbf{J}}{dt}\right)_{\text{rot}} + \boldsymbol{\omega} \times \mathbf{J} = \boldsymbol{\omega} \times \mathbf{J}
$$

$$
= \begin{vmatrix} \mathbf{e}_x & \mathbf{e}_y & \mathbf{e}_z \\ \omega \sin\theta & 0 & \omega \cos\theta \\ \dfrac{ma^2}{12}\omega \sin\theta & 0 & \dfrac{ma^2}{6}\omega \cos\theta \end{vmatrix}
$$

$$
= -\frac{ma^2}{12}\omega^2 \sin\theta \cos\theta\, \mathbf{e}_y.
$$

The torque can be expressed in terms of components in the laboratory frame:

$$\mathbf{M} = -\frac{ma^2}{12}\omega^2 \sin\theta \cos\theta(-\sin\omega t \mathbf{e}_{x'} + \cos\omega t \mathbf{e}_{y'}).$$

This can also be obtained by differentiating $\mathbf{L}$ in the laboratory frame:

$$\mathbf{M} = \left(\frac{d\mathbf{J}}{dt}\right)_{\text{lab}}.$$

1171

A thin flat rectangular plate, of mass M and sides a by $2a$, rotates with constant angular velocity ω about an axle through two diagonal corners, as shown in Fig. 1.131. The axle is supported at the corners of the plate by bearings which can exert forces only on the axle. Ignoring gravitational and frictional forces find the force exerted by each bearing on the axle as a function of time.

(Princeton)

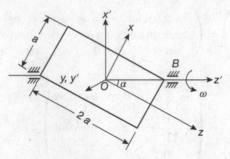

Fig. 1.131

Sol: Use a coordinate frame attached to the plate with the origin at the center of mass O, the y-axis along the normal, and the z-axis parallel to the long side of the rectangle, as shown in Fig. 1.131. Then the x-, y- and z-axes are the principal axes with principal moments of inertia

$$I_{xx} = \frac{Ma^2}{12}, \quad I_{yy} = \frac{5Ma^2}{12}, \quad I_{zz} = \frac{4Ma^2}{12}.$$

Let z' denote the axis of rotation and α the angle between the z- and z'-axes. The angular momentum of the plate is

$$\mathbf{L} = \frac{Ma^2}{12}\begin{pmatrix} 1 & 0 & 0 \\ 0 & 5 & 0 \\ 0 & 0 & 4 \end{pmatrix}\begin{pmatrix} \omega\sin\alpha \\ 0 \\ \omega\cos\alpha \end{pmatrix} = \frac{Ma^2\omega}{12}\begin{pmatrix} \sin\alpha \\ 0 \\ 4\cos\alpha \end{pmatrix}.$$

The torque on the axle of the plate is then

$$\tau = \left(\frac{d\mathbf{L}}{dt}\right)_{\text{fixed}} = \left(\frac{d\mathbf{L}}{dt}\right)_{\text{rot}} + \omega \times \mathbf{L} = \omega \times \mathbf{L}$$

$$= \frac{Ma^2\omega^2}{12} \begin{vmatrix} \mathbf{e}_x & \mathbf{e}_y & \mathbf{e}_z \\ \sin\alpha & 0 & \cos\alpha \\ \sin\alpha & 0 & 4\cos\alpha \end{vmatrix}$$

$$= \frac{Ma^2\omega^2}{4}\cos\alpha\sin\alpha\,\mathbf{e}_y = -\frac{Ma^2\omega^2}{10}\mathbf{e}_y,$$

as $\sin\alpha = \dfrac{1}{\sqrt{5}}$, $\cos\alpha = \dfrac{2}{\sqrt{5}}$. Let N_A, N_B be the constraint forces exerted by the bearings on the axle at A, B respectively. Rotate the coordinate frame $Oxyz$ about the y-axes so that the z and z'-axis coincide. The new coordinate axes are the x'-axis, y'-axis which is identical with the y-axis, and z'-axis shown in Fig. 1.131. As the center of mass is stationary, we have

$$N_{Ax'} + N_{Bx'} = 0, \qquad N_{Ay'} + N_{By'} = 0.$$

Considering the torque about O we have

$$N_{Bx'}d - N_{Ax'}d = -\frac{Ma^2\omega^2}{10}, \qquad N_{Ay'}d - N_{By'}d = 0,$$

where $d = \dfrac{1}{2}\overline{AB} = \dfrac{\sqrt{5}}{2}a$. The above equations give

$$N_{Ay'} = N_{By'} = 0$$

$$N_{Ax'} = \frac{Ma^2\omega^2}{20d} = \frac{Ma\omega^2}{10\sqrt{5}},$$

$$N_{Bx'} = -\frac{Ma^2\omega^2}{20d} = -\frac{Ma\omega^2}{10\sqrt{5}}.$$

These forces are fixed in the rotating frame. In a fixed coordinate frame they rotate with angular velocity ω. In a fixed frame $Ox''y''z''$ with the same z'-axis and the x''-axis coinciding with the x'-axis at $t = 0$,

$$N_{Ax''} = \frac{Ma\omega^2}{10\sqrt{5}}\cos\omega t, \qquad N_{Ay''} = -\frac{Ma\omega^2}{10\sqrt{5}}\sin\omega t,$$

$$N_{Bx''} = -\frac{Ma\omega^2}{10\sqrt{5}}\cos\omega t, \qquad N_{By''} = \frac{Ma\omega^2}{10\sqrt{5}}\sin\omega t.$$

1172

A homogeneous thin rod of mass M and length b is attached by means of an inextensible cord to a spring whose spring constant is k. The cord passes over a very small and smooth pulley fixed at P. The rod is free to rotate about A without friction throughout the angular range $-\pi < \theta \le \pi$ (Fig. 1.132). When $c = 0$ the spring has its natural length. It is assumed that $b < a$ and that gravity acts downward.

   a. Find the values of θ for which the system is in static equilibrium, and determine in each case if the equilibrium is stable, unstable or neutral.

   b. Find the frequencies for small oscillations about the points of stable equilibrium.

(Note: line PA is parallel to $\mathbf{g}$).

(SUNY, Buffalo)

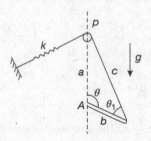

Fig. 1.132

Sol:

   a. Take the direction pointing out of the paper as the positive direction of the torques. The torque about point A due to gravity is

$$L_g = -\frac{Mgb}{2} \sin \theta,$$

and that caused by the restoring force due to the spring is $L_k = kc\, b \sin \theta_1$, where θ_1 is the angle formed by the rod with the rope, or, using the sine theorem

$$\frac{c}{\sin \theta} = \frac{a}{\sin \theta_1},$$
$$L_k = kba \sin \theta.$$

For equilibrium, we require $L_g + L_k = 0$, or $ka \sin \theta = \dfrac{Mg}{2} \sin \theta$.

i.   If $ka = Mg/2$, the equilibrium condition is satisfied for all θ and the equilibrium is neutral.

ii. If $ka < Mg/2$, the equilibrium condition is satisfied if $\theta = 0$ or $\theta = \pi$. Consider the equilibrium at $\theta = 0$. Let $\theta = 0 \pm \varepsilon$, where $\varepsilon > 0$ is a small angle. Then

$$L = L_k + L_g \approx \mp \left(\frac{Mg}{2} - ka \right) \varepsilon.$$

Thus
$$L < 0 \text{ for } \theta = +\varepsilon,$$
$$L > 0 \text{ for } \theta = -\varepsilon.$$

Hence L tends to increase ε in both cases and the equilibrium is unstable. For the equilibrium at $\theta = \pi \pm \varepsilon$, we have

$$L = \pm \left(\frac{Mg}{2} - ka \right) \varepsilon.$$

Then
$$L < 0 \text{ for } \theta = \pi - \varepsilon,$$
$$L > 0 \text{ for } \theta = \pi + \varepsilon.$$

In the case L tends to reduce ε and the equilibrium is stable.

iii. If $ka > Mg/2$, the situation is opposite to that of (ii). Hence in this case $\theta = 0$ is a position of stable equilibrium and $\theta = \pi$ is a position of unstable equilibrium.

b.   Take the case of $ka > Mg/2$ where $\theta = 0$ is a position of stable equilibrium. Let $\theta = \varepsilon$ where ε is a small angle. The equation of motion is

$$b \left(ka - \frac{Mg}{2} \right) \sin \varepsilon = -\frac{Mb^2}{3} \ddot{\varepsilon},$$

or for small oscillations

$$b(2ka - Mg)\varepsilon + \frac{2Mb^2}{3} \ddot{\varepsilon} = 0.$$

Hence the frequency of oscillation is

$$f = \frac{1}{2\pi} \sqrt{\frac{3(2ka - Mg)}{2Mb}}.$$

Similarly in the case $ka < Mg/2$, the frequency of small oscillations about th
position of stable equilibrium at $\theta = \pi$ is

$$f = \frac{1}{2\pi} \sqrt{\frac{3(Mg - 2ka)}{2Mb}}.$$

1173

A thin ring of mass M and radius R is pivoted at P on a frictionless table, a
shown in Fig. 1.133. A bug of mass m runs along the ring with speed v with
respect to the ring. The bug starts from the pivot with the ring at rest. Hov
fast is the bug moving with respect to the table when it reaches the diametri-
cally opposite point on the ring (point X)?

(MIT)

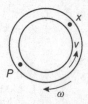

Fig. 1.133

Sol: The moment of inertia of the ring with respect to the pivot P is

$$I = MR^2 + MR^2 = 2MR^2.$$

When the bug reaches point X, its velocity with respect to the table is $v - 2R\omega$
and the angular momentum of the ring about P is

$$J = 2MR^2\omega,$$

where ω is the angular velocity of the ring about P at that instant. Initially the total
angular momentum of the ring and bug about P is zero. Conservation of angular
momentum then gives

$$2MR^2\omega - 2mR(v - 2R\omega) = 0,$$

or

$$\omega = \frac{mv}{R(M + 2m)}.$$

The velocity of the bug at point X with respect to the table is

$$v - 2R\omega = \frac{Mv}{M + 2m}.$$

1174

A cone of height h and base radius R is constrained to rotate about its vertical axis, as shown in Fig. 1.134. A thin, straight groove is cut in the surface of the cone from apex to base as shown. The cone is set rotating with initial angular velocity ω_0 around its axis and a small (point-like) bead of mass m is released at the top of the frictionless groove and is permitted to slide down under gravity. Assume that the bead stays in the groove, and that the moment of inertia of the cone about its axis is I_0.

a. What is the angular velocity of the cone when the bead reaches the bottom?

b. Find the speed of the bead in the laboratory just as it leaves the cone.

(*MIT*)

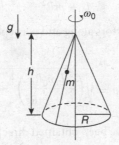

Fig. 1.134

Sol:

a. As the total angular momentum of the system is conserved, the angular velocity ω of the cone at the time when the bead reaches the bottom satisfies the relation

$$I_0\omega_0 = (I_0 + mR^2)\omega.$$

Hence

$$\omega = \frac{I_0\omega_0}{I_0 + mR^2}.$$

b. As the energy of the system is conserved, the velocity v of the bead when it reaches the bottom satisfies

$$\frac{1}{2}I_0\omega_0^2 + mgh = \frac{1}{2}mv^2 + \frac{1}{2}I_0\omega^2.$$

with

$$v^2 = v_\parallel^2 + v_\perp^2 = v_\parallel^2 + R^2\omega^2,$$

where $v_\parallel$ is velocity of the bead parallel to the groove and $v_\perp$ is that perpendicular to the groove. Thus

$$\frac{1}{2}mv_\parallel^2 = \frac{1}{2}I_0\omega_0^2 + mgh - \frac{1}{2}I_0\omega^2 - \frac{1}{2}mR^2\omega^2,$$

giving

$$v_\parallel^2 = I_0\omega_0^2 - \frac{(I_0 + mR^2)}{m}\frac{I_0^2\omega^2}{(I_0 + mR^2)^2} + 2gh = \frac{I_0\omega_0^2 R^2}{I_0 + mR^2} + 2gh.$$

Hence the velocity of the bead when it reaches the bottom is

$$\mathbf{v} = v_\perp \mathbf{i} + v_\parallel \mathbf{j}$$

$$= \frac{I_0\omega_0 R}{I_0 + mR^2}\mathbf{i} + \sqrt{\frac{I_0\omega_0^2 R^2}{I_0 + mR^2} + 2gh}\,\mathbf{j},$$

i and j being unit vectors along and perpendicular to the groove respectively, with magnitude

$$v = \sqrt{\left(\frac{I_0\omega_0 R}{I_0 + mR^2}\right)^2 + \frac{I_0\omega_0^2 R^2}{I_0 + mR^2} + 2gh}.$$

This speed could have been obtained directly by substituting the expression for ω in the energy equation.

1175

A thin uniform disc, radius a and mass m, is rotating freely on a frictionless bearing with uniform angular velocity ω about a fixed vertical axis passing through its center, and inclined at angle α to the symmetry axis of the disc. What is the magnitude and direction of the torque, and of the net force acting between the disc and the axis?

(Columbia)

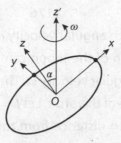

Fig. 1.135

Sol: Take a coordinate frame $Oxyz$ attached to the disc with the origin at its center O, the z-axis along the normal to the disc, and the x-axis in the plane of the z-axis and the axis of rotation z', as shown in Fig. 1.135. The x-, y- and z-axes are the principal axes of the disc with principal moments of inertia

$$I_x = \frac{1}{4}ma^2, \quad I_y = \frac{1}{4}ma^2, \quad I_z = \frac{1}{2}ma^2.$$

The angular momentum about O is

$$\mathbf{L} = \begin{pmatrix} \frac{1}{4}ma^2 & 0 & 0 \\ 0 & \frac{1}{4}ma^2 & 0 \\ 0 & 0 & \frac{1}{2}ma^2 \end{pmatrix} \begin{pmatrix} \omega \sin \alpha \\ 0 \\ \omega \cos \alpha \end{pmatrix}$$

$$= \frac{1}{4}mu^2 (\omega \sin \alpha \mathbf{e}_x + 2\omega \cos \alpha \mathbf{e}_z).$$

Hence the torque is

$$\mathbf{M} = \left(\frac{d\mathbf{L}}{dt}\right)_{\text{fixed}} = \left(\frac{d\mathbf{L}}{dt}\right)_{\text{rot}} + \omega \times \mathbf{L} = \omega \times \mathbf{L}$$

$$= (\omega \sin \alpha \mathbf{e}_x + \omega \cos \alpha \mathbf{e}_z) \times \frac{1}{4}ma^2\omega(\sin \alpha \mathbf{e}_x + 2\cos \alpha \mathbf{e}_z)$$

$$= -\frac{1}{4}ma^2\omega \sin \alpha \cos \alpha \mathbf{e}_y.$$

The torque is in the plane of the disc and is perpendicular to the plane formed by the normal to the disc and the axis of rotation. It rotates with the disc. As the center of mass of the disc is stationary, the net force on the disc is zero.

1176

A moon of mass m orbits with angular velocity ω around a planet of mass M Assume $m \ll M$. The rotation of the moon can be neglected but the plane rotates about its axis with angular velocity Ω. The axis of rotation of the plane is perpendicular to the plane of the orbit. Let I = moment of the inertia of the planet about its axis and D = distance from the moon to the center of the planet.

a. Find expressions for the total angular momentum L of the system about its center of mass and for the total energy E. Eliminate D from both these expressions.

b. Generally the two angular velocities ω and Ω are unequal. Suppose there is a mechanism such as tidal friction which can reduce E if $\omega \neq \Omega$ but conserves angular momentum. By examining the behavior of E as a function of ω, show that there is a range of initial conditions such that eventually $\omega = \Omega$ and a stable final configuration obtains.

Famous examples of this effect occur in the orbits of the moons of Mercury and Venus. (However, it is the lighter body whose rotation is relevant in these examples.)

(Princeton)

Sol:

a. As $M \gg m$, the position of the planet can be considered to be fixed in space. The total angular momentum about the center of mass and the total energy of the system of moon and planet are then

$$L = I\Omega + mD^2\omega,$$
$$E = \frac{1}{2}I\Omega^2 + \frac{1}{2}mD^2\omega^2 - \frac{GMm}{D}.$$

Considering the gravitational attraction between the two bodies we have

$$\frac{GMm}{D^2} = mD\omega^2,$$

or

$$D = \left(\frac{GM}{\omega^2}\right)^{\frac{1}{3}}.$$

Substituting this in the above gives

$$L = I\Omega + m\left(\frac{G^2M^2}{\omega}\right)^{\frac{1}{3}},$$

$$E = \frac{1}{2}I\Omega^2 - \frac{m}{2}(GM\omega)^{\frac{2}{3}}.$$

(1)

b. As angular momentum is to be conserved, $dL = 0$, giving

$$\frac{d\Omega}{d\omega} = \frac{mD^2}{3I}.$$

For a configuration to be stable, the corresponding energy must be a minimum. Differentiating (1), we have

$$dE = I\Omega d\Omega - \frac{m}{3}(GM)^{\frac{3}{2}}\omega^{\frac{-1}{3}}d\omega$$

$$= \frac{mD^2}{3}(\Omega - \omega)\,d\omega,$$

$$\frac{d^2E}{d\omega^2} = \frac{2mD}{3}(\Omega - \omega)\frac{dD}{d\omega} + \frac{mD^2}{3}\left(\frac{d\Omega}{d\omega} - 1\right)$$

$$= \frac{mD^2}{9}\left(\frac{mD^2}{I} + 1 - \frac{4\Omega}{\omega}\right).$$

Hence for the configuration to be stable, we require that

$$\Omega \approx \omega,$$

and furthermore that

$$\frac{mD^2}{I} + 1 > \frac{4\Omega}{\omega}.$$

This latter condition can be satisfied by a range of initial conditions.

1177

A pendulum consists of a uniform rigid rod of length L; mass M, a bug of mass $M/3$ which can crawl along the rod. The rod is pivoted at one end and swing in a vertical plane. Initially the bug is at the pivot-end of the rod, which is at rest at an angle θ_0 ($\theta_0 \ll 1$ rad) from the vertical as shown in Fig. 1.136, is released. For $t > 0$ the bug crawls slowly with constant speed V along the rod towards the bottom end of the rod.

a. Find the frequency ω of the swing of the pendulum when the bug has crawled a distance l along the rod.

b. Find the amplitude of the swing of the pendulum when the bug has crawled to the bottom end of the rod ($l = L$).

c. How slowly must the bug crawl in order that your answers for part (a) and (b) be valid?

(Wisconsin)

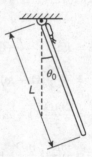

Fig. 1.136

Sol:

a. When the bug has crawled a distance l, the moment of inertia of the rod and bug about the pivot is

$$I = \frac{1}{3}ML^2 + \frac{1}{3}Ml^2 = \frac{1}{3}M(L^2 + l^2).$$

The equation of motion of the pendulum is

$$\frac{d}{dt}(I\dot{\theta}) = -Mg\frac{L}{2}\sin\theta - \frac{1}{3}Mgl\sin\theta,$$

or

$$\frac{1}{3}M(L^2 + l^2)\ddot{\theta} + \frac{2}{3}Mll\dot{\theta} = -Mg\,\sin\theta\left(\frac{L}{2} + \frac{l}{3}\right).$$

For small oscillations it becomes

$$\ddot{\theta} + \frac{2l\dot{l}\dot{\theta}}{L^2 + l^2} + \frac{g\left(l + \dfrac{3L}{2}\right)\theta}{L^2 + l^2} = 0.$$

If the bug crawls so slowly that the change in l in a period of oscillation is negligible, i.e. $\dot{l} = v \ll l\omega$, we can ignore the second term and write

$$\ddot{\theta} + \frac{g(2l + 3L)}{2(L^2 + l^2)}\theta = 0$$

Hence the angular frequency of oscillation ω is

$$\omega = \sqrt{\frac{g(2l + 3L)}{2(L^2 + l^2)}}.$$

b. Consider the motion of the bug along the rod,

$$\frac{M}{3}(\ddot{l} - l\dot{\theta}^2) = \frac{Mg\,\cos\theta}{3} - f,$$

where f is the force exerted on the bug by the rod. As the bug crawls with constant speed, $\ddot{l} = 0$. Also for small oscillations, $\cos\theta \approx 1 - \dfrac{\theta^2}{2}$. The above gives

$$f = \frac{Mg}{3} - \frac{Mg}{6}\theta^2 + \frac{Ml\dot{\theta}^2}{3}.$$

The work done by f as the bug crawls a distance dl is then

$$dW = -f\,dl = -\frac{Mg}{3}dl + \frac{M}{3}\left(\frac{g\theta^2}{2} - l\dot{\theta}^2\right)dl,$$

which is stored as energy of the system. The first term on the right-hand side is the change in the potential energy of the bug, while the second term is the change in the energy of oscillation E of the system,

$$dE = \frac{M}{3}\left(\frac{g\theta^2}{2} - l\dot{\theta}^2\right)dl.$$

Under the condition $\dot{l} \ll l\omega$, l hardly changes in a period of oscillation and can be taken to be constant. For each l, when we consider a full period, the kinematic quantities in the above equation can be replaced by their average values

$$dE = \frac{M}{3}\left(\frac{g\overline{\theta^2}}{2} - l\overline{\dot{\theta}^{\,2}}\right) dl.$$

Now, in single harmonic oscillations the potential and kinetic energy are equal on average, so that

$$\overline{T} = \frac{1}{2} \cdot \frac{M}{3}(L^2 + l^2)\,\dot{\theta}^2 = \frac{E}{2},$$

$$\overline{V} = \frac{MgL}{2}(1 - \overline{\cos\theta}) + \frac{Mgl}{3}(1 - \overline{\cos\theta})$$

$$= \frac{Mg}{2}\left(\frac{L}{2} + \frac{l}{3}\right)\overline{\theta^2} = \frac{E}{2};$$

or

$$\overline{\dot{\theta}^{\,2}} = \frac{3E}{M(L^2 + l^2)},$$

$$\overline{\theta^2} = \frac{6E}{Mg(3L + 2l)}.$$

Substituting these in the energy equation we have

$$\frac{dE}{E} = \left(\frac{1}{3L + 2l} - \frac{l}{L^2 + l^2}\right) dl,$$

or

$$\ln E = \frac{1}{2}\ln\left(\frac{3L + 2l}{L^2 + l^2}\right) + K,$$

where K is a constant. Initially, $l = 0$, $E = E_0$, i.e.

$$\ln E_0 = \frac{1}{2}\ln\left(\frac{3}{L}\right) + K,$$

and we thus have

$$\ln \frac{E}{E_0} = \frac{1}{2}\ln\left[\frac{(3L + 2l)L}{3(L^2 + l^2)}\right].$$

When $l = L$,

$$\ln \frac{E}{E_0} = \frac{1}{2} \ln \frac{5}{6},$$

i.e.

$$E = \sqrt{\frac{5}{6}} E_0.$$

θ is equal to the amplitude when $\dot{\theta} = 0$ i.e. $T = 0$ and $E = V$. When $l = L$, the amplitude θ_{max} is given by

$$\frac{1}{2} Mg \left(\frac{L}{2} + \frac{L}{3} \right) \theta^2_{\text{max}} = E.$$

When $l = 0$, we have

$$\frac{1}{2} Mg \cdot \frac{L}{2} \theta_0^2 = E_0.$$

Then as $E = \sqrt{\frac{5}{6}} E_0$, the above expressions give

$$\theta_{\text{max}} = \left(\frac{3}{10} \right)^{\frac{1}{4}} \theta_0.$$

c. We have neglected the radial velocity of the bug as compared with its tangential velocity: $\dot{l} \ll l\omega$. This is the condition that must be assumed for the above to be valid.

1178

A rectangular lamina of side "*a*" and side "*b*" has a superficial density distribution given by $\sigma = x^2 + y^2 + xy + 1$, where "*x*" and "*y*" are measured from one corner. The mass of the lamina is M. Determine the coordinates of the center of mass.

Sol: Let the corner from which measurements are made be the origin of the coordinate system. The mass of an elemental area $dxdy$ is

$$dm = (x^2 + y^2 + xy + 1) \, dxdy.$$

The coordinates of the center of mass is then

$$x_{cm} = \frac{\displaystyle\iint_0^{x=a,y=b} (x^2 + xy + y^2 + 1)\,x\,dx\,dy}{M} = \frac{1}{M}\int_0^b \left(\frac{a^4}{4} + \frac{a^3 y}{3} + \frac{a^2 y^2}{2} + \frac{a^2}{2} \right) dy$$

$$= \frac{1}{M}\left(\frac{a^4 b}{4} + \frac{a^3 b^2}{6} + \frac{a^2 b^3}{6} + \frac{a^2 b}{2} \right).$$

$$y_{cm} = \frac{\displaystyle\iint_0^{x=a,y=b} (x^2 + xy + y^2 + 1)\,y\,dx\,dy}{M} = \frac{1}{M}\int_0^b \left(\frac{a^3 y}{3} + \frac{a^2 y^2}{2} + ay^3 + ay \right) dy$$

$$= \frac{1}{M}\left(\frac{a^3 b^2}{6} + \frac{a^2 b^3}{6} + \frac{ab^4}{4} + \frac{ab^2}{2} \right).$$

1179

A yo-yo of mass M lies on a smooth horizontal table as shown in Fig. 1.137 The moment of inertia about the center may be taken as $\frac{1}{2}MA^2$. A string is pulled with force F from the inner radius B as indicated in Fig. 1.138.

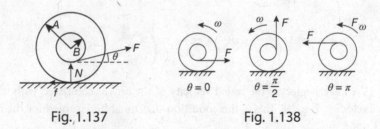

Fig. 1.137   Fig. 1.138

a. In what direction will the yo-yo roll if $\theta = 0, \pi/2, \pi$?

b. For what value of θ will the yo-yo slide without rolling independent of the roughness (coefficient of friction) of the table or the magnitude of F?

c. At what angle θ will the yo-yo roll, independent of the smoothness of the table?

(Columbia)

Sol: Assume that the yo-yo is at rest before the application of the force F.

a. As there is no friction acting on the yo-yo, the direction of rolling is only determined by the direction of the torque of the applied force F about its center. The direction of rolling is shown in Fig. 1.137 for $\theta = 0$, $\pi/2$ or π.

b. The friction acting on the yo-yo is $f = \mu N$, where N is the normal reaction of the table, as shown in Fig. 1.138. The yo-yo will slide without rolling if

$$FB = \mu NA.$$

The acceleration a of the center of mass of the yo-yo is given by

$$F\cos\theta - \mu N = Ma.$$

Thus

$$\cos\theta = \frac{Ma}{F} + \frac{B}{A}.$$

If this condition is satisfied, θ is independent of μ. It still depends on F unless $a = 0$, i.e. no motion.

c. Let the acceleration of the center of mass of the yo-yo and its angular acceleration about the center be a and α respectively. We have (Fig. 1.138)

$$F\cos\theta - f = Ma,$$
$$fA - FB = \frac{1}{2}MA^2\alpha.$$

For rolling without slipping, $a = -A\alpha$. Eliminating α and a gives

$$f = \frac{2F}{A}\left(B - \frac{1}{2}A\cos\theta\right).$$

As

$$f \le \mu N = \mu(Mg - F = \sin\theta),$$

for the yo-yo to roll without slipping irrespective of the smoothness of the table, i.e. independent of μ, we require

$$\sin\theta = \frac{Mg}{F}, \qquad \cos\theta = \frac{2B}{A},$$

or

$$\tan\theta = \frac{A\,Mg}{2B\,F}.$$

Thus we require that, first of all, $2B < A$, $Mg < F$. Then two values of θ, one positive and one negative, with the same $|\sin\theta|$ are possible.

1180

A bowling ball of uniform density is thrown along a horizontal alley with initial velocity v_0 in such a way that it initially slides without rolling. The ball has mass m, coefficient of static friction μ_s and coefficient of sliding friction μ_d with the floor. Ignore the effect of air friction.

Compute the velocity of the ball when it begins to roll without sliding.

(Princeton)

Sol: When the bowling ball slides without rolling the friction $f = \mu_d mg$ gives rise to an acceleration

$$a = -\frac{f}{m} = -\mu_d g.$$

The moment of f gives rise to an angular acceleration α given by

$$fR = \frac{2}{5}MR^2\alpha,$$

as the ball has a moment of inertia $\frac{2}{5}mR^2$ about an axis through its center, R being its radius. Suppose at time t the ball begins to roll without sliding.

We require

$$R\alpha t = v_0 + at,$$

giving

$$t = \frac{v_0}{R\alpha - a} = \frac{2mv_0}{7f} = \frac{2v_0}{7\mu_d g}.$$

The velocity of the ball when this happens is

$$v = v_0 + at = v_0 - \mu_d gt = \frac{5}{7}v_0.$$

1181

A coin spinning about its axis of symmetry with angular frequency ω is set down on a horizontal surface (Fig. 1.139). After it stops slipping, with what velocity does it roll away?

(Wisconsin)

Sol: Take coordinates as shown in Fig. 1.140. Before the coin stops slipping, the frictional force is $f = \mu mg$, where μ is the coefficient of sliding friction. Let x_c be the x coordinate of the center of mass of the coin. The equations of motion of the coin before it stops slipping are

$$m\ddot{x}_c = -\mu mg,$$
$$I\ddot{\theta} = -\mu mgR,$$

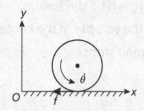

Fig. 1.139   Fig. 1.140

where m and R are respectively the mass and radius of the coin, and $I = \frac{1}{2}mR^2$. Integrating and using initial conditions $\dot{x}_c = 0$, $\dot{\theta} = \omega$ at $t = 0$, we have

$$\dot{x}_c = -\mu gt,$$
$$\dot{\theta} = \omega - \frac{2\mu gt}{R}.$$

When the coin rolls without slipping, we have

$$\dot{x}_c = -\dot{\theta}R.$$

Suppose this happens at time t, then the above give

$$-\mu gt = -\omega R + 2\mu gt$$

or

$$t = \frac{\omega R}{3\mu g}.$$

At this time, the velocity of the center of mass of the coin is

$$\dot{x}_c = -\mu gt = -\frac{1}{3}\omega R,$$

which is the velocity with which the coin rolls away without slipping.

1182

A wheel of mass M and radius R is projected along a horizontal surface with an initial linear velocity V_0 and an initial angular velocity ω_0 as shown in Fig. 1.141, so it starts sliding along the surface (ω_0 tends to produce rolling in the direction opposite to V_0). Let the coefficient of friction between the wheel and the surface be μ.

a. How long is it till the sliding ceases?

b. What is the velocity of the center of mass of the wheel at the time when the slipping stops?

(*Columbia*)

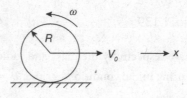

Fig. 1.141

Sol:

a. Take the positive x direction as towards the right and the angular velocity $\dot{\theta}$ as positive when the wheel rotates clockwise. Assume the wheel has moment of inertia $\frac{1}{2}MR^2$ about the axle. We then have two equations of motion:

$$M\ddot{x} = -\mu Mg,$$
$$\frac{1}{2}MR^2\ddot{\theta} = \mu MgR.$$

Making use of the initial conditions $\dot{x}_c = V_0$, $\dot{\theta}_c = -\omega_0$ at $t = 0$ we obtain by integration

$$\dot{x} = V_0 - \mu gt,$$
$$\dot{\theta} = -\omega_0 + \frac{2\mu gt}{R}.$$

Let T be the time when sliding ceases. Then at T

$$\dot{x} = R\dot{\theta},$$

or

$$V_0 - \mu gT = -R\omega_0 + 2\mu gT,$$

giving

$$T = \frac{V_0 + R\omega_0}{3\mu g}.$$

b. The velocity of the center of mass of the wheel at the time when slipping stops is

$$\dot{x} = V_0 - \mu gT = \frac{1}{3}(2V_0 - R\omega_0).$$

1183

A thin hollow cylinder of radius R and mass M slides across a frictionless floor with speed V_0. Initially the cylinder is spinning backward with angular velocity $\omega_0 = 2V_0/R$ as shown in Fig. 1.142. The cylinder passes onto a rough area and continues moving in a straight line. Due to friction, it eventually rolls. What is the final velocity $\mathbf{V}_f$?

(MIT)

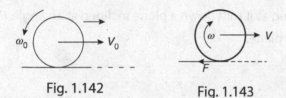

Fig. 1.142   Fig. 1.143

Sol: Suppose the cylinder enters the rough area at time $t = 0$ and starts to roll without slipping at time $t = t_0$. At $0 < t < t_0$ the equations of motion of the cylinder are (Fig. 1.143)

$$-f = M\frac{dV}{dt},$$

$$fR = I\frac{d\omega}{dt}$$

with $I = MR^2$. Integrating we obtain

$$M(V_f - V_0) = -\int_0^{t_0} f\,dt,$$

or

$$I[\omega_f - (-\omega_0)] = R\int_0^{t_0} f\,dt,$$

giving

$$I(\omega_f + \omega_0) = MR(V_0 - V_f).$$

The cylinder rolls without slipping at $t = t_0$, when $V_f = \omega_f R$. We are also given $\omega_0 R = 2V_0$. The last equation then gives

$$V_f = -\frac{1}{2}V_0.$$

Hence the cylinder will eventually move backward with a speed $\frac{1}{2}V_0$.

1184

Calculate the minimum coefficient of friction necessary to keep a thin circular ring from sliding as it rolls down a plane inclined at an angle θ with respect to the horizontal plane.

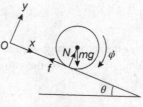

Fig. 1.144

(Wisconsin)

Sol: Use coordinates as shown in Fig. 1.144 and write down the equations of motion for the ring:

$$m\ddot{x} = mg\sin\theta - f, \qquad I\ddot{\varphi} = fR,$$

where m and R are the mass and radius of the ring respectively, $I = mR^2$ is the moment of inertia of the ring about its axis of symmetry and f is the static friction on the ring. The above equations combine to give

$$\ddot{x} + R\ddot{\varphi} = g \sin \theta.$$

The condition for no sliding is $R\dot{\varphi} = \dot{x}$, or $R\ddot{\varphi} = \ddot{x}$, giving

$$\ddot{x} = \frac{1}{2}g \sin \theta.$$

Hence

$$f = mg \sin \theta - m\ddot{x} = \frac{1}{2}mg \sin \theta.$$

The normal reaction of the inclined plane is $N = mg \cos \theta$, and for no slipping we require $f < \mu N$, or

$$\frac{1}{2}mg \sin \theta < \mu mg \cos \theta,$$

i.e.

$$\frac{1}{2}\tan \theta < \mu.$$

Hence the minimum coefficient of friction necessary to keep the ring from slipping is $\mu = \frac{1}{2}\tan \theta$.

1185

A solid uniform cylinder of mass m, radius R is placed on a plane inclined at angle θ relative to the horizontal as shown in Fig. 1.145. Let g denote the usual acceleration due to gravity, and let a be the acceleration along the incline of the axis of the cylinder. The coefficient of friction between cylinder and plane is μ.

For θ less than some critical angle θ_c, the cylinder will roll down the incline without slipping.

a. What is the angle θ_c?

b. For $\theta < \theta_c$, what is the acceleration a?

(CUSPEA)

Sol: Let f denote the frictional force and α the angular acceleration about the axis of the cylinder. The equations of motion are

Fig. 1.145

$$mg \sin \theta - f = ma,$$
$$fR = I\alpha,$$

with

$$I = \frac{1}{2}MR^2.$$

a.   If there is no slipping, we require $a = R\alpha$, $f < \mu N$, where N, the normal reaction of the inclined plane, equals $mg \cos \theta$. The equations of motion give

$$f = \frac{1}{3}mg \sin \theta.$$

Hence we require

$$\mu mg \cos \theta > \frac{1}{3}mg \sin \theta,$$

or

$$3\mu > \tan \theta.$$

Let $\tan \theta_c = 3\mu$. Then we require $\tan \theta < \tan \theta_c$ for no slipping. Therefore the critical angle is $\theta_c = \arctan 3\mu$.

b.   For $\theta < \theta_c$, the cylinder rolls without slipping and the above gives

$$a = g \sin \theta - \frac{f}{m} = \frac{2}{3}g \sin \theta.$$

1186

A wheel of radius r, mass m, and moment of inertia $I = mR^2$ is pulled along a horizontal surface by application of a horizontal force **F** to a rope unwinding from an axle of radius b as shown in Fig. 1.146. You may assume there is a frictional force between the wheel and the surface such that the wheel rolls without slipping. In the expression $I = mR^2$ the quantity R is a constant with dimensions of length.

a. What is the linear acceleration of the wheel?

b. Calculate the frictional force that acts on the wheel.

(Wisconsin)

Sol: Let x be the displacement of the center of mass of the wheel along the horizontal direction and θ the angular displacement of the wheel from an initial direction through its center of mass.

a. The equations of motion of the wheel are (Fig. 1.146)

$$m\ddot{x} = F - f,$$
$$I\ddot{\theta} = Fb + fr.$$

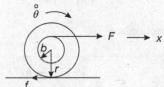

Fig. 1.146

The constraint for no sliding is $\dot{x} = r\dot{\theta}$ or $\ddot{x} = r\ddot{\theta}$. Hence

$$\frac{mR^2}{r}\ddot{x} = Fb + (F - m\ddot{x})r,$$

or

$$\ddot{x} = \frac{F(b + r)r}{m(R^2 + r^2)},$$

which is the linear acceleration of the wheel.

b. The frictional force is

$$f = F - m\ddot{x}$$

$$= F\left[1 - \frac{(b+r)r}{R^2 + r^2}\right] = \frac{F(R^2 - br)}{R^2 + r^2}.$$

1187

A flat disc of mass $m = 1.8$ kg and radius $r = 0.2$ m lies on a frictionless horizontal table. A string wound around the cylindrical surface of the disc exerts a force of 3 Newtons in the northerly direction (Fig. 1.147). Find the acceleration (magnitude and direction) of the center of mass a and the angular acceleration α about the center of mass. Is $a = r\alpha$? Explain.

(*Wisconsin*)

Fig. 1.147

Sol: The equations of motion are

$$f = ma,$$
$$fr = I\alpha,$$

where $I = mr^2/2$, giving

$$a = \frac{f}{m} = 1.7 \text{ m/s},$$

$$\alpha = \frac{2f}{mr} = 17 \text{ rad/s}.$$

The direction of a is the same as that of f. It is seen that $a \neq \alpha r$. This is because as the disc lies on its flat surface the two motions are not related even though they are due to the same force.

1188

A body of mass m is attached to a string of length l and rotated in a horizontal circle of radius r. Show that if the speed with which the body is rotated is doubled, the radius of the circle and the angle the string makes with the vertical is also doubled pvided the length is much longer than the radius of the circle the body traces.

Sol: Resolving the forces vertically and horizontally

$$T\cos\theta = mg \quad \text{and} \quad T\sin\theta = \frac{mv^2}{r}.$$

giving

$$\tan\theta = \frac{v^2}{rg} = \frac{v^2}{lg\sin\theta}$$

$$v^2 = lg\sin\theta\tan\theta = lg\left(\frac{r}{l}\right)\left(\frac{r}{\sqrt{l^2 - r^2}}\right)$$

$$= \frac{gr^2}{l}\left(1 - \frac{r^2}{l^2}\right)^{-\frac{1}{2}} \approx \frac{gr^2}{l}\left(1 + \frac{r^2}{2l^2}\right) \approx \frac{gr^2}{l}$$

since $r \ll l$.

When v is doubled

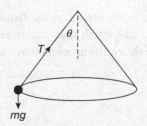

$$mg$$

Fig. 1.148

$$4v^2 = \frac{gr'^2}{l}$$

Or $$r' = 2r$$

Again when $r \ll l$

$$\sin\theta \approx \theta = \frac{r}{l}\text{ and }\sin\theta' = \theta' = \frac{2r}{l}$$

Or $$\theta' = 2\theta$$

1189

Two uniform discs in a vertical plane of masses M_1 and M_2 with radii R_1 and R respectively have a thread wound about their circumferences, and are thu connected as shown in Fig. 1.149.

The first disc has fixed frictionless horizontal axis of rotation through its cen ter. Set up the equations to determine the acceleration of the center of mas: of the second disc if it falls freely.

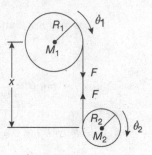

Fig. 1.149

(You need not solve the equations.)

(*Wisconsin*)

Sol: Let F be the tension in the thread, x_1 the distance of the center of mass of disc 2 from that of disc 1, and $\dot{\theta}_1$, $\dot{\theta}_2$ the angular velocities of the discs, as shown in Fig. 1.149. We have the equations of motion

$$M_2\ddot{x} = M_2 g - F,$$
$$I_1\ddot{\theta}_1 = FR_1,$$
$$I_2\ddot{\theta}_2 = FR_2,$$

where $I_1 = m_1 R_1^2/2$, $I_2 = m_2 R_2^2/2$. W$_2$ also have the constraint

$$\dot{x} = R_1\dot{\theta}_1 + R_2\dot{\theta}_2,$$

or

$$\ddot{x} = R_1\ddot{\theta}_1 + R_2\ddot{\theta}_2.$$

From the four equations the unknowns $\ddot{\theta}_1$, $\ddot{\theta}_2$, $\ddot{x}$ and F can be determined.

1190

A yo-yo of mass M is composed of 2 large disks of radius R and thickness t separated by a distance t with a shaft of radius r. Assume a uniform density throughout. Find the tension in the massless string as the yo-yo descends under the influence of gravity.

(*Wisconsin*)

Sol: Let the density of the yo-yo be ρ, then its moment of inertia and mass are respectively

$$I = 2 \cdot \frac{1}{2}\pi t\rho R^4 + \frac{1}{2}\pi t\rho r^4,$$

$$M = 2\pi t\rho R^2 + \pi t\rho r^2,$$

whence

$$I = \frac{1}{2}M\left(\frac{2R^4 + r^4}{2R^2 + r^2}\right).$$

The equations of motion of the yo-yo are

$$M\ddot{x} = Mg - F,$$
$$I\ddot{\theta} = Fr,$$

where F is the tension in the string. We also have the constraint $\ddot{x} = r\ddot{\theta}$. From the above we obtain

$$F = \frac{IMg}{I + Mr^2} = \frac{(2R^4 + r^4)Mg}{2R^4 + 4R^2r^2 + 3r^4}.$$

1191

A sphere of mass M and radius R $\left(I = \frac{2}{5}MR^2\right)$ rests on the platform of a truck. The truck starts from rest and has a constant acceleration A. Assuming that the sphere rolls without slipping, find the acceleration of the center of mass of the ball relative to the truck.

(*Wisconsin*)

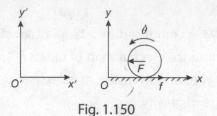

Fig. 1.150

Sol: Let Oxy and $O'x'y'$ be coordinate frames attached to the truck and fixed in space respectively with the x- and x'-axes along the horizontal as shown in Fig. 1.150. Denoting $\overline{O'O} = \xi$, we have for the center of mass of the sphere

$$x' = x + \xi, \quad \text{or} \quad \ddot{x}' = \ddot{x} + \ddot{\xi}.$$

As the force acting on the sphere is the friction f, Newton's second law gives writing A for $\ddot{\xi}$,

$$f = M\ddot{x}' = M\ddot{x} + MA,$$

or

$$M\ddot{x} = f - MA.$$

Thus in the moving frame there is a fictitious force $F = -MA$ acting on the sphere through the center of mass, in addition to the friction f. Considering the torque about the center of mass, we have

$$I\ddot{\theta} = fR$$

with $I = \dfrac{2}{5} MR^2$. We also have the constraint for no slipping, $\dot{x} = -R\dot{\theta}$, or $\ddot{x} = -R\ddot{\theta}$. These three equations give $\ddot{x} = -\dfrac{5}{7} A$, which is the acceleration of the center of mass of the sphere relative to the truck.

1192

Referring to Fig. 1.151, find the minimum height h (above the top position in the loop) that will permit a spherical ball of radius r (which rolls without slipping) to maintain constant contact with the rail of the loop. (The moment of inertia of a sphere about the center is $\dfrac{2}{5} mr^2$.)

(Wisconsin)

Sol: Conservation of mechanical energy requires that the kinetic energy of the sphere at the top position in the loop is equal to the decrease mgh in potential energy as it falls from the initial position to this position. The kinetic energy of the sphere is composed of two parts: the translational kinetic energy of the sphere and the rotational kinetic energy of the sphere

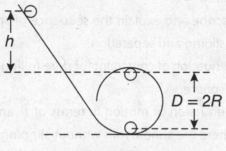

Fig. 1.151

about its center of mass. Let m, T, v, ω be respectively the mass, kinetic energy, velocity of the center of mass, and angular velocity about the center of mass of the sphere. Then

$$T = \frac{1}{2}mv^2 + \frac{1}{2}I\omega^2$$

with $I = \frac{2}{5}mr^2$. As the sphere rolls without slipping, $v = \omega r$ and

$$T = \frac{1}{2}\left(mv^2 + \frac{2}{5}mr^2\frac{v^2}{r^2}\right) = \frac{7}{10}mv^2.$$

In the critical case, the force exerted by the loop on the sphere is zero when the latter reaches the top of the loop. In other words, the centripetal force needed for the circular motion of the sphere is supplied entirely by gravity:

$$\frac{mv^2}{R} = mg,$$

whence $v^2 = Rg$ and

$$T = \frac{7}{10}mRg = mgh.$$

Hence $h = 7R/10$ is the minimum initial height required.

1193

A sphere of radius b is at rest at $\theta = 0$ upon a fixed sphere of radius $a > b$. The upper sphere is moved slightly to roll under the influence of gravity a shown in Fig. 1.152. The coefficient of static friction is $\mu_s > 0$, the coefficien of sliding friction is $\mu = 0$.

   a. Briefly describe and explain the sequence of sphere motions in terms of rolling, sliding and separation.

   b. Write the equation of constraint for pure rolling of the upper sphere on the lower sphere.

   c. Write the equation of motion in terms of $\ddot{\theta}$ and θ for the part of the motion where the sphere rolls without slipping.

   d. Find a related equation between $\ddot{\theta}$ and θ.

   e. Solve this equation for $\theta(t)$, assuming $0 < \theta(0) \ll \theta(t)$. You may wish to use the relation

$$\int \frac{dx}{\sin\left(\frac{x}{2}\right)} = 2 \ln \tan\left(\frac{x}{4}\right).$$

(MIT)

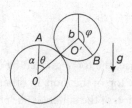

Fig. 1.152

Sol:

   a. At first the upper sphere rolls without slipping, the angular velocity becoming larger and the normal pressure on it smaller with increasing θ. When the condition for pure rolling is not satisfied, the sphere begins to slide and finally when the centripetal force is not large enough to maintain the circular motion of the upper sphere, it will separate from the lower sphere.

b. Suppose initially O, A, O', B are on the same vertical line. As the upper sphere rolls by an angle φ, its center has traveled through a path $\overline{OO'}\theta$, as shown in Fig. 1.152. Hence the condition for pure rolling is

$$(a + b)\theta = b\varphi.$$

c. The equations of motion of the upper sphere are

$$m(a + b)\ddot{\theta} = mg \sin \theta - f,$$

$$I\ddot{\varphi} = \frac{2}{5}mb^2\ddot{\varphi} = fb,$$

where f is the static friction on the sphere. When the sphere rolls without slipping, we have from (b)

$$(a + b)\ddot{\theta} = b\ddot{\varphi}.$$

Then the equations of motion give

$$\ddot{\theta} = \frac{5g \sin \theta}{7(a + b)}.$$

d. As

$$\ddot{\theta} = \frac{1}{2}\frac{d\dot{\theta}^2}{d\theta},$$

the last equation gives

$$\dot{\theta}^2 = -\frac{10g \cos \theta}{7(a + b)} + K.$$

With $\dot{\theta} = 0$ at $\theta = 0$, $K = \frac{10g}{7(a + b)}$. Hence

$$\dot{\theta}^2 = \frac{10g(1 - \cos \theta)}{7(a + b)}.$$

e. As

$$\frac{d\theta}{dt} = \sqrt{\frac{10g(1 - \cos \theta)}{7(a + b)}} = \sqrt{\frac{20g}{7(a + b)}} \sin\frac{\theta}{2},$$

we have, with $\theta_0 = \theta(0)$ at $t = 0$,

$$\int_{\theta_0}^{\theta} \frac{d\theta}{\sin \dfrac{\theta}{2}} = \sqrt{\frac{20g}{7(a+b)}} \int_0^t dt$$

or

$$\ln \left(\frac{\tan \dfrac{\theta}{4}}{\tan \dfrac{\theta_0}{4}} \right) = \alpha t,$$

where $\alpha = \sqrt{\dfrac{5g}{7(a+b)}}$. Hence

$$\theta = 4 \arctan \left(e^{\alpha t} \tan \frac{\theta_0}{4} \right),$$

valid for the part of the motion where the sphere rolls without slipping.

1194

A sphere of mass m, radius a, and moment of inertia $\frac{2}{5} ma^2$ rolls without slipping

from its initial position at rest atop a fixed cylinder of radius b (see Fig. 1.153).

a. Determine the angle $\theta_{\max}$ at which the sphere leaves the cylinder.

b. What are the components of the velocity of the sphere's center at the instant it leaves the cylinder?

(Wisconsin)

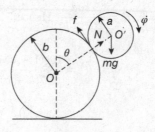

Fig. 1.153

Sol:

a. The forces on the sphere are as shown in Fig. 1.153. The equations of motion for the center of mass of the sphere are

$$m(a+b)\ddot{\theta} = mg \sin \theta - f, \qquad (1)$$

$$m(a+b)\dot{\theta}^2 = mg \cos \theta - N, \qquad (2)$$

and that for the rotation of the sphere is

$$\frac{2}{5}ma^2\ddot{\varphi} = fa. \tag{3}$$

The condition for it to roll without slipping is

$$(a + b)\,\dot{\theta} = a\dot{\varphi}, \quad \text{or} \quad (a + b)\,\ddot{\theta} = a\ddot{\varphi}. \tag{4}$$

From (3) and (4), we found

$$f = \frac{2}{5}m(a + b)\,\ddot{\theta}.$$

Substitution in (1) gives

$$\ddot{\theta} = \frac{5g\,\sin\theta}{7(a + b)}.$$

As $\theta = \dot{\theta} = 0$ at $t = 0$ and $\ddot{\theta} = \frac{1}{2}\frac{d\dot{\theta}^2}{d\theta}$, it gives

$$\dot{\theta}^2 = \frac{10g(1 - \cos\theta)}{7(a + b)}.$$

Substitution in (2) gives

$$N = mg\,\cos\theta - \frac{10}{7}mg(1 - \cos\theta) = mg\left(\frac{17\,\cos\theta}{7} - \frac{10}{7}\right).$$

After the sphere leaves the cylinder, $N = 0$. We assume that the coefficient of friction is large enough for the period of both rolling and slipping which occurs before the sphere leaves the cylinder to be negligible. Then at the instant N becomes zero, $\theta = \theta_{\max}$ given by

$$\cos\theta_{\max} = \frac{10}{17}.$$

b. At that instant the velocity of the center of the sphere has magnitude

$$v = (a + b)\,\dot{\theta} = \sqrt{\frac{10}{17}g(a + b)},$$

and is parallel to the tangential direction of the cylinder at the point where $\theta = \theta_{\max}$.

1195

In Fig. 1.154, the ball on the left rolls horizontally without slipping at speed V toward an identical ball initially at rest. Each ball is a uniform sphere of mass M. Assuming that all the frictional forces are small enough to have a negligible effect during the instant of collision, and that the instantaneous collision is perfectly elastic, calculate:

a. The velocity of each ball a long enough time after the collision when each ball is again rolling without slipping.

b. The fraction of the initial energy transformed by the frictional forces to thermal energy.

The moment of inertia of a sphere of mass M, radius R about its center is $\frac{2}{5}MR^2$.

(CUSPEA)

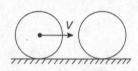

Fig. 1.154

Sol:

a. Before the collision

$$V_1 = V, \quad V_2 = 0, \quad \omega_1 = \frac{V}{R}, \quad \omega_2 = 0.$$

During the collision, as friction can be neglected, the forces with which the balls interact are directed through the centers so that the angular momentum about the center of each ball is conserved. Thus

$$\omega'_1 = \omega_1, \quad \omega'_2 = 0.$$

As the collision is elastic, conservation of translational momentum and that of kinetic energy then require

$$V'_1 = 0, \quad V'_2 = V_1 = V.$$

In the above, single primes denote quantities immediately after the collision. After some time, the balls again roll without slipping. Let the quantities at this time be denoted by double primes. The positive directions of these quantities are shown in Fig. 1.155.

The angular momentum of each ball about some fixed point in the plane of motion is conserved. Consider the angular momentum of each ball about the point of contact with the horizontal plane.

For ball 1,

$$MRV_1' + I\omega_1' = MRV_1'' + I\omega_1'',$$

or

$$\frac{IV}{R} = \left(MR + \frac{I}{R} \right) V_1'',$$

giving

$$V_1'' = \frac{V}{\dfrac{MR^2}{I} + 1} = \frac{2}{7} V.$$

For ball 2,

$$MRV_2' + I\omega_2' = MRV_2' + I\omega_2'',$$

or

$$MRV = \left(MR + \frac{I}{R} \right) V_2'',$$

giving

$$V_2'' = \frac{V}{1 + \dfrac{I}{MR^2}} = \frac{5}{7} V.$$

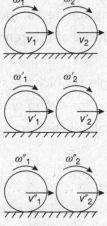

Fig. 1.155

b. The initial and final energies of the system are

$$W_i = \frac{1}{2}MV_1^2 + \frac{1}{2}I\omega_1^2 = \frac{1}{2}\left(MV^2 + \frac{2}{5}MR^2 \cdot \frac{V^2}{R^2}\right)$$

$$= \frac{1}{2}MV^2 \cdot \frac{7}{5},$$

$$W_f = \frac{1}{2}M(V''^2_1 + V''^2_2) + \frac{1}{2}I(\omega''^2_1 + \omega''^2_2)$$

$$= \frac{1}{2}M\left[V''^2_1 + V''^2_2 + \frac{2}{5}(V''^2_1 + V''^2_2)\right]$$

$$= \frac{1}{2}MV^2 \cdot \frac{7}{5} \cdot \frac{29}{49}.$$

Hence the loss of energy is

$$W_i - W_f = \frac{1}{2} \cdot \frac{7}{5}MV^2 \cdot \frac{20}{49},$$

and the fractional loss is $\frac{20}{49}$.

1196

A small homogeneous sphere of mass m and radius r rolls without sliding on the outer surface of a larger stationary sphere of radius R as shown in Fig. 1.156. Let θ be the polar angle of the small sphere with respect to a coordinate system with origin at the center of the large sphere and z-axis vertical. The smaller sphere starts from rest at the top of the larger sphere ($\theta = 0$).

a. Calculate the velocity of the center of the small sphere as a function of θ.

b. Calculate the angle at which the small sphere flies off the large one.

c. If one now allows for sliding with a coefficient of friction μ, at what point will the small sphere start to slide?

(Columbia)

Sol:

a. As the sum of the kinetic and potential energies of the small sphere is a constant of the motion when it rolls without sliding, we have

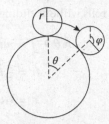

Fig. 1.156

$$\frac{1}{2}mv^2 + \frac{1}{2}\cdot\frac{2}{5}mr^2\cdot\dot{\varphi}^2 + mg(R+r)\cos\theta = mg(R+r)$$

with $v = r\dot{\varphi} = (R+r)\dot{\theta}$, whence

$$\dot{\theta} = \sqrt{\frac{10}{7}\frac{(1-\cos\theta)g}{(R+r)}}.$$

The velocity of the center of the small sphere is

$$v = (R+r)\dot{\theta} = \sqrt{\frac{10}{7}(R+r)(1-\cos\theta)g}.$$

b.   At the moment of flying off, the support force on the small sphere $N = 0$. From the force equation

$$mg\cos\theta - N = \frac{mv^2}{R+r},$$

we find the angle θ_c at which the small sphere flies off the large sphere as given by

$$\cos\theta_c = \frac{10}{17}.$$

Thus

$$\theta_c = \arccos\left(\frac{10}{17}\right).$$

Note that this derivation applies only for sufficiently large coefficient of friction.

c. When the small sphere rolls without sliding, we have

$$mg \sin \theta - f = m\dot{v},$$

$$fr = \frac{2}{5} mr^2 \ddot{\varphi},$$

$$v = (R + r)\dot{\theta} = \dot{\varphi},$$

where f is the frictional force on the sphere. From these we find

$$f = \frac{2}{7} mg \sin \theta.$$

At the moment when the sphere starts to slide, the frictional force is

$$f = \mu N,$$

i.e.

$$\frac{2}{7} mg \sin \theta = \mu \left(mg \cos \theta - \frac{mv^2}{R + r} \right).$$

Then, using the expression for v from (a), we have

$$2 \sin \theta = 17\mu \cos \theta - 10\mu.$$

Solving this we find that the angle θ_s at which the small sphere starts to slide is given by

$$\cos \theta_s = \frac{170\mu^2 \pm \sqrt{756\mu^2 + 4}}{289\mu^2 + 4}.$$

However, we require that $\theta_c > \theta_s$, or $\cos \theta_s > \cos \theta_c$. Where this can be satisfied by the value of μ, we generally have to take the upper sign. Hence

$$\theta_s = \arccos \left(\frac{170\mu^2 + \sqrt{756\mu^2 + 4}}{289\mu^2 + 4} \right).$$

1197

A spherical ball of radius r is inside a vertical circular loop of radius $(R + r)$ as shown in Fig. 1.157. Consider two cases (i) rolling without sliding (ii) frictionless sliding without rolling.

a. In each case what minimum velocity v_1 must the sphere have at the bottom of the loop so as not to fall at the top?

b. For a 10% smaller v_1 and the sliding case, where on the loop will falling begin?

(*Columbia*)

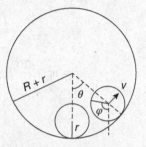

Fig. 1.157

Sol:

a. For rolling without sliding, $R\theta = r\varphi$. Hence

$$\omega = \dot{\varphi} = \frac{R\dot{\theta}}{r} = \frac{v}{r},$$

where v is velocity of the center of the ball. In order that the ball does not fall at the top of the loop, the force N_t the loop exerts on the ball at the top must be such that

$$N_t = \frac{mv^2}{R} - mg \geq 0.$$

Thus we require that

$$v^2 \geq Rg.$$

The minimum velocity v_t that satisfies such condition is $v_t^2 = Rg$ and the corresponding kinetic energy is

$$T_t = \frac{1}{2}mv_t^2 + \frac{1}{2} \cdot \frac{2}{5}mr^2 \cdot \omega^2 = \frac{7}{10}mv_t^2.$$

At the bottom of the loop, if the ball has the required minimum velocity v_1, we have

$$T_b = T_t + V_t,$$

i.e.

$$\frac{7}{10} mv_1^2 = \frac{7}{10} mv_t^2 + 2mRg,$$

giving

$$v_1^2 = v_t^2 + \frac{20}{7} Rg = \frac{27}{7} Rg,$$

or

$$v_1 = \sqrt{\frac{27}{7} Rg}.$$

(ii) For sliding without rolling, we still require that $v^2 \geq Rg$ at the top of the loop, i.e. the minimum velocity at the top is given by

$$v_t^2 = Rg,$$

and the corresponding kinetic energy is

$$T_t = \frac{1}{2} mv_t^2.$$

Thus we have

$$\frac{1}{2} mv_1^2 = \frac{1}{2} mv_t^2 + 2mgR,$$

giving

$$v_1^2 = 5Rg,$$

or

$$v_1 = \sqrt{5Rg}.$$

b. Suppose falling begins at θ. At that moment the velocity v of the center of the ball is given by

$$\frac{1}{2} m(0.9v_1)^2 = \frac{1}{2} mv^2 + mg(R - R\cos\theta),$$

and

$$N - mg \cos \theta = \frac{mv^2}{R}$$

with

$$N = 0.$$

These equations give

$$3Rg \cos \theta = 2Rg - 0.81v_1^2 = -2.05Rg,$$

i.e.

$$\cos \theta = -0.683,$$

or

$$\theta = 133.1°,$$

1198

A body of mass *m* is projected at an angle *β* to the horizontal with a velocity *u*. It hits a plane inclined at an angle *α* at the point A as shown in the figure. Obtain an expression for the time it takes to hit the plane. If the angle *β* is 60° and *α* is 45°, determine the *x* and *y* components of velocity of the body when it hits the plane.

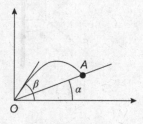

Fig. 1. 158

Sol: Let the coordinates of the body on the inclined plane be *(x, y)*. Then

$$x = ut \cos \beta$$

$$y = ut \sin \beta - \frac{gt^2}{2}$$

Now

$$\tan \alpha = \frac{y}{x} = \tan \beta - \frac{gt}{2u \cos \beta}$$

Reorganizing we get

$$t = \frac{2(\tan \beta - \tan \alpha)u \cos \beta}{g}$$

Substituting the values of α and β

$$t = \frac{(\sqrt{3} - 1)u}{g}$$

The x and y components of the velocity of the body when it hits the plane are

$$v_{ox} = u \cos \beta = \frac{u}{2} \text{ or } 0.5u$$

and $v_{oy} = u \sin \beta - gt = \frac{\sqrt{3}}{2}u - (\sqrt{3} - 1)u = \frac{2 - \sqrt{3}}{2}u$ or $0.134u$

1199

A block of mass m is pressed against a wall and kept by a force F applied at an angle $\theta < 90°$. Find an expression for this force. At what angle would the force exerted be a minimum? The coefficient of friction between the block and the wall is μ.

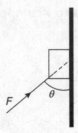

Fig. 1. 159

Sol:

$$mg - F \cos \theta = \mu F \sin \theta$$

$$F = \frac{mg}{\mu \sin \theta + \cos \theta}$$

For F to be minimum

$$\frac{dF}{d\theta} = 0 \text{ and } \frac{d^2F}{d\theta^2} > 0$$

$$\frac{dF}{d\theta} = \frac{-mg(\mu \cos \theta - \sin \theta)}{(\mu \sin\theta + \cos\theta)^2} = 0 \text{ or } \tan \theta = \mu$$

$$\frac{d^2F}{d\theta^2} = -mg\frac{d}{d\theta}\left((\mu\sin\theta + \cos\theta)^{-2}(\mu\cos\theta - \sin\theta)\right)$$

$$= -mg\{-2\mu\cos\theta(\mu\sin\theta + \cos\theta)^{-3}(\mu\cos\theta - \sin\theta)$$

$$+ (\mu\sin\theta + \cos\theta)^{-2}(-\mu\sin\theta) + (2)\sin\theta(\mu\sin\theta + \cos\theta)^{-3}$$

$$(\mu\cos\theta - \sin\theta) + (\mu\sin\theta + \cos\theta)^{-2}(-\cos\theta)\}$$

$$= -mg\{-2(\mu\sin\theta + \cos\theta)^{-3}(\mu\cos\theta - \sin\theta)(\mu\cos\theta - \sin\theta)$$

$$+ (\mu\sin\theta + \cos\theta)^{-2}(-\mu\sin\theta - \cos\theta)\}$$

$$= -mg\{-2\cos^2\theta(\mu\sin\theta + \cos\theta)^{-3}(\mu - \tan\theta)(\mu - \tan\theta)$$

$$+ \cos^{-1}\theta(\mu\tan\theta + 1)^{-2}(-\mu\tan\theta - 1)\}$$

$$= 0 - mg\frac{-\mu^2 - 1}{\cos\theta(\mu^2 + 1)^2} = \frac{mg}{(\mu^2 + 1)\cos\theta}$$

$$\tan\theta = \mu \text{ or } \tan(180 - \theta) = -\mu$$

$$\text{then } \cos(180 - \theta) = \frac{1}{(\mu^2 + 1)^{\frac{1}{2}}} = \cos\theta$$

$$\frac{d^2F}{d\theta^2} = \frac{mg}{(\mu^2 + 1)(\mu^2 + 1)^{\frac{1}{2}}} = \frac{mg}{(\mu^2 + 1)^{\frac{3}{2}}} > 0$$

F has a minimum value when $\tan\theta = \mu$

1200

Two long uniform rods A and B each 1 m long and of masses 1 kg (A) and 2 kg (B) lie parallel to each other on a frictionless horizontal plane (x, y). Rod B is initially at rest at $y = 0, x = 0$ to $x = 1$ m. Rod A is moving at 10 m/s in the positive y direction, and it extends from $x = (-1 + \varepsilon)$ m to $x = \varepsilon$ m, $(\varepsilon \ll 1$ m) as shown in Fig. 1.160. Rod A reaches $y = 0$ at $t = 0$ and collides elastically with B. Find the subsequent motion of the rods, ignoring the possibility of subsequent collisions. Check for equality of energy before and after collision.

(Columbia)

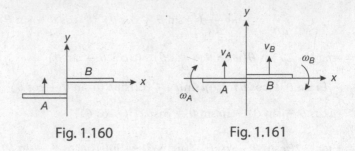

Fig. 1.160   Fig. 1.161

Sol: Let I be the impulse rod A exerts on rod B during the collision. Its direction is the direction of the motion of A, i.e. the positive y direction. Let v_A, ω_A, v_B, ω_B, be the velocity of the center of mass and the angular velocity about the center of mass of A and B respectively, as shown in Fig. 1.161. Denoting the masses of A, B by m_A, m_B respectively, we have

$$-I = m_A(v_A - 10),$$

$$\frac{1}{2}I = \frac{1}{12}m_A\omega_A,$$

$$I = m_B v_B,$$

$$\frac{1}{2}I = \frac{1}{12}m_B\omega_B.$$

The condition of elastic collision means that the relative velocity of the points of collision remains the same in magnitude but reverses in direction:

$$\left(v_B + \frac{1}{2}\omega_B\right) - \left(v_A - \frac{1}{2}\omega_A\right) = 10.$$

The above equations give

$$I = \frac{5m_A m_B}{m_A + m_B} = \frac{10}{3}\,\text{Ns},$$

$$v_A = 10 - \frac{I}{m_A} = \frac{20}{3}\,\text{m/s};$$

$$\omega_A = \frac{6I}{m_A} = 20\,\text{rad/s},$$

$$v_B = \frac{I}{m_B} = \frac{5}{3}\,\text{m/s},$$

$$\omega_B = \frac{6I}{m_B} = 10\,\text{rad/s}$$

for the subsequent motion. The energy of the two rods before collision is

$$E_i = \frac{1}{2} \cdot 1 \cdot 10^2 = 50 \, \text{J}$$

and after collision is

$$E_f = \frac{1}{2} m_A v_A^2 + \frac{1}{2} m_B v_B^2 + \frac{1}{2} \cdot \frac{1}{12} m_A \omega_A^2 + \frac{1}{2} \cdot \frac{1}{12} m_B \omega_B^2$$

$$= \frac{225}{9} + \frac{300}{12} = 50 \, \text{J} = E_i.$$

Hence the equality of energy holds

1201

A billiard ball of radius R and mass M is struck with a horizontal cue stick at a height h above the billiard table as shown in Fig. 1.162. Given that the moment of inertia of a sphere is $\frac{2}{5} MR^2$, find the value of h for which the ball will roll without slipping.

(Wisconsin)

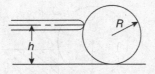

Fig. 1.162

Sol: Suppose that f is the impact force on the ball exerted by the stick and that it acts for a time Δt causing a change of momentum of the ball of $M\Delta v$ and a change of its angular momentum about the center of mass of $I\Delta\omega$. We have the equations of motion

$$M\Delta v = f\Delta t,$$
$$I\Delta\omega = f(h - R)\Delta t$$

with $I = \frac{2}{5} MR^2$, which yield

$$\Delta v = \frac{2R^2 \Delta\omega}{5(h - R)}.$$

As the ball is at rest initially, the velocity of its center of mass and the angular velocity after impact satisfy

$$v = \frac{2R^2 \omega}{5(h - R)}.$$

The ball will roll without slipping if $v = R\omega$. Hence we require

$$5(h - R) = 2R,$$

or

$$h = \frac{7}{5}R.$$

1202

A uniform solid ball of radius a rolling with velocity u on a level surface collides inelastically with a step of height $h < a$, as shown in Fig. 1.163. Find, in terms of h and a, the minimum velocity for which the ball will "trip" up over the step. Assume that no slipping occurs at the impact point, and remember that the moment of inertia of a solid sphere with respect to an axis through its center is $\frac{2}{5} Ma^2$.

(Wisconsin)

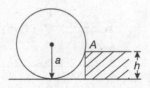

Fig. 1.163

Sol: Let ω and ω', J and J' be the angular velocity of the ball with respect to its center of mass and its angular momentum about the point of impact A before and after collision with the step, respectively. We have

$$J = mv(a - h) + \frac{2}{5} ma^2 \omega = \frac{7}{5} mva - mvh$$

as $v = a\omega$ for rolling without slipping, and

$$J' = \left(\frac{2}{5} ma^2 + ma^2 \right) \omega' = \frac{7}{5} ma^2 \omega'$$

as the center of mass of the ball is momentarily at rest after the collision. Conservation of angular momentum requires

$$\frac{7}{5}ma^2\omega' = \frac{7}{5}mva - mvh,$$

yielding

$$\omega' = \left(1 - \frac{5h}{7a}\right)\frac{v}{a}.$$

In order that the ball can just trip up over the step, its kinetic energy must be sufficient to provide for the increase in potential energy:

$$\frac{1}{2}I'\omega'^2 = mgh,$$

where $I' = \frac{2}{5}ma^2 + ma^2 = \frac{7}{5}ma^2$ is the moment of inertia of the ball about a horizontal axis through A. Hence the minimum velocity required is given by

$$\frac{7}{10}ma^2\left(1 - \frac{5h}{7a}\right)^2\left(\frac{v}{a}\right)^2 = mgh,$$

yielding

$$v = \frac{a\sqrt{70gh}}{7a - 5h}.$$

1203

A parked truck has its rear door wide open as shown in the plane view in Fig. 1.164(a). At time $t = 0$ the truck starts to accelerate with constant acceleration a. The door will begin to close, and at a later time t the door will be passing through the position shown in Fig. 1.164(b) such that the door makes an angle θ with its original orientation. You may assume that the door has mass m uniformly distributed along its length L.

a. Using θ and its time derivatives to describe the motion, write down dynamic equations relating the two components, $F_{\parallel}$ and $F_{\perp}$, of the force exerted on the door at the hinge to the kinematic quantities. $F_{\parallel}$ is the component of the force parallel to the door in the plane of the diagram and $F_{\perp}$ is the component perpendicular to the door.

b. Express $\ddot{\theta} = d^2\theta/dt^2$, $F_{\parallel}$ and $F_{\perp}$ in terms of θ, m, L and a.

c. Write down, but do not attempt to integrate, an expression for the total time elapsed from the start of acceleration to the closing of the door.

(MIT

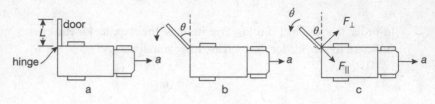

Fig. 1.164

Sol:

a. In a frame attached to the accelerating truck, the center of mass of the door has components of acceleration $\frac{1}{2}L\ddot{\theta}$ perpendicular to the door and $-\frac{1}{2}L\dot{\theta}^2$ parallel to the door. The directions of $F_{\parallel}$ and $F_{\perp}$ are as shown in Fig. 1.164(c). In this frame a fictitious force $-ma$ acting at the center of mass is included in the equations of motion:

$$F_{\perp} - ma \cos\theta = -\frac{1}{2}mL\ddot{\theta},$$

$$F_{\parallel} - ma \sin\theta = \frac{1}{2}mL\dot{\theta}^2,$$

$$\frac{1}{2}LF_{\perp} = \frac{1}{12}mL^2\ddot{\theta},$$

where $\frac{1}{12}mL^2$ is the moment of inertia of the door about an axis perpendicular to the top edge of the door through the center of mass.

b. The above equations give

$$\ddot{\theta} = \frac{3a \cos\theta}{2L},$$

$$F_{\perp} = \frac{1}{4}ma \cos\theta.$$

As $\ddot{\theta} = \frac{1}{2}\frac{d\dot{\theta}^2}{d\theta}$, integrating the expression for $\ddot{\theta}$ and noting that $\theta = \dot{\theta} = 0$ initially we have

$$\dot{\theta}^2 = \frac{3a \sin \theta}{L},$$

whence

$$F_{\parallel} = ma \sin \theta + \frac{3}{2} ma \sin \theta = \frac{5}{2} ma \sin \theta.$$

c. As

$$\frac{d\theta}{dt} = \sqrt{\frac{3a \sin \theta}{L}},$$

the total time elapsed from start of acceleration to the closing of the door is

$$t = \int_0^\pi \sqrt{\frac{L}{3a \sin \theta}} \, d\theta.$$

1204

Consider a solid cylinder of mass m and radius r sliding without rolling down the smooth inclined face of a wedge of mass M that is free to move on a horizontal plane without friction (Fig. 1.165).

a. How far has the wedge moved by the time the cylinder has descended from rest a vertical distance h?

b. Now suppose that the cylinder is free to roll down the wedge without slipping. How far does the wedge move in this case?

c. In which case does the cylinder reach the bottom faster? How does this depend on the radius of the cylinder?

(UC, Berkeley)

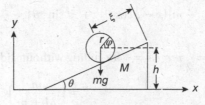

Fig. 1.165

Sol:

a. Let ξ be the distance of the center of mass of the cylinder from its initial position. In a fixed coordinate frame, let x be the horizontal coordinate of the center of mass of the wedge. The horizontal component of the velocity of the cylinder in the fixed frame is $\dot{x} - \dot{\xi}\cos\theta$. As the total momentum of the system in the x direction is conserved, we have, since the system is initially at rest,

$$M\dot{x} + m(\dot{x} - \dot{\xi}\cos\theta) = 0,$$

giving

$$(M + m)\dot{x} = m\dot{\xi}\cos\theta.$$

Without loss of generality we set $\xi = x = 0$ at $t = 0$. Integration of the above then gives

$$(M + m)x = m\xi\cos\theta.$$

When the cylinder has descended a vertical distance h, it has moved a distance $\xi = \dfrac{h}{\sin\theta}$, and the wedge has moved a distance

$$x = \frac{m\xi}{M+m}\cos\theta = \frac{mh}{M+m}\cot\theta.$$

b. If the cylinder is allowed to roll, conservation of the horizontal component of the total linear momentum of the system still holds. It follows that the result obtained in (a) is also valid here.

c. Conservation of the total mechanical energy of the system holds for both cases. As the center of mass of the cylinder has velocity $(\dot{x} - \dot{\xi}\cos\theta, -\dot{\xi}\sin\theta)$ and that of the wedge has velocity $(\dot{x}, 0)$, we have for the sliding cylinder,

$$\frac{1}{2}m[(\dot{x} - \dot{\xi}\cos\theta)^2 + \dot{\xi}^2\sin^2\theta] + \frac{1}{2}M\dot{x}^2 = mg\xi\sin\theta,$$

and for the rolling cylinder,

$$\frac{1}{2}m[(\dot{x} - \dot{\xi}\cos\theta)^2 + \dot{\xi}^2\sin^2\theta] + \frac{1}{2}I\dot{\varphi}^2 + \frac{1}{2}M\dot{x}^2 = mg\xi\sin\theta$$

with $I = \dfrac{1}{2}mr^2$, $\dot{\varphi} = \dfrac{\dot{\xi}}{r}$ for rolling without sliding. As

$$\dot{x} = \left(\frac{m}{M+m}\right)\dot{\xi}\cos\theta,$$

the above respectively reduce to

$$\frac{m}{2(M + m)}(M + m \sin^2 \theta)\dot{\xi}^2 = mg\xi \sin \theta,$$

$$\frac{m}{4(M + m)}[3M + m(1 + 2 \sin^2 \theta)]\dot{\xi}^2 = mg\xi \sin \theta.$$

These equations have the form $\dot{\xi} = b\sqrt{\xi}$. As $\xi = 0$ at $t = 0$, integration gives $t = \frac{2}{b}\sqrt{\xi}$. Hence for the same $\xi = \frac{h}{\sin \theta}$, $t \propto \frac{1}{b}$. As

$$3M + m(1 + 2 \sin^2 \theta) - 2(M + m \sin^2 \theta) = M + m > 0,$$

the sliding cylinder will take a shorter time to reach the bottom.

1205

A stepladder consists of two legs held together by a hinge at the top and a horizontal rope near the bottom, and it rests on a horizontal surface at 60° as shown in Fig. 1.166. If the rope is suddenly cut, what is the acceleration of the hinge at that instant? Assume the legs to be uniform, identical to each other, and neglect all friction.

(UC, Berkeley)

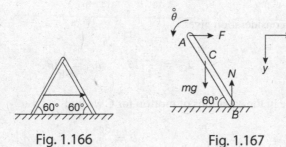

Fig. 1.166   Fig. 1.167

Sol: Consider the instant when the horizontal rope is suddenly cut.

By symmetry the forces which the two legs exert on each other at the hinge A are horizontal and the acceleration of A, $\mathbf{a}_A$, is vertically downward.

Consider one leg of the stepladder. The forces acting on it are as shown in Fig. 1.167. Let l be the length of the leg and $\mathbf{a}_C$ the acceleration of its center of mass C at the instant the rope is cut. We have

$$mg - N = ma_{Cy},$$

$$F = ma_{Cx},$$

$$\frac{1}{2} Nl \cos 60° - \frac{1}{2} Fl \sin 60° = I\ddot{\theta}$$

with $I = \frac{1}{12} ml^2$, or

$$N - \sqrt{3}F = \frac{1}{3} ml\ddot{\theta}.$$

The velocity of A in terms of the velocity of C are given by

$$\dot{x}_A = \dot{x}_C - \frac{1}{2} l\dot{\theta} \sin 60°, \qquad \dot{y}_A = \dot{y}_C + \frac{1}{2} l\dot{\theta} \cos 60°.$$

Hence $\mathbf{a}_A$, which is in the y direction, has components

$$0 = a_{Cx} - \frac{\sqrt{3}}{4} l\ddot{\theta},$$

$$a_A = a_{Cy} + \frac{1}{4} l\ddot{\theta}.$$

Consider now the acceleration $\mathbf{a}_B$ of point B. At the instant the rope is cut it has only a horizontal component. Thus $a_{By} = 0$, i.e.

$$a_{Cy} - \frac{1}{2} l\ddot{\theta} \cos 60° = a_{Cy} - \frac{1}{4} l\ddot{\theta} = 0.$$

The above consideration gives

$$a_{Cx} = \frac{\sqrt{3}}{4} l\ddot{\theta}, \qquad a_{Cy} = \frac{1}{4} l\ddot{\theta}.$$

Using these in the equations of motion for C we find

$$\ddot{\theta} = \frac{3g}{4l},$$

which gives the acceleration of the hinge as

$$a_A = \frac{1}{2} l\ddot{\theta} = \frac{3}{8} g,$$

directed vertically downward.

1206

A particle of mass m and speed v collides elastically with the end of a uniform thin rod of mass M as shown in Fig. 1.168. After the collision, m is stationary. Calculate M.

(MIT)

Fig. 1.168

Sol: Let v_c be the velocity of the center of mass of the rod and ω the angular velocity of the rod about the center of mass. Conservation of momentum and that of energy of this system give

$$mv = Mv_c,$$

$$\frac{1}{2}mv^2 = \frac{1}{2}Mv_c^2 + \frac{1}{2}I\omega^2$$

with $I = \frac{1}{12}Ml^2$, l being the length of the rod. Conservation of the angular momentum of the system about a fixed point located at the center of the rod before collision gives

$$\frac{1}{2}lmv = I\omega.$$

The above equations give

$$M = 4m.$$

1207

A uniform thin cylindrical rod of length L and mass m is supported at its ends by two massless springs with spring constants k_1 and k_2. In equilibrium the rod is horizontal, as shown in Fig. 1.169. You are asked to consider small-amplitude motion about equilibrium under circumstances where the springs can move only vertically.

a. First consider the special case $k_1 = k_2$. Find the eigenfrequencies of the normal modes and describe the corresponding normal mode motions. Here you might well be guided by intuitive reasoning.

b. Now consider the general case where k_1 and k_2 are not necessarily equal. Find the normal mode eigenfrequencies.

(Princeton)

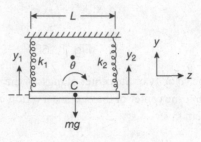

Fig. 1.169

Sol:

a. Let y_1 and y_2 be the vertical displacements from the equilibrium position of the two ends of the rod as shown in Fig. 1.169. As the displacement of the center of mass C is $\frac{1}{2}(y_1 + y_2)$, its equation of translational motion is

$$\frac{1}{2}m(\ddot{y}_1 + \ddot{y}_2) = -k_1y_1 - k_2y_2.$$

For small-amplitude rotation about the center of mass, we have

$$I\ddot{\theta} = -\frac{1}{2}L(k_1y_1 - k_2y_2)$$

with $I = \frac{1}{12}mL^2$, $\theta \approx \frac{y_1 - y_2}{L}$. For $k_1 = k_2 = k$, the equations of motion reduce to

$$\ddot{y}_1 + \ddot{y}_2 = -\frac{2k}{m}(y_1 + y_2),$$

$$\ddot{y}_1 - \ddot{y}_2 = -\frac{6k}{m}(y_1 - y_2).$$

Hence there exist two normal modes.

i. *Symmetric mode*

$$y_s = y_1 + y_2$$

with eigenfrequency $\omega_s = \sqrt{\dfrac{2k}{m}}$. This mode corresponds to vertical harmonic oscillation of the rod as a whole.

ii. *Asymmetric mode*

$$y_a = y_1 - y_2$$

with eigenfrequency $\omega_a = \sqrt{\dfrac{6k}{m}}$. This mode corresponds to harmonic oscillation about a horizontal axis perpendicular to the rod and through its center of mass.

b. For the general case $k_1 \neq k_2$, let $y_1 = A_1 e^{i\omega t}$, $y_2 = A_2 e^{i\omega t}$, where ω is the eigenfrequency of oscillation. The equations of motion now give

$$\left(k_1 - \frac{1}{2}m\omega^2\right)A_1 + \left(k_2 - \frac{1}{2}m\omega^2\right)A_2 = 0,$$

$$\left(\frac{I\omega^2}{L} - \frac{1}{2}Lk_1\right)A_1 + \left(\frac{1}{2}Lk_2 - \frac{I\omega^2}{L}\right)A_2 = 0.$$

For a non-zero solution we require

$$\begin{vmatrix} k_1 - \dfrac{1}{2}m\omega^2 & k_2 - \dfrac{1}{2}m\omega^2 \\[2mm] \dfrac{I\omega^2}{L} - \dfrac{1}{2}Lk_1 & \dfrac{1}{2}Lk_2 - \dfrac{I\omega^2}{L} \end{vmatrix} = 0,$$

i.e.

$$\frac{Im\omega^4}{L} - \left(\frac{I}{L} + \frac{1}{4}mL\right)(k_1 + k_2)\omega^2 + Lk_1k_2 = 0,$$

or

$$m^2\omega^4 - 4m(k_1 + k_2)\omega^2 + 12k_1k_2 = 0.$$

Solving for ω^2 we obtain the eigenfrequencies

$$\omega = \sqrt{\frac{2}{m}\left[(k_1 + k_2) \pm \sqrt{(k_1^2 - k_1k_2 + k_2^2)}\right]}.$$

Note that for $k_1 = k_2 = k$, this expression gives $\omega = \sqrt{\dfrac{2k}{m}}, \sqrt{\dfrac{6k}{m}}$, in agreement with (a).

1208

A rigid wheel has principal moments of inertial $I_1 = I_2 \neq I_3$ about its body fixed principal axes $\hat{x}_1$, $\hat{x}_2$ and $\hat{x}_3$, as shown in Fig. 1.170. The wheel i attached at its center of mass to a bearing which allows frictionless rotatior about one space-fixed axis. The wheel is "dynamically balanced", i.e. it car rotate at constant $\omega \neq 0$ and exert no torque on its bearing. What condition must the components of ω satisfy? Sketch the permitted motion(s).

(*MIT*

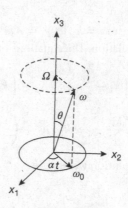

Fig. 1.170

Sol: Setting $I_1 = I_2 = I$ in Euler's equations

$$I_1 \dot{\omega}_1 + (I_3 - I_2)\omega_3\omega_2 = 0, \tag{1}$$

$$I_2 \dot{\omega}_2 + (I_1 - I_3)\omega_1\omega_3 = 0, \tag{2}$$

$$I_3 \dot{\omega}_3 + (I_2 - I_1)\omega_2\omega_1 = 0, \tag{3}$$

we see that (3) can be readily integrated to give

$$\omega_3 = \text{constant} = \Omega, \text{ say.}$$

We then rewrite (1) and (2) as

$$\dot{\omega}_1 = -\left(\frac{I_3 - I}{I}\right)\Omega\omega_2,$$

$$\dot{\omega}_2 = \left(\frac{I_3 - I}{I}\right)\Omega\omega_1,$$

which are the conditions that must be satisfied. Differentiating these equations gives

$$\ddot{\omega}_1 = -\left(\frac{I_3 - I}{I}\Omega\right)\dot{\omega}_2 = -\left(\frac{I_3 - I}{I}\Omega\right)^2 \omega_1 = -\alpha^2\omega_1,$$

$$\ddot{\omega}_2 = -\left(\frac{I_3 - I}{I}\Omega\right)^2 \omega_2 = -\alpha^2\omega_2,$$

where $\alpha = \left(\frac{I_3 - I}{I}\right)\Omega$. The general solution is

$$\omega_1 = \omega_0 \cos(\alpha t + \varepsilon), \qquad \omega_2 = \omega_0 \sin(\alpha t + \varepsilon).$$

Hence the total angular velocity has magnitude

$$\omega = \sqrt{\Omega^2 + \omega_1^2 + \omega_2^2} = \sqrt{\Omega^2 + \omega_0^2},$$

which is a constant. As $\omega_3 = \Omega$ is a constant the total angular velocity vector $\boldsymbol{\omega}$ makes a constant angle θ with the $\hat{\mathbf{x}}_3$-axis as shown in Fig. 1.170. Furthermore the plane of $\boldsymbol{\omega}$ and $\hat{\mathbf{x}}_3$ rotates about the $\hat{\mathbf{x}}_3$-axis with an angular velocity α, or a period

$$\frac{2\pi}{\alpha} = \frac{2\pi I}{(I_3 - I)\Omega}.$$

The motion, which is the only one allowed, is sketched in Fig. 1.170.

1209

A rigid body is in space. All external influences (including gravity) are negligible.

a. Use Newton's law to show that angular momentum is conserved; mention any assumptions made.

b. Suppose the center of mass of the body is at rest in an inertial frame. Must its axis of rotation have a fixed direction? Justify your answer briefly.

(UC, Berkeley)

Sol:

a. The angular momentum of a rigid body about a fixed point O is defined as

$$\mathbf{L} = \sum_i^n \mathbf{r}_i \times m_i \dot{\mathbf{r}}_i,$$

where $\mathbf{r}_i$ is the radius vector from O of a particle m_i of the rigid body, which consists of n particles. As there are no external forces, only internal forces act and according to Newton's second law

$$m_i \ddot{\mathbf{r}}_i = \sum_{j \neq i}^{n} \mathbf{F}_{ij},$$

where $\mathbf{F}_{ij}$ is the force acting on m_i by particle m_j of the rigid body. Consider

$$\sum_{i}^{n} \mathbf{r}_i \times m_i \ddot{\mathbf{r}}_i = \sum_{i}^{n} \sum_{j \neq i}^{n} \mathbf{r}_i \times \mathbf{F}_{ij}. \tag{1}$$

By Newton's third law, the internal forces $\mathbf{F}_{ij}$ occur in pairs such that

$$\mathbf{F}_{ij} = -\mathbf{F}_{ji},$$

both acting along the same line joining the two particles. This means that the double-summation on the right-hand side of (1) consists of sums like

$$\mathbf{r}_i \times \mathbf{F}_{ij} + \mathbf{r}_j \times \mathbf{F}_{ji}.$$

As shown in Fig. 1.171, each such sum adds up to zero. Hence

$$\sum_{r}^{n} \mathbf{r}_i \times m \ddot{\mathbf{r}}_i = 0.$$

Then

$$\dot{\mathbf{L}} = \sum_{i}^{n} \dot{\mathbf{r}}_i \times m \dot{\mathbf{r}}_i + \sum_{i}^{n} \mathbf{r}_i \times m \ddot{\mathbf{r}}_i = 0,$$

or

$$\mathbf{L} = \text{constant}.$$

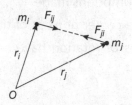

Fig. 1.171

That is, the angular momentum of a rigid body about an arbitrary point is conserved.

b. The above argument holds also for a point fixed in an inertial frame, so that the angular momentum **L** of the body about the center of mass is a constant vector in the inertial frame. However, the angular velocity ω of the body about the center of mass need not be in the same direction as **L**. Only when the axis of rotation is along a principal axis of the body is ω parallel to **L**. Hence, in general the axis of rotation is not fixed even though the direction of **L** is.

1210

The trash can beside the Physics Department mailboxes has a conical-shaped lid which is supported by a pivot at the center. Suppose you tip the cone of the lid and spin it rapidly with spin velocity ω about the symmetry axis of the cone (Fig. 1.172). Does the lid precess in the same or opposite sense to the spin direction of ω? Document your answer with appropriate formula and vector diagram.

(Wisconsin)

Sol: The torque of gravity about O is

$$\mathbf{M} = \overline{OC} \times m\mathbf{g}.$$

In a fixed frame we have

$$\frac{d\mathbf{L}}{dt} = \mathbf{M} = -\frac{\overline{OC}\mathbf{L}}{L} \times m\mathbf{g} = \frac{\overline{OC}}{L}m\mathbf{g} \times \mathbf{L} = \omega_p \times \mathbf{L},$$

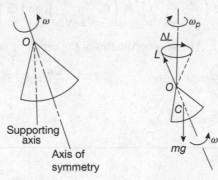

Fig. 1.172   Fig. 1.173

where $\omega_p = \dfrac{\overline{OC}mg}{L}$. Thus **L** and hence the axis of symmetry of the lip precess with

angular velocity $\omega_p = \dfrac{\overline{OC}mg}{L}$ about the vertical axis in a sense opposite to that of

the spin, as shown in Fig. 1.173.

1211

Masses m, $2m$, $3m$, and $4m$ are placed in the vertices of a regular tetrahedron of side a shown in the figure. Determine the position of the center of mass of the system.

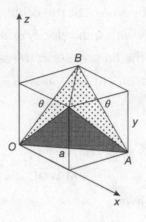

Fig. 1.174

Sol: Taking one of the vertex to be at the origin, the coordinates of the vertices are

$$O: (0, 0, 0), \quad A: \left(\frac{a}{\sqrt{2}}, \frac{a}{\sqrt{2}}, 0\right), \quad B: \left(0, \frac{a}{\sqrt{2}}, \frac{a}{\sqrt{2}}\right), \text{ and } C: \left(\frac{a}{\sqrt{2}}, 0, \frac{a}{\sqrt{2}}\right).$$

Let (x_{cm}, y_{cm}, z_{cm}) be the coordinates of the center of mass of the system. Then,

$$x_{cm} = \frac{2a + 4a}{10\sqrt{2}} = 0.3\sqrt{2}a; \qquad y_{cm} = \frac{2a + 3a}{10\sqrt{2}} = 0.25\sqrt{2}a;$$

$$z_{cm} = \frac{3a + 4a}{10\sqrt{2}} = 0.35\sqrt{2}a$$

1212

We consider an ideal free gyro, i.e. a rotationally symmetric rigid body (with principal moments of inertia $I_1 = I_2 < I_3$) so suspended that it can rotate freely about its center of gravity, and move under the influence of no torque. Let

$\omega(t)$ be the instantaneous angular velocity vector, and let $\mathbf{L}(t)$ be the instantaneous angular momentum. Let the unit vector $\mathbf{u}(t)$ point along the symmetry axis of the body (associated with the moment of inertia I_3). These vectors are in an inertial frame with respect to which the body rotates. Derive expressions for $\mathbf{L}(t)$, $\omega(t)$, and $\mathbf{u}(t)$ in terms of initial values $\mathbf{u}_0 = \mathbf{u}(0)$ and $\omega_0 = \omega(0)$.

(UC, Berkeley)

Sol: Let $t = 0$ be an instant when $\mathbf{L}$, ω and the axis of symmetry of the gyro, $\mathbf{u}$, are coplanar. Use a fixed coordinate frame $Oxyz$ with origin at the center of mass of the gyro which at $t = 0$ has the z-axis along the angular momentum vector $\mathbf{L}$ and the y-axis perpendicular to the angular velocity ω_0. Also use a rotating coordinate frame $Ox'y'z'$ attached to the gyro such that the z'-axis coincides with the axis of symmetry and the x'-axis is in the plane of z'- and z-axes at $t = 0$. The relation between the two frames is shown in Fig. 1.175, which also defines the Eulerian angles θ, φ, ψ. Note that initially the y'-and y-axes coincide and $\psi_0 = \varphi_0 = 0$.

As seen from Fig. 1.175, the angular velocity $\omega(t)$ of the gyro can be expressed in the rotating frame in terms of the Eulerian angles as

$$\omega_{x'} = \dot{\theta} \sin \psi - \dot{\varphi} \sin \theta \cos \psi,$$
$$\omega_{y'} = \dot{\theta} \cos \psi + \dot{\varphi} \sin \theta \sin \psi,$$
$$\omega_{z'} = \dot{\varphi} \cos \theta + \dot{\psi}$$

Since the x'-, y'- and z'-axes are principal axes, $\mathbf{L}$ can be expressed as

$$\mathbf{L} = I_1 \omega_{x'} \mathbf{i}' + I_1 \omega_{y'} \mathbf{j}' + I_3 \omega_{z'} \mathbf{k}' \qquad (1)$$

for $I_1 = I_2$. As there is no torque acting on the gyro, $\mathbf{L} = $ constant and is along the z-axis. Furthermore, the Euler equation

$$I_3 \dot{\omega}_{z'} - (I_1 - I_2)\omega_{x'} \omega_{y'} = 0$$

gives for $I = I_2$,

$$\omega_{z'} = \text{constant} = \omega_{0z'}.$$

As

$$L = \sqrt{I_1^2(\omega_{x'}^2 + \omega_{y'}^2) + I_3^2 \omega_{0z'}^2} = \text{constant},$$

we have

$$\omega_{x'}^2 + \omega_{y'}^2 = \text{constant} = \omega_{0x'}^2 + \omega_{0y'}^2 = \omega_{0x'}^2$$

since $\omega_{0y'} = \omega_{0y} = 0$. Hence

$$\mathbf{L} = \sqrt{I_1^2 \omega_{0x'}^2 + I_3^2 \omega_{0z'}^2}.$$

It can also be expressed in terms of the Eulerian angles as (Fig. 1.175)

$$\mathbf{L} = -L \sin\theta \cos\psi \mathbf{i}' + L \sin\theta \sin\psi \mathbf{j}' + L \cos\theta \mathbf{k}'$$

in the rotating frame. Comparing this with (1) we find

$$L \cos\theta = I_3 \omega_{0z'},$$

showing that $\cos\theta = \text{constant} = \cos\theta_0$, say, and thus $\dot{\theta} = 0$. Furthermore,

$$-L \sin\theta \cos\psi = I_1 \omega_{x'} = -I_1 \dot{\varphi} \sin\theta \cos\psi,$$

giving

$$\dot{\varphi} = \frac{L}{I} = \text{constant}.$$

Similarly,

$$L \cos\theta = I_3 \omega_{z'} = I_3(\dot{\varphi}\cos\theta + \dot{\psi}),$$

giving

$$\dot{\psi} = \frac{L\cos\theta}{I_3} - \frac{L\cos\theta}{I_1} = \left(1 - \frac{I_3}{I_1}\right)\frac{L\cos\theta}{I_3} = \left(1 - \frac{I_3}{I_1}\right)\omega_{0z'} = \text{constant}.$$

What the above means is that the motion of the free, symmetric gyro consists of two steady motions: a spin of angular velocity $\dot{\psi}$ about the axis of symmetry and a precession of angular velocity $\dot{\varphi}$ about the constant angular momentum vector $\mathbf{L}$.

Consider now the unit vector $\mathbf{u}(t)$, which is along the axis of symmetry, in the fixed frame (Fig. 1.176):

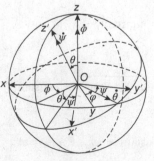

Fig. 1.175

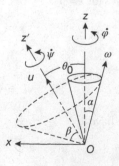

Fig. 1.176

$$\mathbf{u}(t) = \sin\theta\cos\varphi\mathbf{i} + \sin\theta\sin\varphi\mathbf{j} + \cos\theta\mathbf{k}.$$

As $\theta = \theta_0$ and at $t = 0$, $\varphi = 0$, we have

$$\mathbf{u}(0) = \sin\theta_0\mathbf{i} + \cos\theta_0\mathbf{k},$$

and, as $\varphi = \dot{\varphi}t$,

$$\mathbf{u}(t) = u_{0x}\cos(\dot{\varphi}t)\mathbf{i} + u_{0x}\sin(\dot{\varphi}t)\mathbf{j} + u_{0z}\mathbf{k}.$$

Consider the angular velocity $\boldsymbol{\omega}$. In the rotating frame, we have for time t

$$\boldsymbol{\omega} = (-\dot{\varphi}\sin\theta_0\cos\psi, \dot{\varphi}\sin\theta_0\sin\psi, \dot{\varphi}\cos\theta_0 + \dot{\psi}),$$

as $\theta = \theta_0$, $\dot{\theta} = 0$, and for time $t = 0$

$$\boldsymbol{\omega}_0 = (-\dot{\varphi}\sin\theta_0, 0, \dot{\varphi}\cos\theta_0 + \dot{\psi}) = (\omega_{0x'}, 0, \omega_{0z'}).$$

Thus

$$\boldsymbol{\omega} = (\omega_{0x'}\cos\psi, -\omega_{0x'}\sin\psi, \omega_{0z'})$$

with

$$\omega = \sqrt{\omega_{0x'}^2 + \omega_{0z'}^2} = \omega_0.$$

Hence $\boldsymbol{\omega}$ has a constant magnitude. It makes an angle α with the z-axis given by

$$\cos\alpha = \frac{\omega_z}{\omega} = \frac{\boldsymbol{\omega}\cdot\mathbf{L}}{\omega L}$$

$$= \frac{1}{\omega}(\dot{\varphi}\sin^2\theta_0\cos^2\psi + \dot{\varphi}\sin^2\theta_0\sin^2\psi + \dot{\varphi}\cos^2\theta_0 + \dot{\psi}\cos\theta_0)$$

$$= \frac{\dot{\varphi} + \dot{\psi}\cos\theta_0}{\omega},$$

which is a constant as $\dot{\varphi}, \dot{\psi}, \omega$ are all constants. It makes an angle β with the z'-axis given by

$$\cos\beta = \frac{\omega_{z'}}{\omega} = \frac{\boldsymbol{\omega}\cdot\mathbf{k}'}{\omega} = \frac{\dot{\varphi}\cos\theta_0 + \dot{\psi}}{\omega},$$

which is also a constant. In the fixed frame,

$$\omega = (\dot{\psi}\sin\theta\cos\varphi - \dot{\theta}\sin\varphi, \dot{\psi}\sin\theta\sin\varphi - \dot{\theta}\cos\varphi, \dot{\psi}\cos\theta + \dot{\varphi})$$
$$= (\dot{\psi}\sin\theta_0\cos\varphi, \dot{\psi}\sin\theta_0\sin\varphi, \dot{\psi}\cos\theta_0 + \dot{\varphi})$$

as $\theta = \theta_0$, $\dot{\theta} = 0$. At $t = 0$, $\varphi = \psi = 0$ so that

$$\omega_0 = (\dot{\psi}\sin\theta_0, 0, \dot{\psi}\cos\theta_0 + \dot{\varphi}).$$

Hence

$$\omega(t) = \omega_{0x}\cos(\dot{\varphi}t)\mathbf{i} + \omega_{0x}\sin(\dot{\varphi}t)\mathbf{j} + \omega_{0z}\mathbf{k}.$$

The precessions of ω about $\mathbf{L}$ and $\mathbf{u}$ are depicted in Fig. 1.176. Note that $\mathbf{u}$ itself precesses about $\mathbf{L}$.

1213

Let I_1, I_2, I_3 be the principal moments of inertia (relative to the center of mass) of a rigid body and suppose these moments are all different with $I_1 > I_2 > I_3$. If the body in free space is set to spin around one of the principal axes, it will continue spinning about that axis. However, we are concerned about the stability. What happens if the initial spin axis is very close to, but not exactly aligned with, a principal axis? Stability implies that the spin axis never wanders far from that principal axis. One finds that the motion is in fact stable for the principal axes corresponding to I_1 and I_3, the largest and the smallest moments of inertia. Explain this analytically using Euler's equations.

(CUSPEA)

Sol: Let ω_1, ω_2, ω_3 be the components of the angular velocity along the principal axes. Then, using Euler's equations for zero torque

$$I_1\dot{\omega}_1 - \omega_2\omega_3(I_2 - I_3) = 0,$$
$$I_2\dot{\omega}_2 - \omega_3\omega_1(I_3 - I_1) = 0,$$
$$I_3\dot{\omega}_3 - \omega_1\omega_2(I_1 - I_2) = 0,$$

we consider the following cases.

i. Suppose initially ω directs almost parallel to the x-axis, i.e. $\omega_1 \gg \omega_2, \omega_3$. If ω_2, ω_3 remain small in the subsequent rotation, the motion is stable. As $|\omega| = $ constant and $\omega = \sqrt{\omega_1^2 + \omega_2^2 + \omega_3^2} \approx \omega_1$, we can take ω_1 to be constant to first order. Then

$$\ddot{\omega}_2 = \frac{\omega_1(I_3 - I_1)}{I_2}\,\dot{\omega}_3 = \frac{\omega_1^2(I_3 - I_1)(I_1 - I_2)}{I_2 I_3}\,\omega_2,$$

$$\ddot{\omega}_3 = \frac{\omega_1^2(I_1 - I_2)(I_3 - I_1)}{I_3 I_2}\,\omega_3.$$

As $I_1 > I_2 > I_3$, the coefficients on the right-hand side of the above are both negative and the equations of motion have the form of that of a harmonic oscillator. Thus ω_2 and ω_3 will oscillate about same equilibrium values and remain small. Hence the motion is stable. The same conclusion is drawn if ω is initially almost parallel to the z-axis.

ii. If ω initially is almost parallel to the y-axis. The same consideration gives

$$\ddot{\omega}_1 = \frac{\omega_2^2(I_2 - I_3)(I_1 - I_2)}{I_1 I_3}\,\omega_1,$$

$$\ddot{\omega}_3 = \frac{\omega_2^2(I_1 - I_2)(I_2 - I_3)}{I_3 I_1}\,\omega_3.$$

As $I_2 > I_3$, $I_1 > I_2$, the coefficients on the right-hand side are both positive and the motion is unstable at least in first-order approximation.

1214

A spherical ball of mass m, radius R and uniform density is attached to a massless rigid rod of length l in such a way that the ball may spin around the rod. The ball is in a uniform gravitational field, say that of the earth. Supposing the ball and the rod rotate about the z-axis without nutation (i.e. θ is fixed), the angular velocity of the rod and ball about the z-axis is ω, and the ball spins about the rod with angular velocity Ω. Give the relation between ω and Ω (you may assume $R/l \ll 1$ though this is not necessary for the form of the solution). Does the ball move in a right-handed or a left-handed sense about the z-axis?

(Columbia)

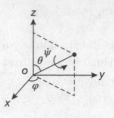

Fig. 1.177

Sol: The orientation of the ball can be described in terms of the Eulerian angles θ, φ, ψ (Problem **1212**). As there is no nutation, $\dot{\theta} = 0$. The angular momentum of the ball about the origin O (Fig. 1.177) is

$$\mathbf{L} = \frac{2}{5} mR^2 \dot{\varphi} \mathbf{e}_r + l \sin\theta \cdot \dot{\varphi} m l \sin\theta \mathbf{e}_z$$

$$= \frac{2}{5} mR^2 \Omega \mathbf{e}_r + m l^2 \sin^2\theta \omega \mathbf{e}_z$$

in cylindrical coordinates. As $\mathbf{e}_z$ is fixed, θ is a constant, we have

$$\frac{d\mathbf{L}}{dt} = \frac{2}{5} mR^2 \Omega \frac{d\mathbf{e}_r}{dt} = \mathbf{M},$$

where M is the torque due to gravity. As

$$\frac{d\mathbf{e}_r}{dt} = \dot{\theta}\mathbf{e}_\theta + \dot{\varphi}\sin\theta\mathbf{e}_\varphi = \omega\sin\theta\mathbf{e}_\varphi,$$

the above becomes

$$\frac{2}{5} mR^2 \Omega\omega\sin\theta\mathbf{e}_\varphi = l\mathbf{e}_r \times mg(-\mathbf{e}_z) = lmg\sin\theta\mathbf{e}_\varphi.$$

Hence

$$\omega = \frac{5lg}{2R^2\Omega}.$$

As $\dot{\varphi} \equiv \omega > 0$, the ball moves in a right-handed sense about the z-axis.

1215

A gyroscope at latitude 45°N is mounted on bearings in such a way that the axis of spin is constrained to be horizontal but otherwise no torques occur in the bearings. Taking into account the rotation of the earth, show that an orientation with the axis of spin along the local north-south is stable and find the period for small oscillations of the spin axis about this direction. Assume that the rotor can be approximated by a thin circular ring (i.e. the spokes and other parts are of negligible mass). (In working out this problem it is simpler when writing the angular velocity of the rotor

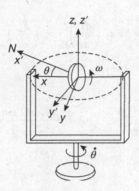

Fig. 1.178

about the x-axis (Fig. 1.178) to lump together the spin term and the term due to rotation of the earth).

(*UC, Berkeley*)

Sol: Use an inertial frame $Ox'y'z'$ fixed with respect to a distant star which, at the instant under consideration, has the origin O at the center of mass of the rotor, the z'-axis pointing vertically up and the x'-axis pointing north, and a rotating frame $Oxyz$ attached to the earth with the same z-axis but with the x-axis at that instant along the spin axis of the rotor as shown in Fig. 1.178. Denote the spin angular velocity by ω, and the moments of inertia about the x-, y-, z-axes, which are the principal axes of the gyroscope, by C, A, A respectively. The angular momentum then has components

$$(C\omega, 0, A\dot{\theta})$$

in the rotating frame, and

$$(C\omega \cos \theta, C\omega \sin \theta, A\dot{\theta})$$

in the fixed frame. Note that the z component which is the same in both frame is contributed by the precession. In the fixed frame, the earth's rotational angula velocity at latitude $\lambda = 45°$N has components

$$\Omega(\cos 45°, 0, \sin 45°) = \frac{\Omega}{\sqrt{2}}(1, 0, 1).$$

Also, the only torques are those that constrain the spin axis to the horizontal s that

$$M_{z'} = 0.$$

As

$$\mathbf{M} = \left(\frac{d\mathbf{L}}{dt}\right)_{\text{fix}} = \left(\frac{d\mathbf{L}}{dt}\right)_{\text{rot}} + \mathbf{\Omega} \times \mathbf{L} = 0,$$

its z component is

$$A\ddot{\theta} + \frac{C\omega\Omega \sin\theta}{\sqrt{2}} = 0,$$

or

$$A\ddot{\theta} + \frac{C\omega\Omega}{\sqrt{2}}\theta = 0$$

for small θ. Note that for $\mathbf{\Omega} \times \mathbf{L}$ we have resolved the vectors in the fixed frame. The last equation shows that the spin axis oscillates harmonically about the local north-south direction with angular frequency

$$\omega' = \sqrt{\frac{C\omega\Omega}{\sqrt{2}A}}$$

and the orientation is stable. The period is

$$T = \frac{2\pi}{\omega} = 2\pi\sqrt{\frac{\sqrt{2}A}{C\omega\Omega}}.$$

If the rotor is approximated by a thin circular ring of mass M and radius R, we have

$$C = MR^2, \quad A = \frac{MR^2}{2}, \quad T = \frac{2\pi}{\sqrt{\sqrt{2}\omega\Omega}}.$$

1216

A thin disk of mass M and radius A is connected by two springs of spring constant k to two fixed points on a frictionless table top. The disk is free to rotate but it is constrained to move in a plane. Each spring has an unstretched length of l_0, and initially both are stretched to length $l > l_0$ in the equilibrium position, as shown in Fig. 1.179. What are the frequencies

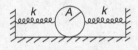

Fig. 1.179

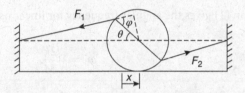

Fig. 1.180

of the normal modes of oscillation for small vibrations? Sketch the motion for each mode.

(Princeton)

Sol: The motion of the disk is confined to the vertical plane. Let the displacement of the center of mass from equilibrium be x and the angular displacement be θ, as shown in Fig. 1.180. To first order in θ, the restoring forces are

$$F_1 = k(l + x - l_0), \quad F_2 = k(l - x - l_0).$$

The equations of motion are then

$$M\ddot{x} = F_2 - F_1 = -2kx,$$

or

$$\ddot{x} + \frac{2k}{M}x = 0, \tag{1}$$

and

$$I\ddot{\theta} = (F_2 + F_1)A\sin\varphi,$$

where $I = \frac{1}{2}MA^2$ and φ is given by

$$\frac{\sin(\pi - \varphi)}{l + A + x} = \frac{\sin\theta}{l + x},$$

or

$$\sin\varphi \approx \left(\frac{l + A}{l}\right)\sin\theta \approx \left(\frac{l + A}{l}\right)\theta,$$

i.e.

$$\ddot{\theta} + \frac{4k(l - l_0)(l + A)}{MlA}\theta = 0. \tag{2}$$

Equation (1) gives the angular frequency for linear oscillation,

$$\omega_1 = \sqrt{\frac{2k}{M}}.$$

Equation (2) gives the angular frequency for rotational oscillation,

$$\omega_2 = \sqrt{\frac{4k(l - l_0)(l + A)}{MlA}}.$$

The normal mode fequencies of small oscillations are therefore

$$\frac{\omega_1}{2\pi}, \quad \frac{\omega_2}{2\pi},$$

and the motions of the two normal modes are as shown in Fig. 1.181.

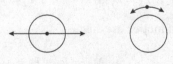

Fig. 1.181

1217

A simple symmetrical top consists of a disk of mass M and radius r mounted on the center C of the massless cylindrical rod of length l and radius a as shown in Fig. 1.182. The top is rotated with large angular velocity $\omega(t)$ and is placed at an angle θ to the vertical on a horizontal surface with a small coefficient of friction. Neglect nutation and assume that the rate of slowing of $\omega(t)$ is small in one period of procession.

   a. Describe the entire subsequent motion of the top.

   b. Compute the angular frequency of the (slow) precession.

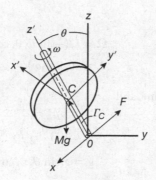

Fig. 1.182

   c. Estimate the time required before the axis of the top becomes vertical.

(UC, Berkeley)

Sol:

   a. The motion of the top consists mainly of three components:

      1. spinning with angular velocity ω about its axis of symmetry,

      2. a slow procession Ω about the vertical axis due to gravity,

      3. motion of the axis of symmetry to come to the vertical gradually due to the effect of the frictional torque.

   b. Use two coordinate frames with origin O as shown in Fig. 1.182: a fixed frame $Oxyz$ with the z-axis along the upward vertical, and a rotating frame $Ox'y'z'$ with the z'-axis along the axis of symmetry in the same direction as the spin angular velocity ω, both the x- and x'-axes being taken in the plane of the z- and z'-axes at the instant under consideration. We have

$$\left(\frac{d\mathbf{L}}{dt}\right)_{\text{fix}} = \left(\frac{d\mathbf{L}}{dt}\right)_{\text{rot}} + \mathbf{\Omega} \times \mathbf{L}.$$

Under the condition that the spin angular velocity ω is very large, the total angular momentum can be taken to be approximately

$$\mathbf{L} = I_3\omega\mathbf{k}'.$$

Further, as ω does not change appreciably in a period of precession, $\left(\dfrac{d\mathbf{L}}{dt}\right)_{\text{rot}} \approx 0$. We then have

$$\left(\frac{d\mathbf{L}}{dt}\right)_{\text{fix}} = \mathbf{\Omega} \times \mathbf{L} = \mathbf{r}_C \times M\mathbf{g}$$

with

$$\mathbf{\Omega} = \Omega(-\sin\theta, 0, \cos\theta),$$
$$\mathbf{L} = (0, 0, I_3\omega)$$

in the rotating frame, and

$$\mathbf{r}_C = \left(0, 0, \frac{l}{2}\right),$$
$$\mathbf{g} = g(\sin\theta, 0, \cos\theta).$$

In the fixed frame, the above gives

$$I_3\omega\Omega\,\sin\theta = \frac{1}{2}Mlg\,\sin\theta,$$

i.e.

$$\Omega = \frac{Mlg}{2I_3\omega} = \frac{lg}{r^2\omega},$$

as

$$I_3 = \frac{1}{2}Mr^2.$$

c.   When the axis of symmetry makes an angle θ with the vertical, the frictional force f on the contact point of the rod with the ground is approximately μMg. Actually only the left edge of the bottom end touches the ground. The frictional force is opposite to the slipping velocity of the contact point and has the direction shown in Fig. 1.182. This force causes an acceleration of the center of mass C of the top and generates a torque about C at the same time. Neglecting any specific condition of the rod, we can take the torque about C as approximately

$$\tau \approx \mu Mg \cdot \frac{1}{2} l\mathbf{j}.$$

This torque changes the magnitude of the angle θ and causes the axis of symmetry to eventually become vertical.

When the axis is vertical, the bottom of the rod contacts the ground evenly so that the frictional force is distributed symmetrically. The total torque about C due to friction is then zero. Actually the torque of the frictional force about the z'-axis (relating to the z' component of $\mathbf{L}$) does not vanish altogether, but as the rod is so thin the torque is quite small and causes ω to decrease only slowly. We have for the frictional torque approximately

$$\frac{d\mathbf{L}_C}{dt} = \frac{1}{2}\mu Mgl\mathbf{j},$$

i.e.

$$-\dot{\theta}I_3\omega = \frac{1}{2}\mu Mgl,$$

or

$$\frac{d\theta}{dt} = -\frac{\mu gl}{r^2\omega},$$

which gives

$$t = -\int_{\theta}^{0} \frac{r^2\omega}{\mu gl}\, d\theta = \frac{r^2\omega}{\mu gl}\theta.$$

1218

A heavy symmetrical top with one point fixed is precessing at a steady angular velocity Ω about the vertical axis z. What is the minimum spin ω' about its symmetrical axis z' (z' is inclined at an angle θ with respect to the z-axis)? The top has mass m and its center of gravity is at a distance h from the fixed point. Use the coordinate systems indicated in Fig. 1.183, with the axes z, z', x and x' in the same plane at the time under consideration and assume $I_1 = I_2$.

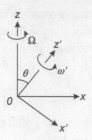

Fig. 1.183

(SUNY, Buffalo

Sol: Referring to the Eulerian angles defined in Problem **1212**, the torque due to gravity is in a direction perpendicular to the xz-plane and in the rotating frame $Ox'y'z'$ attached to the top has components

$$mgh \sin\theta \sin\psi, \qquad mgh \sin\theta \cos\psi, \qquad 0.$$

Euler's equations, which apply in the rotating frame, are, for $I_1 = I_2$,

$$I_1\dot{\omega}_{x'} - (I_1 - I_3)\omega_{y'}\omega_{z'} = mgh \sin\theta \sin\psi,$$

$$I_1\dot{\omega}_{y'} - (I_3 - I_1)\omega_{x'}\omega_{z'} = mgh \sin\theta \cos\psi,$$

$$I_3\dot{\omega}_{z'} = 0.$$

The angular velocity vector ω in the rotating frame has components

$$-\dot{\varphi}\sin\theta \cos\psi, \qquad \dot{\varphi}\sin\theta \sin\psi, \qquad \dot{\varphi}\cos\theta + \dot{\psi}$$

as $\dot{\theta} = 0$. So writing Ω for $\dot{\varphi}$ and noting that $\ddot{\varphi} = 0$ for steady precession, the first Euler's equation becomes

$$\Omega^2(I_1 - I_3)\cos\theta - \Omega I_3\dot{\psi} + mgh = 0,$$

giving

$$\omega' \equiv \dot{\psi} = \frac{mgh + (I_1 - I_3)\Omega^2 \cos\theta}{I_3\Omega}.$$

However for Ω to be real we require that

$$I_3^2\omega'^2 - 4(I_1 - I_3)mgh \cos\theta \geq 0,$$

or

$$\omega' \geq \frac{1}{I_3}\sqrt{4(I_1 - I_3)mgh \cos\theta}.$$

1219

The game of "Jacks" is played with metal pieces that can be approximated by six masses on orthogonal axes of length *l* with total mass *M*, as shown in Fig 1.184.

a. If you spin the jack around one of the axes so that there is a steady precession around the vertical (Fig. 1.185) what is the relation between the spin velocity *s*, the precession rate, and the angle θ between the vertical and the rotation axis of the jack?

b. What must the spin velocity be for the jack to spin stably around a vertical axis (i.e. $\theta = 0$)?

(Princeton)

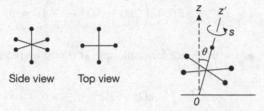

Fig. 1.184   Fig. 1.185

Sol: Use fixed frame *Oxyz* and rotating frame *Ox′y′z′* as in Problem **1212** with *O* at the point of contact with the ground and the latter frame attached to the jack. The *z*-axis is along the upward vertical and the *z′*-axis is along the axis of spin as shown in Fig. 1.185. The moments of inertia about the *x′*−, *y′*− and *z′*-axes are

$$I_1 = I_2 = 4ml^2 + 6ml^2 = 10ml^2,$$
$$I_3 = 4ml^2,$$

with $m = \dfrac{M}{6}$.

a. In the rotating frame, the torque due to gravity has components

$$6mgl \sin\theta \sin\psi, \qquad 6mgl \sin\theta \cos\psi, \qquad 0,$$

and the angular velocity ω has components

$$\dot{\theta} \sin\psi - \dot{\varphi} \sin\theta \cos\psi, \qquad \dot{\theta} \cos\psi + \dot{\varphi} \sin\theta \sin\psi, \qquad \dot{\psi} + \dot{\varphi} \cos\theta.$$

Euler's equations then give

$$5\dot\omega_{x'} - 3\omega_{y'}\,\omega_{z'} = \frac{3g}{l}\sin\theta\sin\psi, \tag{1}$$

$$5\dot\omega_{y'} + 3\omega_{z'}\,\omega_{x'} = \frac{3g}{l}\sin\theta\cos\psi, \tag{2}$$

$$4\dot\omega_{z'} = 0.$$

The last equation gives

$$\omega_{z'} = \dot\psi + \dot\varphi\cos\theta = s + \Omega\cos\theta = \text{constant}.$$

where Ω is the precession rate.

b. If the spin axis is nearly vertical, $\theta \approx 0$ and we take the approximation $\sin\theta \approx \theta$, $\cos\theta \approx 1$. Then $\sin\psi \times (1) + \cos\psi \times (2)$ gives

$$5\ddot\theta + \left(2\Omega s - 3\Omega^2 - \frac{3g}{l}\right)\theta = 0$$

with $\Omega = \dot\varphi$, $s = \dot\psi$. Hence for stable spin at $\theta = 0$ we require

$$2\Omega s - 3\Omega^2 - \frac{3g}{l} > 0,$$

or

$$s > \frac{3\Omega}{2} + \frac{3g}{2l\Omega}.$$

1220

A propeller-driven airplane flies in a circle, counterclockwise when viewed from above, with a constant angular velocity χ with respect to an inertial frame. Its propeller turns at a constant angular velocity $d\psi/dt$ clockwise as seen by the pilot.

a. For a flat, four-bladed propeller, what relations exist among the moments of inertia?

b. Find the magnitude and direction of the torque that must be applied to the propeller shaft by the bearings to maintain level flight in a circle.

(UC, Berkeley)

Sol:

a. Take a fixed frame $Oxyz$ at the instantaneous position of the center of the propeller with the z-axis pointing vertically up and a rotating frame

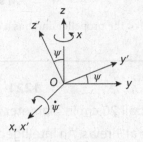

Fig. 1.186

$Ox'y'z'$ fixed to the propeller such that, the x'-axis is along the spin axis, and the z'-axis is along a propeller blade, the x-axis being taken to coincide with the x'-axes at the instant under consideration, as shown in Fig. 1.186. The rotating coordinate axes are then the principal axes with moments of inertia

$$I_2 = I_3 = I,$$

and $I_1 = 2I$ by the perpendicular axis theorem. The angular velocity has components in the rotating frame of

$$\dot{\psi}, \quad \chi \sin \psi, \quad \chi \cos \psi,$$

where $\psi = \dot{\psi}t$. Euler's equations of motion

$$I_1 \dot{\omega}_{x'} - (I_2 - I_3)\omega_y \omega_{z'} = M_{x'},$$
$$I_2 \dot{\omega}_{y'} - (I_3 - I_1)\omega_z \omega_{x'} = M_{y'},$$
$$I_3 \dot{\omega}_{z'} - (I_1 - I_2)\omega_x \omega_{y'} = M_{z'}$$

then give for the torque **M** exerted on the propeller shaft

$$M_{x'} = 0,$$

as $\dot{\psi} = $ constant and $I_2 = I_3$,

$$M_{y'} = 2I\dot{\psi}\chi \cos(\dot{\psi}t),$$
$$M_{z'} = -2I\dot{\psi}\chi \sin(\dot{\psi}t),$$

as $\chi = $ constant. Hence

$$M = 2I\dot{\psi}\chi$$

and as

$$M_{x'} = 0,$$
$$M_{y'} = M \cos \psi,$$
$$M_{z'} = -M \sin \psi,$$

M is in the plane of the propeller and has a direction along the y-axis of the fixe‹ frame.

1221

A perfectly uniform ball 20 cm in diameter and with a density of 5 g/cm³ i‹ rotating in free space at 1 rev/s. An intelligent flea of 10^{-3} g resides in a smal (massless) house fixed to the ball's surface at a rotational pole as shown ir Fig. 1.187. The flea decides to move the equator to the house by walking quickly to a latitude of 45° and waiting the proper length of time. How long should it wait? Indicate how you obtain this answer.

Note: Neglect the small precession associated with the motion of the flea or the surface of the ball.

(Princeton)

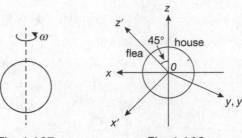

Fig. 1.187   Fig. 1.188

Sol: After the flea moves to a position of latitude 45°, the angular velocity ω no longer coincides with a principal axis of the system. This causes the ball to precess. As the mass of the flea is much smaller than that of the ball, the center of mass of the system can be taken to be at the center of the ball O. Use a fixed frame $Oxyz$ with the z-axis along the original direction of ω and a rotating frame $Ox'y'z'$ attached to the ball with the z'-axis through the new position of the flea, with both the x- and x'-axes in the plane of the z- and z'-axes at $t = 0$ as shown in Fig. 1.188. As the system is in free space, there is no external force. We assume that the flea moves so quickly to the new position that ω remains the same at $t = 0$ as for $t < 0$.

The rotating axes are the new principal axes. Let the corresponding moments of inertia be I_1, I_2 and I_3 with $I_1 = I_2$ for symmetry. Euler's equations are then

$$I_1\dot{\omega}_{x'} - (I_1 - I_3)\omega_{y'}\omega_{z'} = 0, \tag{1}$$

$$I_1\dot{\omega}_{y'} - (I_3 - I_1)\omega_{z'}\omega_{x'} = 0, \tag{2}$$

$$I_3\dot{\omega}_{z'} = 0. \tag{3}$$

Equation (3) shows that

$$\omega_{z'} = \text{constant} = \omega_{0z'}.$$

Equations (2) and (3) then give

$$\ddot{\omega}_{x'} + \Omega^2\omega_{x'} = 0$$

with $\Omega = \left|\dfrac{I_3 - I_1}{I_1}\right|\omega_{0z'}$. Its solution is

$$\omega_{x'} = A\cos(\Omega t + \phi),$$

where A and ϕ are constants. Equation (2) then gives

$$\omega_{y'} = A\sin(\Omega t + \phi).$$

Initially, ω has components in the rotating frame

$$\omega_{0x'} = -\frac{\omega}{\sqrt{2}}, \quad \omega_{0y'} = 0, \quad \omega_{0z'} = \frac{\omega}{\sqrt{2}}.$$

These give

$$A = \frac{\omega}{\sqrt{2}}, \quad \phi = \pi.$$

Hence at time t, ω has components

$$\omega_x = \frac{\omega}{\sqrt{2}}\cos(\Omega t + \pi),$$

$$\omega_y = \frac{\omega}{\sqrt{2}}\sin(\Omega t + \pi),$$

$$\omega_z = \frac{\omega}{\sqrt{2}}.$$

Thus, both the magnitude and the z' component of ω are constant, and the angular velocity vector ω describes a cone in the rigid body with axis along the z'-axis. In other words, ω precesses about the z'-axis with an angular rate

$$\Omega = \left| \frac{I_3 - I_1}{I_1} \right| \frac{\omega}{\sqrt{2}}.$$

For the equator at be at the flea house, the angular velocity ω must be midwa between the x'- and z'-axes, i.e.

$$\omega_{x'} = \frac{\omega}{\sqrt{2}}, \qquad \omega_{y'} = 0, \qquad \omega_{z'} = \frac{\omega}{\sqrt{2}}.$$

This means that $\Omega t = \pi$, or that the time required is

$$\frac{\pi}{\Omega} = \frac{\sqrt{2}\pi}{\omega} \left| \frac{I_1}{I_3 - I_1} \right| \approx \frac{\sqrt{2}\pi 2MR^2}{\omega \ 5mR^2}$$

$$= \frac{2\sqrt{2}\pi}{5\omega m}\left(\frac{4}{3}\pi R^3 \rho\right) = \frac{4\sqrt{2}}{3}\pi \times 10^6 = 6 \times 10^6 \text{s}.$$

1222

A mountaineer is caught between two cliffs as shown. The coefficient of static friction between the shoe and the wall is μ_1 and that between the clothing and the wall is μ_2. If the mountaineer is just held up by the frictional force and the normal force exerted by the mountaineer is N, determine the fraction of the body weight held up by each of the frictional forces.

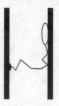

Fig. 1.189

Sol:

$$Mg = (\mu_1 + \mu_1)N$$

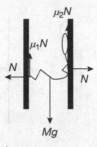

Fig. 1.190

Fraction of the weight held up by the frictional force of the shoe is

$$\frac{\mu_1 N}{Mg} = \frac{\mu_1}{\mu_1 + \mu_2}$$

and that by the clothing is

$$\frac{\mu_2}{\mu_1 + \mu_2}$$

1223

A uniform rod of length $2a$ and mass M is rotated with constant angular velocity ω in a horizontal circle of center B and radius b. The rod is hinged at A so that it can move freely only in the vertical plane containing it. The angle between the vertical and the rod is θ as shown in Fig. 1.191. The earth's gravitational field is in the vertical direction.

  a. Compute the kinetic and potential energies of the rod as a function of θ, θ and ω.

  b. Find a general expression for the possible equilibrium positions of the rod.

  c. Solve the expression found in part (b) by a graphical technique to find the equilibrium positions in each quadrant of θ between 0 and 2π.

  d. Which of these equilibrium positions are stable? Unstable? For each quadrant of θ how does the existence of the equilibrium position(s) depend on the parameters ω, b and a?

  e. For each quadrant of θ make a force diagram to verify qualitatively the existence and nature of the equilibrium positions.

(MIT)

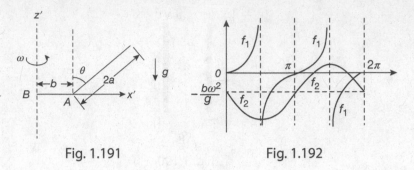

Fig. 1.191   Fig. 1.192

Sol: Use a coordinate frame $Ox'y'z'$ such that the origin O coincides with B, the z'-axis is along the axis of rotation of angular velocity ω and the x'-axis is in the vertical plane containing the z'-axis and the rod.

a.   In this rotating frame the kinetic energy of the system is

$$T = \frac{1}{2} I_A \dot{\theta}^2 = \frac{1}{2} \cdot \frac{4}{3} ma^2 \cdot \dot{\theta}^2 = \frac{2}{3} ma^2 \dot{\theta}^2.$$

The potential energy consists of two parts, a centrifugal potential and a gravitational potential. In the rotating frame, a fictitious centrifugal force $m\omega^2 x'$ must be introduced on every mass point m, corresponding to a potential $-\frac{1}{2} mx'^2 \omega^2$. For the entire system this fictitious centrifugal potential is $-\frac{1}{2} I_z \omega^2$, where $I_{z'} = \frac{1}{3} ma^2 \sin^2 \theta + m(b + a \sin \theta)^2$. As the gravitational potential is $mga \cos \theta$, we have

$$V = -\frac{1}{2} m \left[\frac{1}{3} a^2 \sin^2 \theta + (b + a \sin \theta)^2 \right] \omega^2 + mga \cos \theta.$$

b.   For equilibrium, $\frac{dV}{d\theta} = 0$, which gives the equation for possible equilibrium positions of the rod,

$$-m\omega^2 \left(b + \frac{4}{3} a \sin \theta \right) \cdot a \cdot \cos \theta - mga \sin \theta = 0,$$

or

$$\tan \theta = -\frac{a\omega^2}{g} \left(\frac{b}{a} + \frac{4}{3} \sin \theta \right).$$

c.   Let the left-hand side of the above equation be f_1 and the right-hand side be f_2 and draw these curves in Fig. 1.192. The equilibrium positions are given by their intersections. It can be seen that one equilibrium position occurs in each of the second and fourth quadrants of θ. In the third quadrant, $f_1 = \tan \theta$ is positive, and

$$f_2 = \left(-\frac{a}{g} + \frac{4a}{3g} |\sin \theta| \right) \omega^2$$

as $\sin \theta$ is negative. It is seen that only if f_2 is positive and sufficiently large can there be one or two equilibrium positions, otherwise there will be none.

d. For an equilibrium position to be stable, we require that

$$\frac{d^2 V}{d\theta^2} > 0$$

at that position. As

$$\frac{dV}{d\theta} = -m\omega^2 \left(b + \frac{4}{3} a \sin \theta \right) a \cos \theta - mga \sin \theta,$$

we require that

$$\frac{d^2 V}{d\theta^2} = -\frac{4}{3} ma^2 \omega^2 (\cos^2 \theta - \sin^2 \theta) + mab\omega^2 \sin \theta - mga \cos \theta$$

$$= \frac{ma \cos^2 \theta}{\sin \theta} \left(\frac{4}{3} a\omega^2 \sin \theta \, \tan^2 \theta + b\omega^2 \tan^2 \theta + b\omega^2 \right)$$

$$= \frac{ma \cos^2 \theta}{\sin \theta} (-g \tan^3 \theta + b\omega^2) > 0$$

for an equilibrium position θ to be stable.

When θ is in the second quadrant $\left[\frac{\pi}{2}, \pi \right]$, as $\sin \theta > 0$, $\tan \theta < 0$, we have

$$\frac{d^2 V}{d\theta^2} > 0$$

and the equilibrium is stable.

When θ is in the fourth quadrant $\left[\frac{3\pi}{2}, 2\pi \right]$, as $\sin \theta < 0$, $\tan \theta < 0$, we have

$$\frac{d^2 V}{d\theta^2} < 0$$

and the equilibrium is unstable.

When θ is in the third quadrant $\left[\pi, \dfrac{3\pi}{2}\right]$, we write

$$\frac{d^2V}{d\theta^2} = \frac{ma\cos^2\theta}{\sin\theta}\left(\frac{4}{3}a\omega^2\sin\theta\tan^2\theta + b\omega^2\sec^2\theta\right)$$

$$= \frac{ma\omega^2}{\sin\theta}\left(\frac{4}{3}a\sin^3\theta + b\right) = -\frac{ma\omega^2}{|\sin\theta|}\left(b - \frac{4}{3}a|\sin\theta|^3\right)$$

as $\sin\theta < 0$. Then if $b < \dfrac{4}{3}a|\sin\theta|^3$, the equilibrium is stable, and i

$b > \dfrac{4}{3}a|\sin\theta|^3$, the equilibrium is unstable.

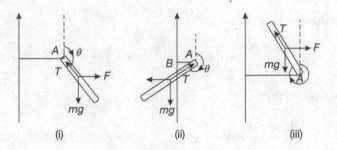

Fig. 1.193

e. The force diagram for each equilibrium situation is shown in Fig. 1.193, where (i), (ii) and (iii), are for the second, third and fourth quadrants respectively, with T and F denoting the support force by the hinge and the fictitious centrifugal force. By considering a small deviation $\delta\theta$ from equilibrium, we see that (i) is stable and (iii) is unstable, while for (ii) the situation is more complicated; whether it is stable or not depends on the relative values of the parameters.

DYNAMICS OF DEFORMABLE BODIES (1224–1272)

1224

A string is stretched between two rigid supports 100 cm apart. In the frequency range between 100 and 350 cps only the following frequencies can be excited: 160, 240, 320 cps. What is the wavelength of each of these modes of vibration?

(*Wisconsin*)

Sol: As the two ends of the string are fixed, we have $n\lambda = 2L$, where L is the length of the string and n an integer. Let the wavelengths corresponding to frequencies 160, 240, 320 Hz be $\lambda_0, \lambda_1, \lambda_2$ respectively. Then

$$n\lambda_0 = (n+1)\lambda_1 = (n+2)\lambda_2 = 200,$$
$$160\lambda_0 = 240\lambda_1 = 320\lambda_2.$$

Hence $n = 2$, and

$$\lambda_0 = 100 \text{ cm}, \quad \lambda_1 = 67 \text{ cm}, \quad \lambda_2 = 50 \text{ cm}.$$

1225

a. Give the equation which relates the fundamental frequency of a string to the physical and geometrical properties of the string.

b. Derive your result from Newton's equations by analysing what happens to a small section of the string.

(*Wisconsin*)

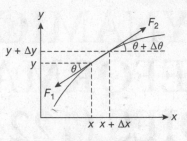

Fig. 1.194

Sol:

a. Let ω be the fundamental frequency of a string of length l, linear density ρ and tension F. The equation relating F, l and ρ is

$$\omega = \frac{\pi}{l}\sqrt{\frac{F}{\rho}}.$$

b. Consider a small length Δl of a string along the x direction undergoing small oscillations and let F_1, F_2 be the tensions at its two ends, as shown in Fig. 1.194. For small oscillations, $\theta \approx 0$ and $\Delta \theta$ is a second-order small quality. Furthermore as there is no x motion, we can take the x-component of the net force on Δl to be zero. Thus

$$f_x = F_2 \cos(\theta + \Delta\theta) - F_1 \cos\theta$$
$$\approx (F_2 - F_1)\cos\theta - F_2\Delta\theta \sin\theta$$
$$\approx F_2 - F_1 = 0,$$

or $F_2 \approx F_1$. Then

$$f_y = F\sin(\theta + \Delta\theta) - F\sin\theta \approx F\frac{d\sin\theta}{d\theta}\Delta\theta$$
$$= F\cos\theta\frac{d\theta}{dx}\Delta x \approx F\frac{d\theta}{dx}\Delta x.$$

For small θ,

$$\theta \approx \frac{dy}{dx}, \quad \frac{d\theta}{dx} = \frac{d^2y}{dx^2},$$

and the above becomes

$$\rho \Delta l \frac{\partial^2 y}{\partial t^2} = F \frac{\partial^2 y}{\partial x^2} \Delta x$$

by Newton's second law. As $\Delta l \approx \Delta x$, this gives

$$\frac{\partial^2 y}{\partial x^2} - \frac{\rho}{F} \frac{\partial^2 y}{\partial x^2} = 0,$$

which is the equation for a wave with velocity of propagation

$$v = \sqrt{\frac{F}{\rho}}.$$

For the fundamental mode in a string of length l with the two ends fixed, the wavelength λ is given by $l = \lambda/2$. Hence the fundamental angular frequency is

$$\omega = \frac{2\pi v}{\lambda} = \frac{\pi v}{l} = \frac{\pi}{l}\sqrt{\frac{F}{\rho}}.$$

1226

A violin string on a violin is of length L and can be considered to be fastened at both ends. The fundamental of the open string has a frequency f_0. The violinist bows the string at a distance $L/4$ from one end and touches the string lightly at the midpoint.

a. Under these conditions, what is the lowest frequency he can excite? Sketch the shape of the string.

b. What is the frequency of the first overtone under these conditions?

(Wisconsin)

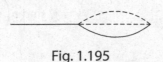

Fig. 1.195

Sol:

a. For the open string, the wavelength λ_0 corresponding to the fundamental frequency f_0 is given by $\lambda_0/2 = L$. When the violinist bows at $L/4$ from one end and touches the string at $L/2$, the former point is a node and the latter point an antinode so that $\lambda_0 = L$. Hence the string has the shape shown in Fig. 1.195 and as $f_0 \propto 1/\lambda_0$, the fundamental frequency is $2f_0$.

b. The frequency of the first overtone is $4f_0$.

1227

A guitar string is 80 cm long and has a fundamental frequency of 400 Hz. In its fundamental mode the maximum displacement is 2 cm at the middle. If the tension in the string is 10^6 dynes, what is the maximum of that component of the force on the end support which is perpendicular to the equilibrium position of the string?

(Wisconsin)

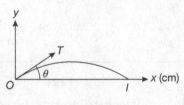

Fig. 1.196

Sol: Use Cartesian coordinates with the x-axis along the equilibrium position of the string and the origin at one of its fixed ends. Then the two fixed ends are at $x = 0$ and $x = l = 80$ cm, as shown in Fig. 1.196. At $x = 0$, the y-component of the force on the support is

$$F_y = T \sin \theta \simeq T\theta \simeq T\frac{\partial y}{\partial x},$$

where T is the tension in the string. The guitar string has a sinusoidal form

$$y = y_0 \sin \left[\omega \left(t - \frac{x}{v} \right) \right]$$

with $\omega = \dfrac{2\pi v}{\lambda} = \dfrac{2\pi v}{2l} = \dfrac{\pi v}{80}$, $y_0 = 2$ cm. Thus

$$y = 2 \sin \left(\omega t - \frac{\pi x}{80} \right) \text{ cm.}$$

Hence at $x = 0$,

$$F_y = -\frac{2\pi T}{80}\cos(\omega t)$$

and

$$F_{y\,\text{max}} = \frac{\pi T}{40} = 7.85 \times 10^4 \text{ dynes.}$$

1228

A transverse traveling sinusoidal wave on a long stretched wire of mass per unit length ρ has frequency ω and wave speed c. The maximum amplitude is y_0, where $y_0 \ll \lambda$. The wave travels toward increasing x.

a. Write an expression for the amplitude y as a function of t and x, where x is distance measured along the wire.

b. What is the energy density (energy/unit length)?

c. What is the power transmitted along the wire?

d. If the wave is generated by a mechanical device at $x = 0$, find the transverse force $F_y(t)$ that it exerts on the wire.

(Wisconsin)

Sol:

a. $y = y_0 \sin\left[\omega\left(t - \frac{x}{c}\right)\right].$

b. Every point the wave travels through undergoes simple harmonic motion. Consider an element of the wire from x to $x + \Delta x$. The mechanical energy of the element is the sum of its potential and kinetic energies and is a constant equal to the maximum of its kinetic energy. As

$$\dot{y} = \omega y_0 \cos\left[\omega\left(t - \frac{x}{c}\right)\right],$$

the maximum vibrational velocity of the element is ωy_0 and its total mechanical energy is

$$\frac{1}{2}\Delta m \cdot \omega^2 y_0^2 = \frac{1}{2}\rho\omega^2 y_0^2 \Delta x.$$

Hence the energy per unit length of the string is

$$E = \frac{1}{2}\rho\omega^2 y_0^2.$$

c. As the wave travels at a speed c, the energy that passes through a point on th string in time t is Ect. Hence the power transmitted is

$$\frac{1}{2}\rho c \omega^2 y_0^2.$$

d. The tension T in the string is given by $c = \sqrt{\dfrac{T}{\rho}}$ (Problem **1225**). The transvers force the mechanical device exerts on the wire at $x = 0$ is (Problem **1227**)

$$F_y(t) = -T\left(\frac{\partial y}{\partial x}\right)_{x=0} = \rho c \omega y_0 \cos(\omega t).$$

1229

A violin string, 0.5 m long, has a fundamental frequency of 200 Hz.

a. At what speed does a transverse pulse travel on this string?

b. Draw a pulse before and after reflection from one end of the string.

c. Show a sketch of the string in the next two higher modes of oscillation and give the frequency of each mode.

(Wisconsin)

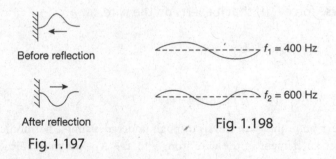

Before reflection

After reflection

Fig. 1.197

$f_1 = 400$ Hz

$f_2 = 600$ Hz

Fig. 1.198

Sol:

a. For a string of length l fastened at both ends, the wavelength λ of the fundamental mode is given by $\lambda/2 = l$. Hence

$$v = \lambda\nu = 2l\nu = 2 \times 0.5 \times 200 = 200 \text{ m/s}.$$

b. Figure 1.197 shows the shape of a pulse before and after reflection from one end of the string.

c. The frequencies of the next two higher modes are 400 Hz and 600 Hz. The corresponding shapes of the string are as shown in Fig. 1.198.

1230

A piano string of length l is fixed at both ends. The string has a linear mass density σ and is under tension T.

a. Find the allowed solutions for the vibrations of the string. What are the allowed frequencies and wavelengths?

b. At time $t = 0$ the string is pulled a distance s from equilibrium position at its midpoint so that it forms an isosceles triangle and is then released ($s \ll l$, see Fig. 1.199). Find the ensuing motion of the string, using the Fourier analysis method.

(Columbia)

Sol:

a. The vibration of the string is described by the wave equation (Problem **1225**)

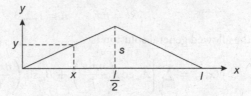

Fig. 1.199

$$\frac{\partial^2 y}{\partial x^2} - \frac{\sigma}{T}\frac{\partial^2 y}{\partial t^2} = 0,$$

subject to the conditions

$$y(0, t) = y(l, t) = 0$$

for all t. Let

$$y(x, t) = X(t)A(t)$$

and obtain from the above

$$\frac{1}{X}\frac{d^2 X}{dx^2} = \frac{\sigma}{TA}\frac{d^2 A}{dt^2}.$$

As the left-hand side depends only on x and the right-hand side only on t, each must be equal to a constant; let it be $-k^2$. We then have the ordinary differential equations

$$\frac{d^2X}{dx^2} + k^2X = 0,$$

$$\frac{d^2A}{dt^2} + v^2k^2A = 0,$$

where $v = \sqrt{\dfrac{T}{\sigma}}$.

Solutions of the above equations are respectively

$$X(x) = c_1 \cos(kx) + c_2 \sin(kx),$$
$$A(t) = b_1 \cos(vkt) + b_2 \sin(vkt).$$

With the boundary conditions $X(0) = X(l) = 0$, we have

$$c_1 = 0, \qquad c_2 \sin(kl) = 0.$$

As c_1 and c_2 cannot both be zero (otherwise $y(x, t)$ would be identically zero), we have to choose $\sin(kl) = 0$ or

$$kl = n\pi, \qquad n = 1, 2, 3 \dots.$$

Thus the allowed general solution is

$$y(x, t) = \sum_{n=1}^{\infty} \left[A_n \cos\left(\frac{n\pi vt}{l}\right) + B_n \sin\left(\frac{n\pi vt}{l}\right) \right] \sin\left(\frac{n\pi x}{l}\right),$$

where we have replaced the integration constants b_1c_2 by A_n and b_2c_2 by B_n for integer n. Each term in the general solution is an allowed solution corresponding to an allowed mode. The period for the nth mode is given by

$$\frac{n\pi v}{l}T_n = 2\pi,$$

the frequency being

$$\nu_n = \frac{1}{T_n} = \frac{n}{2l}v = \frac{n}{2l}\sqrt{\frac{T}{\sigma}},$$

and the wavelength being

$$\lambda_n = \frac{v}{\nu_n} = \frac{2l}{n}.$$

b.   Initially the string is as shown in Fig. 1.199. As

$$\frac{y}{x} = \frac{y}{l - x} = \frac{s}{\dfrac{l}{2}},$$

the initial conditions are

$$y(x, 0) = \begin{cases} \dfrac{2sx}{l} & \text{for} \quad 0 \leq x \leq \dfrac{l}{2}, \\[3mm] \dfrac{2s(l - x)}{l} & \text{for} \quad \dfrac{l}{2} \leq x \leq l. \end{cases}$$

Furthermore, the string is initially at rest so

$$\left(\frac{\partial y}{\partial t}\right)_{t=0} = 0.$$

Hence

$$B_n = 0,$$

and the A_n are given by

$$y(x, 0) = \sum_{n=1}^{\infty} A_n \sin\left(\frac{n\pi x}{l}\right).$$

Multiply both sides by $\sin(m\pi x/l)$ and integrate from 0 to l:

$$\int_0^l y(x, 0) \sin\left(\frac{m\pi x}{l}\right) dx = \sum_{n=1}^{\infty} A_n \int_0^l \sin\left(\frac{n\pi x}{l}\right) \sin\left(\frac{m\pi x}{l}\right) dx$$

$$= A_m \int_0^l \sin^2\left(\frac{m\pi x}{l}\right) dx$$

$$= \frac{A_m l}{\pi} \int_0^\pi \sin^2 \xi\, d\xi = \frac{1}{2} A_m l.$$

Hence

$$A_m = \frac{2}{l} \int_0^l y(x, 0) \sin\left(\frac{m\pi x}{l}\right) dx$$

$$= \frac{2}{l} \left[\frac{2s}{l} \int_0^{\frac{l}{2}} x \sin\left(\frac{m\pi x}{l}\right) dx + 2s \int_{\frac{l}{2}}^l \left(1 - \frac{x}{l}\right) \sin\left(\frac{m\pi x}{l}\right) dx\right]$$

$$= \frac{8s}{(m\pi)^2} \sin\left(\frac{m\pi}{2}\right),$$

use having been made of the formulae

$$\int_0^\pi \sin(mx)\sin(nx)\,dx = \frac{1}{2}\pi\delta_{mn},$$

$$\int x\sin(ax)\,dx = \frac{1}{a^2}\sin(ax) - \frac{x}{a}\cos(ax).$$

Thus the motion of the string is described by

$$y(x,t) = \sum_{n=1}^\infty \frac{8s}{(n\pi)^2}\sin\left(\frac{n\pi}{2}\right)\cos\left(\frac{n\pi vt}{l}\right)\sin\left(\frac{n\pi x}{l}\right)$$

with $v = \sqrt{\dfrac{T}{\sigma}}$.

1231

A spring of rest length X and force constant k has a mass m. One end is fixed and the other end is attached to a mass M. The orientation is horizontal, and M moves on a frictionless surface.

a. Derive a wave equation for longitudinal oscillation of this system.

b. Find the frequency of the lowest mode as a function of mass for the case where M and k are finite and $m \ll M$.

(Princeton)

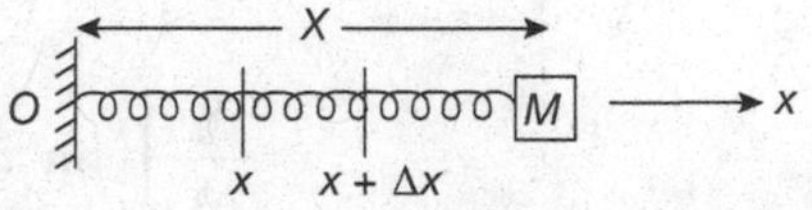

Fig. 1.200

Sol:

a. Take the x-axis along the length of the spring with origin at the fixed end and consider a section of length Δx extending from x to $x + \Delta x$ as shown in Fig. 1.200. Then as M moves to the right, the point x moves to $x + \xi$ and the point $x + \Delta x$ moves to $x + \Delta x + \xi + \Delta\xi$ as shown in Fig. 1.201.

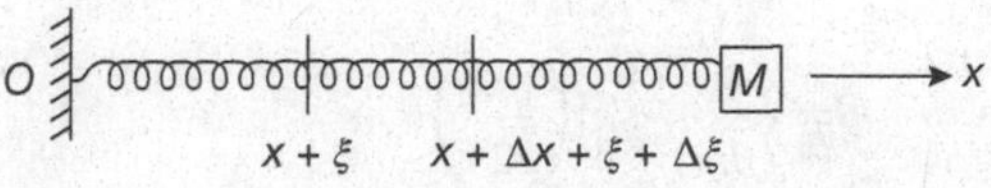

Fig. 1.201

Let σ be the Young's modulus of the spring. The restoring force F is given by $F = a\sigma\Delta l/l$, where a is the area of the cross section of the spring and $\Delta l/l$ is the extension per unit length. Write K_0 for $a\sigma$. The net force on the section under consideration is

$$F_{x+\Delta x} - F_x = K_0\left(\frac{\partial\xi}{\partial x}\right)_{x+\Delta x} - K_0\left(\frac{\partial\xi}{\partial x}\right)_x,$$

which by Newton's second law is equal to $\rho\Delta x(\partial^2\xi/\partial t^2)_x$, ρ being the mass per unit length of the spring, assumed constant for small extensions. Thus

$$\left(\frac{\partial^2\xi}{\partial t^2}\right)_x = \frac{K_0}{\rho}\frac{\left(\frac{\partial\xi}{\partial x}\right)_{x+\Delta x} - \left(\frac{\partial\xi}{\partial x}\right)_x}{\Delta x} = \frac{K_0}{\rho}\left(\frac{\partial^2\xi}{\partial x^2}\right)_x,$$

or

$$\frac{\partial^2\xi}{\partial x^2} - \frac{m}{K_0 X}\frac{\partial^2\xi}{\partial t^2} = 0.$$

This is the equation for propagation of longitudinal waves along the spring and gives the velocity of propagation as

$$v = \sqrt{\frac{K_0 X}{m}} = X\sqrt{\frac{k}{m}},$$

as $k = K_0/X$ by definition.

b. Try a solution of the form

$$\xi(x, t) = \xi_0(x)\cos(\omega t + \varphi),$$

where ω, φ are constants. Substitution in the wave equation gives

$$\frac{\partial^2\xi_0}{\partial x^2} + K^2\xi_0 = 0,$$

where

$$K^2 = \frac{m\omega^2}{K_0 X} = \frac{\omega^2}{v^2}.$$

Its general solution is

$$\xi_0 = A\sin(Kx) + B\cos(Kx),$$

A, B being constants of integration. The boundary condition $\xi_0 = 0$ at $x = 0$ give $B = 0$. We also have from Newton's second law

$$M\left(\frac{\partial^2 \xi}{\partial t^2}\right)_X = -K_0\left(\frac{\partial \xi}{\partial x}\right)_X,$$

or

$$KX \tan(KX) = \frac{m}{M},$$

which can be solved to give a series of K, and hence of the vibrational frequencies of the spring.

For $m \ll M$ and the lowest frequency, $\tan(KX) \simeq KX$ and the above becomes

$$\frac{\omega^2 m}{k} \approx \frac{m}{M},$$

giving the lowest angular frequency as

$$\omega = \sqrt{\frac{k}{M}}.$$

Note that this is just the vibrational frequency of an oscillator which consists of a massless spring of force constant k with one end fixed and the other end connected to a mass M.

To obtain a more accurate approximate solution, expand

$$\tan(KX) = KX + \frac{1}{3}(KX)^3 + \cdots$$

and retain the first two terms only. We then have

$$(KX)^2 = \frac{m}{M}\left[1 + \frac{1}{3}(KX)^2\right]^{-1} \approx \frac{m}{M}\left[1 - \frac{1}{3}(KX)^2\right],$$

or

$$K^2 = \frac{3m}{(3M + m)X^2},$$

giving

$$\omega = \sqrt{\frac{3k}{3M + m}}.$$

1232

a. Suppose you have a string of uniform mass per unit length ρ and length l held at both ends under tension T. Set up the equation for small transverse oscillations of the string and then find the eigenfrequencies.

b. Now consider the case where the string is free at one end and attached to a vertical pole at the other end, and is rotating about the pole at an angular frequency ω (neglect gravity) as shown in Fig. 1.202. Set up the equation for small transverse oscillations for this case.

c. Find the eigenfrequencies.

(Hint: the equation you get should look familiar in terms of the Legendre polynomials.)

(CUSPEA)

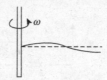

Fig. 1.202

Sol:

a. Consider a section of the string as shown in Fig. 1.203. The y-component of the tension at x is

$$F_y(x) = -T\sin\theta \approx -T\theta \approx -T\left(\frac{\partial y}{\partial x}\right)_x.$$

Simarly at $x + \Delta x$

$$F_y(x + \Delta x) \approx T\left(\frac{\partial y}{\partial x}\right)_{x+\Delta x}.$$

Note that T is constant. Thus

$$F_y(x + \Delta x) - F_y(x) = T\left[\left(\frac{\partial y}{\partial x}\right)_{x+\Delta x} - \left(\frac{\partial y}{\partial x}\right)_x\right] = T\frac{\partial}{\partial x}\left(\frac{\partial y}{\partial x}\right)\cdot\Delta x$$

$$= T\frac{\partial^2 y}{\partial x^2}\cdot\Delta x.$$

The section has length Δx, mass $\rho\Delta x$, and by applying Newton's second law to the section we have

$$\frac{\partial^2 y}{\partial x^2} - \frac{\rho}{T}\frac{\partial^2 y}{\partial t^2} = 0.$$

This is the wave equation for small transverse oscillations, the velocity of propagation being $v = \sqrt{\frac{T}{\rho}}$. The general solution is (Problem **1230**)

$$y(x, t) = \sum_{n=1}^{\infty}\left(A_n \cos\frac{n\pi v t}{l} + B_n \sin\frac{n\pi v t}{l}\right)\sin\frac{n\pi x}{l}.$$

The eigenfrequencies are

$$\nu_n = \frac{\omega_n}{2\pi} = \frac{n\pi v}{2\pi l} = \frac{nv}{2l}.$$

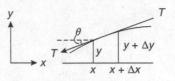

Fig. 1.203

b. Take a rotating frame $Oxyz$ attached to the spring with the y-axis along the axis of rotation and the x-axis along the string. There is a fictitious centrifugal force acting on the string which is balanced by the tension. Consider a section Δx of the string. The difference of tension across its ends is

$$-\Delta T = \rho\Delta x \cdot x\omega^2,$$

whence

$$\frac{dT}{dx} = -\rho\omega^2 x.$$

Integrating and applying the boundary condition $T = 0$ at $x = l$ we find

$$T = \frac{1}{2}\rho\omega^2(l^2 - x^2).$$

Following the procedure of (a) we have

$$F_y(x + \Delta x) - F_y(x) = \left(T\frac{\partial y}{\partial x}\right)_{x+\Delta x} - \left(T\frac{\partial y}{\partial x}\right)_x$$

$$\approx \frac{\partial}{\partial x}\left[\frac{1}{2}\rho\omega^2(l^2 - x^2)\frac{\partial y}{\partial x}\right]\Delta x.$$

Newton's second law gives

$$\rho \Delta x \frac{\partial^2 y}{\partial t^2} = \frac{\partial}{\partial x}\left[\frac{1}{2}\rho\omega^2(l^2 - x^2)\frac{\partial y}{\partial x}\right]\Delta x,$$

or

$$\frac{\partial}{\partial x}\left[(l^2 - x^2)\frac{\partial y}{\partial x}\right] = \frac{2}{\omega^2}\frac{\partial^2 y}{\partial t^2},$$

for small transverse oscillations.

c. Try a solution of the type $y \sim e^{-i\Omega t}$ and let $\xi = \frac{x}{l}$. The above equation then becomes

$$(1 - \xi^2)\frac{\partial^2 y}{d\xi^2} - 2\xi\frac{\partial y}{\partial \xi} + \frac{2\Omega^2}{\omega^2}y = 0,$$

with $0 \leq \xi \leq 1$. This differential equation has finite solutions if

$$\frac{2\Omega^2}{\omega^2} = n(n + 1),$$

n being an integer. The equation is then known as Legendre's differential equation and the solutions are known as Legendre's polynomials. Thus the eigenhequencies are given by

$$\frac{\Omega}{2\pi} = \frac{\omega}{2\pi}\sqrt{\frac{1}{2}n(n + 1)},$$

where $n = 1, 2, 3, ...$ However we still have to satisfy the boundary condition $y = 0$ at $\xi = 0$. This limits the allowable n to odd integers $1, 3, 5, ...$ since Legendre's polynomials $P_n(\xi) = 0$ at $\xi = 0$ only for odd values of n.

1233

A long string of linear density (mass per unit length) μ is under tension T. A point mass m is attached at a particular point of the string. A wave of angular frequency ω traveling along the string is incident from the left.

a. Calculate what fraction of the incident energy is reflected by the mass m.

b. Suppose that the point mass m is replaced by a string of linear density $\mu_m \gg \mu$ and short length l such that $l = m/\mu_m$. For what range of l values (for fixed m) does the answer for (a) remain approximately correct?

(CUSPEA)

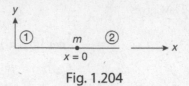

Fig. 1.204

Sol: Divide the space into two regions with separation at the location of m, which i taken to be the origin of the x-axis, as shown in Fig. 1.204. In region 1, let the wav function be

$$y^{(1)} = e^{ikx} + Ae^{-ikx},$$

where $k = \omega/v$, $v = \sqrt{T/\mu}$ being the velocity of the wave (Problem **1225**), and the second term of the right-hand side represents the reflected wave. In region 2 we have

$$y^{(2)} = Be^{ikx}.$$

At $x = 0$, where the mass m is located, we require that

$$y^{(1)} = y^{(2)}$$

i.e.

$$1 + A = B. \tag{1}$$

Furthermore, considering the forces on the point mass m we have

$$m\frac{\partial^2 y}{\partial t^2} = T\frac{\partial y^{(2)}}{\partial x} - T\frac{\partial y^{(1)}}{\partial x},$$

where for y we can use either $y^{(1)}$ or $y^{(2)}$. Then

$$-m\omega^2 B = ikT(B - 1 + A). \tag{2}$$

Solving (1) and (2) we have

$$A = \frac{-m\omega^2}{2ikT + m\omega^2},$$

$$B = \frac{2ikT}{2ikT + m\omega^2}.$$

Therefore the fraction of the incident energy that is reflected is

$$|A|^2 = \frac{m^2\omega^4}{4k^2T^2 + m^2\omega^4}.$$

b. The calculation in (a) still applies provided $l \ll \lambda$, where λ is the wavelength, being

$$\frac{2\pi}{k} = \frac{2\pi\upsilon}{\omega} = \frac{2\pi}{\omega}\sqrt{\frac{T}{\mu}}.$$

Hence the condition that the answer to (a) remains approximately correct is

$$l \ll \frac{2\pi}{\omega}\sqrt{\frac{T}{\mu}}.$$

1234

A perfectly flexible string with uniform linear mass density ρ and length L is hanging from a fixed support with its bottom end free, as shown in Fig. 1.205.

a. Derive the partial differential equation describing small transverse (in one plane) oscillations of the string, and from it, the differential equation for the form of the normal modes.

b. Solve this differential equation using standard (power series) methods (the trick for transforming it into Bessel's equation is not what is wanted), and, using approximate numerical methods, solve for the frequency of the lowest normal mode.

(Princeton)

Fig. 1.205

Sol:

a. Use coordinate frame $Oxyz$ as shown in Fig. 1.205. following the procedure of Problem **1232**, we have, by Newton's second law, for a section Δx of the string

$$\rho\Delta x \frac{\partial^2 y}{\partial t^2} = \left(T\frac{\partial y}{\partial x}\right)_{x+\Delta x} - \left(T\frac{\partial y}{\partial x}\right)_x,$$

or

$$\rho \frac{\partial^2 y}{\partial t^2} = \frac{\partial}{\partial x}\left(T \frac{\partial y}{\partial x}\right).$$

The tension T in the string at x is related to the gravity by

$$T = \int_x^L \rho g \, dx = \rho g(L - x),$$

so the above equation becomes

$$\frac{\partial^2 y}{\partial t^2} = g \frac{\partial}{\partial x}\left[(L - x)\frac{\partial y}{\partial x}\right].$$

This is the partial differential equation for small transverse oscillations of the string. Applying the method of separation of variables by putting

$$y(x, t) = \xi(x)\tau(t),$$

we obtain

$$\frac{1}{g\tau}\frac{d^2\tau}{dt^2} = \frac{1}{\xi}\frac{d}{dx}\left[(L - x)\frac{d\xi}{dx}\right].$$

As the left-hand side depends on t alone and the right-hand side depends on x alone, each must be equal to a constant, say $-\lambda$, λ being a positive number. We thus have the equivalent ordinary differential equations

$$\frac{d}{dx}\left[(L - x)\frac{d\xi}{dx}\right] + \lambda\xi = 0,$$

$$\frac{d^2\tau}{dt^2} + \lambda g\tau = 0.$$

The boundary conditions are

$$y(0, t) = 0, \qquad y(L, t) = \text{finite},$$

i.e.

$$\xi(0) = 0, \qquad \xi(L) = \text{finite}.$$

b. The ξ-equation can be written as

$$(x - L)\xi'' + \xi' - \lambda\xi = 0.$$

As $x = L$ is a regular singular point, the equation has a solution of the form

$$\xi = \sum_{n=0}^{\infty} a_n(x - L)^n.$$

Then

$$\xi' = \sum_1^\infty na_n(x - L)^{n-1} = \sum_0^\infty (n + 1)a_{n+1}(x - L)^n,$$

$$\xi'' = \sum_1^\infty n(n + 1)a_{n+1}(x - L)^{n-1} = \sum_2^\infty (n - 1)na_n(x - L)^{n-2},$$

$$(x - L)\xi'' = \sum_2^\infty (n - 1)na_n(x - L)^{n-1} = \sum_1^\infty n(n + 1)a_{n+1}(x - L)^n,$$

and the ξ-equation becomes

$$(a_1 - \lambda a_0) + \sum_1^\infty [(n + 1)^2 a_{n+1} - \lambda a_n](x - L)^n = 0.$$

Equating the coefficients of $(x - L)^n$ on both sides of the equation, we find

$$a_1 = \lambda a_0, \quad a_{n+1} = \frac{\lambda}{(n + 1)^2}a_n.$$

Hence

$$a_2 = \frac{\lambda}{2^2}a_1 = \frac{\lambda^2}{2^2}a_0,$$

$$a_3 = \frac{\lambda}{3^2}a_2 = \frac{\lambda^3}{(3 \cdot 2)^2}a_0,$$

$$\ldots\ldots$$

$$a_n = \frac{\lambda^n}{(n!)^2}a_0,$$

$$\ldots\ldots,$$

giving

$$\xi(x) = a_0 \sum_0^\infty \frac{\lambda^n}{(n!)^2}(x - L)^n.$$

The boundary condition $\xi(0) = 0$ then yields

$$f(\lambda L) = \sum_0^\infty \frac{(-1)^n(\lambda L)^n}{(n!)^2} = 1 - \lambda L + \frac{1}{4}\lambda^2 L^2 - \ldots = 0.$$

This equation can be solved to find the roots λL, which then give the frequencies of the various modes, $\sqrt{\lambda g}/2\pi$, according to the τ-equation.

For an approximate solution we retain only the terms up to $n = 2$ in $f(\lambda L)$:

$$f(\lambda L) \approx 1 - \lambda L + \frac{1}{4}(\lambda L)^2,$$

Newton's approximate method gives a better approximate root of $f(\lambda L) = 0$, α_{k+} if we input an approximate root α_k by calculating

$$\alpha_{k+1} = \alpha_k - \frac{f(\alpha_k)}{f'(\alpha_k)}.$$

As

$$f'(\lambda L) \approx -1 + \frac{1}{2}\lambda L,$$

if we take $\alpha_1 = 0$, then

$$\alpha_2 = 1, \qquad\qquad f(\alpha_2) \approx 0.25,$$

$$\alpha_3 = 1 - \frac{0.25}{-0.5} = 1.5, \quad f(\alpha_3) \approx 0.625.$$

As $f(\alpha_3)$ is quite close to zero we can consider $\alpha_3 = 1.5$ as the smallest positive root. Thus

$$\lambda_{\min} = \frac{1.5}{L}.$$

for the lowest mode. Then for this mode

$$\tau = A\cos(\sqrt{\lambda g}\, t) + B\sin(\sqrt{\lambda g}\, t)$$

and the frequency is

$$\nu_{\min} \approx \frac{1}{2\pi}\sqrt{\frac{3g}{2L}}.$$

1235

A common lecture demonstration is as follows: hold or clamp a one meter long thin aluminium bar at the center, strike one end longitudinally (i.e. parallel to the axis of the bar) with a hammer, and the result is a sound wave of frequency 2500 Hz.

a.   From this experiment, calculate the speed of sound in air.

b.   Rom this experiment, calculate the speed of sound in aluminium.

c.   Where might you hold the bar to excite a frequency of 3750 Hz? Explain. Does it matter which end of the bar is struck? Explain.

d.   Suppose you hold the bar at the center as before, but strike the bar transverse to its length, rather longitudinally. Qualitatively explain why the resultant sound wave is of lower frequency than before.

(UC, Berkeley)

Sol:

a. The point where the bar is struck is an antinode and the point where it is held a node. With the bar held at the center and its one end struck, the wavelength λ is related to its length L by $\lambda = 2L$. Hence the speed of sound propagation in the aluminium bar is

$$v_{Al} = \nu\lambda = 2\nu L = 2 \times 2500 \times 1 = 5000 \text{ m/s}.$$

The speed of sound in a solid is

$$v = \sqrt{\frac{Y}{\rho}},$$

where Y is the Young's modulus of its material and ρ its density. The speed of sound in a fluid is

$$v = \sqrt{\frac{M}{\rho}},$$

where M is its bulk modulus and ρ its density. For adiabatic compression of a gas, $M = \gamma p$, where p is its pressure and γ the ratio of its principal specific heats; $\gamma = 1.4$ for air, a diatomic gas. Hence

$$\frac{v_{air}}{v_{Al}} = \sqrt{\frac{1.4 p \rho_{Al}}{Y \rho_{air}}}.$$

With

$$p = 1.013 \times 10^6 \text{ dyn/cm}^2 \quad \text{(standard atmosphere)},$$
$$Y = 7.05 \times 10^{11} \text{ dyn/cm}^2,$$
$$\rho_{Al} = 2.7 \text{ g/cm}^3,$$
$$\rho_{air} = 1.165 \times 10^{-3} \text{ g/cm}^3 \quad \text{(at 30°C)},$$
$$v_{air} = 6.83 \times 10^{-2} \times 5000 = 341 \text{ m/s}.$$

b. $v_{Al} = 5000$ m/s.

c. Suppose the bar is held at distance x from the struck end. We have

$$x = \frac{\lambda}{4} = \frac{v}{4\nu} = \frac{5000}{4 \times 3750} = \frac{1}{3} m.$$

Hence the bar is to be held at $\frac{1}{3}$ m from the struck end. If it is so held but struck at the other end, we would have

$$\frac{2}{3} = \frac{v}{4\nu}$$

and the frequency would become 1875 Hz.

d. If the bar is struck transversely, the wave generated will be transverse, not com pressional, and the velocity of propagation is then given by

$$v = \sqrt{\frac{N}{\rho}},$$

where N is the shear modulus. As the shear modulus of a solid is generally smalle than its bulk modulus, v is now smaller. And as

$$\nu = \frac{v}{2L}$$

the frequency generated is lower.

1236

a. A violin string of length L with linear density μ kg/m and tension T new tons undergoes small oscillations (Fig. 1.206 (a)). Write the solutions for the fundamental and first harmonic, and sketch their x-dependences Give the angular frequency ω_1 of the fundamental and ω_2 of the first harmonic.

b. The left-hand 1/3 of the string is wrapped so as to increase its linear density to 4μ kg/m (Fig. 1.206 (b)). Repeat part (a), i.e. derive and sketch the new fundamental and first harmonic, and express the new ω_1 and ω_2 in terms of the original ω_1 and ω_2 of part (a).

(UC, Berkeley)

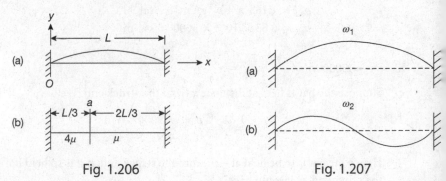

Fig. 1.206   Fig. 1.207

Sol:

a. Use coordinates as shown in Fig. 1.206 (a). The equation of motion for the string is (Problem **1225**)

$$\frac{\partial^2 y}{\partial x^2} - \frac{\mu}{T}\frac{\partial^2 y}{\partial t^2} = 0,$$

from which it is seen that the wave propagating along the string has velocity $v = \sqrt{T/\mu}$. As the two ends of the string are fixed the fundamental mode (Fig. 1.207 (a)) has wavelength λ_1 given by

$$L = \frac{1}{2}\lambda_1.$$

Hence the fundamental angular frequency is

$$\omega_1 = \frac{2\pi v}{\lambda_1} = \frac{\pi}{L}\sqrt{\frac{T}{\mu}}.$$

The solution for the fundamental mode is

$$y_1 = A_1 \sin\left(\frac{\pi x}{L}\right) \cos(\omega_1 t + \varphi_1),$$

where A_1, φ_1 are constants to be determined from the initial conditions. The wavelength for the first harmonic (Fig. 1.207 (b)) is $\lambda_2 = L$. Hence for the first harmonic the angular frequency is

$$\omega_2 = \frac{2\pi v}{\lambda_2} = \frac{2\pi}{L}\sqrt{\frac{T}{\mu}}$$

and the solution is

$$y_2 = A_2 \sin\left(\frac{2\pi x}{L}\right) \cos(\omega_2 t + \varphi_2),$$

where A_2, φ_2 are constants to be determined from the initial conditions.

b. The equations of motion for the two sections are

$$\frac{\partial^2 y}{\partial x^2} - \frac{4\mu}{T}\frac{\partial^2 y}{\partial t^2} = 0, \quad 0 \le x \le \frac{L}{3},$$

$$\frac{\partial^2 y}{\partial x^2} - \frac{\mu}{T}\frac{\partial^2 y}{\partial t^2} = 0, \quad \frac{L}{3} \le x \le L.$$

The boundary conditions are that for all t, $y = 0$ at $x = 0, L$, and y and $\partial y/\partial x$ are continuous at $x = L/3$. Thus the solutions of the equations of motion are

$$y(x, t) = \begin{cases} (A_1 \cos \omega t + B_1 \sin \omega t) \sin\left(\dfrac{\omega}{v_1} x\right), & 0 \le x \le \dfrac{L}{3}, \\[2ex] (A_2 \cos \omega t + B_2 \sin \omega t) \sin\left[\dfrac{\omega}{v_2}(L - x)\right], & \dfrac{L}{3} \le x \le L, \end{cases}$$

with

$$v_1 = \sqrt{\frac{T}{4\mu}}, \quad v_2 = \sqrt{\frac{T}{\mu}} = 2v_1$$

and

$$(A_1 \cos \omega t + B_1 \sin \omega t) \sin\left(\frac{L\omega}{3v_1}\right)$$

$$= (A_2 \cos \omega t + B_2 \sin \omega t) \sin\left(\frac{L\omega}{3v_1}\right),$$

$$\frac{\omega}{v_1}(A_1 \cos \omega t + B_1 \sin \omega t) \cos\left(\frac{L\omega}{3v_1}\right)$$

$$= -\frac{\omega}{2v_1}(A_2 \cos \omega t + B_2 \sin \omega t) \cos\left(\frac{L\omega}{3v_1}\right).$$

Equating separately the coefficients of $\cos \omega t$ and $\sin \omega t$ on the two sides of the last two equations gives

$$A_1 \sin\left(\frac{L\omega}{3v_1}\right) - A_2 \sin\left(\frac{L\omega}{3v_1}\right) = 0,$$

$$A_1 \frac{\omega}{v_1} \cos\left(\frac{L\omega}{3v_1}\right) + A_2 \frac{\omega}{2v_1} \cos\left(\frac{L\omega}{3v_1}\right) = 0,$$

$$B_1 \sin\left(\frac{L\omega}{3v_1}\right) - B_2 \sin\left(\frac{L\omega}{3v_1}\right) = 0,$$

$$B_1 \frac{\omega}{v_1} \cos\left(\frac{L\omega}{3v_1}\right) + B_2 \frac{\omega}{2v_1} \cos\left(\frac{L\omega}{3v_1}\right) = 0.$$

For A_1, A_2, B_1, B_2 not all zero we require

$$\begin{vmatrix} \sin\left(\dfrac{L\omega}{3v_1}\right) & -\sin\left(\dfrac{L\omega}{3v_1}\right) \\[3ex] \dfrac{\omega}{v_1}\cos\left(\dfrac{L\omega}{3v_1}\right) & \dfrac{\omega}{2v_1}\cos\left(\dfrac{L\omega}{3v_1}\right) \end{vmatrix} = \frac{3\omega}{2v_1} \sin\left(\frac{L\omega}{3v_1}\right) \cos\left(\frac{L\omega}{3v_1}\right)$$

$$= \frac{3\omega}{4v_1} \sin\left(\frac{2L\omega}{3v_1}\right) = 0,$$

i.e.

$$\frac{2L\omega}{3v_1} = n\pi, \quad n = 1, 2, 3, \dots .$$

Hence the new fundamental and first harmonic angular frequencies are

$$\omega_1' = \frac{3\pi v_1}{2L} = \frac{3\pi}{4L}\sqrt{\frac{T}{\mu}} = \frac{3}{4}\omega_1,$$

$$\omega_2' = \frac{6\pi v_1}{2L} = \frac{3\pi}{2L}\sqrt{\frac{T}{\mu}} = \frac{3}{2}\omega_1.$$

For the fundamental frequency ω_1',

$$A_2 = A_1, \quad B_2 = B_1.$$

For the first harmonic frequency ω_2',

$$A_2 = -2A_1, \quad B_2 = -2B_1.$$

The corresponding wave forms are sketched in (a) and (b) of Fig. 1.208 respectively.

(a) 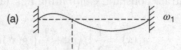 ω_1

(b)  ω_2

Fig. 1.208

1237

A string of infinite length has tension T and linear density σ. At $t = 0$, the deformation of the string is given by the function $f(x)$, and its initial velocity distribution by $g(x)$. What is the motion of the string for $t > 0$?

(Chicago)

Sol: The deformation of the string travels as a wave following the wave equation

$$\frac{\partial^2 y}{\partial x^2} - \frac{1}{v^2}\frac{\partial^2 y}{\partial t^2} = 0$$

with

$$v = \pm\sqrt{\frac{T}{\sigma}}.$$

The general solution is a sum of waves traveling in the $-x$ and $+x$ directions:

$$y = f_1(x + vt) + f_2(x - vt).$$

The initial conditions give

$$f_1(x) + f_2(x) = f(x), \tag{1}$$

$$f_1'(x) - f_2'(x) = \frac{g(x)}{v}, \tag{2}$$

where

$$f_1'(x) = \left(\frac{\partial f_1(\xi)}{\partial \xi}\right)_{t=0}, \qquad f_2'(x) = \left(\frac{\partial f_2(\xi)}{\partial \xi}\right)_{t=0},$$

with $\xi = x + vt$, $\xi = x - vt$ respectively. Integrating (2) gives

$$f_1(x) - f_2(x) = \frac{1}{v}\int^x g(x')\,dx' + C, \tag{3}$$

C being an arbitrary constant. Combining Eqs. (1) and (3) we obtain

$$f_1(x) = \frac{1}{2}\left[f(x) + \frac{1}{v}\int^x g(x')\,dx' + C\right],$$

$$f_2(x) = \frac{1}{2}\left[f(x) - \frac{1}{v}\int^x g(x')\,dx' - C\right].$$

Hence

$$\begin{aligned}
y(x, t) &= f_1(x + vt) + f_2(x - vt) \\
&= \frac{1}{2}\left\{\left[f(x + vt) + \frac{1}{v}\int^{x+vt} g(x')\,dx' + C\right]\right. \\
&\quad + \left.\left[f(x - vt) - \frac{1}{v}\int^{x-vt} g(x')\,dx' - C\right]\right\} \\
&= \frac{1}{2}\left[f(x + vt) + f(x - vt) + \frac{1}{v}\int_{x-vt}^{x+vt} g(x')\,dx'\right].
\end{aligned}$$

1238

A long wave packet with amplitude A composed predominantly of frequencies very near ω_0 propagates on an infinitely long string of linear mass density μ stretched with a tension T as shown in Fig. 1.209. The packet encounters a bead of mass m attached to the string as shown in the sketch.

a. What is the amplitude of the transmitted wave packet?

b. In the hmit of large m and high frequency (large ω_0), how does the amplitude of the transmitted wave depend on ω_0?

(MIT)

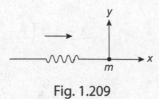

Fig. 1.209

Sol:

a. The equation of motion for the string for small transverse oscillations is a wave equation (Problem **1225**)

$$\frac{\partial^2 y}{\partial x^2} - \frac{\mu}{T}\frac{\partial^2 y}{\partial t^2} = 0,$$

the velocity of wave propagation being $v = \pm\sqrt{T/\mu}$. For waves of angular frequency ω, define wave number

$$k = \frac{\omega}{v} = \omega\sqrt{\frac{\mu}{T}}.$$

For waves with angular frequencies very nearly ω_0, the wave equation has solutions

$$
\begin{aligned}
y_1 &= Ae^{i(kx-\omega_0 t)} + Be^{-i(kx-\omega_0 t)} && \text{for } x < 0,\\
y_2 &= Ce^{i(kx-\omega_0 t)} && \text{for } x > 0,
\end{aligned}
$$

where A, B, C are the amplitudes of the incident, reflected and transmitted waves respectively, and the position of the bead is taken to be the origin of the x-axis. The continuity of the displacement at the boundary requires $y_1 = y_2$ at $x = 0$ for all t, i.e.

$$A + B = C.$$

The equation of the motion of the bead is

$$m\left(\frac{\partial^2 y}{\partial t^2}\right)_{x=0} = -T\sin\theta_1 + T\sin\theta_2$$

$$\approx -T\theta_1 + T\theta_2$$

$$\approx -T\left(\frac{\partial y_1}{\partial x}\right)_{x=0} + T\left(\frac{\partial y_2}{\partial x}\right)_{x=0},$$

where θ_1, θ_2 are the angles the string makes with the x-axis for $x < 0$ and $x >$ respectively as shown in Fig. 1.210. Thus

$$-m\omega_0^2 C = -ikT(A - B) + ikTC,$$

or

$$A - B = \left(1 + \frac{m\omega_0^2}{ikT}\right)C.$$

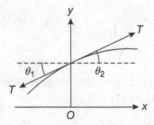

Fig. 1.210

As $A + B = C$ we have

$$C = \frac{2A}{\left(2 + \dfrac{m\omega_0^2}{ikT}\right)},$$

and the amplitude of the transmitted wave is

$$|C| = \sqrt{C^*C} = \frac{2A}{\sqrt{4 + \dfrac{m^2\omega_0^4}{k^2 T^2}}} = \frac{2A}{\sqrt{4 + \dfrac{m^2\omega_0^4}{\mu T}}}.$$

b.   In the limit of large m and large ω_0 we have

$$|C| \approx \frac{2A}{m\omega_0}\sqrt{\mu T} \propto \frac{1}{\omega_0}.$$

1239

A uniform string has length L and mass per unit length ρ. It undergoes small transverse vibration in the (x, y) plane with its endpoints held fixed at $(0, 0)$ and $(L, 0)$ respectively. The tension is K. A velocity-dependent hictional force is present: if a small piece of length δl has transverse velocity v the frictional force is $-kv\delta l$. Using appropriate approximations, the following equations hold for the vibration amplitude $y(x, t)$:

i. $\dfrac{\partial^2 y}{\partial t^2} + a\dfrac{\partial y}{\partial t} = b\dfrac{\partial^2 y}{\partial x^2}$,   **ii.** $y(0, t) = 0 = y(L, t)$.

a. Find the constants a and b in (i). If you cannot do this part, take a and b as given positive constants and go on.

b. Find all solutions of (i) and (ii) which have the product form $y = X(x)T(t)$. You may assume $a^2 < b/L^2$.

c. Suppose $y(x, 0) = 0$,

$$\dot{y}(x, 0) = A\sin\left(\frac{3\pi x}{L}\right) + B\sin\left(\frac{5\pi x}{L}\right).$$

Here A and B are constants. Find $y(x, t)$.

d. Suppose, instead, that $a = 0$ and $\dot{y}(x, 0) = 0$ while

$$y(x, 0) = \begin{cases} Ax, & 0 \le x \le \dfrac{L}{2} \\[2mm] A(L - x), & \dfrac{L}{2} \le x \le L. \end{cases}$$

Find $y(x, t)$.

(UC, Berkeley)

Sol:

a. The frictional force acting on unit length of the string is $-kv = -k\partial y/\partial t$, so the transverse vibration of the string is described by

$$\rho\frac{\partial^2 y}{\partial t^2} = K\frac{\partial^2 y}{\partial x^2} - k\frac{\partial y}{\partial t},$$

or

$$\frac{\partial^2 y}{\partial t^2} + \left(\frac{k}{\rho}\right)\frac{\partial y}{\partial t} = \left(\frac{K}{\rho}\right)\frac{\partial^2 y}{\partial x^2}.$$

Hence $a = k/\rho$, $b = K/\rho$.

b.   Setting $y = X(x)\,T(t)$ and substituting it in the wave equation we obtain

$$\frac{T''}{T} + \frac{aT'}{T} = \frac{bX''}{X}.$$

As the left-hand side depends only on t and the right-hand side depends only on x, each must be equal to a constant, say $-b\lambda^2$. Thus we have

$$X'' + \lambda^2 X = 0,$$
$$T'' + aT' + b\lambda^2 T = 0.$$

Using the boundary conditions

$$y(0, t) = y(L, t) = 0, \text{ i.e. } X(0) = X(L) = 0,$$

we obtain the solutions for the first equation

$$X_n(x) = A_n \sin(\lambda_n x) = A_n \sin\left(\frac{n\pi x}{L}\right),$$

where A_n is a constant and $n = 1, 2, 3, \ldots$. The second equation then becomes

$$T'' + aT' + b\left(\frac{n\pi}{L}\right)^2 T = 0.$$

Letting $T(t) = e^{pt}$ we obtain the characteristic equation

$$p^2 + ap + \frac{n^2\pi^2 b}{L^2} = 0,$$

whose solutions are

$$p_\pm = \frac{-a \pm \sqrt{a^2 - 4n^2\pi^2 b/L^2}}{2} = -\frac{a}{2} \pm i\omega_n,$$

where

$$\omega_n = \sqrt{\frac{n^2\pi^2 b}{L^2} - \frac{a^2}{4}}$$

is real as $b/L^2 > a^2$. Hence the solution of the second equation can be written as

$$T_n = [C_n' \sin(\omega_n t) + D_n' \cos(\omega_n t)] e^{-\frac{at}{2}},$$

and thus

$$y_n = \sin\left(\frac{n\pi x}{L}\right)[C_n \sin(\omega_n t) + D_n \cos(\omega_n t)] e^{-\frac{at}{2}},$$

grouping the constants in each term into one. The general solution of the wave equation is thus

$$y(x, t) = \sum_{n=1}^{\infty} y_n(x, t).$$

c. As $y(x, 0) = 0$, $D_n = 0$ for all n and we have

$$y(x, t) = \sum_{n=1}^{\infty} C_n \sin\left(\frac{n\pi x}{L}\right) \sin(\omega_n t) e^{-\frac{at}{2}}.$$

and

$$\dot{y}(x, t) = \sum_{n=1}^{\infty} C_n \sin\left(\frac{n\pi x}{L}\right) [\omega_n \cos(\omega_n t) - \frac{a}{2} \sin(\omega_n t)] e^{-\frac{at}{2}}.$$

Then as

$$\dot{y}(x, 0) = A \sin\left(\frac{3\pi x}{L}\right) + B \sin\left(\frac{5\pi x}{L}\right) = \sum_{n=1}^{\infty} C_n \sin\left(\frac{n\pi x}{L}\right) \omega_n,$$

we have

$$C_3 = \frac{A}{\omega_3}, \quad C_5 = \frac{B}{\omega_5}$$

and all other $C_n = 0$. Hence

$$y(x, t) = \left[\frac{A}{\omega_3} \sin\left(\frac{3\pi x}{L}\right) \sin(\omega_3 t) + \frac{B}{\omega_5} \sin\left(\frac{5\pi x}{L}\right) \sin(\omega_5 t)\right] e^{-\frac{at}{2}}$$

with

$$\omega_3 = \sqrt{\frac{9\pi^2 b}{L^2} - \frac{a^2}{4}}, \quad \omega_5 = \sqrt{\frac{25\pi^2 b}{L^2} - \frac{a^2}{4}}.$$

d. Starting with the general solution

$$y(x, t) = \sum_{n=1}^{\infty} \sin\left(\frac{n\pi x}{L}\right) [C_n \sin(\omega_n t) + D_n \cos(\omega_n t)] e^{-\frac{at}{2}},$$

we find $C_n = 0$ for all n as $\dot{y}(x, 0) = 0$. Then

$$y(x, 0) = \sum_{n=1}^{\infty} D_n \sin\left(\frac{n\pi x}{L}\right) = \begin{cases} Ax, & 0 \leq x \leq L/2, \\ A(L - x), & L/2 \leq x \leq L. \end{cases}$$

As

$$\int_0^L y(x, 0) \sin\left(\frac{m\pi x}{L}\right) dx = \sum_{n=1}^{\infty} \int_0^L D_n \sin\left(\frac{n\pi x}{L}\right) \sin\left(\frac{m\pi x}{L}\right) dx$$

$$= \frac{LD_m}{2},$$

we have

$$D_m = \frac{2}{L}\int_0^L y(x,0)\sin\left(\frac{m\pi x}{L}\right)dx$$

$$= \frac{2}{L}\int_0^{L/2} Ax\sin\left(\frac{m\pi x}{L}\right)dx + \frac{2}{L}\int_{L/2}^L A(L-x)\sin\left(\frac{m\pi x}{L}\right)dx$$

$$= \frac{4AL}{m^2\pi^2}\sin\left(\frac{m\pi}{2}\right).$$

Note that we have used the formula

$$\int_0^\pi \sin(mx)\sin(nx)\,dx = \frac{\pi}{2}\delta_{mn}$$

in the above. Finally we have

$$y(x,t) = \sum_{n=1}^\infty \left(\frac{4AL}{n^2\pi^2}\right)\sin\left(\frac{n\pi}{2}\right)\sin\left(\frac{n\pi x}{L}\right)\cos(\omega_n t),$$

where

$$\omega_n = \frac{n\pi}{L}\sqrt{b}$$

as $a = 0$.

1240

a. Plot the pressure and air displacement diagrams along a pipe closed at one end for the second mode.

b. What is the frequency of this mode relative to the fundamental?

(Wisconsin)

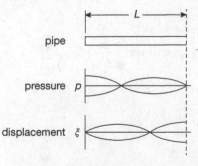

Fig. 1.211

Sol:

a. The pressure and air displacement as functions of distance from the closed end are sketched in Fig. 1.211.

b. For this mode, $L = 3\lambda/4$, while for the fundamental mode, $L = \lambda/4$. Hence if ω_0 is the fundamental frequency, the frequency of this mode is $3\omega_0$.

1241

An organ pipe of length l open on both ends is used in a subsonic wind tunnel to measure the Mach number v/c of air in the tunnel as shown in Fig. 1.212. The pipe when fixed in the tunnel is observed to resonate with a fundamental period t. If $v/c = 1/2$, calculate the ratio of periods t/t_0 where t_0 is the fundamental period of the pipe in still air.

(Wisconsin)

Sol: As the organ pipe is open at both ends, the fundamental wavelength of sound in resonance with it is given by $\lambda/2 = l$. The corresponding period is

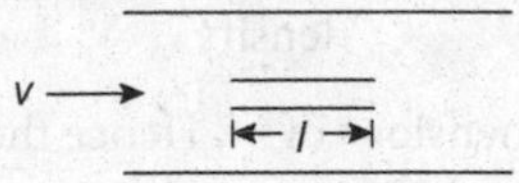

Fig. 1.212

$$t = \frac{\lambda}{v} = \frac{2l}{v},$$

where v is the velocity of sound relative to the pipe.

When the air in the pipe is still, v is equal to the velocity of sound in still air, c, and the fundamental period is

$$t_0 = \frac{2l}{c}.$$

When the air in the pipe moves with velocity $c/2$, the pipe can be considered to move with velocity $-c/2$ in still air. Thus $v = c - (-c/2) = 3c/2$ and the period is

$$t = \frac{2l}{\frac{3c}{2}} = \frac{4l}{3c}.$$

Hence we have the ratio

$$\frac{t}{t_0} = \frac{2}{3}.$$

1242

The speed of sound in a gas is calculated as

$$V = \sqrt{\frac{\text{adiabatic bulk modulus}}{\text{density}}}.$$

a. Show that this is a dimensionally-correct equation.

b. This formula implies that the propagation of sound through air is a quasistatic process. On the other hand, the speed for air is about 340 m/sec at a temperature for which the rms speed of an air molecule is about 500 m/sec. How then can the process be quasistatic?

(Wisconsin)

Sol:

a. The dimensions of the bulk modulus are the same as those of pressure while the adiabatic factor is dimensionless. Thus dimensionally

$$\frac{\text{adiabatic bulk modulus}}{\text{density}} \sim \frac{\text{g/cm} \cdot \text{s}^2}{\text{g/cm}^3} = \text{cm}^2/\text{s}^2,$$

which are the dimensions of v^2. Hence the formula is dimensionally correct.

b. Consider for example sound of frequency 1000 Hz. Its wavelength is about 0.34 m. Although the rms speed of an air molecule is large, its collision mean free path is only of the order of 10^{-5} cm, much smaller than the wavelength of sound. So the motion of the air molecules does not affect sound propagation through air, which is still adiabatic and quasistatic.

1243

A vertical cylindrical pipe, open at the top, can be partially filled with water. Successive resonances of the column with a 512 sec^{-1} tuning fork are observed when the distance from the water surface to the top of the pipe is 15.95 cm, 48.45 cm, and 80.95 cm.

a. Calculate the speed of sound in air.

b. Locate precisely the antinode near the top of the pipe.

c. The above measurements are presented to you by a team of sophomore lab students. How would you criticize their work?

(Wisconsin)

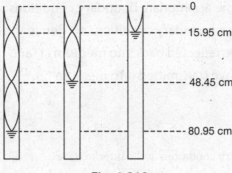

Fig. 1.213

Sol:

a. The wave forms of the successive resonances in the air column are shown in Fig. 1.213. It is seen that for successive resonances, the air columns differ in height by half a wavelength: $d = \lambda/2$. As

$$d = 48.45 - 15.95 = 80.95 - 48.45 = 32.50 \text{ cm},$$
$$\lambda = 2d = 65.00 \text{ cm}.$$

The velocity in air is then

$$v = \lambda\nu = 0.6500 \times 512 = 330 \text{ m/s}.$$

b. As $\lambda/4 = 16.25$ cm and 16.25 cm -15.95 cm $= 0.30$ cm, the uppermost antinode is located at 0.30 cm above the top of the pipe.

c. This method of measuring sound velocity in air is rather inaccurate as the human ear is not sensitive enough to detect precisely small variations in the intensity of sound, and the accuracy of measurement is rather limited. Still, the data obtained are consistent and give a good result. The students ought to be commended for their careful work.

1244

Two media have a planar, impermeable interface as shown in Fig. 1.214. Plane acoustic waves of pressure amplitude A and frequency f are generated in medium (1), directed toward medium (2). Take A and f as given quantities and assume the wave propagation is normal to the interface. Medium (1) has density ρ_1 and sound velocity c_1, while medium (2) has density ρ_2 and sound velocity c_2

a.   What are the appropriate boundary conditions at the interface?

b.   Apply these boundary conditions to derive the pressure amplitude A of the wave reflected back into medium (1) and the pressure amplitude B of the wave transmitted into medium (2).

(CUSPEA

Sol:

a.   The boundary conditions at the interface are

i.   the pressure is continuous,

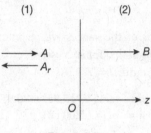

Fig. 1.214

ii.   the component of the rate of fluid displacement perpendicular to the interface is continuous, otherwise the interface would be permeable.

b.   Take the z-axis perpendicular to the interface with the origin on the interface and let the pressure be

$$Ae^{i(\omega t - k_1 z)} \quad \text{for the incident wave,}$$
$$A_r e^{i(\omega t - k_1 z)} \quad \text{for the reflected wave,}$$
$$Be^{i(\omega t - k_2 z)} \quad \text{for the transmitted wave,}$$

with $k_j = \omega/c_j$, c_j being the velocity of sound in the jth medium. The boundary condition (i) gives

$$A + A_r = B. \tag{1}$$

The velocity of sound in a fluid is given by

$$c = \sqrt{\frac{M}{\rho}},$$

where $M = -p(\Delta v/v)^{-1}$ is the bulk modulus, Δv being the change of the original volume v by an excess pressure p. For a compressional wave, Δv is solely longitudinal so that

$$\frac{\Delta v}{v} \approx \frac{\Delta \xi}{\Delta z} \approx \frac{\partial \xi}{\partial z},$$

where ξ is the displacement of fluid layers from their equilibrium positions. Thus

$$p = -\rho c^2 \frac{\partial \xi}{\partial z},$$

or

$$\frac{\partial \xi}{\partial z} = \frac{-p}{\rho c^2} \sim -\frac{e^{i(\omega t \mp kz)}}{\rho c^2}.$$

Integrating we have

$$\xi \sim \pm \frac{e^{i(\omega t \mp kz)}}{ik\rho c^2}.$$

For the three waves we have respectively

$$\xi_A = \frac{1}{\rho_1 c_1^2}\left(\frac{A}{ik_1}\right) e^{i(\omega t - k_1 z)},$$

$$\xi_{A_r} = \frac{-1}{\rho_1 c_1^2}\left(\frac{A_r}{ik_1}\right) e^{i(\omega t + k_1 z)},$$

$$\xi_B = \frac{1}{\rho_2 c_2^2}\left(\frac{B}{ik_2}\right) e^{i(\omega t - k_2 z)},$$

and thus

$$\dot{\xi}_A = \frac{A}{\rho_1 c_1} e^{i(\omega t - k_1 z)},$$

$$\dot{\xi}_{A_r} = \frac{-A_r}{\rho_1 c_1} e^{i(\omega t + k_1 z)},$$

$$\dot{\xi}_B = \frac{B}{\rho_2 c_2} e^{i(\omega t - k_2 z)}.$$

The boundary condition (ii) states that at $z = 0$.

$$\dot{\xi}_A + \dot{\xi}_{A_r} = \dot{\xi}_B,$$

or

$$\frac{A}{\rho_1 c_1} - \frac{A_r}{\rho_1 c_1} = \frac{B}{\rho_2 c_2}. \tag{2}$$

Combining Eqs. (1) and (2) we obtain the amplitudes of the reflected and transmitted pressure waves:

$$A_r = A\frac{\rho_2 c_2 - \rho_1 c_1}{\rho_1 c_1 + \rho_2 c_2},$$

$$B = \frac{2A\rho_2 c_2}{\rho_1 c_1 + \rho_2 c_2}.$$

1245

Let the speed of sound in air be c and the velocity of a source of sound moving through the air be v in the x-direction.

a. For $v < c$: a pulse of sound is emitted at the origin at time $t = 0$. Sketch the relationship of the wavefront at time t to the position of the sound at time t. Label your sketch carefully. Write an equation for the position of the wavefront as seen from the source at time t.

b. For $v > c$: a source emits a continuous signal. Sketch the wavefront set up by the moving source. Indicate on your sketch the construction which leads to your result. Write an equation relating the shape of the wavehont to other known factors in the problem.

(Wisconsin)

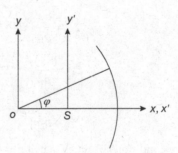

Fig. 1.215

Sol:

a. Let S be the position of the source at time t. Take coordinate frames Oxy, $Sx'y'$ with origins at O and S, the x-, x'-axes along OS, and the y-, y'-axes parallel to each other as shown in Fig. 1.215. We have

$$x' = x - vt, \quad y' = y.$$

The wavehont at time t is given by $x = ct \cos \varphi$, $y = ct \sin \varphi$, with $0 \leq \varphi \leq 2\pi$. Then the wavefront as seen from the source is given by $x' = ct \cos \varphi - vt$, $y' = ct \sin \varphi$.

b. Suppose the source moves from point O to point S in the time interval $t = 0$ to $t = t$ and consider the signals emitted at $t = 0$ and intermediate instants $t_1, t_2, \ldots$, when the source is at $S_1, S_2, \ldots$, with $OS_1 = vt_1$, $OS_2 = vt_2, \ldots$. Each signal will propagate from the point of emission as a spherical wave. At time t, the wavefronts of the signals emitted at $O, S_2, S_2, \ldots$ will have radii $ct, c(t - t_1), c(t - t_2), \ldots$, respectively. As

$$\frac{ct}{vt} = \frac{c(t - t_1)}{v(t - t_1)} = \frac{c(t - t_2)}{v(t - t_2)} = \ldots,$$

all these wavefronts will be enveloped by a cone with vertex at S of semivertex angle θ given by

$$\sin \theta = \frac{ct}{vt} = \frac{c}{v},$$

as shown in Fig. 1.216. Hence the resultant wavefront of the continuous signal is a cone of semi-vertex angle $\arcsin(c/v)$ with the vertex at the moving source.

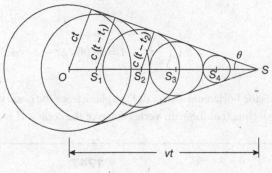

Fig. 1.216

1246

The velocity of sound in the atmosphere is 300 m/s. An airplane is traveling with velocity 600 m/s at an altitude of 8000 m over an observer as shown in Fig. 1.217. How far past the observer will the plane be when he hears the sonic boom?

(Wisconsin)

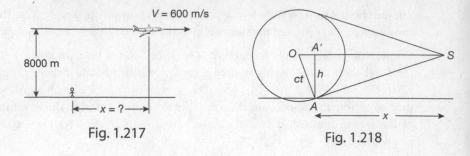

Fig. 1.217   Fig. 1.218

Sol: As the velocity v of the source S is greater than the velocity c of sound propagation, the wavefront is a cone with vertex at the moving source (Problem **1245**).
The observer at A will hear the sonic boom, which was emitted when the source
was at O, when the cone sweeps past him, as shown in Fig. 1.218. The source is
now at S. Let A' be a point on the path of the source directly above A. We have

$$OA \perp AS, \quad OA = ct, \quad OS = vt,$$

and

$$\frac{h}{x} = \frac{ct}{AS} = \frac{ct}{\sqrt{OS^2 - OA^2}} = \frac{c}{\sqrt{v^2 - c^2}}$$

or

$$x = h\sqrt{\left(\frac{v}{c}\right)^2 - 1} = 8000\sqrt{2^2 - 1} = 1.39 \times 10^4\,\text{m}.$$

This is the horizontal distance of the plane from the observer when he hears the sonic
boom. Note that the semi-vertex angle of the cone is $\theta = \arcsin(c/v)$ as required.

1247

It is a curious fact that one occasionally hears sound from a distant source
with startling clarity when the wind is blowing from the source toward the
observer.

- **a.** Show that this effect cannot be explained by "the wind carrying the
  sound along with it", i.e. a uniform wind velocity cannot account for the
  effect.

- **b.** Wind blowing over the ground has a vertical velocity gradient which
  can be well represented near the ground by the formula $v = ky^2$, where
  y is the height above the ground and k is a constant which depends

on the wind speed outside of the boundary layer where the parabolic velocity profile is a good approximation. For a given value of k and of the speed of sound v_s, calculate the distance s, downwind from a sound source, where the maximum enhancement of sound intensity occurs. HINT: You may assume that the sound rays follows low, arc-like paths which are well represented by

$$y = h \sin\left(\frac{\pi x}{s}\right).$$

c. One also notices an enhancement of the transmission of sound over a lake, even for no wind. What is happening in this case?

(Princeton)

Sol:

a. The effect cannot be explained by the wind carrying the sound with it, for across the path of a uniformly moving wind, all observers would then hear the sound with equal clarity. This not being the case the effect is in fact due to refraction of sound brought about by the variation of the sound velocity, with respect to a fixed observer, at different points of the medium. This may arise from two possible causes, temperature gradient or velocity gradient in the moving wind. The velocity of compressional waves in a gas varies with temperature T as $\sqrt{T}$. It also varies if the velocity of the medium itself varies. Refraction of sound changes the direction of its wavefront. Near the surface of the earth, both gradients may be present and the path of sound can bend in different ways, making it possible for a distant observer to hear it with startling clarity.

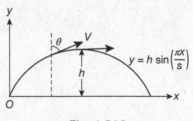

Fig. 1.219

b. Take coordinate axes as shown in Fig. 1.219. It is assumed that the wind velocity near ground is horizontal with a vertical gradient, i.e.

$$v = v_x = ky^2,$$

so the medium can be considered as consisting of horizontal layers with differen[t] sound velocities. The law of refraction is

$$\frac{\sin \theta}{V} = \text{constant},$$

where θ is the angle between the direction of sound propagation in the layer an[d] the vertical, and V is the velocity of sound with respect to the ground. Conside[r] two points on the sound path with variables

$$\theta_1 = \theta, \qquad V_2 = v_s + v_x \sin \theta = v_s + v \sin \theta,$$
$$\theta_2 = \theta + d\theta, \quad V_2 = v_s + (v + dv) \sin(\theta + d\theta).$$

The law of refraction then gives

$$\frac{v_s + (v + dv) \sin(\theta + d\theta)}{v_s + v \sin \theta} = \frac{\sin(\theta + d\theta)}{\sin \theta}.$$

As $\sin(\theta + d\theta) \approx \sin \theta + \cos \theta d\theta$, retaining only the lowest order terms w[e] have

$$\frac{dv}{v_s} = \frac{d \sin \theta}{\sin^2 \theta}.$$

Thus

$$\int_0^h 2ky \frac{dy}{v_s} = \int_{\theta_0}^{\frac{\pi}{2}} \frac{d \sin \theta}{\sin^2 \theta},$$

or

$$\frac{kh^2}{v_s} = \frac{1}{\sin \theta_0} - 1.$$

On the other hand the given sound path yields

$$\cot \theta = \frac{dy}{dx} = \frac{\pi h}{s} \cos \frac{\pi x}{s},$$

or

$$\frac{1}{\sin \theta} = \sqrt{1 + \cot^2 \theta} = \sqrt{1 + \left(\frac{\pi h}{s}\right)^2 \cot^2 \frac{\pi x}{s}},$$

in particular,

$$\frac{1}{\sin \theta_0} = \sqrt{1 + \left(\frac{\pi h}{s}\right)^2}.$$

Substituting this in the above gives the path length s, downwind from the sound source, where maximum enhancement of sound intensity occurs as

$$s = \frac{\pi v_s}{\sqrt{k(2v_s + kh^2)}}$$

c. The speed of sound in a gas varies with absolute temperature T as $\sqrt{T}$. Vertically above a lake, for some range of heights, the temperature increases during daytime and establishes a vertical gradient. So does the speed of sound. Refraction of sound occurs during daytime simar to that described in (b).

1248

Consider a plane standing sound wave of frequency 10^3 Hz in air at 300 K. Suppose the amplitude of the pressure variation associated with this wave is 1 dyn/cm^2 (compared with the ambient pressure of 10^6 dyn/cm^2). Estimate (order of magnitude) the amplitude of the displacement of the air molecules associated with this wave.

(Columbia)

Sol: The longitudinal displacement ξ from equilibrium of a point in a plane stationary compressional wave in the x direction can be expressed as

$$\xi = \xi_0 \sin(kx)e^{-i\omega t},$$

with $k = n\pi/l$, l being the thickness of the gas and $n = 1, 2, \ldots$. The velocity of the wave is

$$v = \sqrt{\frac{M}{\rho}}.$$

Here the bulk modulus M is by definition

$$M = -p\left(\frac{\Delta V}{V}\right)^{-1},$$

p being the excess pressure and V the original volume. Consider a cylinder of the gas of cross-sectional area A and length Δx. We have

$$\frac{\Delta V}{V} = \frac{A\Delta\xi}{A\Delta x} \approx \frac{\partial \xi}{\partial x}.$$

Then

$$p = -M \frac{\partial \xi}{\partial x}$$
$$= -Mk\xi_0 \cos(kx)e^{-i\omega t}$$
$$= -p_0 \cos(kx)e^{-i\omega t},$$

where $p_0 = Mk\xi_0 = \rho v^2 k\xi_0$ is the amplitude of the excess pressure. Hence

$$\xi_0 = \frac{p_0}{\rho v^2 k}.$$

For the lowest mode

$$n = 1, \qquad \lambda = 2l,$$
$$k = \frac{2\pi}{\lambda} = \frac{2\pi\nu}{v},$$

ν being the frequency of the sound wave. Thus

$$\xi_0 = \frac{p_0}{2\pi \rho v \nu}.$$

For an ideal gas

$$p_a V = \frac{m}{M} RT,$$

giving

$$\rho = \frac{m}{V} = \frac{p_a M}{RT},$$

where p_a, T are the ambient pressure and temperature respectively. As $p_0 = 1 \text{ dyn/cm}^2 = 10^{-1} \text{ N/m}^2$, $p_a = 10^6 \text{ dyn/cm}^2 = 10^5 \text{ N/m}^2$, $M = 29 \times 10^{-3}$ kg/mol, $R = 8.31$ J/mol/K, $T = 300$ K, $v = 340$ m/s, $\nu = 10^3$ Hz, we find $\xi_0 = 4 \times 10^{-8}$ m as the amplitude of the displacement of the air molecules.

1249

An acoustical motion detector emits a 50 kHz signal and receives the echo signal. If the echoes have Doppler shift frequency components departing from 50 kHz by more than 100 Hz, a "moving object" is registered. For a sound velocity in air of 330 m/sec, calculate the speed with which an object must move toward (or away from) the detector in order to be registered as a "moving object".

(Wisconsin)

Sol: Consider a source emitting sound of frequency ν. The Doppler effect has it that if an observer moves with velocity v toward the source he will detect the frequency as

$$\nu' = \left(\frac{c+v}{c}\right)\nu,$$

c being the speed of sound propagation. On the other hand, if the source moves with velocity v toward the observer, who is stationary, then

$$\nu' = \left(\frac{c}{c-v}\right)\nu.$$

Thus the object, moving toward the detector, receives a signal of frequency

$$\nu' = \left(\frac{c+v}{c}\right)\nu,$$

and the signal after reflection by the object is detected by the detector as having frequency

$$\nu'' = \left(\frac{c}{c-v}\right)\nu' = \left(\frac{c+v}{c-v}\right)\nu.$$

For the moving object to be registered, we must have, $\nu'' = \nu \pm \Delta\nu$, where $\Delta\nu \geq 10^2$ Hz. Then

$$\nu \pm \Delta\nu = \left(\frac{c+v}{c-v}\right)\nu,$$

or

$$v = \pm\frac{c\Delta\nu}{2\nu\pm\Delta\nu} \approx \pm\frac{c\Delta\nu}{2\nu},$$

as $\Delta\nu \ll \nu$. Hence the object must be moving toward or receding from the detector at

$$v \geq \frac{330 \times 10^2}{2 \times 5 \times 10^4} = 0.330 \text{ m/s}$$

for it to be registered.

1250

A student near a railroad track hears a train's whistle when the train is coming directly toward him and then when it is going directly away. The two observed frequencies are 250 and 200 Hz. Assume the speed of sound in air to be 360 m/s. What is the train's speed?

(Wisconsin)

Sol: Let ν_0, ν_1, ν_2 be respectively the frequency of the whistle emitted by the train and the frequencies heard by the student when the train is coming and when it is moving away. The Doppler effect has it that

$$\nu_1 = \left(\frac{c}{c-v}\right)\nu_0,$$

$$\nu_2 = \left(\frac{c}{c+v}\right)\nu_0,$$

where c is the speed of sound and v is the speed of the train, and thus

$$\frac{\nu_1}{\nu_2} = \frac{c+v}{c-v}.$$

Putting in the data, we have

$$1.25 = \frac{360 + v}{360 - v},$$

or

$$\frac{2.25}{0.25} = \frac{720}{2v},$$

and thus

$$v = \frac{360}{9} = 40 \text{ m/s}.$$

1251

The velocity of blood flow in an artery can be measured using Doppler-shifted ultrasound. Suppose sound with frequency 1.5×10^6 Hz is reflected straight back by blood flowing at 1 m/s. Assuming the velocity of sound in tissue is 1500 m/s and that the sound is incident at a very small angle as shown in Fig. 1.220, calculate the frequency shift between the incident and reflected waves.

(Wisconsin)

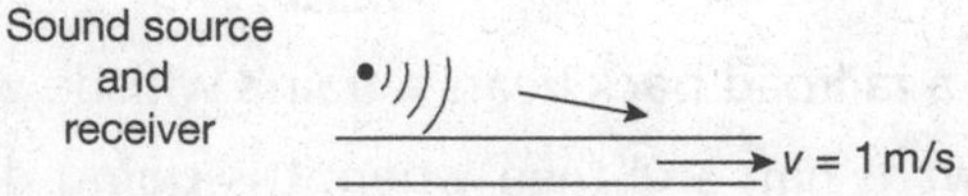

Fig. 1.220

Sol: As the sound is incident at a very small angle, the blood can be considered to be flowing directly away. Then the results of Problem **1249** can be applied with v replaced by $-v$:

$$v'' = \left(\frac{c-v}{c+v}\right)v.$$

The frequency shift is then

$$v'' - v = -\frac{2vv}{c+v} \approx -\frac{2vv}{c} = -2 \times 10^3 \text{ Hz}.$$

1252

A car has front- and back-directed speakers mounted on its roof, and drives toward you with a speed of 50 ft/s, as shown in Fig. 1.221. If the speakers are driven by a 1000 Hz oscillator, what beat frequency will you hear between the direct sound and the echo off a brick building behind the car? (Take the speed of sound as 1000 ft/s.)

(Wisconsin)

Fig. 1.221

Sol: The sound from the back-directed speaker has Doppler frequency

$$v_b = \left(\frac{c}{c+v}\right)v,$$

where c and v are the speeds of sound and the car respectively, and v is the frequency of the sound emitted. As the wall is stationary with respect to the observer, v_b is also the frequency as heard by the latter. The sound from the front-directed speaker has Doppler frequency

$$v_f = \left(\frac{c}{c-v}\right)v.$$

Hence the beat frequency is

$$v_f - v_b = cv\left(\frac{1}{c-v} - \frac{1}{c+v}\right) = \frac{2vcv}{c^2 - v^2} \approx \frac{2vv}{c} = 100 \text{ Hz}.$$

1253

A physics student holds a tuning fork vibrating at 440 Hz and walks at 1.2 m/ away from a wall. Does the echo from the wall have a higher or lower pitch than the tuning fork? What beat frequency does he hear between the fork and the echo? The speed of sound is 330 m/s

(Wisconsin

Sol: As the tuning fork, which emits sound of frequency ν, moves away the wall a speed v, the sound that is incident on the wall has frequency

$$\nu' = \left(\frac{c}{c + v}\right)\nu.$$

Then the student, who is moving away from the wall at speed v, hears the reflected frequency

$$\nu'' = \left(\frac{c - v}{c}\right)\nu' = \left(\frac{c - v}{c + v}\right)\nu.$$

As

$$\nu'' - \nu = -\frac{2v\nu}{c + v} < 0,$$

the echo has a lower frequency. The beat frequency between the fork and the echo is

$$\frac{2v\nu}{c + v} \approx \frac{2v\nu}{c} = 3.2 \text{ Hz}.$$

1254

A rope is attached at one end to a wall and is wrapped around a capstan through an angle θ. If someone pulls on the other end with a force F as shown in Fig. 1.222(a), find the tension in the rope at a point between the wall and the capstan in terms of F, θ and μ_s, the coefficient of friction between the rope and capstan.

(Columbia)

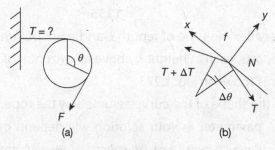

Fig. 1.222

Sol: Consider an element of the rope as shown in Fig. 1.222(b). The forces acting on the element are the tensions T and $T + \Delta T$ at its two ends, the reaction N exerted by the capstan, and the friction f. As the element is in equilibrium we have

$$f + (T + \Delta T) \cos\left(\frac{\Delta \theta}{2}\right) - T \cos\left(\frac{\Delta \theta}{2}\right) = 0,$$

$$N - (T + \Delta T) \sin\left(\frac{\Delta \theta}{2}\right) - T \sin\left(\frac{\Delta \theta}{2}\right) = 0.$$

In first-order approximation the above equations become,

$$f + (T + \Delta T) - T = 0, \quad \text{or} \quad f = -\Delta T,$$

$$N - T\frac{\Delta \theta}{2} - T\frac{\Delta \theta}{2} = 0, \quad \text{or} \quad N = T\Delta \theta.$$

Then as $f = \mu_s N$, we find

$$\frac{\Delta T}{\Delta \theta} = -\mu_s T,$$

or, letting $\Delta \theta \to 0$,

$$\frac{dT}{d\theta} = -\mu_s T.$$

Integrating we have

$$T = Ce^{-\mu_s \theta},$$

where C is a constant. As $T = F$ at $\theta = 0$, $C = F$. Hence

$$T = Fe^{-\mu_s \theta}.$$

1255

A uniform, very flexible rope of length L and mass per unit length ρ is hung from two supports, each at height h above a horizontal plane, separated by distance, $2x_0$, as shown in Fig. 1.223.

  a. Derive the shape of the curve assumed by the rope.

      HINT: A parameter in your solution will depend on a transcendental equation, which need not be solved. However, any differential equations which you encounter should be solved.

  b. Find an expression for the tension in the rope at the supports.

      Suppose the supports are now replaced by frictionless pulleys of negligible size, and a uniform rope of infinite length is hung over the two pulleys (see Fig. 1.223). There is no friction between the rope and the table. In this case the shape of the curve assumed by the rope depends only on a dimensionless parameter $\alpha = h/x_0$.

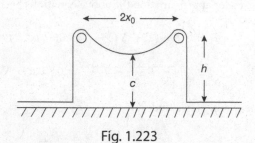

Fig. 1.223

  c. Assuming that the rope hangs in a smooth curve with minimum height c, derive a transcendental equation relating h/c to α.

  d. Find an exact solution for the shape of the rope when $\alpha \ll 1$.

  e. Relate the shape of the rope in parts (c) and (d) to the shape of a soap film stretched between two circular wires of radius h and separation $2x_0$ as shown in Fig. 1.224.

                                                                    (MIT)

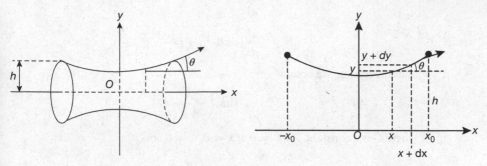

Fig. 1.224   Fig. 1.225

Sol:

a. Use coordinates as shown in Fig. 1.225 and let the tension in the rope be $T = T(x)$. Consider an infinitesimal element between the points x and $x + dx$. Conditions for equilibrium are

$$(T\cos\theta)_{x+dx} - (T\cos\theta)_x = 0,$$

$$(T\sin\theta)_{x+dx} - (T\sin\theta)_x = \rho g\sqrt{(dx)^2 + (dy)^2} = \rho g\sqrt{1 + y'^2}\,dx.$$

The first equation givsses

$$\frac{d(T\cos\theta)}{dx} = 0, \quad \text{or} \quad T\cos\theta = \text{constant} = A, \quad \text{say.}$$

The second equation gives

$$\frac{d(T\sin\theta)}{dx} = \rho g\sqrt{1 + y'^2}.$$

As

$$\tan\theta = \frac{dy}{dx} \equiv y',$$

we have

$$\sin\theta = \frac{y'}{\sqrt{1 + y'^2}}, \quad \cos\theta = \frac{1}{\sqrt{1 + y'^2}}$$

and the above equations become

$$T = A\sqrt{1 + y'^2}, \tag{1}$$

$$Ay'' = \rho g\sqrt{1 + y'^2}. \tag{2}$$

Writing (2) as

$$\frac{1}{\sqrt{1 + y'^2}}\frac{dy'}{dx} = \frac{\rho g}{A},$$

or

$$\frac{d}{dx}(\sinh^{-1}y') = \frac{\rho g}{A},$$

and integrating, we obtain

$$y' = \sinh\left(\frac{\rho g x}{A} + C\right),$$

where C is a constant. As $y' = 0$ at $x = 0$, $C = 0$ and

$$y' = \sinh\left(\frac{\rho g x}{A}\right). \tag{3}$$

Further integrating gives

$$y = \frac{A}{\rho g}\cosh\left(\frac{\rho g x}{A}\right) + B.$$

With the boundary condition $y = h$ at $x = x_0$, we find

$$B = h - \frac{A}{\rho g}\cosh\left(\frac{\rho g x_0}{A}\right).$$

Hence the shape of the rope is described by

$$y(x) = \frac{A}{\rho g}\left[\cosh\left(\frac{\rho g x}{A}\right) - \cosh\left(\frac{\rho g x_0}{A}\right)\right] + h \tag{4}$$

with the constant A yet to be determined. Consider the tensions $T(\pm x_0)$ at the supports $x = \pm x_0$. Their y-components satisfy

$$2T\sin\theta = L\rho g,$$

i.e.

$$\frac{2Ty'}{\sqrt{1 + y'^2}} = L\rho g.$$

Using Eqs. (1) and (3), we can write this as

$$2A\sinh\left(\frac{\rho g x_0}{A}\right) = L\rho g,$$

from which A can be determined. The tensions in the rope at $x = \pm x_0$ are given by (1) to be

$$T(\pm x_0) = A\sqrt{1 + y'^2}\big|_{x=x_0} = A\cosh\left(\frac{\rho g x_0}{A}\right), \tag{5}$$

use having been made of Eq. (3).

c. The tensions $T(\pm x_0)$ in the rope on the two sides of each pulley are equal. Hence

$$T(\pm x_0) = h\rho g,$$

or, by Eq. (5),

$$A \cosh\left(\frac{\rho g x_0}{A}\right) = h\rho g.$$

Substituting this in (4) gives the equation describing the shape of the rope between the pulleys:

$$y(x) = \frac{A}{\rho g} \cosh\left(\frac{\rho g x}{A}\right). \tag{6}$$

Let $y = c$ at $x = 0$, then $c = A/\rho g$. As $y = h$ at $x = x_0$, we have

$$h = c \cosh\left(\frac{x_0}{c}\right),$$

or

$$\frac{h}{c} = \cosh\left(\frac{h}{c\alpha}\right). \tag{7}$$

This equation determines h/c as a function of $\alpha = h/x_0$ only. Equation (6) can be written as

$$\frac{y}{h} = \frac{c}{h} \cosh\left(\frac{h}{c}\frac{x}{h}\right).$$

If we scale the coordinates by h, i.e.

$$\xi = \frac{x}{h}, \quad \eta = \frac{y}{h},$$

we have

$$\eta = \frac{c}{h} \cosh\left(\frac{h}{c}\xi\right).$$

This equation, which describes the shape of the curve, depends only on h/c, which in term depends only on $\alpha = h/x_0$ through Eq. (7).

d. Physically, $c < h$, so that if $\alpha \ll 1$, $\cosh(h/c\alpha) \gg 1$. Then for h to remain finite, we require $c \to 0$ as indicated by Eq. (7). This means that the whole rope is lying on the ground.

e. Let σ be the coefficient of surface tension of the soap. For equilibrium in the horizontal direction at a point (x, y) on the film, we have

$$(\sigma \cdot 2\pi y \cos\theta)_{x+dx} - (\sigma \cdot 2\pi y \cos\theta)_x = 0,$$

or

$$\frac{d}{dx}(2\pi\sigma y\cos\theta) = 0,$$

i.e.

$$y\cos\theta = \text{constant}.$$

Suppose $y = c$ at $x = 0$, then as $\theta = 0$ for $x = 0$, the constant is equal to c. Furthermore, as

$$\cos\theta = \frac{dx}{\sqrt{(dx)^2 + (dy)^2}} = \frac{1}{\sqrt{1 + y'^2}},$$

we have

$$y = c\sqrt{1 + y'^2}$$

or

$$dx = \frac{cdu}{\sqrt{u^2 - 1}}$$

with $u = y/c$. Integrating we have

$$x = c\cos h^{-1}\left(\frac{y}{c}\right) + \text{constant}.$$

As $y = c$ at $x = 0$, the constant is zero. Hence

$$y = c\cosh\left(\frac{x}{c}\right),$$

which is identical with Eq. (6) of part (c).

1256

a. A bounded, axially symmetric body has mass density $\rho(x, y, z) = \rho(r, \theta)$. At large distances from the body its gravitational potential has the form

$$\phi = -\frac{GM}{r} + \frac{f(\theta)}{r^2} + \dots,$$

where

$$M = \int \rho(x', y', z')dx'\, dy'\, dz' = 2\pi \int \rho(r', \theta')r'^2 \sin\theta'\, dr'\, d\theta'$$

is the total mass. Find $f(\theta)$.

b. A small test body has mass density $\sigma(x, y, z)$ and is placed in a gravitational potential $\phi(x, y, z)$. What is its gravitational potential energy?

c. Suppose the body in (a) is spherically symmetric, i.e. $\rho = \rho(r)$, then $\phi = \phi(r)$. Suppose the body is made of gas and supported against its own gravity by a pressure $p(r)$. Denote its radius by R. Some of the following integrals correctly represent the gravitational potential energy of the body, others are incorrect by simple numerical factors (positive or negative). Identify the correct ones and find the missing factors for the others. That is, if U = potential energy/4π, then is

$$U = \begin{cases} (i) & -\displaystyle\int_0^R \rho \frac{d\phi}{dr} r^3 dr \quad ? \\[2ex] (ii) & \dfrac{1}{4\pi G}\displaystyle\int_0^R \left(\frac{d\phi}{dr}\right)^2 r^2 dr \quad ? \\[2ex] (iii) & \dfrac{1}{2}\displaystyle\int_0^R \rho\phi r^2 dr \quad ? \\[2ex] (iv) & -\displaystyle\int_0^R p r^2 dr \quad ? \end{cases}$$

d. The test body in (b) is placed with its center of mass at $(0,0,r_0)$ in a spherically symmetric potential

$$\phi(r) = -\frac{MG}{r}.$$

For large r_0 the gravitational potential energy has the form

$$-\frac{mGM}{r_0} + \frac{d}{r_0^3} + O\left(\frac{1}{r_0^4}\right),$$

where $m = \displaystyle\int \sigma d^3x$. Find d.

(UC, Berkeley)

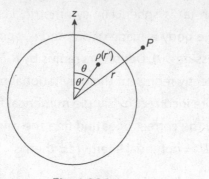

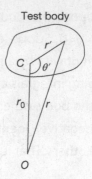

$$\text{Fig. 1.226} \qquad\qquad \text{Fig. 1.227}$$

Sol:

a. As in Fig. 1.226 take z-axis along the axis of symmetry and origin O inside the body. The gravitational potential (potential energy per unit mass of test body) at a distant point P due to the body is

$$\phi = -\int_{V'} \frac{G\rho(\mathbf{r}')}{|\mathbf{r} - \mathbf{r}'|}\, dV',$$

where V' is the volume of the body.

As

$$(\mathbf{r} - \mathbf{r}')^2 = r^2 + r'^2 - 2rr'\cos(\theta' - \theta),$$

for large distances from the body $|\mathbf{r} - \mathbf{r}'|^{-1}$ can be approximated:

$$\frac{1}{|\mathbf{r} - \mathbf{r}'|} = \frac{1}{r}\left[1 + \left(\frac{r'}{r}\right)^2 - \frac{2r'}{r}\cos(\theta - \theta')\right]^{-\frac{1}{2}}$$

$$\approx \frac{1}{r}\left[1 + \frac{r'}{r}\cos(\theta - \theta')\right].$$

Substituting it in the integral gives

$$\phi = -\frac{G}{r}\cdot 2\pi\int \rho(r', \theta')r'^2 \sin\theta'\, dr'\, d\theta'$$

$$-\frac{G}{r^2}\cdot 2\pi\int \rho(r', \theta')r'^3 \sin\theta' \cos(\theta - \theta')\, dr'\, d\theta'.$$

Comparing it with the given form

$$\phi \approx -\frac{GM}{r} + \frac{f(\theta)}{r^2},$$

we find

$$f(\theta) = -2\pi G \int \rho(r', \theta') r'^3 \sin\theta' \cos(\theta - \theta') dr' d\theta'.$$

b. In a gravitational potential $\phi(x, y, z)$ the potential energy of a test body with mass density $\sigma(x, y, z)$ and volume V is

$$W = \int_V \sigma(x, y, z)\phi(x, y, z) dV.$$

c. For a closed system of mass density ρ and volume V the gravitational energy is

$$W = \frac{1}{2} \int_V \rho\phi dV.$$

Then for a spherically symmetric gaseous body of radius R we have

$$W = \frac{1}{2} \int_0^R \rho(r)\phi(r) 4\pi r^2 dr,$$

taking the origin at its center. Thus

$$U = \frac{W}{4\pi} = \frac{1}{2} \int_0^R \rho\phi r^2 dr. \tag{1}$$

Thus integral (iii) is correct.

Consider a spherical shell of the gaseous body of radius r and thickness Δr. As the body is supported against its own gravity by pressure p, we require that for equilibrium

$$4\pi r^2 [p(r) - p(r + \Delta r)] - 4\pi r^2 \rho \frac{d\phi}{dr} \Delta r = 0,$$

or

$$\frac{dp}{dr} = -\rho \frac{d\phi}{dr}.$$

Poissons's equation for attracting masses is

$$\nabla^2 \phi = 4\pi G\rho,$$

or, for spherical symmetry,

$$\frac{1}{r^2} \frac{d}{dr}\left(r^2 \frac{d\phi}{dr}\right) = 4\pi G\rho,$$

giving

$$\rho = \frac{1}{4\pi G} \cdot \frac{1}{r^2} \frac{d}{dr}\left(r^2 \frac{d\phi}{dr}\right).$$

Hence

$$\frac{dp}{dr} = -\frac{1}{4\pi G} \cdot \frac{1}{r^2}\frac{d}{dr}\left(r^2\frac{d\phi}{dr}\right)\frac{d\phi}{dr}.$$

Outside the spherical body, p is zero and $dp/dr = 0$. Hence

$$\left(\frac{d\phi}{dr}\right)_{r=R} = 0.$$

Equation (1) can then be written as

$$\begin{aligned}
U &= \frac{1}{8\pi G}\int_0^R \phi\frac{d}{dr}\left(r^2\frac{d\phi}{dr}\right)dr \\
&= \frac{1}{8\pi G}\left[r^2\phi\frac{d\phi}{dr}\bigg|_0^R - \int_0^R r^2\left(\frac{d\phi}{dr}\right)^2 dr\right] \\
&= -\frac{1}{8\pi G}\int_0^R r^2\left(\frac{d\phi}{dr}\right)^2 dr.
\end{aligned}$$

(2

Thus integral (ii) has to be multiplied by a factor $-\frac{1}{2}$. Consider now integral (i) It can be written as

$$\begin{aligned}
&-\frac{1}{4\pi G}\int_0^R r\frac{d}{dr}\left(r^2\frac{d\phi}{dr}\right)\cdot\frac{d\phi}{dr}dr \\
&= -\frac{1}{4\pi G}\left[r^3\left(\frac{d\phi}{dr}\right)^2\bigg|_0^R - \int_0^R r^2\frac{d\phi}{dr}\cdot d\left(r\frac{d\phi}{dr}\right)\right] \\
&= \frac{1}{8\pi G}\int_0^R r\,d\left(r\frac{d\phi}{dr}\right)^2 \\
&= \frac{1}{8\pi G}\left[r^3\left(\frac{d\phi}{dr}\right)^2\bigg|_0^R - \int_0^R r^2\left(\frac{d\phi}{dr}\right)^2 dr\right] \\
&= -\frac{1}{8\pi G}\int_0^R r^2\left(\frac{d\phi}{dr}\right)^2 dr,
\end{aligned}$$

which is the same as Eq. (2). Hence integral (i) is correct.

Integral (iv) can be written as

$$-\frac{1}{3}\int_0^R p\,dr^3 = -\frac{1}{3}\left[pr^3 \Big|_0^R - \int_0^R r^3 \frac{dp}{dr}\,dr \right]$$

$$= \frac{1}{3}\int_0^R r^3 \frac{dp}{dr}\,dr$$

$$= -\frac{1}{3}\int_0^R \rho r^3 \frac{d\phi}{dr}\,dr.$$

Compared with integral (i), which is correct, it has to be multiplied by a factor 3.

d. Let C be the center of mass of the test body, and consider a volume element dV' at radius vector $\mathbf{r}'$ from C as shown in Fig. 1.227. We have

$$\mathbf{r} = \mathbf{r}_0 + \mathbf{r}',$$

or

$$r^2 = r_0^2 + r'^2 + r_0 r' \cos\theta',$$

giving

$$r^{-1} = r_0^{-1}\left[1 + \frac{r'}{r_0}\cos\theta' + \left(\frac{r'}{r_0}\right)^2 \right]^{-\frac{1}{2}} \approx r^{-1}\left[1 - \frac{r'}{2r_0}\cos\theta' + O\left(\frac{r'^2}{r_0^2}\right) \right].$$

The gravitational potential energy of the test body is

$$W = \int_{V'} \sigma\phi\,dV' = -\int_{V'} \sigma(\mathbf{r}')\frac{GM}{r_0}\left(1 - \frac{r'}{2r_0}\cos\theta' + O\left(\frac{r'}{r_0}\right)^2 \right)dV',$$

Where V' is the volume of the test body. Use spherical coordinates (r', θ', φ') with origin at C, we have

$$dV' = r'^2 \sin\theta'\,dr'\,d\theta'\,d\varphi'$$

and can write the above as

$$W = -\frac{GMm}{r_0} + \frac{GM}{2r_0^2}\int \sigma(r', \theta', \varphi')r'^3 \sin 2\theta'\,dr'\,d\theta'\,d\varphi' + O\left(\frac{1}{r_0^3}\right).$$

Hence

$$d = \frac{GMr_0}{2}\int \sigma(r', \theta', \varphi')r'^3 \sin 2\theta'\,dr'\,d\theta'\,d\varphi'.$$

1257

A beam of seasoned oak, 2 in × 4 in in cross section is built into a concrete wa⸱
so as to extend out 6 ft, as shown in Fig. 1.228. It is oriented so as to suppor⸱
the load L with the least amount of bending. The elastic hmit for oak is a stres⸱
of 7900 lb/in^2. The modulus of elasticity, I (dp/dl), is 1.62×10^6 lb/in^2. What i⸱
the largest load L that can be supported without permanently deforming the⸱
beam and what is the displacement of the point P under this load? In working
this problem make reasonable approximations, including that it is adequate
to equate the radius of the curvature of the beam to $-(d^2y/dx^2)^{-1}$ instead o⸱
the exact expression.

(*UC, Berkeley*⸱

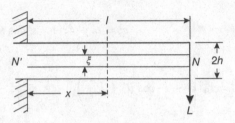

Fig. 1.228

Sol: Neglect shear stresses and assume pure bending. During bending, the upper
fibers will be extended while the lower fibers are compressed, and there is a neutral plane $N'N$ which remains unstrained. Consider fibers a distance ξ from $N'N$
as shown in Fig. 1.228. Let the radius of curvature of $N'N$ be r and that of the
fibers under consideration $r + \xi$. The latter thus suffer a longitudinal strain

$$\frac{(r + \xi) - r}{r} = \frac{\xi}{r}.$$

Consider a cross section A of the beam at x. The longitudinal stress at ξ from the
neutral axis in which the cross section intersects the neutral plane is

$$T(\xi) = \frac{E\xi}{r},$$

where E is the Young's modulus of the material. The total moment of the longitudinal stresses about the neutral axis is

$$M(x) = \int T\xi dA = \frac{E}{r}\int \xi^2 dA = \frac{EI}{r}. \tag{1}$$

I is the moment of inertia of the cross-sectional area about the neutral axis. The maximum bending moment occurs at the cross section $x = 0$ and the maximum stress occurs at the upper and lower boundaries. As

$$T = \frac{E\xi}{r} = \frac{M(x)\xi}{I},$$

$$T_{\max} = \frac{M(0)h}{I} = \frac{Llh}{I}.$$

For least bending the beam should be mounted so that its height is $2h = 4$ in and width $\omega = 2$ in. Thus

$$I = \int \xi^2 dA = \omega \int_{-h}^{h} \xi^2 d\xi = \frac{2}{3}\omega h^3 = \frac{32}{3} \text{ in}^4.$$

With $l = 72$ in, limiting stress $T_{\max} = 7900$ lb/in^2, this gives the maximum load as

$$L = \frac{7900 \times 32}{3 \times 72 \times 2} = 585 \text{ lb}.$$

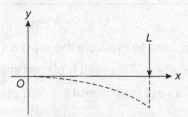

Fig. 1.229

Figure 1.229 shows the bending of the neutral plane $N'N$. Equation (1) gives

$$-\frac{d^2y}{dx^2} \approx \frac{1}{r} = \frac{M(x)}{EI} = \frac{L(l-x)}{EI}.$$

Integrating and noting that $dy/dx = 0$ at $x = 0$, we have

$$\frac{dy}{dx} = -\frac{L}{EI}\left(lx - \frac{x^2}{2}\right).$$

Further integration with $y = 0$ at $x = 0$ gives

$$y = \frac{L}{EI}\left(\frac{x^3}{6} - \frac{lx^2}{2}\right).$$

The displacement of the point P is therefore

$$\frac{L}{EI}\left(\frac{l^3}{6} - \frac{l^3}{2}\right) = -\frac{Ll^3}{3EI} = -4.21 \text{ in.}$$

1258

Many elementary textbooks quote Pascal's principle for hydrostatics as "any change in the pressure of a confined fluid is transmitted undiminished and instantaneously to all other parts of the fluid". Is this a violation of relativity? Explain clearly what "instantaneously" must mean here.

(Wisconsin)

Sol: Pascal's principle does not really violate relativity. It assumes the fluid to be incompressible, which is a simplified model and does not correspond to a real fluid.

A change in the pressure at a point of a fluid is transmitted throughout the fluid with the speed of sound. As the size of an ordinary container is very small compared with the distance traversed by sound in a short time, the change in pressure appears to be transmitted to all parts of the fluid instantaneously.

1259

A beam balance is used to measure the mass m_1 of a solid of volume V_1 which has a very low density ρ_1. This solid is placed in the left-hand balance pan and metal weights of a very high density ρ_2 are placed in the right-hand pan to achieve balance.

  a. If the balancing is first carried out in air and then the balance casing is evacuated, will the apparatus remain balanced? If not, which pan will go down?

  b. Determine the percentage error (if any) in the measured mass m_1 when the balancing is carried out in air (density of air air $= \rho_A$).

(Wisconsin)

Sol:

  a. The apparatus will not remain balanced after the balance casing is evacuated. The left-hand pan, which carries a lower density solid, and hence an object of a larger volume, will go down, as it had been supported more by air in the earlier balancing.

b. Let the true and apparent masses of the solid be m and m_1 respectively. Then

$$mg - \frac{m}{\rho_1} \rho_A g = m_1 g - \frac{m_1}{\rho_2} \rho_A g,$$

or

$$m - m_1 = \left(\frac{m}{\rho_1} - \frac{m_1}{\rho_2} \right) \rho_A \approx \frac{m_1}{\rho_1} \rho_A,$$

i.e.

$$\frac{\Delta m_1}{m_1} \approx \frac{\rho_A}{\rho_1}.$$

1260

A bucket of water is rotated at a constant angular velocity ω about its symmetry axis. Determine the shape of the surface of the water after everything has settled down.

(MIT)

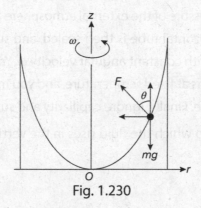

Fig. 1.230

Sol: Consider a particle of water of mass m at the surface. Two forces act on it: a force **F** normal to the surface due to neighboring water particles, and gravity mg, as shown in Fig. 1.230. As it moves in a circular orbit with constant angular velocity ω, in a cylindrical coordinate system with origin at the lowest point of the surface we have

$$F \cos \theta = mg,$$
$$F \sin \theta = m\omega^2 r,$$

where θ is the angle formed by the normal to the water surface and the z-axi
Hence

$$\tan \theta = \frac{\omega^2 r}{g}.$$

As $\tan \theta$ is the slope of the curve representing the shape of the surface,

$$\frac{dz}{dr} = \frac{\omega^2 r}{g},$$

giving

$$z = \frac{\omega^2 r^2}{2g},$$

as $z = 0$ for $r = 0$. Hence the surface is a paraboloid generated by rotating th
above parabola about the z-axis.

1261

A device consisting of a thin vertical tube and wide horizontal tube joined
together in the way shown in Fig. 1.231 is immersed in a fluid of density ρ_f
The density and pressure of the external atmosphere are ρ_a and p_a respectively
The end of the horizontal tube is then sealed, and subsequently the device is
rotated as shown with constant angular velocity ω. You may treat the air everywhere as an ideal gas at fixed temperature, and you may ignore the variation of
density with altitude. Finally, ignore capillarity and surface friction.

Find the height h to which the fluid rises in the vertical tube to second order
in ω.

(Princeton)

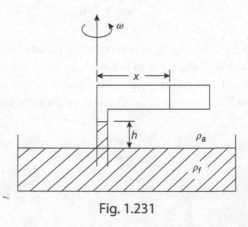

Fig. 1.231

Sol: The pressure p and density ρ of the air in the horizontal tube are not uniform. Consider a vertical layer of the air of thickness dx at distance x from the axis of rotation as shown in Fig. 1.231. As the tube is rotating with angular velocity ω, we have

$$[p(x + dx) - p(x)]A = \omega^2 x \rho A dx,$$

A being the cross-sectional area of the tube, or

$$\frac{dp}{dx} = \omega^2 x \rho.$$

Treating air as an ideal gas of molecular weight M, we have

$$pV = \frac{m}{M}RT,$$

or

$$\rho = \frac{pM}{RT},$$

where R is the gas constant. Hence

$$d\rho = \frac{M}{RT} dp$$

and

$$\frac{d\rho}{\rho} = \frac{M\omega^2}{RT} x dx.$$

Integration of the above gives

$$\ln\left(\frac{\rho}{\rho_0}\right) = \frac{M\omega^2}{2RT} x^2,$$

where ρ_0 is the density of the air $x = 0$. Thus

$$\rho = \rho_0 e^{\alpha x^2}$$

With $\alpha = M\omega^2/2RT$. ρ_0 can be determined by considering the total mass of the air in the tube:

$$\int_0^L \rho S dx = \rho_a SL,$$

i.e.

$$\rho_0 \int_0^L e^{\alpha x^2} dx = \rho_a L.$$

For moderate ω, α is a small number. As

$$e^{\alpha x^2} = 1 + \alpha x^2 + \frac{\alpha^2 x^4}{2!} + \ldots \approx 1 + \alpha x^2,$$

the above becomes approximately

$$\rho_0 L \left(1 + \frac{\alpha L^2}{3}\right) \approx \rho_a L,$$

or

$$\rho_0 \approx \left(1 - \frac{\alpha L^2}{3}\right)\rho_a.$$

As p is proportional ρ to since the temperature is assumed the same everywhere we have the pressure at $x = 0$ as

$$p_0 = \left(1 - \frac{\alpha L^2}{3}\right)p_a.$$

Consider now the liquid in the thin vertical tube. For equilibrium we have

$$p_a = p_0 + gh\rho_f,$$

or

$$\frac{\alpha L^2}{3}p_a = gh\rho_f,$$

giving

$$h = \frac{M\omega^2 L^2 p_a}{6RTg\rho_f} = \frac{\omega^2 L^2}{6g} \cdot \frac{\rho_a}{\rho_f}.$$

1262

A cylindrical container of circular cross section, radius R, is so supported that it can rotate about its vertical axis. It is first filled with liquid (assumed to be incompressible) of density ρ to a level h above its flat bottom. The cylinder is then set in rotation with angular velocity ω about its axis. The angular velocity is kept constant, and we wait for while until steady state is achieved. It is assumed that the liquid does not overflow, and it is also assumed that no portion of the bottom is "dry".

a. Find the equation for the upper surface of the liquid.

b. Find an expression for the pressure $p(z)$ on the cylindrical surface at height z above the bottom.

c. Find an expression for the pressure $p_0(z)$ along the axis at height above the bottom.

d. Is the fluid flow as viewed by stationary observer irrotational? The liquid is, of course, subject to the influence of gravity, and we assume that the normal atmospheric pressure p_a prevails in the environment.

(UC, Berkeley)

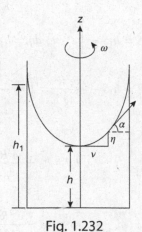

Fig. 1.232

Sol:

a. Consider vertical plane containing the axis of rotation. Let α be the angle made by the tangent to the upper surface of the liquid with the horizontal at point distance ξ from the rotational axis and height η above the lowest point of the upper surface, as shown in Fig. 1.232. Following Problem **1260** we have

$$\tan \alpha = \frac{d\eta}{d\xi} = \frac{\omega^2 \xi}{g}.$$

Its integration gives the parabola

$$\eta = \frac{\omega^2 \xi^2}{2g}.$$

The upper surface is obtained by rotating this parabola about the axis of rotation.

b. The upper surface of the liquid is an isobaric surface with pressure equal to the atmosphere pressure p_a. Note that each such revolving parabola in the liquid is a isobaric surface, the difference in pressure between it and the upper surface being determined by the distance between the two surfaces along the rotational axis. Let h be the height of the lowest point of the upper surfaces above the bottom of the container. The height of the highest point of the upper surface above the bottom is then

$$h_1 = h + \frac{\omega^2 R^2}{2g}.$$

If $S = \pi R^2$ and h_0 is the height of the liquid when it is not rotating, the total volume of the liquid is

$$h_0 S = h_1 S - \int_0^{\frac{\omega^2 R^2}{2g}} \pi \xi^2 d\eta = h_1 S - \int_0^R \frac{\pi \omega^2}{g} \xi^3 d\xi$$

$$= h_1 S - \frac{\pi \omega^2 R^4}{4g} = \left(h_1 - \frac{\omega^2 R^2}{4g} \right) \pi R^2,$$

giving

$$h_1 = h_0 + \frac{\omega^2 R^2}{4g},$$

and hence

$$h = h_0 - \frac{\omega^2 R^2}{4g}.$$

The pressure on the cylindrical surface at height z above the bottom is therefore

$$p(z) = p_a + (h_1 - z)\rho g = p_a + \left(h_0 + \frac{\omega^2 R^2}{4g} - z \right) \rho g.$$

c. The pressure along the axis at height z above the bottom is

$$p_0(z) = p_a + (h - z)\rho g = p_a + \left(h_0 - \frac{\omega^2 R^2}{4g} - z \right) \rho g.$$

d.

$$\nabla \times \mathbf{v} = \nabla \times (\boldsymbol{\omega} \times \mathbf{r}) \times \begin{vmatrix} \mathbf{i} & \mathbf{j} & \mathbf{k} \\ 0 & 0 & \omega \\ x & y & z \end{vmatrix}$$

$$= \nabla \times (-\omega y \mathbf{i} + \omega x \mathbf{j})$$

$$= \begin{vmatrix} \mathbf{i} & \mathbf{j} & \mathbf{k} \\ \dfrac{\partial}{\partial x} & \dfrac{\partial}{\partial y} & \dfrac{\partial}{\partial z} \\ -\omega y & \omega x & 0 \end{vmatrix} = 2\omega \mathbf{k}.$$

As $\nabla \times \mathbf{v} \neq 0$, the fluid flow is rotational.

1263

Given that the angular diameter of the moon and that of the sun are nearly equal and that the tides raised by the moon are about twice as high as those raised by the sun, what statement can you derive about the relative densities of the sun and moon?

(UC, Berkeley)

Sol: Let R_e, R_m, R_s be the radii, M_e, M_m, M_s the masses of the earth, moon and sun, and denote by h_m, h_s the heights of the tides raised by the moon and sun at a point on earth, and by D_m, D_s the distances of the moon and sun from the center of the earth, respectively. The disturbing effect of the moon at a point on the earth's surface may be represented by a potential which is approximately

$$\frac{3}{2} \frac{G M_m R_e^2}{D_m^3} \left(\frac{1}{3} - \cos\theta \right),$$

where θ is the moon's zenith distance at that point. This being equal to $g h_m$, where

$$g = \frac{G M_e}{R_e^2}$$

is the acceleration due to the earth's gravity, we have

$$h_m = \frac{3}{2} \frac{R_e^4 M_m}{D_m^3 M_e} \left(\frac{1}{3} - \cos^2\theta \right).$$

For the same zenith distance,

$$\frac{h_m}{h_s} = \left(\frac{D_s}{D_m} \right)^3 \frac{M_m}{M_s} = \left(\frac{D_s}{D_m} \right)^3 \left(\frac{R_m}{R_s} \right)^3 \frac{\rho_m}{\rho_s},$$

with ρ_m, ρ_s denoting the average densities of the moon and sun respectively. As the angular diameters of the moon and sun as seen from the earth are approximately equal, we have

$$\frac{R_m}{D_m} = \frac{R_s}{D_s},$$

and hence

$$\frac{\rho_m}{\rho_s} = \frac{h_m}{h_s} = 2,$$

which is the density of the moon relative to that of the sun.

1264

A hypothetical material out of which an astronomical object is formed has an equation of state

$$p = \frac{1}{2} K \rho^2,$$

where p is the pressure and ρ the mass density.

a. Show that for this material, under conditions of hydrostatic equilibrium, there is a linear relation between the density and the gravitational potential. The algebraic sign of the proportionality term is important.

b. Write a differential equation satisfied at hydrostatic equilibrium by the density. What boundary conditions or other physical constraints should be applied?

c. Assuming spherical symmetry, find the radius of the astronomical object at equilibrium.

(UC, Berkeley)

Sol:

a. Suppose the fluid is acted upon by an external force **F** per unit volume. Consider the surfaces normal to the x-axis of a volume element $d\tau = dxdydz$ of the fluid. At equilibrium **F** is balanced by the pressure in the fluid, thus

$$F_x d\tau = [p(x + dx) - p(x)] dydz = \frac{\partial p}{\partial x} d\tau,$$

i.e.

$$F_x = \frac{\partial p}{\partial x}$$

or

$$\mathbf{F} = \nabla p.$$

Then if **f** is the external force per unit mass of the fluid, we have

$$\mathbf{f} = \frac{1}{\rho} \nabla p.$$

As p is given by the equation of state, we have

$$\nabla p = K\rho\nabla\rho,$$

and

$$\mathbf{f} = K\nabla\rho.$$

If the external force is due to gravitational potential ϕ, then

$$\mathbf{f} = -\nabla\phi.$$

A comparison with the above gives

$$\nabla\phi + K\nabla\rho = 0,$$

or

$$\phi + K\rho = \text{constant}.$$

Hence ϕ and ρ are related linearly.

b. Poisson's equation

$$\nabla^2\phi = 4\pi G\rho$$

then gives

$$\nabla^2\rho + \frac{4\pi G\rho}{K} = 0.$$

This is the differential equation that has to be satisfied by the density at equilibrium. The boundary condition is that ρ is zero at the edge of the astronomical object.

c. For spherical symmetry use spherical coordinates with origin at the center of the object. The last equation then becomes

$$\frac{d^2\rho}{dr^2} + \frac{2}{r}\frac{d\rho}{dr} + \frac{4\pi G\rho}{K} = 0.$$

Let $u = \rho r, \omega^2 = 4\pi G/K$ and write the above as

$$\frac{d^2u}{dr^2} + \omega^2 u = 0,$$

which has solution

$$u = u_0 \sin(\omega r + \beta),$$

giving

$$\rho = \frac{r_0\rho_0}{r}\sin(\omega r + \beta),$$

where r_0, ρ_0 and β are constants. The boundary condition $\rho = 0$ at $r = R$, where R is the radius of the astronomical object, requires

$$\omega R + \beta = n\pi, \quad n = 1, 2, 3, \ldots.$$

However, the density ρ must be positive so that $\omega r + \beta \le \pi$. This means that $n = 1$ and $\omega R + \beta = \pi$. Consider

$$\mathbf{f} = K\nabla\rho$$

$$= K\left[-\frac{r_0\rho_0}{r^2}\sin(\omega r + \beta) + \frac{r_0\rho_0}{r}\omega\cos(\omega r + \beta)\right]\mathbf{e}_r$$

$$= -K\frac{\rho_0 r_0}{r^2}\cos(\omega r + \beta)[\tan(\omega r + \beta) - \omega r]\mathbf{e}_r.$$

Due to symmetry we require $\mathbf{f} = 0$ at $r = 0$. This means that as $r \to 0$

$$\tan(\omega r + \beta) - \omega r = \beta + \frac{1}{3}(\omega r + \beta)^3 + \dots \to 0.$$

Hence $\beta = 0$ and $\omega R = \pi$, giving the radius as

$$R = \sqrt{\frac{\pi K}{4G}}.$$

1265

Consider a self-gravitating slab of fluid matter in hydrostatic equilibrium of total thickness $2h$ and infinite lateral extent (in the x and y directions). The slab is uniform such that the density $\rho(z)$ is a function of z only, and the matter distribution is furthermore symmetric about the midplane $z = 0$. Derive an expression for the pressure p in this midplane in terms of the quantity

$$\sigma = \int_0^h \rho(z)\,dz$$

without making any assumption about the equation of state.

(UC, Berkeley)

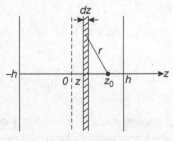

Fig. 1.233

Sol: In hydrostatic equilibrium the applied force on unit mass of the fluid is (Problem **1264**)

$$\mathbf{f} = \frac{1}{\rho}\nabla p.$$

As there is variation only in the z-direction,

$$f = \frac{1}{\rho}\frac{dp}{dz}. \tag{1}$$

Consider the gravitational force acting on unit mass at a point at z_0, as shown in Fig. 1.233, by a layer of the fluid of thickness dz at z:

$$-\int_0^\infty \frac{G\rho(z)dz \cdot 2\pi r dr}{r^2 + (z_0 - z)^2} \cdot \frac{z_0 - z}{\sqrt{r^2 + (z_0 - z)^2}}$$

$$= -2\pi G\rho(z)(z_0 - z)dz\int_0^\infty \frac{r dr}{[r^2 + (z_0 - z)^2]^{\frac{3}{2}}}$$

$$= -2\pi G\rho(z)(z_0 - z)\left[\frac{-1}{\sqrt{r^2 + (z_0 - z)^2}}\right]_0^\infty dz$$

$$= \frac{-2\pi G\rho(z)(z_0 - z)dz}{|z_0 - z|}.$$

The total gravitational force acting on the unit mass at $z = z_0$ is

$$f(z_0) = -2\pi G\left[\int_{-h}^{z_0}\rho(z)dz - \int_{z_0}^{h}\rho(z)dz\right]$$

$$= -2\pi G\int_{-z_0}^{z_0}\rho(z)dz$$

as $\rho(z)$ is symmetric with respect to the plane $z = 0$. Applying Eq. (1) to the point $z = z_0$ and integrating, we have

$$p(h) - p(0) = \int_0^h dp(z_0) = -2\pi G\int_0^h \rho(z_0)dz_0\int_{-z_0}^{z_0}\rho(z)dz.$$

This gives for symmetric $\rho(z)$

$$p(h) - p(0) = -4\pi G\int_0^h \rho(z_0)dz_0\int_0^{z_0}\rho(z)dz.$$

Setting $\varphi(z_0) = \int_0^{z_0} \rho(z)\,dz$, we have $d\varphi/dz_0 = \rho(z_0)$ and

$$\int_0^h \rho(z_0)\,dz_0 \int_0^{z_0} \rho(z)\,dz = \int_0^h \rho(z_0)\varphi(z_0)\,\frac{d\varphi}{\rho(z_0)} = \int_0^\sigma \varphi\,d\varphi = \frac{\sigma^2}{2},$$

where $\sigma = \int_0^h \rho(z)\,dz$. Using the boundary condition $p(h) = 0$ we finally obtain

$$p(0) = 2\pi G\sigma^2.$$

1266

a. A boat of mass M is floating in a (deep) tank of water with vertical side walls. A rock of mass m is dropped into the boat. How much does the water level in the tank rise? If the rock misses the boat and falls into the water, how much does the water level rise then?

[You may assume any reasonable shapes for the tank, boat and rock, i: you require.]

b. A U-tube with arms of different cross-sectional areas A_1, A_2 is filled with an incompressible liquid to a height d, as shown in Fig. 1.234. Air is blown impulsively into one end of the tube. Describe quantitatively the subsequent motion of the liquid. You may neglect surface tension effects and the viscosity of the fluid.

(UC, Berkeley)

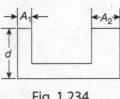

Fig. 1.234

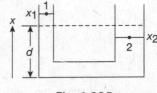

Fig. 1.235

Sol:

a. Let ρ_w and ρ_r be the densities of water and the rock, S_t and S_h the horizontal cross-sectional areas of the tank and boat, respectively. With the rock in the boat, the boat will sink a distance (from water surface) Δh such that an additional buoyancy is made available of magnitude

$$mg = \rho_\omega S_b \Delta h g,$$

giving

$$\Delta h = \frac{m}{\rho_\omega S_b}.$$

This will cause the water level in the tank to rise by ΔH given by

$$S_t \Delta H = S_b \Delta h,$$

or

$$\Delta H = \frac{m}{\rho_\omega S_t}.$$

If the rock misses the boat and falls into the water, it drops to the bottom of the tank. This increases the "water" volume by m/ρ_r, which then causes the water level in the tank to rise a height

$$\Delta H = \frac{m}{\rho_r S_t}.$$

b. The motion of the fluid is irrotational and non-steady, and is described by Bernoulli's equation of the form

$$\frac{1}{2}\rho v^2 + p + U - \rho\frac{\partial\phi}{\partial t} = \text{constant},$$

which holds for all points of the fluid at any given time t. Here U is the potential of the external force $\mathbf{F}$ defined by $\mathbf{F} = -\nabla U$, and ϕ is the velocity potential defined by $\mathbf{v} = -\nabla\phi$. Consider two surface points 1, 2, one on each arm of the vessel, at distances x_1, x_2 from the equilibrium level d, as shown in Fig. 1.235. Bernoulli's equation gives

$$\frac{1}{2}\rho v_1^2 + p_1 + U_1 - \rho\frac{\partial\phi_1}{\partial t} = \frac{1}{2}\rho v_2^2 + p_2 + U_2 - \rho\frac{\partial\phi_2}{\partial t}$$

with

$$p_1 = p_2 = \text{atmospheric pressure},$$

$$U_1 = (d + x_1)\rho g, \quad U_2 = (d - x_2)\rho g,$$

$$v_1 = \dot{x}_1, \quad v_2 = \dot{x}_2,$$

$$\frac{\partial\phi_1}{\partial t} = -\int_0^{d+x_1}\frac{\partial v_1}{\partial t}\,dx \approx -\ddot{x}_1 d,$$

$$\frac{\partial\phi_2}{\partial t} = -\int_0^{d-x_1} -\frac{\partial v_2}{\partial t}\,dx \approx \ddot{x}_2 d,$$

retaining only first order terms of the small quantities x_1, x_2 and their time derivations. In the same approximation, Bernoulli's equation becomes

$$(\dot{x}_1 + \dot{x}_2) + \frac{g}{d}(x_1 + x_2) = 0.$$

Making use of the continuity equation

$$A_1 x_1 = A_2 x_2,$$

we have

$$\dot{x}_1 + \frac{g x_1}{d} = 0,$$

$$\dot{x}_2 + \frac{g x_2}{d} = 0.$$

Hence the subsequent motion of the liquid is that of harmonic vibration with angular frequency $\omega = \sqrt{g/d}$.

1267

A space station is made from a large cylinder of radius R_0 filled with air. The cylinder spins about its symmetry axis at angular speed ω to provide acceleration at the rim equal to the gravitational acceleration g at the earth's surface

If the temperature T is constant inside the station, what is the ratio of air pressure at the center to the pressure at the rim?

(MIT)

Sol: Consider a cylindrical shell of air of radius r and thickness Δr. The pressure difference across its curved surfaces provides the centripetal force for the rotating air. Thus

$$[p(r + \Delta r) - p(\Delta r)]2\pi rl = \omega^2 r \cdot 2\pi rl\rho\Delta r,$$

where ρ is the density of the air and l is the length of the cylinder, giving

$$\frac{dp}{dr} = \rho\omega^2 r.$$

The air follows the equation of state of an ideal gas

$$pV = \frac{m}{M}RT,$$

or

$$\rho = \frac{M}{RT}p,$$

where T and M are the absolute temperature and molecular weight of air and R is the gas constant. Hence

$$\frac{dp}{dr} = \frac{M\omega^2}{RT} pr.$$

Integrating we have

$$\int_{p(0)}^{p(R_0)} \frac{dp}{p} = \frac{M\omega^2}{RT} \int_0^{R_0} r\, dr,$$

i.e.

$$\ln\left[\frac{p(R_0)}{p(0)}\right] = \frac{M\omega^2 R_0^2}{2RT} = \frac{MR_0 g}{2RT},$$

as the acceleration at the rim, $\omega^2 R_0$, is equal to g. Hence the ratio of the pressures is

$$\frac{p(0)}{p(R_0)} = \exp\left(-\frac{MR_0 g}{2RT}\right).$$

1268

Calculate the surface figure of revolution describing the equatorial bulge attained by a slowly rotating planet. Assume that the planet is composed entirely of an incompressible liquid of density ρ and total mass M that rotates with uniform angular velocity ω. When rotating, the equilibrium distance from the center of the planet to its poles is R_p.

a. Write down the equation of hydrostatic equilibrium for this problem.
b. Solve for the pressure near the surface of the planet using the crude approximation that the gravitational field near the surface can be written as $-GMr/r^2$.
c. Find an equation for the surface of the planet.
d. If the equatorial bulge ($R_e - R_p$) is a small fraction of the planetary radius, find an approximation to the expression obtained in (c) to describe the deviation of the surface from sphericity.
e. For the case of earth ($R_p = 6400$ km, $M = 6 \times 10^{24}$ kg) make a numerical estimate of the height of the equatorial bulge.

(MIT)

Sol:

a. Use coordinate as shown in Fig. 1.236 and consider a point P in the planet. I equilibrium the external forces are balanced by the pressure force per unit volume

$$\nabla p = \left(\frac{\partial p}{\partial r}, \frac{\partial p}{r \partial \theta}, 0 \right),$$

in spherical coordinates with the assumption that the planet is symmetric with respect to the axis of rotation.

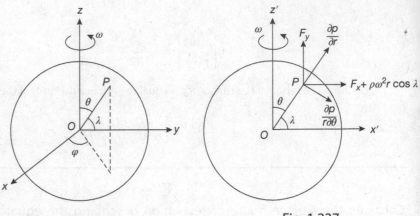

Fig. 1.236   Fig. 1.237

Now use a rotating coordinate frame attached to the planet such that the z'-axis coincides with the axis of rotation and the $x'z'$-plane contains OP. In this frame a fictitious centrifugal force per unit volume, $\rho \omega^2 r \cos \lambda$, where $\lambda = \frac{\pi}{2} - \theta$ is the latitude, has to the introduced. Let $\mathbf{F}$ be the gravitational force per unit volume. Then the forces involved are as shown in Fig. 1.237. As $d\theta = -d\lambda$, we can write $\partial p / \partial \theta = -\partial p / \partial \lambda$ and have, in the x' and z' directions,

$$\frac{\partial p}{\partial r} \cos \lambda - \frac{\partial p}{r \partial \lambda} \sin \lambda = F_{x'} + \rho \omega^2 r \cos \lambda, \tag{1}$$

$$\frac{\partial p}{\partial r} \sin \lambda + \frac{\partial p}{r \partial \lambda} \cos \lambda = F_{y'}. \tag{2}$$

b. The gravitational force per unit volume at P, as given, has components

$$F_{x'} = -\frac{\rho G M \cos \lambda}{r^2}, \quad F_{y'} = -\frac{\rho G M \sin \lambda}{r^2}.$$

Substitution in Eq. (2) gives

$$\frac{\partial p}{r \partial \lambda} = -\left(\frac{\rho G M}{r^2} + \frac{\partial p}{\partial r} \right) \tan \lambda,$$

which, with Eq. (1), then gives

$$\frac{\partial p}{\partial r} = \rho\omega^2 r \cos^2 \lambda - \frac{\rho GM}{r^2}.$$

As $p = 0$ at $r = R$, its integration gives

$$p = \frac{1}{2}(r^2 - R^2)\rho\omega^2 \cos^2 \lambda + \left(\frac{1}{r} - \frac{1}{R}\right)\rho GM$$

For a point a depth h under the surface at latitude λ, we have, as $r = R - h$ with $h \ll R$,

$$r^2 - R^2 \approx -2Rh, \quad \frac{1}{r} - \frac{1}{R} \approx \frac{h}{R^2}$$

and

$$p \approx \left(-R\omega^2 \cos^2 \lambda + \frac{GM}{R^2}\right)\rho h.$$

c. The surface of the planet is an equipotential surface. The potential (potential energy per unit mass) at the surface due to gravitational force is

$$U = -\frac{GM}{R} + \text{constant}.$$

The potential ϕ due to the centrifugal force is given by

$$-\nabla\phi = \left(-\frac{\partial\phi}{\partial r}, \frac{\partial\phi}{r\partial\lambda}\right) = (\omega^2 r \cos^2 \lambda, \omega^2 r \cos \lambda \sin \lambda).$$

Thus

$$\frac{\partial\phi}{\partial r} = -\omega^2 r \cos^2 \lambda,$$

or

$$\phi = -\frac{1}{2}\omega^2 r^2 \cos^2 \lambda + f(\lambda).$$

As

$$\frac{\partial\phi}{r\partial\lambda} = \omega^2 r \cos \lambda \sin \lambda + \frac{1}{r}f'(\lambda) = \omega^2 r \cos \lambda \sin \lambda,$$

$$f'(x) = 0 \quad \text{or} \quad f(x) = \text{constant}.$$

Hence for the surface we have

$$-\frac{1}{2}\omega^2 R^2 \cos^2 \lambda - \frac{Gm}{R} = \text{constant}.$$

At the poles, $\lambda = \pm\frac{\pi}{2}, R = R_p$. We thus have

$$\frac{1}{2}\omega^2\cos^2\lambda R^3 - \frac{GM}{R_p}R + GM = 0.$$

d. At the equator, $\lambda = 0, R = R_e$, and the above equation becomes

$$\omega^2 R_e^3 = 2GM\left(\frac{R_e - R_p}{R_p}\right).$$

The deviation of the surface from sphericity is therefore

$$\frac{R_e - R_p}{R_p} = \frac{\omega^2 R_e^3}{2GM}.$$

e. For the earth,

$$R = 6400 \text{ km}, \quad R_e \approx R_p = 6400 \text{ km}, \quad M = 6 \times 10^{24} \text{ kg},$$

$$\omega = \frac{2\pi}{24 \times 3600}\text{ s}^{-1}, \quad G = 6.67 \times 10^{-11} \text{ Nm}^2/\text{kg}^2,$$

we have

$$R_e - R_x = 11 \text{ km}.$$

1269

The compressibility K of a gas or liquid is defined as $K = -(dV/V)/dp$, where $-dV$ is the volume decrease due to a pressure increase dp. Air (at STP) has about 15,000 times greater compressibility than water.

a. Derive the formula for the velocity v of sound waves, $1/v^2 = K\rho$, where ρ is the mass density. Use any method you wish. (A simple model will suffice.)

b. The velocity of sound in air (at STP) is about 330 m/s. Sound velocity in water is about 1470 m/s. Suppose you have water filled with a homogeneous mixture of tiny air bubbles (very small compared with sound wavelengths in air) that occupy only 1% of the volume. Neglect the effect the bubbles have on the mass density of the mixture (compared with pure water). Find the compressibility K of the mixture, and thus find v for the mixture. Compare the numerical value of v for the given 1% volume fraction with v for pure water or air.

(UC, Berkeley)

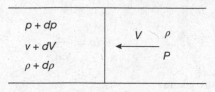

Fig. 1.238

Sol:

a. Without loss of generality, we can consider the problem in one dimension and suppose the front of the compressed region, i.e. the wavefront, propagates from left to right at speed v. For convenience we use coordinates such that the compressed region is at rest, then the gas particles in the region not reached by the wave will move from left to right at speed v in this frame. Let the pressure and density in the latter region be p and ρ respectively. When the particles enter the compressed area, their velocity changes to $v + dv$, pressure changes to $p + dp$, and density changes to $\rho + d\rho$, as shown in Fig. 1.238. The mass of gas passing through a unit area of the wavefront is

$$pv = (\rho + d\rho)(v + dv),$$

yielding, to first-order quantities,

$$v d\rho = -\rho dv.$$

The change of momentum per unit time crossing the unit area is

$$(\rho + d\rho)(v + dv) \cdot (v + dv) - \rho v \cdot v = v^2 d\rho + 2\rho v dv.$$

By Newton's second law this corresponds to the excess pressure of the right-hand side over the left-hand side. Thus

$$v^2 d\rho + 2\rho v dv = p - (p + dp) = -dp.$$

The above two equations give

$$v^2 d\rho = dp.$$

For a given mass m of the gas,

$$m = \rho V$$

or

$$dp = -\rho \frac{dV}{V}.$$

Hence

$$v^2 d\rho = -v^2 \rho \frac{dV}{V} = dp,$$

or

$$v^2 = -\left(\frac{dV}{V}\right)^{-1}\frac{dp}{\rho} = \frac{1}{K\rho},$$

i.e.

$$v = \frac{1}{\sqrt{K\rho}}.$$

b. For the mixture given,

$$K = -\frac{dV}{Vdp} = -\frac{dV_1 + dV_2}{Vdp} = \frac{K_1 V_1 + K_2 V_2}{V}.$$

For water and air we have respectively

$$v_1^2 = \frac{1}{K_1\rho_1}, \quad v_2^2 = \frac{1}{K_2\rho_2},$$

and so

$$\frac{K_2}{K_1} = \left(\frac{v_1}{v_2}\right)^2 \frac{\rho_1}{\rho_2}$$

$$= \left(\frac{1470}{330}\right)^2 \frac{1}{1.293 \times 10^{-3}}$$

$$= 1.53 \times 10^4.$$

Hence, for the mixture,

$$K = \left(\frac{V_1}{V} + \frac{K_2 V_2}{K_1 V}\right) K_1 = (0.99 + 153)K_1 \approx 154K_1,$$

$$v = \frac{1}{\sqrt{K\rho}} \approx \frac{1}{\sqrt{154 K_1\rho_1}} = \frac{1470}{\sqrt{154}} = 118 \text{ m/s},$$

which is much less than the velocity of sound in pure water or air.

1270

Consider the spherically symmetric expansion of a homogeneous, self-gravitating gas with negligible pressure. The initial conditions of expansion are unspecified; instead, you are given that when the density is ρ_0, a fluid element at a radius R_0 from the origin has a velocity v_0.

a. Find $v(R)$.

b. Describe the ultimate fate of the gas in terms of v_0, R_0 and ρ_0.

(UC, Berkeley)

Sol:

a. Consider the motion of a unit mass at the surface of the gas, conservation of mechanical energy gives

$$\frac{1}{2}v_0^2 - \frac{GM}{R_0} = \frac{1}{2}v^2 - \frac{GM}{R},$$

where $M = 4\pi\rho_0 R_0^3/3$ is the total mass of the gas. Hence the speed of the unit mass when the radius of the volume of gas is R is

$$v = \sqrt{v_0^2 + \frac{8}{3}\pi G\rho_0 R_0^3\left(\frac{1}{R} - \frac{1}{R_0}\right)}.$$

b. As R increases, v decreases, and finally $v = 0$ and expansion stops when the radius becomes

$$R = \left[\frac{1}{R_0} - \frac{3v_0^2}{8\pi G\rho_0 R_0^3}\right]^{-1}.$$

1271

An incompressible fluid of mass density ρ, viscosity η is pumped in steady-state laminar flow through a circular pipe of internal radius R and length L. The pressure at the inlet end is p_1, the pressure at the exit is p_2, $p_1 > p_2$.

Let Q be the mass of fluid that flows through the pipe per unit time. Compute Q.

(CUSPEA)

Sol: Use cylindrical coordinates (r, φ, z) with the z-axis along the axis of the pipe. For laminar flow the velocity $\mathbf{v}$ of the fluid has components

$$v_r = v_\phi = 0, \quad v_z = v.$$

Furthermore, because of symmetry, $v = v(r)$. Then in the Navier-Stokes equation

$$\rho\frac{\partial\mathbf{v}}{\partial t} + \rho(\mathbf{v}\cdot\nabla)\mathbf{v} - \eta\nabla^2\mathbf{v} + \nabla p = \mathbf{F},$$

$\partial\mathbf{v}/\partial t = 0$ for steady-state motion,

$$(\mathbf{v}\cdot\nabla)\mathbf{v} = v_z\frac{\partial v_z}{\partial z}\mathbf{e}_z = 0,$$

as $v = v_z(r)$, and the external force per unit volume $\mathbf{F}$ is zero provided gravity can be neglected, we have

$$\nabla p = \eta\nabla^2\mathbf{v}.$$

This becomes

$$\frac{\partial p}{\partial z} = \frac{\eta}{r}\frac{\partial}{\partial r}\left(r\frac{\partial v}{\partial r}\right), \quad \frac{\partial p}{\partial r} = \frac{\partial p}{\partial \varphi} = 0$$

for $\mathbf{v} = v(r)\mathbf{e}_z$. As the right-hand side of the first equation depends on r while p is a function of z, either side must be a constant, which is

$$\frac{\partial p}{\partial z} = \frac{p_2 - p_1}{L} = -\frac{\Delta p}{L},$$

where $\Delta p = p_1 - p_2$. Hence

$$\frac{d}{dr}\left(r\frac{dv}{dr}\right) = -\left(\frac{\Delta p}{\eta L}\right)r.$$

Integration gives

$$v = -\frac{1}{4}\left(\frac{\Delta p}{\eta L}\right)r^2 + C_1 \ln r + C_2,$$

C_1, C_2 being constants. As

$$v(r) = \text{finite}, \quad v(R) = 0,$$

we require

$$C_1 = 0, \quad C_2 = \frac{1}{4}\left(\frac{\Delta p}{\eta L}\right)R^2.$$

Hence

$$v = \frac{\Delta p}{4\eta L}(R^2 - r^2).$$

The mass of the fluid flowing through the pipe per unit time is then

$$Q = \rho\int_0^R v \cdot 2\pi r dr = \frac{\pi\rho R^4 \Delta p}{8\eta L}.$$

1272

A sphere of radius R moves with uniform velocity $\mathbf{u}$ in an incompressible ($\nabla \cdot \mathbf{v}(x) = 0$, $\mathbf{v}(x)$ being the velocity of the fluid), non-viscous, ideal fluid.

a. Determine the velocity $\mathbf{v}$ of the fluid passing any point on the surface of the sphere.

b. Calculate the pressure distribution over the surface of the sphere.

c. What is the force necessary to keep the sphere in uniform motion?

(Columbia)

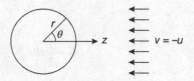

Fig. 1.239

Sol: We can consider the sphere as being at rest while the fluid flows past it with velocity $v = -u$ as shown in Fig. 1.239. Use spherical coordinates (r, θ, φ) with origin at the center of the sphere such that the velocity of the fluid is in the direction $\theta = \pi$. Define the velocity potential ϕ by

$$\mathbf{v} = -\nabla\phi.$$

The incompressibility of the fluid means that

$$\nabla \cdot \mathbf{v} = -\nabla^2\phi = 0.$$

Thus ϕ satisfies Laplace's equation. The boundary conditions are

$$-\left(\frac{\partial \phi}{\partial r}\right)_R = -u \cos\theta$$

as the surface of the sphere is impenetrable, and

$$\phi = 0 \quad \text{for} \quad r \to \infty$$

as $v = -u = $ constant at large distances from the sphere.

The general solution of Laplace's equation is

$$\phi = \sum_{n=0}^{\infty} \sum_{m=0}^{n} (a_{nm}r^n + b_{nm}r^{-n-1})P_n^m(\cos\theta)e^{\pm im\varphi}.$$

As the geometry is cylindrically symmetric, ϕ is independent of φ and we have to take $m = 0$. Thus we have

$$\phi = \sum_{n=0}^{\infty} (a_n r^n + b_n r^{-n-1})P_n(\cos\theta),$$

where $P_n(\cos\theta)$ are the Legendre polynomials, and a_n, b_n are arbitrary constants. As $\phi = 0$ for $r \to \infty$, we require $a_n = 0$. As

$$\left(\frac{\partial \phi}{\partial r}\right)_R = \sum_{n=0}^{\infty} (-n-1)b_n R^{-n-2} P_n(\cos \theta) = u P_1(\cos \theta)$$

we require

$$b_n = 0 \quad \text{for all} \quad n \neq 1$$

and

$$b_1 = -\frac{1}{2} u R^3.$$

Hence

$$\phi = -\frac{u R^3}{2r^2} \cos \theta.$$

a. At a point (R, θ) on the sphere the velocity of the fluid is

$$\mathbf{v} = -\nabla \phi = \left[-\frac{\partial \phi}{\partial r} \mathbf{e}_r - \frac{\partial \phi}{r \partial \theta} \mathbf{e}_\theta \right]_{r=R}$$

$$= -\frac{1}{2} u R^3 \left[\frac{2}{r^3} \cos \theta \mathbf{e}_r + \frac{\sin \theta}{r^3} \mathbf{e}_\theta \right]_{r=R}$$

$$= -u \cos \theta \mathbf{e}_r - \frac{1}{2} u \sin \theta \mathbf{e}_\theta.$$

b. Bernoulli's equation for the irrotational steady flow of a nonviscous, incompressible fluid is

$$\frac{1}{2} \rho v^2 + p + U = \text{constant},$$

where $U = $ constant if there is no external force. Consider a point (R, θ) on the surface of the sphere and a point at infinity, where the pressure is p_0. Then

$$\frac{\rho}{2} u^2 \left(\cos^2 \theta + \frac{1}{4} \sin^2 \theta \right) + p = \frac{1}{2} \rho u^2 + p_0,$$

or

$$p(R, \theta) = \frac{3}{8} \rho u^2 \sin^2 \theta + p_0.$$

This gives the distribution of the pressure over the surface of the sphere.

c. The net total force exerted by the pressure on the sphere is in the direction of u and has magnitude

$$F_z = \int p \cos \theta \, dS$$

$$= \int_0^\pi \left(\frac{3}{8} \rho u^2 \sin^2 \theta + p_0 \right) \cdot 2\pi R \sin \theta \cdot \cos \theta \cdot R d\theta$$

$$= 0.$$

Hence no force is required to keep the sphere in uniform motion. This can be anticipated as the sphere moves uniformly without friction.

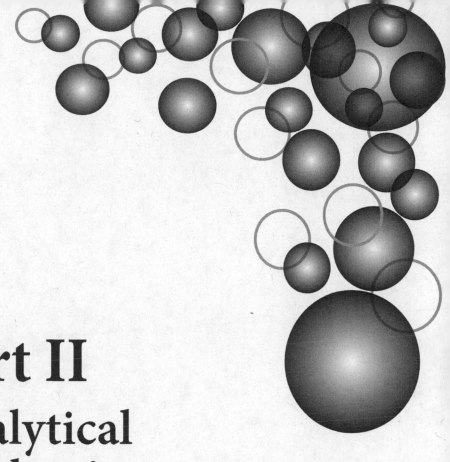

Part II
Analytical Mechanics

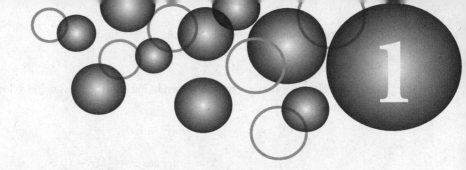

LAGRANGE'S EQUATIONS (2001–2027)

2001

A massless spring of rest length l_0 (with no tension) has a point mass m connected to one end and the other end fixed so the spring hangs in the gravity field as shown in Fig. 2.1. The motion of the system is only in one vertical plane.

  a. Write down the Lagrangian.

  b. Find Lagrange's equations using variables $\theta, \lambda = (r - r_0)/r_0$, where r_0 is the rest length (hanging with mass m). Use $\omega_s^2 = k/m$, $\omega_p^2 = g/r_0$.

  c. Discuss the lowest order approximation to the motion when λ and θ are small with the initial conditions $\theta = 0$, $\dot\lambda = 0$, $\lambda = A$, $\dot\theta = \omega_p B$ at $t = 0$. A and B are constants.

  d. Discuss the next order approximation to the motion. Under what conditions will the λ motion resonate? Can this be realized physically?

(Wisconsin)

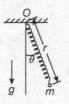

Fig. 2.1

429

Sol:

a. In polar coordinates (r, θ) as shown in Fig. 2.1, the mass m has velocity $\mathbf{v} = (\dot{r}, r\dot{\theta})$. Thus

$$T = \frac{1}{2}m(\dot{r}^2 + r^2\dot{\theta}^2),$$

$$V = -mgr\cos\theta + \frac{1}{2}k(r - l_0)^2,$$

k being the spring constant. The Lagrangian of m is therefore

$$L = T - V = \frac{1}{2}m(\dot{r}^2 + r^2\dot{\theta}^2) + mgr\cos\theta - \frac{1}{2}k(r - l_0)^2.$$

b.

$$\frac{d}{dt}\left(\frac{\partial L}{\partial \dot{r}}\right) - \frac{\partial L}{\partial r} = 0$$

gives

$$m\ddot{r} - mr\dot{\theta}^2 - mg\cos\theta + k(r - l_0) = 0.$$

$$\frac{d}{dt}\left(\frac{\partial L}{\partial \dot{\theta}}\right) - \frac{\partial L}{\partial \theta} = 0$$

gives

$$mr^2\ddot{\theta} + 2mr\dot{r}\dot{\theta} + mgr\sin\theta = 0.$$

The rest length of the spring with mass m hanging, r_0, is given by Hooke's law

$$k(r_0 - l_0) = mg.$$

Thus with $\lambda = (r - r_0)/r_0$ we have

$$r - l_0 = \lambda r_0 + \frac{mg}{k},$$

$$r = r_0(1 + \lambda), \quad \dot{r} = r_0\dot{\lambda}, \quad \ddot{r} = r_0\ddot{\lambda},$$

and the equations of motion become

$$\ddot{\lambda} + \frac{k\lambda}{m} - (1 + \lambda)\dot{\theta}^2 + \frac{g}{r_0}(1 - \cos\theta) = 0,$$

$$(1 + \lambda)\ddot{\theta} + 2\dot{\lambda}\dot{\theta} + \frac{g}{r_0}\sin\theta = 0;$$

or, with $\omega_s^2 = \frac{k}{m}$, $\omega_p^2 = \frac{g}{r_0}$,

$$\ddot{\lambda} + (\omega_s^2 - \dot{\theta}^2)\lambda - \dot{\theta}^2 + \omega_p^2(1 - \cos\theta) = 0,$$

$$(1 + \lambda)\ddot{\theta} + 2\dot{\lambda}\dot{\theta} + \omega_p^2\sin\theta = 0.$$

c. When λ and θ are small, we can neglect second order quantities in $\theta, \lambda, \dot\theta, \dot\lambda$, and the equations of motion reduce to

$$\ddot\lambda + \omega_s^2\lambda = 0,$$
$$\ddot\theta + \omega_p^2\theta = 0$$

in the lowest order approximation. For the given initial conditions, we find

$$\lambda = A\cos(\omega_s t),$$
$$\theta = B\sin(\omega_p t).$$

Thus λ and θ each oscillates sinusoidally with angular frequencies ω_s and ω_p respectively, the two oscillations differing in phase by $\pi/2$.

d. If we retain also terms of the second order, the equations become

$$\ddot\lambda + \omega_s^2\lambda = \dot\theta^2 - \frac{1}{2}\omega_p^2\theta^2,$$

$$(1 + \lambda)\ddot\theta + 2\dot\lambda\dot\theta + \omega_p^2\theta = 0.$$

Using the results of the lowest order approximation, the first equation above can be approximated as

$$\ddot\lambda + \omega_s^2\lambda \approx \frac{1}{2}B^2\omega_p^2[2\cos^2(\omega_p t) - \sin^2(\omega_p t)]$$

$$= \frac{1}{4}B^2\omega_p^2[3\cos(2\omega_p t) + 1].$$

Thus λ may resonate if $\omega_s = 2\omega_p$. However this is unlikely to realize physically since as the amplitude of λ increases toward a resonance the lowest order approximation no longer holds and higher order effects will take place. Furthermore the nonlinear properties of the spring will also come into play, invalidating the original simplified model.

2002

A disk of mass *M* and radius *R* slides without friction on a horizontal surface. Another disk of mass *m* and radius *r* is pinned through its center to a point off the center of the first disk by a distance *b*, so that it can rotate without friction on the first disk as shown in Fig. 2.2. Describe the motion and identify its constants.

(Wisconsin)

Sol: Take generalized coordinates as follows: *x, y*, the coordinates of the center of mass of the larger disk, θ, the angle of rotation of the larger disk and ϕ the angle

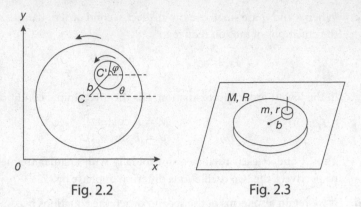

Fig. 2.2   Fig. 2.3

of rotation of the smaller disk as shown in Fig. 2.3. The center of mass of the smaller disk has coordinates

$$x + b\cos\theta, \quad y + b\sin\theta$$

and velocity components

$$\dot{x} - b\dot{\theta}\sin\theta, \quad \dot{y} + b\dot{\theta}\cos\theta.$$

Hence the total kinetic energy of the system of the two disks is

$$T = \frac{1}{2}M(\dot{x}^2 + \dot{y}^2) + \frac{1}{4}MR^2\dot{\theta}^2$$
$$+ \frac{1}{2}m[(\dot{x} - b\dot{\theta}\sin\theta)^2 + (\dot{y} + b\dot{\theta}\cos\theta)^2] + \frac{1}{4}mr^2\dot{\varphi}^2$$

and the Lagrangian is

$$L = T - V = T$$
$$= \frac{1}{2}M(\dot{x}^2 + \dot{y}^2) + \frac{1}{4}MR^2\dot{\theta}^2$$
$$+ \frac{1}{2}m[\dot{x}^2 + \dot{y}^2 + b^2\dot{\theta}^2 - 2b\dot{x}\dot{\theta}\sin\theta + 2b\dot{y}\dot{\theta}\cos\theta] + \frac{1}{4}mr^2\dot{\varphi}^2.$$

Consider Lagrange's equations

$$\frac{d}{dt}\left(\frac{\partial L}{\partial \dot{q}_i}\right) - \frac{\partial L}{\partial q_i} = 0.$$

As

$$\frac{\partial L}{\partial x} = 0, \quad \frac{d}{dt}\left(\frac{\partial L}{\partial \dot{x}}\right) - \frac{\partial L}{\partial x} = \frac{d}{dt}\left(\frac{\partial L}{\partial \dot{x}}\right) = 0,$$

we have

$$\frac{\partial L}{\partial \dot{x}} = \text{constant},$$

or

$$(M + m)\dot{x} - mb\dot{\theta}\sin\theta = \text{constant.} \tag{1}$$

As $\partial L/\partial y = 0$, we have $\partial L/\partial\dot{y} = \text{constant}$, or

$$(M + m)\dot{y} + mb\dot{\theta}\cos\theta = \text{constant.} \tag{2}$$

As $\partial L/\partial\varphi = 0$, we have $\partial L/\partial\dot{\varphi} = \text{constant}$, or

$$\dot{\varphi} = \text{constant.} \tag{3}$$

As

$$\frac{\partial L}{\partial\theta} = -mb\dot{x}\dot{\theta}\cos\theta - mb\dot{y}\dot{\theta}\sin\theta,$$

$$\frac{\partial L}{\partial\dot{\theta}} = \frac{1}{2}MR^2\dot{\theta} + mb^2\dot{\theta} - mb\dot{x}\sin\theta + mb\dot{y}\cos\theta,$$

we have the equation of motion

$$\frac{1}{2}MR^2\ddot{\theta} + mb^2\ddot{\theta} = mb\ddot{x}\sin\theta + mb\ddot{y}\cos\theta = 0. \tag{4}$$

Equations (1)–(4) describe the motion of the system. Since $V = 0$ and $T + V = \text{constant}$ as there is no external force, the total kinetic energy of the system, T, is a conserved quantity. Conservation of the angular momentum about the center of mass of the system requires that, as $\dot{\varphi} = \text{constant}$, $\dot{\theta} = \text{constant}$ too.

2003

A uniform solid cylinder of radius R and mass M rests on a horizontal plane and an identical cylinder rests on it, touching it along the highest generator as shown in Fig 2.4. The upper cylinder is given an infinitesimal displacement so that both cylinders roll without slipping.

  a. What is the Lagrangian of the system?
  b. What are the constants of the motion?
  c. Show that as long as the cylinders remain in contact

$$\dot{\theta}^2 = \frac{12g(1 - \cos\theta)}{R(17 + 4\cos\theta - 4\cos^2\theta)},$$

where θ is the angle which the plane containing the axes makes with the vertical.

(Wisconsin)

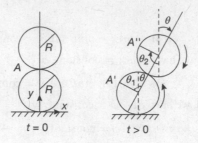

Fig. 2.4

Sol:

a. The system possesses two degrees of freedom so that two generalized coordinates are required. For these we use θ_1, the angle of rotation of the lower cylinder, and θ, the angle made by the plane containing the two axes of the cylinders and the vertical.

Initially the plane containing the two axes of the cylinders is vertical. At a later time, this plane makes an angle θ with the vertical. The original point of contact, A now moves to A' on the lower cylinder and to A'' on the upper cylinder. With the angles so defined we have from Fig. 2.4

$$\theta_1 + \theta = \theta_2 - \theta,$$

or

$$\theta_2 = \theta_1 + 2\theta.$$

Taking Cartesian coordinates (x, y) in the vertical plane normal to the axes of the cylinders and through their centers of mass, as shown in Fig. 2.4, we have, at $t > 0$, for the lower cylinder

$$x_1 = -R\theta_1, \qquad y_1 = R,$$

and for the upper cylinder

$$x_2 = x_1 + 2R \sin \theta,$$
$$y_2 = 3R - 2(R - \cos \theta) = R + 2R \cos \theta.$$

The corresponding velocity components are

$$\dot{x}_1 = -R\dot{\theta}_1, \qquad \dot{y}_1 = 0,$$
$$\dot{x}_2 = -R\dot{\theta}_1 + 2R\dot{\theta} \cos \theta, \qquad \dot{y}_2 = -2R\dot{\theta} \sin \theta.$$

The kinetic energy of the lower cylinder is thus

$$T_1 = \frac{1}{2} M \dot{x}_1^2 + \frac{1}{4} MR^2 \dot{\theta}_1^2 = \frac{3}{4} MR^2 \dot{\theta}_1^2,$$

and that of the upper cylinder is

$$T_2 = \frac{1}{2} M(\dot{x}_2^2 + \dot{y}_2^2) + \frac{1}{4} MR^2 \dot{\theta}_2^2$$

$$= \frac{1}{2} MR^2 (\dot{\theta}_1^2 - 4\dot{\theta}_1 \dot{\theta} \cos\theta + 4\dot{\theta}^2) + \frac{1}{4} MR^2 (\dot{\theta}_1^2 + 4\dot{\theta}_1 \dot{\theta} + 4\dot{\theta}^2)$$

$$= \frac{1}{4} MR^2 [3\dot{\theta}_1^2 + 4\dot{\theta}_1 \dot{\theta}(1 - 2\cos\theta) + 12\dot{\theta}^2].$$

The potential energy of the system, taking the horizontal plane as level of reference, is

$$V = Mg(y_l + y_2) = 2MR(1 + \cos\theta)g.$$

Hence the Lagrangian of the system is

$$L = T - V$$

$$= \frac{1}{2} MR^2 [3\dot{\theta}_1^2 + 2\dot{\theta}_1 \dot{\theta}(1 - 2\cos\theta) + 6\dot{\theta}^2] - 2MR(1 + \cos\theta)g.$$

b. As only gravity is involved, the total mechanical energy of the system is a constant of the motion:

$$E = T + V$$

$$= \frac{1}{2} MR^2 [3\dot{\theta}_1^2 + 2\dot{\theta}_1 \dot{\theta}(1 - 2\cos\theta) + 6\dot{\theta}^2] + 2MR(1 + \cos\theta)g$$

$$= \text{constant}.$$

Furthermore, if $\partial L / \partial q_i = 0$, Lagrange's equation

$$\frac{d}{dt}\left(\frac{\partial L}{\partial \dot{q}_i}\right) - \frac{\partial L}{\partial q_i} = 0$$

requires that $\partial L / \partial \dot{q}_i$ is conserved. For the system under consideration, $\partial L / \partial \theta_1 = 0$ so that

$$\frac{\partial L}{\partial \dot{\theta}_1} = MR^2 [3\dot{\theta}_1 + \dot{\theta}(1 - 2\cos\theta)] = \text{constant}.$$

c. As long as the cylinders remain in contact the results of (b) hold. Initially, $\theta = 0$, $\dot{\theta}_1 = \dot{\theta} = 0$, so that

$$\frac{1}{2} MR^2 [3\dot{\theta}_1^2 + 2\dot{\theta}_1 \dot{\theta}(1 - 2\cos\theta) + 6\dot{\theta}^2] + 2MR(1 + \cos\theta)g = 4MRg,$$

$$MR^2 [3\dot{\theta}_1 + \dot{\theta}(1 - 2\cos\theta)] = 0.$$

These combine to give

$$\dot{\theta}^2[18 - (1 - 2\cos\theta)^2] = \frac{12}{R}(1 - \cos\theta)g,$$

i.e.

$$\dot{\theta}^2 = \frac{12(1 - \cos\theta)g}{R(17 + 4\cos\theta - 4\cos^2\theta)}.$$

2004

Two particles of the same mass m are constrained to slide along a thin rod o
mass M and length L, which is itself free to move in any manner. Two identi
cal springs link the particles with the central point of the rod. Consider only
motions of this system in which the lengths of the springs (i.e. the distance:
of the two particles from the center of the rod) are equal. Taking this to be ar
isolated system in space, find equations of motion for it and solve them (up tc
the point of doing integrations). Describe qualitatively the motion.

(Wisconsin

Sol: Use a fixed Cartesian coordinate frame, and a moving frame with origin at the
midpoint O of the rod and its Cartesian axes parallel to those of the former
respectively. Let (r, θ, φ) be the spherical coordinates of a point referring to the
moving frame, as shown in Fig. 2.5. Then the point O has coordinates (x, y, z)
in the fixed frame and the two masses have spherical coordinates (r, θ, φ) and
$(-r, \theta, \varphi)$ in the moving frame.

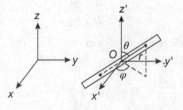

Fig. 2.5

The kinetic energy of a system is equal to the kinetic energy it would have if all its
mass were concentrated at the center of mass plus the kinetic energy of motion
about the center of mass. As O is the center of mass of the system, we have

$$T = \frac{1}{2}(M + 2m)(\dot{x}^2 + \dot{y}^2 + \dot{z}^2) + m(\dot{r}^2 + r^2\dot{\theta}^2 + r^2\dot{\varphi}^2\sin^2\theta) + T_{\text{rot}},$$

where T_{rot} is the rotational kinetic energy of the rod. The angular velocity of the rod about O is

$$\omega = \dot{\varphi} \cos \theta \mathbf{e}_r - \dot{\varphi} \sin \theta \mathbf{e}_\theta - \dot{\theta} \mathbf{e}_\varphi,$$

resolved along its principal axes, the corresponding moments of inertia being

$$I_r = 0, \qquad I_\theta = \frac{1}{12} ML^2, \qquad I_\varphi = \frac{1}{12} ML^2.$$

Hence

$$T_{\text{rot}} = \frac{1}{2}(I_r \omega_r^2 + I_\theta \omega_\theta^2 + I_\varphi \omega_\varphi^2)$$

$$= \frac{1}{24} ML^2 (\dot{\theta}^2 + \dot{\varphi}^2 \sin^2 \theta).$$

The system is in free space so the only potential energy is that due to the action of the springs,

$$V = 2 \cdot \frac{1}{2} K(r - r_0)^2 = K(r - r_0)^2,$$

where K and r_0 are the spring constant and the natural length of each spring respectively. Hence

$$L = T - V$$

$$= \frac{1}{2}(M + 2m)(\dot{x}^2 + \dot{y}^2 + \dot{z}^2) + m(\dot{r}^2 + r^2\dot{\theta}^2 + r^2\dot{\varphi}^2 \sin^2 \theta$$

$$+ \frac{1}{24} ML^2 (\dot{\theta}^2 + \dot{\varphi}^2 \sin^2 \theta) - K(r - r_0)^2.$$

Lagrange's equations

$$\frac{d}{dt}\left(\frac{\partial L}{\partial \dot{q}_i}\right) - \frac{\partial L}{\partial q_i} = 0$$

then give the following constants of motion:

$$(M + 2m)\dot{x} = \text{constant},$$
$$(M + 2m)\dot{y} = \text{constant},$$
$$(M + 2m)\dot{z} = \text{constant},$$
$$\left(2mr^2 + \frac{1}{12} ML^2\right)\dot{\varphi} \sin^2 \theta = \text{constant}.$$

The first three equations show that the velocity $(\dot{x}, \dot{y}, \dot{z})$ of the center of mass of the system is a constant vector. Thus the center of mass moves in a uniform rectilinear motion with whatever velocity it had initially. The last equation shows that the component of the angular momentum about the z'-axis is a constant of the motion. Since the axis has been arbitrarily chosen, this means that the angular momentum is conserved.

Lagrange's equations also give the following equations of motion:

$$\ddot{r} - r\dot{\theta}^2 - r\dot{\varphi}^2 \sin^2\theta + \frac{K}{m}(r - r_0) = 0$$

$$\left(r^2 + \frac{ML^2}{24m}\right)\ddot{\theta} + 2r\dot{r}\dot{\theta} - \left(r^2 + \frac{ML^2}{24m}\right)\dot{\varphi}^2 \sin\theta\cos\theta = 0.$$

These and the φ equation above describe the motion about the center of mass o the system.

Thus under the constraint that the two masses m slide along the rod symmetri cally with respect to the midpoint O, the motion of the center of mass O of the system is a uniform rectilinear motion, and the motion of the system about O i such that the total angular momentum about O is conserved.

2005

A rectangle coordinate system with axes x, y, z is rotating relative to ar inertial frame with constant angular velocity ω about the z-axis. A parti cle of mass m moves under a force whose potential is $V(x, y, z)$. Set up the Lagrange equations of motion in the coordinate system x, y, z. Show that their equations are the same as those for a particle in a fixed coordinate system acted on by the force $-\nabla V$ and a force derivable from a velocity-dependent potential U. Find U.

(Wisconsin)

Sol: Let the inertial frame have the same origin as the rotating frame and axes x', y', z'. Denote the velocities of the particle in the two frames by $\mathbf{v}$ and $\mathbf{v}'$. As

$$\mathbf{v}' = \mathbf{v} + \boldsymbol{\omega} \times \mathbf{r}$$

with

$$\boldsymbol{\omega} = (0, 0, \omega), \quad \mathbf{r} = (x, y, z), \quad \mathbf{v} = (\dot{x}, \dot{y}, \dot{z}),$$

we have

$$v'^2 = v^2 + 2\mathbf{v} \cdot \boldsymbol{\omega} \times \mathbf{r} + (\boldsymbol{\omega} \times \mathbf{r})^2$$
$$= \dot{x}^2 + \dot{y}^2 + \dot{z}^2 + 2\omega(x\dot{y} - \dot{x}y) + \omega^2(x^2 + y^2),$$

and the Lagrangian of the particle in the inertial frame, expressed in quantities referring to the rotating frame,

$$L = T - V$$
$$= \frac{1}{2}m(\dot{x}^2 + \dot{y}^2 + \dot{z}^2) + m\omega(x\dot{y} - \dot{x}y) + \frac{1}{2}m\omega^2(x^2 + y^2) - V.$$

Lagrange's equations

$$\frac{d}{dt}\left(\frac{\partial L}{\partial \dot{q}_i}\right) - \frac{\partial L}{\partial q_i} = 0$$

then give

$$m\ddot{x} - 2m\omega\dot{y} - m\omega^2 x + \frac{\partial V}{\partial x} = 0,$$

$$m\ddot{y} + 2m\omega\dot{x} - m\omega^2 y + \frac{\partial V}{\partial y} = 0,$$

$$m\ddot{z} + \frac{\partial V}{\partial z} = 0.$$

For a particle of mass m moving in a fixed frame (x,y,z) under a force $-\nabla V$ and an additional velocity-dependent potential U, the Lagrangian is

$$L = \frac{1}{2}m(\dot{x}^2 + \dot{y}^2 + \dot{z}^2) - V - U.$$

A comparison of this with the Lagrangian obtained previously gives

$$U = -m\omega(x\dot{y} - \dot{x}y) - \frac{1}{2}m\omega^2(x^2 + y^2).$$

This Lagrangian would obviously give rise to the same equations of motion.

2006

a. Show that the moment of inertia of a thin rod about its center of mass is $ml^2/12$.

b. A long thin tube of negligible mass is pivoted so that it may rotate without friction in a horizontal plane. A thin rod of mass M and length l slides without friction in the tube. Choose a suitable set of coordinates and write Lagrange's equations for this system.

c. Initially the rod is centered over the pivot and the tube is rotating with angular velocity ω_0. Show that the rod is unstable in this position, and describe its subsequent motion if it is disturbed slightly. What are the radial and angular velocities of the rod after a long time? (Assume the tube is long enough that the rod is still inside.)

(Wisconsin)

Sol:

a. By definition the moment of inertia is

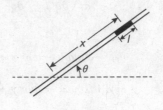

Fig. 2.6

$$I = \sum_i R_i^2 \Delta m_i = \int_{-\frac{l}{2}}^{\frac{l}{2}} x^2 \rho dx = \frac{1}{12}\rho l^3 = \frac{1}{12}ml^2.$$

b. Take the angle θ between the thin tube and a fixed horizontal line through the pivot and the distance x of the center of mass of the thin rod from the pivot of the tube, as shown in Fig. 2.6, as the generalized coordinates. We have

$$T = \frac{1}{2}M(\dot{x}^2 + x^2\dot{\theta}^2) + \frac{1}{24}Ml^2\dot{\theta}^2, \quad V = 0,$$

and the Lagrangian

$$L = \frac{1}{2}M(\dot{x}^2 + x^2\dot{\theta}^2) + \frac{1}{24}Ml^2\dot{\theta}^2.$$

Lagrange's equations then give

$$\ddot{x} = x\dot{\theta}^2$$

$$M\left(x^2 + \frac{1}{12}l^2\right)\dot{\theta} = \text{constant} = C, \quad \text{say.}$$

c. The initial conditions $x = 0$, $\dot{\theta} = \omega_0$ give

$$C = \frac{1}{2}Ml^2\omega_0,$$

i.e.

$$\dot{\theta} = \frac{l^2\omega_0}{12x^2 + l^2}.$$

We then have

$$\ddot{x} = \frac{1}{2}\frac{d\dot{x}^2}{dx} = \frac{l^4\omega_0^2 x}{(12x^2 + l^2)^2}.$$

Integrating we obtain, as initially $x = 0$, $\dot{x} = 0$,

$$\dot{x}^2 = \frac{l^2\omega_0^2 x^2}{12x^2 + l^2}.$$

It is noted that the speed of the rod in the tube,

$$\dot{x} = \frac{l\omega_0}{\sqrt{12 + \dfrac{l^2}{x^2}}},$$

increases as its distance from the initial position increases. Thus the rod is unstable at the initial position. For $t \to \infty$, $x \to \infty$, $\dot{\theta} \to 0$ and $\dot{x} \to l\omega_0/\sqrt{12}$. Hence, after a long time, the rotation will slow down to zero while the speed of the rod in the tube will tend to an upper limit. The distance x however will be ever increasing.

2007

A block of mass M is rigidly connected to a massless circular track of radius a on a frictionless horizontal table as shown in Fig. 2.7. A particle of mass m is confined to move without friction on the circular track which is vertical.

  a. Set up the Lagrangian, using θ as one coordinate.

  b. Find the equations of motion.

  c. In the limit of small angles, solve that equations of motion for θ as a function of time.

(*Wisconsin*)

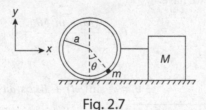

Fig. 2.7

Sol:

  a. As the motion of the system is confined to a vertical plane, use a fixed coordinate frame x, y and choose the x coordinate of the center of the circular track and the angle θ giving the location of m on the circular track as the generalized coordinates as shown in Fig. 2.7. The coordinates of the mass m are then $(x + a \sin\theta, -a \cos\theta)$. As M is rigidly connected to the circular track its velocity is $(\dot{x}, 0)$. Hence the Lagrangian is

$$L = T - V = \frac{1}{2}M\dot{x}^2 + \frac{1}{2}m[(\dot{x} + a\dot{\theta}\cos\theta)^2 + a^2\dot{\theta}^2\sin^2\theta] + mga\cos\theta$$

$$= \frac{1}{2}M\dot{x}^2 + \frac{1}{2}m[\dot{x}^2 + a^2\dot{\theta}^2 + 2a\dot{x}\dot{\theta}\cos\theta] + mga\cos\theta.$$

b. As

$$\frac{\partial L}{\partial x} = 0,$$

$$\frac{\partial L}{\partial \dot{x}} = M\dot{x} + m\dot{x} + ma\dot{\theta}\cos\theta,$$

$$\frac{\partial L}{\partial \theta} = -ma\dot{x}\dot{\theta}\sin\theta - mga\sin\theta,$$

$$\frac{\partial L}{\partial \dot{\theta}} = ma^2\dot{\theta} + ma\dot{x}\cos\theta,$$

Lagrange's equations

$$\frac{d}{dt}\left(\frac{\partial L}{\partial \dot{q}_i}\right) - \frac{\partial L}{\partial q_i} = 0$$

give

$$(M + m)\ddot{x} + ma\ddot{\theta}\cos\theta - ma\dot{\theta}^2\sin\theta = 0,$$

$$a\ddot{\theta} + \ddot{x}\cos\theta + g\sin\theta = 0.$$

c. For small oscillations, θ and $\dot{\theta}$ are small quantities of the 1st order. Neglecting higher order terms the equations of motion become

$$(M + m)\ddot{x} + ma\ddot{\theta} = 0,$$
$$a\ddot{\theta} + \ddot{x} + g\theta = 0.$$

Eliminating $\ddot{x}$ we have

$$\ddot{\theta} + \frac{(M + m)g\theta}{Ma} = 0.$$

Hence

$$\theta = A\sin(\omega t) + B\cos(\omega t),$$

where $\omega = \sqrt{(M + m)g/Ma}$ is the angular frequency of oscillation and A and B are constants to be determined from the initial conditions.

2008

Consider a particle of mass m moving in a bound orbit with potential $V(r) = -k/r$. Using polar coordinates in the plane of the orbit:

a. Find p_r and p_θ as functions of r, θ, $\dot{r}$ and $\dot{\theta}$. Is either one constant?

b. Using the virial theorem show that

$$J_r + J_\theta = \oint \frac{k}{r} dt,$$

where

$$J_r = \oint p_r dr,$$

$$J_\theta = \oint p_\theta d\theta.$$

c. Show that

$$(J_r + J_\theta) = \sqrt{\frac{-2\pi^2 mk^2}{E}},$$

using

$$\int_{r_-}^{r_+} \frac{dr}{\sqrt{-r^2 + ar - b}} = \pi, \quad r_\pm = \frac{1}{2}(a \pm \sqrt{a^2 - 4b}).$$

d. Using the results of (c) show that the period of the orbit is the same for the r and θ motions, namely,

$$\tau = \pi k \sqrt{\frac{m}{-2E^3}}.$$

(Wisconsin)

Sol:

a. We have

$$L = T - V = \frac{1}{2}m(\dot{r}^2 + r^2\dot{\theta}^2) + \frac{k}{r}.$$

The generalized momenta are

$$p_r = \frac{\partial L}{\partial \dot{r}} = m\dot{r},$$

$$p_\theta = \frac{\partial L}{\partial \dot{\theta}} = mr^2\dot{\theta}.$$

As there is no θ in L, $p_\theta = mr^2\dot{\theta}$ is a constant of the motion.

b.

$$J_r + J_\theta = \oint m\dot{r}dr + \oint mr^2\dot{\theta}d\theta$$

$$= \oint m\dot{r}^2 dt + \oint mr^2\dot{\theta}^2 dt$$

$$= \oint m(\dot{r}^2 + r^2\dot{\theta}^2)dt$$

$$= 2\oint T dt = 2\overline{T}\tau,$$

where τ is the period and $\overline{T}$ is the average kinetic energy of the particle over one period. For a particle moving in a bound orbit in the field of an inverse-square law force the virial theorem takes the form

$$\overline{T} = -\frac{1}{2}\overline{V}.$$

Thus

$$J_r + J_\theta = -\overline{V}\tau = \oint \frac{k}{r}dt.$$

c. The total energy of the particle

$$E = T + V = \frac{1}{2}m(\dot{r}^2 + r^2\dot{\theta}^2) - \frac{k}{r} = \frac{1}{2}m\left(\dot{r}^2 + \frac{h^2}{r^2}\right) - \frac{k}{r},$$

where $h = r^2\dot{\theta} = p_\theta/m =$ constant, is a constant. The above gives

$$\dot{r}^2 = \frac{2}{m}\left(E + \frac{k}{r}\right) - \frac{h^2}{r^2},$$

or

$$\dot{r} = \pm\frac{1}{r}\sqrt{\frac{2Er^2}{m} + \frac{2kr}{m} - h^2}$$

$$= \pm\sqrt{\frac{-2E}{m}} \cdot \frac{1}{r}\sqrt{-r^2 - \frac{kr}{E} + \frac{mh^2}{2E}},$$

where it should be noted that $E < 0$ for bound orbits.

For a bound orbit, $r_- \leq r \leq r_+$. The extreme values of r are given by $\dot{r} = 0$, i.e.

$$r^2 + \frac{kr}{E} - \frac{mh^2}{2E} = 0.$$

Writing this as

$$r^2 - ar + b = 0,$$

where a, b are positive numbers $a = -k/E$, $b = -mh^2/2E$, we have

$$r_\pm = \frac{1}{2}(a \pm \sqrt{a^2 - 4b}).$$

Then

$$J_r + J_\theta = \oint \frac{k}{r}\, dt = \oint \frac{k}{r\dot{r}}\, dr$$

$$= 2\int_{r_-}^{r_+} \frac{k\,dr}{\sqrt{\dfrac{-2E}{m}} \cdot \sqrt{-r^2 - \dfrac{kr}{E} + \dfrac{mh^2}{2E}}}$$

$$= 2k\sqrt{\frac{m}{-2E}} \int_{r_-}^{r_+} \frac{dr}{\sqrt{-r^2 - \dfrac{kr}{E} + \dfrac{mh^2}{2E}}}$$

$$= 2k\sqrt{\frac{m}{-2E}} \int_{r_-}^{r_+} \frac{dr}{\sqrt{-r^2 + ar - b}}$$

$$= 2\pi k\sqrt{\frac{m}{-2E}} = \sqrt{\frac{2\pi^2 mk^2}{-E}},$$

using the value given for the integral.

d. As E is a constant, we have

$$-E = -\bar{E} = -(\bar{T} + \bar{V}) = -(\bar{T} - 2\bar{T}) = \bar{T},$$

or

$$-E = \frac{1}{\tau} \oint T\, dt$$

$$= \frac{1}{2\tau}(J_r + J_\theta) = \frac{1}{2\tau}\sqrt{\frac{2\pi^2 mk^2}{-E}},$$

giving

$$\tau = \pi k\sqrt{\frac{m}{-2E^3}}.$$

2009

Two identical discs of mass M and radius R are supported by three identical torsion bars, as shown in Fig. 2.8, whose restoring torque is $\tau = -k\theta$ where $k =$ given torsion constant for length l and twist angle θ. The discs are free to rotate about the vertical axis of the torsion bars with displacements θ_1, θ_2 from equilibrium position. Neglect moment of inertia of the torsion bars. For initial conditions $\theta_1(0) = 0$, $\theta_2(0) = 0$, $\dot{\theta}_1(0) = 0$, $\dot{\theta}_2(0) = \Omega =$ given constant, how long does it take for disc 1 to get all the kinetic energy? You may leave this in the form of an implicit function.

(UC, Berkeley)

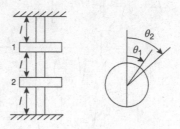

Fig. 2.8

Sol: If I is the moment if inertia of each disc, the Lagrangian of the system is

$$L = \frac{1}{2}I(\dot{\theta}_1^2 + \dot{\theta}_2^2) - \frac{1}{2}k[\theta_1^2 + \theta_2^2 + (\theta_1 - \theta_2)^2].$$

The two Lagrange's equations are

$$I\ddot{\theta}_1 + k(2\theta_1 - \theta_2) = 0,$$
$$I\ddot{\theta}_2 + k(2\theta_2 - \theta_1) = 0.$$

These combine to give

$$I(\ddot{\theta}_1 + \ddot{\theta}_2) + k(\theta_1 + \theta_2) = 0,$$
$$I(\ddot{\theta}_1 - \ddot{\theta}_2) + 3k(\theta_1 - \theta_2) = 0.$$

The solutions are respectively

$$\theta_1 + \theta_2 = A_+ \sin\left(\sqrt{\frac{k}{I}}t + \varphi_+\right),$$

$$\theta_1 - \theta_2 = A_- \sin\left(\sqrt{\frac{3k}{I}}t + \varphi_-\right).$$

The initial conditions

$$\theta_1 + \theta_2 = 0, \quad \theta_1 - \theta_2 = 0 \quad \text{at} \quad t = 0$$

give $\varphi_+ = \varphi_- = 0$. The conditions

$$\dot{\theta}_1 + \dot{\theta}_2 = \Omega, \quad \dot{\theta}_1 - \dot{\theta}_2 = -\Omega \quad \text{at} \quad t = 0$$

give

$$A_+ = \Omega\sqrt{\frac{I}{k}}, \quad A_- = -\Omega\sqrt{\frac{I}{3k}}.$$

Hence

$$\theta_2 = \frac{1}{2}\Omega\left[\sqrt{\frac{I}{k}}\sin\left(\sqrt{\frac{k}{I}}t\right) + \sqrt{\frac{I}{3k}}\sin\left(\sqrt{\frac{3k}{I}}t\right)\right],$$

$$\dot{\theta}_2 = \frac{1}{2}\Omega\left[\cos\left(\sqrt{\frac{k}{I}}t\right) + \cos\left(\sqrt{\frac{3k}{I}}t\right)\right].$$

Only when $\dot{\theta}_2 = 0$, i.e. after a time t given by

$$\cos\left(\sqrt{\frac{k}{I}}t\right) = -\cos\left(\sqrt{\frac{3k}{I}}t\right),$$

will disc 1 get all the kinetic energy.

It should be noted that the kinetic energy of the system is not a constant. When t satisfies the above equation, disk 1 does take all the kinetic energy of the system at that time. However, this kinetic energy varies from time to time this happens.

2010

A thin, uniform rod of length $2L$ and mass M is suspended from a massless string of length l tied to a nail. As shown in Fig. 2.9, a horizontal force F is applied to the rod's free end.

Write the Lagrange equations for this system. For very short times (so that all angles are small) determine the angles that the string and the rod make with the vertical. Start from rest at $t = 0$. Draw a diagram to illustrate the initial motion of the rod.

(UC, Berkeley)

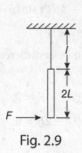

Fig. 2.9

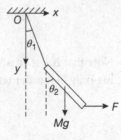

Fig. 2.10

Sol: As the applied force F is horizontal and initially the string and rod are ve[r]-
tical, the motion is confined to a vertical plane. Take Cartesian coordi[-]
nates as shown in Fig. 2.10 and denote the angles made by the string an[d]
the rod with the vertical by θ_1, θ_2 respectively. The center of mass of the ro[d]
has coordinates $(l \sin \theta_1 + L \sin \theta_2, -l \cos \theta_1 - L \cos \theta_2)$ and thus velocit[y]
$(l\dot{\theta}_1 \cos \theta_1 + L\dot{\theta}_2 \cos \theta_2, l\dot{\theta}_1 \sin \theta_1 + L\dot{\theta}_2 \sin \theta_2)$. Its moment of inertia about [a]
perpendicular axis through its center is $ML^2/3$. Hence its kinetic energy is

$$T = \frac{1}{2} M[l^2 \dot{\theta}_1^2 + L^2 \dot{\theta}_2^2 + 2Ll\dot{\theta}_1\dot{\theta}_2 \cos(\theta_1 - \theta_2)] + \frac{1}{6} ML^2 \dot{\theta}_2^2$$

and its potential energy is

$$V = -Mg(l \cos \theta_1 + L \cos \theta_2).$$

The potential U of the horizontal force is by definition

$$U = -\int \mathbf{F} \cdot d\mathbf{r} = -F(l \sin \theta_1 + 2L \sin \theta_2).$$

The Lagrangian is therefore

$$L = T - V - U$$

$$= \frac{1}{2} M[l^2 \dot{\theta}_1^2 + L^2 \dot{\theta}_2^2 + 2Ll\dot{\theta}_1\dot{\theta}_2 \cos(\theta_1 - \theta_2)] + \frac{1}{6} ML^2 \dot{\theta}_2^2$$

$$+ Mg(l \cos \theta_1 + L \cos \theta_2) + F(l \sin \theta_1 + 2L \sin \theta_2).$$

Lagrange's equations

$$\frac{d}{dt}\left(\frac{\partial L}{\partial \dot{q}_i}\right) - \frac{\partial L}{\partial q_i} = 0$$

then give

$$Ml\ddot{\theta}_1 + ML\ddot{\theta}_2 \cos(\theta_1 - \theta_2) + ML\dot{\theta}_2^2 \sin(\theta_1 - \theta_2)$$
$$+ Mg \sin \theta_1 - F \cos \theta_1 = 0,$$

$$\frac{4}{3} ML\ddot{\theta}_2 + Ml\ddot{\theta}_1 \cos(\theta_1 - \theta_2) - Ml\dot{\theta}_1^2 \sin(\theta_1 - \theta_2)$$
$$+ Mg \sin \theta_2 - 2F \cos \theta_2 = 0.$$

Note that if F is small so that $\theta_1, \theta_2, \dot{\theta}_1, \dot{\theta}_2$ can be considered small then, retaining only first order terms, the above become

$$Ml\ddot{\theta}_1 + ML\ddot{\theta}_2 + Mg\theta_1 - F = 0,$$

$$\frac{4}{3} ML\ddot{\theta}_2 + Ml\ddot{\theta}_1 + Mg\theta_2 - 2F = 0.$$

The motion starts from rest at $t = 0$. For a very short time Δt afterwards, the force can be considered as giving rise to a horizontal impulse $F\Delta t$ and an impulsive torque $FL\Delta t$ about the center of mass of the rod. We then have

$$F\Delta t = M(l\dot{\theta}_1 \cos\theta_1 + L\dot{\theta}_2 \cos\theta_2)$$
$$\approx Ml\dot{\theta}_1 + ML\dot{\theta}_2,$$

as the angle θ_1, θ_2 are still small, and

$$FL\Delta t = \frac{1}{3}ML^2\dot{\theta}^2.$$

Eliminating $F\Delta t$ from the above, we have

$$\dot{\theta}_1 \approx -\frac{2L}{3l}\dot{\theta}_2.$$

As $\theta_1 = \overline{\dot{\theta}_1}\Delta t \approx \dot{\theta}_1 \Delta t/2$, $\theta_2 \approx \dot{\theta}_2 \Delta t/2$, the above gives

$$\theta_1 \approx -\frac{2L}{3l}\theta_2.$$

The initial configuration of the system is shown in Fig. 2.11.

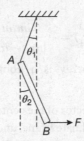

Fig. 2.11

2011

Consider a binary star system.

a. Write the Lagrangian for the system in terms of the Cartesian coordinates of the two stars $\mathbf{r}_1$ and $\mathbf{r}_2$.

b. Show that the potential energy is a homogeneous function of the coordinates of degree –1, i.e.

$$V(\alpha\mathbf{r}_1, \alpha\mathbf{r}_2) = \alpha^{-1}V(\mathbf{r}_1, \mathbf{r}_2),$$

where α is a real scaling parameter.

c. Find a transformation which leaves the Lagrangian the same up to multiplication constant (thereby leaving the physics unchanged) and thus find Kepler's third law relating the period of revolution of the system to the size of its orbit.

(Chicago

Sol:

a. Let r_1, r_2 be the radius vectors of the binary stars, masses m_1, m_2 respectively from the origin of a fixed coordinate frame. Then

$$T = \frac{1}{2} m_1 |\dot{r}_1|^2 + \frac{1}{2} m_2 |\dot{r}_2|^2, \qquad V = -\frac{Gm_1 m_2}{|r_1 - r_2|},$$

and the Lagrangian is

$$L = T - V = \frac{1}{2}(m_1 |\dot{r}_1|^2 + m_2 |\dot{r}_2|^2) + \frac{Gm_1 m_2}{|r_1 - r_2|}.$$

b.

$$V(\alpha r_1, \alpha r_2) = -\frac{Gm_1 m_2}{|\alpha r_1 - \alpha r_2|} = -\frac{1}{\alpha}\frac{Gm_1 m_2}{|r_1 - r_2|} = \frac{1}{\alpha} V(r_1, r_2),$$

i.e. the potential energy is a homogeneous function of the coordinates of degree –1.

c. Let $\mathbf{R}$ be the radius vector of the center of mass of the binary system from the origin of the fixed coordinate frame, and r'_1, r'_2 be the radius vectors of m_1, m_2 from the center of mass respectively. By definition

$$(m_1 + m_2)\mathbf{R} = m_2 r_1 + m_2 r_2,$$
$$r_1 = \mathbf{R} + r'_1, \qquad r_2 = \mathbf{R} + r'_2,$$

so

$$r'_1 = \frac{m_2 r}{m_1 + m_2}, \qquad r'_2 = -\frac{m_1 r}{m_1 + m_2},$$

where $r = r_1 - r_2 = r'_1 - r'_2$.

We can then write the Lagrangian as

$$L = \frac{m_1 + m_2}{2}|\dot{\mathbf{R}}|^2 + \frac{m_1 m_2}{2(m_2 + m_2)}|\dot{r}|^2 + \frac{Gm_1 m_2}{|r|}.$$

As L does not depend on $\mathbf{R} = (x, y, z)$ explicitly, $\partial L/\partial \dot{x}, \partial L/\partial \dot{y}, \partial L/\partial \dot{z}$ and hence $(m_1 + m_2)\dot{\mathbf{R}}$ are constant. Therefore the first term of L, which is the kinetic energy of the system as a whole, is constant. This terms can be neglected when we are interested only in the internal motion of the system.

Thus

$$L = \left(\frac{m_1 m_2}{m_1 + m_2}\right)\left[\frac{1}{2}|\dot{\mathbf{r}}|^2 + \frac{G(M_1 + M_2)}{|\mathbf{r}|}\right]$$

$$= \left(\frac{m_1}{m_1 + m_2}\right)\left[\frac{1}{2}m_2|\dot{\mathbf{r}}|^2 + \frac{Gm_2(m_1 + m_2)}{|\mathbf{r}|}\right]$$

$$= \left(\frac{m_2}{m_1 + m_2}\right)\left[\frac{1}{2}m_1|\dot{\mathbf{r}}|^2 + \frac{Gm_1(m_1 + m_2)}{|\mathbf{r}|}\right],$$

which may be consider as the Lagrangian, apart from a multiplicative constant, of the motion of one star in the gravitational field of a fixed star of mass $m_1 + m_2$. Let m_1 be this "moving" star and consider its centripetal force:

$$m_1 r\dot{\theta}^2 = \frac{Gm_1(m_1 + m_2)}{r^2},$$

or

$$\frac{T^2}{r^3} = \frac{4\pi^2}{G(m_1 + m_2)},$$

where $T = 2\pi/\dot{\theta}$ is the period of m_1 about m_2, which is Kepler's third law. The same is of course true for the motion of m_2 about m_1.

2012

Two thin beams of mass m and length l are connected by a frictionless hinge and a thread. The system rests on a smooth surface in the way shown in Fig. 2.12. At $t = 0$ the thread is cut. In the following you may neglect the thread and the mass of the hinge.

a. Find the speed of the hinge when it hits the floor.

b. Find the time it takes for the hinge to hit the floor, expressing this in terms of a concrete integral which you need not evaluate explicitly.

(Princeton)

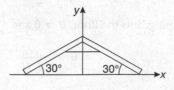

Fig. 2.12

Sol:

a. Due to symmetry, the hinge will fall vertically. Take coordinates as shown in Fig
    2.12 and let θ be the angle each beam makes with the floor. Then the centers o
    mass of the beams have coordinates

$$x_1 = \frac{1}{2} l \cos \theta, \qquad y_1 = \frac{1}{2} l \sin \theta,$$

$$x_2 = -\frac{1}{2} l \cos \theta, \qquad y_2 = \frac{1}{2} l \sin \theta,$$

and velocity components

$$\dot{x}_1 = -\frac{1}{2} l \dot{\theta} \sin \theta, \qquad \dot{y}_1 = \frac{1}{2} l \dot{\theta} \cos \theta,$$

$$\dot{x}_2 = \frac{1}{2} l \dot{\theta} \sin \theta, \qquad \dot{y}_2 = \frac{1}{2} l \dot{\theta} \cos \theta.$$

Each beam has a moment of inertial $ml^2/12$ about a horizontal axis through it
center of mass. The Lagrangian of the system is

$$L = T - V$$

$$= \frac{1}{4} m l^2 \dot{\theta}^2 + \frac{1}{12} m l^2 \dot{\theta}^2 - mgl \sin \theta$$

$$= \frac{1}{3} m l^2 \dot{\theta}^2 - mgl \sin \theta.$$

Lagrange's equation

$$\frac{d}{dt}\left(\frac{\partial L}{\partial \dot{\theta}}\right) - \frac{\partial L}{\partial \theta} = 0$$

Then gives

$$\ddot{\theta} + \frac{3g}{2l} \cos \theta = 0.$$

As

$$\ddot{\theta} = \frac{1}{2}\frac{d\dot{\theta}^2}{d\theta} \qquad \text{and} \qquad \dot{\theta} = 0 \quad \text{when} \quad \theta = 30^\circ,$$

the above integrates to

$$\dot{\theta}^2 = \frac{3g}{2l}(1 - 2\sin\theta).$$

Hence when the hinge hits the floor, $\theta = 0$ and

$$\dot{\theta} = -\sqrt{\frac{3g}{2l}},$$

i.e.

$$|\mathbf{v}| = |l\dot{\theta}| = \sqrt{\frac{3gl}{2}}.$$

b.   The time taken for the hinge to hit the floor is given by

$$t = \int_{30°}^{0} \frac{d\theta}{\dot{\theta}} = \int_{30°}^{0} \frac{d\theta}{-\sqrt{\frac{3g}{2l}(1 - 2\sin\theta)}}$$

$$= \sqrt{\frac{2l}{3g}} \int_{0}^{30°} \frac{d\theta}{1 - 2\sin\theta}.$$

2013

A uniform rod of the length L and mass M moves in the vertical xz-plane, one of its end-points. A being subject to the constraint $z = x\tan\alpha$ ($\alpha = $ constant inclination to the horizontal x-axis). Derive the Lagrangian equations of motion in terms of the generalized coordinates $q_1 = s$ and $q_2 = \theta$ (see Fig. 2.13). Use these to determine if a pure translational motion ($\theta = $ constant) is possible and, if so, for which values of θ.

(Princeton)

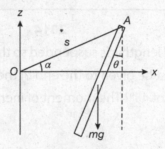

Fig. 2.13

Sol: The coordinates and velocity components of the center of mass of the rod are

$$x = s\cos\alpha - \frac{1}{2}L\sin\theta, \qquad z = s\sin\alpha - \frac{1}{2}L\cos\theta,$$

$$\dot{x} = \dot{s}\cos\alpha - \frac{1}{2}L\dot{\theta}\cos\theta, \qquad \dot{z} = \dot{s}\sin\alpha + \frac{1}{2}L\dot{\theta}\sin\theta,$$

and the moment of inertia of the rod about a perpendicular axis through the center of mass is $ML^2/12$, so the Lagrangian is

$$L = T - V$$

$$= \frac{1}{2}M(\dot{x}^2 + \dot{z}^2) + \frac{1}{24}ML^2\dot{\theta}^2 - Mgz$$

$$= \frac{1}{2}M[\dot{s}^2 - L\dot{s}\dot{\theta}\cos(\theta + \alpha)] + \frac{1}{6}ML^2\dot{\theta}^2 - Mg\left(s\sin\alpha - \frac{1}{2}L\cos\theta\right).$$

Lagrange's equations

$$\frac{d}{dt}\left(\frac{\partial L}{\partial \dot{q}_i}\right) - \frac{\partial L}{\partial q_i} = 0$$

then give

$$\ddot{s} - \frac{1}{2}L\ddot{\theta}\cos(\theta + \alpha) + \frac{1}{2}L\dot{\theta}^2\sin(\theta + \alpha) + g\sin\alpha = 0,$$

$$\ddot{s}\cos(\theta + \alpha) - \frac{2}{3}L\ddot{\theta} - g\sin\theta = 0.$$

If the motion is pure translational, $\theta = $ constant, $\dot{\theta} = 0$, $\ddot{\theta} = 0$ and the above become

$$\ddot{s} + g\sin\alpha = 0,$$
$$\ddot{s}\cos(\theta + \alpha) - g\sin\theta = 0.$$

Eliminating $\ddot{s}$ gives

$$\sin\alpha\cos(\theta + \alpha) = -\sin\theta,$$

or

$$\theta = -\alpha.$$

2014

The rod of mass m and length L is suspended so that it can execute a complete circle. What speed must be given to the end of the rod so that it oscillates with an angular amplitude of 45°? The moment of inertia of the rod about one end is $I = \frac{1}{3}mL^2$.

Sol: Using conservation of energy,

$$\frac{1}{2}I\omega^2 = mgh$$

$$\frac{1}{2}\left(\frac{1}{3}mL^2\right)\omega^2 = mg(L - L\cos\theta)$$

$$\omega^2 = \frac{6g(1 - \cos 45°)}{L}$$

$$\omega^2 = \frac{6g\left(1 - \frac{\sqrt{2}}{2}\right)}{L}$$

$$\omega^2 = \frac{g(6 - 3\sqrt{2})}{L}$$

$$\omega = \sqrt{\frac{g(6 - 3\sqrt{2})}{L}}$$

$$v = \sqrt{gL(6 - 3\sqrt{2})}$$

2015

A spring pendulum consists of a mass m attached to one end of a massless spring with spring constant k. The other end of the spring is tied to a fixed support. When no weight is on the spring, its length is l. Assume that the motion of the system is confined to a vertical plane. Derive the equations of motion. Solve the equations of motion in the approximation of small angular and radial displacements from equilibrium.

(SUNY, Buffalo)

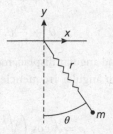

Fig. 2.14

Sol: Use coordinates as shown in Fig. 2.14. The mass m has coordinates $(r \sin \theta, -r \cos \theta)$ and velocity components $(r\dot{\theta} \cos \theta + \dot{r} \sin \theta, r\dot{\theta} \sin \theta - \dot{r} \cos \theta)$ and hence kinetic energy

$$T = \frac{1}{2} m(\dot{r}^2 + r^2 \dot{\theta}^2),$$

and potential energy

$$V = \frac{1}{2} k(r - l)^2 - mgr \cos \theta.$$

The Lagrangian is therefore

$$L = T - V = \frac{1}{2} m(\dot{r}^2 + r^2 \dot{\theta}^2) - \frac{1}{2} k(r - l)^2 + mgr \cos \theta.$$

Lagrange's equations

$$\frac{d}{dt}\left(\frac{\partial L}{\partial \dot{q}_i}\right) - \frac{\partial L}{\partial q_i} = 0$$

then give the equations of motion

$$m\ddot{r} - mr\dot{\theta}^2 + k(r - l) - mg\cos\theta = 0,$$
$$r\ddot{\theta} + 2\dot{r}\dot{\theta} + g\sin\theta = 0.$$

The equilibrium position in polar coordinates (r_0, θ_0) is given by $\ddot{r} = \ddot{\theta} = 0$
$\dot{r} = \dot{\theta} = 0$, namely,

$$\theta_0 = 0, \qquad r_0 = l + \frac{mg}{k}.$$

For small oscillations about equilibrium, θ is a small angle. Let $\rho = r - r_0$ with
$\rho \ll r_0$ and write the equations of motion as

$$m\ddot{\rho} - m(r_0 + \rho)\dot{\theta}^2 + k\rho = 0,$$
$$(r_0 + \rho)\ddot{\theta} + 2\dot{\rho}\dot{\theta} + g\theta = 0,$$

or, neglecting higher order terms of the small quantities $\rho, \dot{\rho}, \dot{\theta}$,

$$\ddot{\rho} + \frac{k}{m}\rho = 0,$$

$$\ddot{\theta} + \frac{g}{r_0}\theta = 0.$$

Thus both the radial and angular displacements execute simple harmonic motion
about equilibrium with angular frequencies $\sqrt{k/m}, \sqrt{g/r_0}$ respectively. The solutions are

$$\rho = A\cos\left(\sqrt{\frac{k}{m}}t + \varphi_1\right),$$

or

$$r = l + \frac{mg}{k} + A\cos\left(\sqrt{\frac{k}{m}}t + \varphi_1\right),$$

and

$$\theta = B\cos\left(\sqrt{\frac{kg}{kl + mg}}t + \varphi_2\right),$$

where the constants $A, \varphi_1, B, \varphi_2$ are to be determined from the initial conditions.

2016

A particle is constrained to be in a plane. It is attracted to a fixed point P in this
plane; the force is always directed exactly at P and is inversely proportional to
the square of the distance from P.

a. Using polar coordinates, write the Lagrangian of this particle.

b. Write Lagrangian equations for this particle and find at least one first integral.

(SUNY, Buffalo)

Sol:

a. Choose polar coordinates with origin at P in the plane in which the particle is constrained to move. The force acting on the particle is

$$\mathbf{F} = -\frac{k\mathbf{r}}{r^3},$$

k being a positive constant. Its potential energy with respect to infinity is

$$V = -\int_\infty^r \mathbf{F} \cdot d\mathbf{r} = -\frac{k}{r}.$$

The kinetic energy of the particle is

$$T = \frac{1}{2}m(\dot{r}^2 + r^2\dot{\theta}^2).$$

Hence the Lagrangian is

$$L = T - V = \frac{1}{2}m(\dot{r}^2 + r^2\dot{\theta}^2) + \frac{k}{r}.$$

b. Lagrange's equations

$$\frac{d}{dt}\left(\frac{\partial L}{\partial \dot{q_i}}\right) - \frac{\partial L}{\partial q_i} = 0$$

then give the equations of motion

$$m\ddot{r} + \frac{k}{r^2} = 0, \quad \frac{d}{dt}(mr^2\dot{\theta}) = 0.$$

The second equation gives immediately a first integral

$$mr^2\dot{\theta} = \text{constant},$$

which means that the angular momentum with respect to P is conserved.

2017

Consider two particles interacting by way of a central force (potential = $V(r)$ where $\mathbf{r}$ is the relative position vector).

a. Obtain the Lagrangian in the center of mass system and show that the energy and angular momentum are conserved. Prove that the motion is in a plane and satisfies Kepler's second law (that **r** sweeps out equa areas in equal times).

b. Suppose that the potential is $V = kr^2/2$, where k is a positive constan and that the total energy E is known. Find expressions for the minimun and maximum values that r will have in the course of the motion.

(SUNY, Buffalo

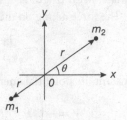

Fig. 2.15

Sol: As the forces acting on the particles always direct along the line of separation, the motion is confined to whatever plane the particles initially move in. Use pola coordinates in this plane as shown in Fig. 2.15 with origin at the center of mass o the particles. By definition of the center of mass,

$$m_1\mathbf{r}_1 + m_2\mathbf{r}_2 = 0,$$

i.e.

$$m_1\mathbf{r}_1 = -m_2\mathbf{r}_2,$$

or

$$m_1 r_1 = m_2 r_2$$

for the magnitudes.

a. The kinetic energy of the particle are

$$T = \frac{m_1}{2}|\dot{\mathbf{r}}_1|^2 + \frac{m_2}{2}|\dot{\mathbf{r}}_2|^2$$

$$= \frac{m_2^2}{2m_1}|\dot{\mathbf{r}}_2|^2 + \frac{m_2}{2}|\dot{\mathbf{r}}_2|^2$$

$$= \frac{m_2(m_1 + m_2)}{2m_1}|\dot{\mathbf{r}}_2|^2 = \frac{m_2^2}{2\mu}|\dot{\mathbf{r}}_2|^2,$$

where $\mu = m_1 m_2/(m_1 + m_2)$ is the reduced mass of the system. The potential energy is

$$V(r_1 + r_2) = V\left(\frac{m_2 r_2}{m_1} + r_2\right) = V\left(\frac{m_2 r_2}{\mu}\right).$$

Hence the Lagrangian is

$$L = T - V = \frac{m_2^2}{2\mu}(\dot{r}_2^2 + r_2^2\dot{\theta}^2) - V\left(\frac{m_2 r_2}{\mu}\right),$$

using r_2 and θ as the generalized coordinates.

The Lagrangian L does not depend on t explicitly. So

$$\frac{dL}{dt} = \sum_j \left(\frac{\partial L}{\partial \dot{q}_j}\frac{d\dot{q}_j}{dt} + \frac{\partial L}{\partial q_j}\dot{q}_j\right)$$

$$= \sum_j \left[\frac{\partial L}{\partial \dot{q}_j}\frac{d\dot{q}_j}{dt} + \frac{d}{dt}\left(\frac{\partial L}{\partial \dot{q}_j}\right)\dot{q}_j\right]$$

$$= \frac{d}{dt}\sum_j \frac{\partial L}{\partial \dot{q}_j}\dot{q}_j,$$

use having been made of Lagrange's equations. Hence

$$\sum_j \frac{\partial L}{\partial \dot{q}_j}\dot{q}_j - L = \text{constant}.$$

In the present case,

$$\frac{\partial L}{\partial \dot{r}_2}\dot{r}_2 = \frac{m_2^2\dot{r}_2^2}{\mu}, \qquad \frac{\partial L}{\partial \dot{\theta}}\dot{\theta} = \frac{m_2^2 r_2^2\dot{\theta}^2}{\mu},$$

and the above gives

$$\frac{m_2^2}{\mu}(\dot{r}_2^2 + r_2^2\dot{\theta}^2) - T + V = \frac{m_2^2}{2\mu}(\dot{r}_2^2 + r_2^2\dot{\theta}^2) + V$$

$$= T + V = \text{constant},$$

showing that the total energy is conserved. Note that this proof is possible because V does not depend on the velocities explicitly.

As L does not depend on θ explicitly, Lagrange's equation gives

$$\frac{\partial L}{\partial \dot{\theta}} = \frac{m_2^2 r_2^2\dot{\theta}}{\mu} = \text{constant} = J, \text{ say}.$$

The angular momentum of the system about the center of mass is

$$m_2 r_2^2\dot{\theta} + m_1 r_1^2\dot{\theta} = \frac{(m_1 + m_2)m_2 r_2^2\dot{\theta}}{m_1} = \frac{m_2^2 r_2^2\dot{\theta}}{\mu} = J.$$

Hence the angular momentum is conserved. The above also implies

$$r^2\dot{\theta} = (r_1 + r_2)^2\dot{\theta} = \left(\frac{m_1 + m_2}{m_1}\right)^2 r_2^2\,\dot{\theta} = \frac{m_2^2 r_2^2\dot{\theta}}{\mu^2} = \text{constant},$$

i.e.

$$\frac{r^2 \Delta \theta}{\Delta t} = \frac{2 \Delta S}{\Delta t} = \text{constant},$$

where ΔS is the area swept out by $\mathbf{r}$ in time Δt. Thus Kepler's second law is satisfied.

b. The total energy

$$E = T + V = \frac{m_2^2}{2\mu}(\dot{r}_2^2 + r_2^2\dot{\theta}^2) + \frac{1}{2}kr^2$$

$$= \frac{m_2^2 \dot{r}_2^2}{2\mu} + \frac{J^2\mu}{2m_2^2 r_2^2}\frac{1}{r_2^2} + \frac{1}{2}kr^2$$

can be written as

$$E = \frac{1}{2}\mu\dot{r}^2 + \frac{J^2}{2\mu}\frac{1}{r^2} + \frac{1}{2}kr^2.$$

When r is a maximum or minimum, $\dot{r} = 0$. Hence the extreme values of r are given by

$$k\mu r^4 - 2E\mu r^2 + J^2 = 0.$$

2018

A particle is attracted to a force center by a force which varies inversely as the cube of its distance from the center. Derive the equations of motion and solve them for the orbits. Discuss how the nature of the orbits depends on the parameters of the system.

(SUNY, Buffalo)

Sol: As the particle moves under a central force its motion is confined to a plane. We use polar coordinates in this plane with origin at the force center. For the force

$$\mathbf{F} = -\frac{k\mathbf{r}}{r^4},$$

where k is a positive constant, the potential energy is

$$V(r) = -\int_\infty^r \mathbf{F} \cdot d\mathbf{r} = -\frac{k}{2r^2}.$$

Hence the Lagrangian is

$$L = T - V = \frac{1}{2}m(\dot{r}^2 + r^2\dot{\theta}^2) + \frac{k}{2r^2}.$$

Lagrange's equations

$$\frac{d}{dt}\left(\frac{\partial L}{\partial \dot{q}_i}\right) - \frac{\partial L}{\partial \dot{q}_i} = 0$$

then give

$$mr^2\dot{\theta} = b, \quad \text{(a constant)},$$

$$m\ddot{r} - mr\dot{\theta}^2 + \frac{k}{r^3} = 0.$$

Let $u = \frac{1}{r}$. The first equation becomes

$$\dot{\theta} = \frac{bu^2}{m}.$$

As

$$\dot{r} = \frac{d}{dt}\left(\frac{1}{u}\right) = -\frac{1}{u^2}\frac{du}{d\theta}\dot{\theta} = -\frac{b}{m}\frac{du}{d\theta},$$

$$\ddot{r} = -\frac{b}{m}\frac{d^2u}{d\theta^2}\dot{\theta} = -\frac{b^2u^2}{m^2}\frac{d^2u}{d\theta^2},$$

the second equation becomes

$$\frac{d^2u}{d\theta^2} + \left(1 - \frac{mk}{b^2}\right)u = 0.$$

Hence, if $b^2 > mk$,

$$u = \frac{1}{r_0}\cos\left[\sqrt{1 - \frac{mk}{b^2}}(\theta - \theta_0)\right],$$

i.e.

$$r\cos\left[\sqrt{1 - \frac{mk}{b^2}}(\theta - \theta_0)\right] = r_0;$$

if $b^2 < mk$,

$$u = \frac{1}{r_0}\cosh\left[\sqrt{\frac{mk}{b^2} - 1}(\theta - \theta_0)\right],$$

i.e.

$$r\cosh\left[\sqrt{\frac{mk}{b^2} - 1}(\theta - \theta_0)\right] = r_0.$$

Here (r_0, θ_0) is a point on the orbit.

2019

Assume the Lagrangian for a certain one-dimensional motion is given by

$$L = e^{\gamma t}\left(\frac{1}{2}m\dot{q}^2 - \frac{1}{2}kq^2\right),$$

where γ, m, k are positive constants. What is the Lagrange's equation? Are there any constants of motion? How would you describe the motion? Suppose a point transformation is made to another generalized coordinate S given by

$$S = \exp\left(\frac{\gamma t}{2}\right)q.$$

What is the Lagrangian in terms of S? Lagrange's equation? Constants of motion? How would you describe the relationship between the two solutions?

(SUNY, Buffalo)

Sol: Lagrange's equation

$$\frac{d}{dt}\left(\frac{\partial L}{\partial \dot{q}}\right) - \frac{\partial L}{\partial q} = 0$$

gives

$$e^{\gamma t}(m\ddot{q} + \gamma m\dot{q} + kq) = 0,$$

or

$$\ddot{q} + \gamma\dot{q} + \frac{kq}{m} = 0.$$

As L contains q, t explicitly, there is no constant of motion.
Try solutions of the form $q \sim e^{\alpha t}$. Substitution gives

$$\alpha^2 + \gamma\alpha + \frac{k}{m} = 0,$$

whose solutions are

$$\alpha = -\frac{\gamma}{2} \pm \sqrt{\left(\frac{\gamma}{2}\right)^2 - \frac{k}{m}}.$$

Write this as $\alpha = -\frac{\gamma}{2} \pm b$ and consider the three possible cases.

i. $\frac{\gamma}{2} > \sqrt{\frac{k}{m}}$. b is imaginary; let it be $i\beta$. The general solution is

$$q = e^{-\frac{\gamma t}{2}}(Ae^{i\beta t} + Be^{-i\beta t}),$$

or

$$q = e^{-\frac{\gamma t}{2}} 2(A' \cos \beta t + B' \sin \beta t),$$

A, B, A', B' being constants. Thus the motion is oscillatory with attenuating amplitude.

ii. $\dfrac{\gamma}{2} > \sqrt{\dfrac{k}{m}}$. $b = 0$ and we have

$$q = q_0 e^{-\frac{\gamma t}{2}},$$

showing that the motion is non-oscillatory with q attenuating from the value q_0 at $t = 0$.

iii. $\dfrac{\gamma}{2} > \sqrt{\dfrac{k}{m}}$. $b = 0$ and

$$q = e^{-\frac{\gamma t}{2}}(Ce^{bt} + De^{-bt}),$$

C and D being constants. This motion is also non-oscillatory and time-attenuating.

The three cases can be characterized as underdamped, critically damped and overdamped.

If we include the time factor in the generalized coordinate by a point transformation

$$S = e^{\frac{\gamma t}{2}} q, \quad \text{i.e.} \quad q = e^{-\frac{\gamma t}{2}} S,$$

the Lagrangian becomes

$$L = \frac{1}{2} m \left(\dot{S} - \frac{1}{2} \gamma S \right)^2 - \frac{1}{2} k S^2.$$

Lagrange's equation then gives the equation of motion

$$\ddot{S} + \omega^2 S = 0$$

with $\omega^2 = k/m - (\gamma/2)^2$. As $\ddot{S} = \dfrac{1}{2} \dfrac{d\dot{S}^2}{dS}$, integration gives

$$\dot{S}^2 + \omega^2 S^2 = \text{constant}.$$

Hence there is now a constant of motion. Physically, however, the situation is not altered. As $S, \dot{S}$ both contain t implicitly, this constant actually changes with time.

For $\gamma/2 < \sqrt{k/m}, \omega^2$ is positive, i.e. ω is real, and the equation of motion in S describes a simple harmonic motion without damping. For $\gamma/2 = \sqrt{k/m}, \omega = 0$ and the motion in S is uniform. For $\gamma/2 > \sqrt{k/m}, \omega$ is imaginary and the motion is non-oscillatory with time attenuation. However, as noted above, S contains a hidden attenuating factor $\exp(-\gamma t/2)$ which causes time attenuation in all the three cases.

We may conclude that both sets of solutions describe identical physica situations but in the second set the attenuating time factor $\exp(-\gamma t/2)$ absorbed in the generalized coordinates and the treatment proceeds as if were nonexistent.

2020

A bead of mass m slides without friction on a rotating wire hoop of radius whose axis of rotation is through a vertical diameter as shown in Fig. 2.16. The constant angular velocity of the hoop is ω.

a. Write the Lagrangian for the system and find any constants of the motion that may exist.

b. Locate the positions of equilibrium of the bead for $\omega < \omega_c$ and $\omega > \omega_c$ where $\omega_c = \sqrt{g/a}$.

Which of these positions of equilibrium are stable and unstable?

c. Calculate the oscillation frequencies of small amplitude vibrations about the points of stable equilibrium.

(UC, Berkeley)

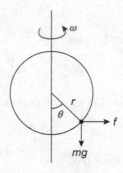

Fig. 2.16

Sol:

a. Use a rotating polar coordinate frame attached to the loop as shown in Fig. 2.16. In this frame, in additional to the gravitational force on the mass, mg, a fictitious centrifugal force $\mathbf{f}$ as shown has to be introduced. In polar coordinates

$$\mathbf{f} = (m\omega^2 r \sin^2 \theta, m\omega^2 r \sin \theta \cos \theta),$$
$$\mathbf{mg} = (mg \cos \theta, -mg \sin \theta).$$

f can be expressed in terms of a potential V_f by

$$\mathbf{f} = -\nabla V_f = \left(-\frac{\partial V_f}{\partial r}, -\frac{\partial V_f}{\gamma \partial \theta} \right),$$

i.e.

$$V_f = -\frac{1}{2} m r^2 \omega^2 \sin^2 \theta.$$

Similarly the gravitational potential is

$$V_g = -mgr \cos \theta.$$

The particle velocity is $(\dot{r}, r\dot{\theta})$. With the constraint $r = a$, the Lagrangian is

$$L = T - V = \frac{1}{2} m a^2 \dot{\theta}^2 + \frac{1}{2} m a^2 \omega^2 \sin^2 \theta + mga \cos \theta.$$

As $\partial L/\partial t = 0$ and V does not contain $\dot{\theta}$ explicitly, $\dot{\theta} \partial L/\partial \dot{\theta} - L = \text{constant}$ (Problem **2017**). Hence

$$\frac{1}{2} m a^2 \dot{\theta}^2 - \frac{1}{2} m a^2 \omega^2 \sin^2 \theta - mga \cos \theta = \text{constant},$$

which means that $T + V = \text{constant}$.

b. Lagrange's equation gives the equation of motion

$$a\ddot{\theta} - a\omega^2 \sin \theta \cos \theta + g \sin \theta = 0.$$

At a position of equilibrium, $\ddot{\theta} = 0$, so

$$\sin \theta \, (a\omega^2 \cos \theta - g) = 0,$$

or

$$a \sin \theta \, (\omega^2 \cos \theta - \omega_c^2) = 0$$

with $\omega_c^2 = \frac{g}{a}$.

If $\omega < \omega_c$, $\omega^2 \cos \theta < \omega_c^2$ and hence $\sin \theta = 0$, and we have two equilibrium positions at $\theta = 0, \pi$.

If $\omega > \omega_c$, we have in addition to the above positions an equilibrium position at

$$\cos \theta = \frac{\omega_c^2}{\omega^2} = \frac{g}{a\omega^2}.$$

c. Suppose θ_0 is an equilibrium position and let $\theta = \theta_0 + \alpha$, where α is a small quantity. The equation of motion reduces, retaining up to first order terms, to

$$a\ddot{\alpha} + (g \cos \theta_0 - a\omega^2 \cos 2\theta_0)\alpha - a\omega^2 \sin \theta_0 \cos \theta_0 + g \sin \theta_0 = 0,$$

or, as $\ddot{\theta} = 0$ at $\theta = \theta_0$,

$$\ddot{\alpha} + \left(\frac{g}{a} \cos \theta_0 - \omega^2 \cos 2\theta_0 \right)\alpha = 0.$$

If $\omega < \sqrt{g/a}$, the coefficient of α is positive for the equilibrium at $\theta_0 = 0$. So thi is a position of stable equilibrium. The coefficient is negative for the equilibriur at $\theta_0 = \pi$, showing that it is an unstable equilibrium.

If $\omega > \sqrt{g/a}$, equilibrium also occurs at $\cos \theta_0 = g/a\omega^2$. In this case the coeffi cient of α is

$$\frac{g}{a} \cos \theta_0 - 2\omega^2 \cos^2 \theta_0 + \omega^2 = \frac{1}{\omega^2}\left(\omega^4 - \frac{g^2}{a^2} \right) > 0,$$

so that the equilibrium is a stable one.

d. The angular frequency of small vibrations about a point of stable equilibrium is

$$\omega' = \sqrt{\frac{g}{a} \cos \theta_0 - \omega^2 \cos 2\theta_0}$$

$$= \begin{cases} \sqrt{\dfrac{g}{a} - \omega^2}, & \text{at } \theta_0 = 0, \\[4mm] \dfrac{1}{\omega}\sqrt{\omega^4 - \left(\dfrac{g}{a}\right)^2} & \text{at } \theta_0 = \cos^{-1}\left(\dfrac{g}{a\omega^2}\right). \end{cases}$$

2021

Particles of mass m_1 and m_2, connected by a light spring with spring constant k, are at rest on a frictionless horizontal surface.

a. An impulse I of very short duration is delivered to m_1. The direction of the impulse is from m_1 to m_2. How far will m_2 move before coming to rest for the first time?

b. Is it possible by delivering a short impulse to m_1 to have the system move from rest so that it is rotating without oscillation? Explain.

(UC, Berkeley)

Sol:

a. Take the initial position of m_1 as origin and the direction from m_1 to m_2 as the positive direction of the x-axis. The Lagrangian of the system is

$$L = T - V = \frac{1}{2} m_1 \dot{x}_1^2 + \frac{1}{2} m_2 \dot{x}_2^2 - \frac{1}{2} k(x_2 - x_1 - l)^2,$$

where l is the natural length of the spring, being equal to $x_2 - x_1$ at $t = 0$. Lagrange's equations

$$\frac{d}{dt}\left(\frac{\partial L}{\partial \dot{q}_i} \right) - \frac{\partial L}{\partial q_i} = 0$$

give

$$m_1\ddot{x}_1 = k(x_2 - x_1 - l),$$
$$m_2\ddot{x}_2 = -k(x_2 - x_1 - l).'$$

from the above, we obtain

$$\ddot{x}_2 - \ddot{x}_1 = -\frac{k(m_1 + m_2)(x_2 - x_1 - l)}{m_1 m_2},$$

or, by setting $u = x_2 - x_1 - l$, $\omega^2 = k(m_1 + m_2)/m_1 m_2$,

$$\ddot{u} + \omega^2 u = 0.$$

The general solution is

$$u = a\cos(\omega t + \alpha),$$

giving

$$x_2 - x_1 - l = a\cos(\omega t + \alpha),$$

where a and α are constants. The initial conditions $x_1 = 0$, $x_2 = l$, $\dot{x}_1 = I/m_1$, $\dot{x}_2 = 0$ at $t = 0$ then give

$$a\cos\alpha = 0,$$

$$a\omega\sin\alpha = \frac{I}{m_1},$$

with solution

$$\alpha = \frac{\pi}{2}, \qquad a = \frac{I}{m_1\omega}.$$

Hence

$$x_2 - x_1 = l + \frac{I}{m_1\omega}\cos\left(\omega t + \frac{\pi}{2}\right). \tag{1}$$

Conservation of momentum gives

$$m_1\dot{x}_1 + m_2\dot{x}_2 = I.$$

Integrating and applying the initial conditions we obtain

$$m_1 x_1 + m_2 x_2 = m_2 l + It.$$

This and Eq. (1) together give

$$x_2 = l + \frac{It}{m_1 + m_2} - \frac{I\sin(\omega t)}{(m_1 + m_2)\omega},$$

and thus

$$\dot{x}_2 = \frac{I}{m_1 + m_2} - \frac{I\cos(\omega t)}{m_1 + m_2}.$$

When m_2 comes to rest for the first time, $\dot{x}_2 = 0$, and the above gives $\cos(\omega t) =$ for the first time. Hence when m_2 comes to rest for the first time,

$$t = \frac{2\pi}{\omega}.$$

At that time m_2 has moved a distance

$$x_2 - l = \frac{2\pi I}{\omega(m_1 + m_2)} = 2\pi I \sqrt{\frac{m_1 m_2}{k}}(m_1 + m_2)^{-3/2}.$$

b. If the impulse given to m_1 has a component perpendicular to the line joining the two particles the system will rotate about the center of mass, in addition to the linear motion of the center of mass. In a rotating frame with origin at the center of mass and the x-axis along the line joining the two particles, there will be (fictitious) centrifugal forces acting on the particles in addition to the restoring force of the string. At the positions of the particles where the forces are in equilibrium the particles have maximum velocities on account of energy conservation (Problem **2017**). Hence oscillations will always occur, besides the rotation of the system as a whole.

2022

A sphere of mass M and radius R rolls without slipping down a triangular block of mass m that is free to move on a frictionless horizontal surface, as shown in Fig. 2.17.

a. Find the Lagrangian and state Lagrange's equations for this system subject to the force of gravity at the surface of the earth.

b. Find the motion of the system by integrating Lagrange's equation, given that all objects are initially at rest and the sphere's center is at a distance H above the surface.

(UC, Berkeley)

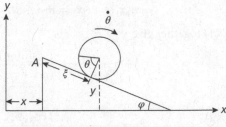

Fig. 2.17

Sol:

a. Use a fixed coordinate frame as shown in Fig. 2.17 and let θ be the angle of rotation of the sphere. As the sphere rolls without slipping down the inclined plane, its center will have coordinates

$$(x + (\xi_0 + R\theta) \cos \varphi, H - R\theta \sin \varphi)$$

and velocity

$$(\dot{x} + R\dot{\theta} \cos \varphi, -R\dot{\theta} \sin \varphi).$$

Note that at $t = 0$, $x = 0$, $\theta = 0$, $\xi = \xi_0$, $y = H$, $\dot{x} = \dot{\theta} = 0$. Then the Lagrangian is

$$L = T - V = \frac{1}{2} m\dot{x}^2 + \frac{1}{2} M(\dot{x}^2 + R^2\dot{\theta}^2 + 2R\dot{x}\dot{\theta} \cos \varphi)$$
$$+ \frac{1}{5} MR^2\dot{\theta}^2 - Mg(H - R\theta \sin \varphi).$$

Lagrange's equations

$$\frac{d}{dt}\left(\frac{\partial L}{\partial \dot{q}_i}\right) - \frac{\partial L}{\partial q_i} = 0$$

give

$$(m + M)\ddot{x} + MR\ddot{\theta} \cos \varphi = 0,$$
$$\ddot{x} \cos \varphi + \frac{7}{5} R\ddot{\theta} - g \sin \varphi = 0.$$

b. Eliminating $\ddot{x}$ from the above gives

$$\left(\frac{7}{5} - \frac{M \cos^2 \varphi}{m + M}\right)\ddot{\theta} = \frac{g \sin \varphi}{R},$$

or, on integration and use of initial conditions,

$$\theta = \frac{5(m + M) \sin \varphi}{2[7(m + M) - 5M \cos^2 \varphi]} \frac{gt^2}{R},$$

and thus

$$x = -\frac{MR \cos \varphi}{m + M} \theta = -\frac{5M \sin(2\varphi)}{4[7(m + M) - 5M \cos^2 \varphi]} gt^2.$$

Note that, as the sphere rolls down the plane, the block moves to the left as expected from momentum conservation.

2023

Two mass points m_1 and $m_2(m_1 \neq m_2)$ are connected by a string of length passing through a hole in a horizontal table. The string and mass points move without friction with m_1 on the table and m_2 free to move in a vertical line.

a. What initial velocity must m_1 be given so that m_2 will remain motionless a distance d below the surface of the table?

b. If m_2 is slightly displaced in a vertical direction, small oscillations ensue. Use Lagrange's equations to find the period of these oscillations.

(UC, Berkeley)

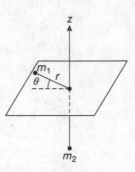

Fig. 2.18

Sol:

a. m_1 must have a velocity v perpendicular to the string such that the centripetal force on it is equal to the gravitational force on m_2:

$$\frac{m_1 v^2}{l - d} = m_2 g,$$

or

$$v = \sqrt{\frac{m_2(l - d)g}{m_1}}.$$

b. Use a frame of polar coordinates fixed in the horizontal table as shown in Fig. 2.18. m_2 has z-coordinate $-(l - r)$ and thus velocity $\dot{r}$. The Lagrangian of the system is then

$$L = T - V = \frac{1}{2} m_1 (\dot{r}^2 + r^2 \dot{\theta}^2) + \frac{1}{2} m_2 \dot{r}^2 + m_2 g(l - r).$$

Lagrange's equations give

$$m_1 r^2 \dot{\theta} = \text{constant},$$

$$(m_1 + m_2)\ddot{r} - m_1 r \dot{\theta}^2 + m_2 g = 0.$$

At $t = 0$, $r = l - d, v = \sqrt{m_2(l - d)g/m_1} = v_0$, say, so

$$\dot{\theta}_0 = \frac{v_0}{l - d} = \sqrt{\frac{m_2}{m_1}\frac{g}{l - d}}.$$

Hence

$$m_1 r^2 \dot{\theta} = m_1(l - d)^2\dot{\theta}_0 = m_1\sqrt{\frac{m_2}{m_1}(l - d)^3 g},$$

giving

$$r\dot{\theta}^2 = \frac{r^4\dot{\theta}^2}{r^3} = \frac{m_2}{m_1}\left(\frac{l - d}{r}\right)^3 g$$

and

$$(m_1 + m_2)\ddot{r} - m_2\left(\frac{l - d}{r}\right)^3 g + m_2 g = 0.$$

Let $r = (l - d) + \rho$, where $\rho \ll (l - d)$. Then

$$\ddot{r} = \ddot{\rho}, \quad r^{-3} = (l - d)^{-3}\left(1 + \frac{\rho}{l - d}\right)^{-3} \approx (l - d)^{-3}\left(1 - \frac{3\rho}{l - d}\right)$$

and the above equation becomes

$$\ddot{\rho} + \frac{3m_2 g}{(m_1 + m_2)(l - d)}\rho = 0.$$

Hence ρ oscillates about O, i.e. r oscillates about the value $l - d$, with angular frequency

$$\omega = \sqrt{\frac{3m_2 g}{(m_1 + m_2)(l - d)}},$$

or period

$$T = 2\pi\sqrt{\frac{(m_1 + m_2)(l - d)}{3m_2 g}}.$$

2024

Two rods *AB* and *BC*, each of length *a* and mass *m*, are frictionlessly joined at *B* and lie on a frictionless horizontal table. Initially the two rods (i.e. point *A, B, C*) are collinear. An impulse $\overline{P}$ is applied at point *A* in a direction perpendicular to the line *ABC*. Find the motion of the rods immediately after the impulse is applied.

(*Columbia*)

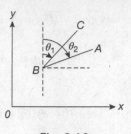

Fig. 2.19

Sol: As the two rods AB, BC are freely joined at B, take coordinates as shown i̇
Fig. 2.19 and let the coordinates of B be (x, y). Then the center of mass of B(
has coordinates

$$\left(x + \frac{1}{2} a \sin \theta_1, y + \frac{1}{2} a \cos \theta_1 \right)$$

and velocity

$$\left(\dot{x} + \frac{1}{2} a \dot{\theta}_1 \cos \theta_1, \dot{y} - \frac{1}{2} a \dot{\theta}_1 \sin \theta_1 \right),$$

and that of AB has coordinates

$$\left(x + \frac{1}{2} a \sin \theta_2, y + \frac{1}{2} a \cos \theta_2 \right)$$

and velocity

$$\left(\dot{x} + \frac{1}{2} a \dot{\theta}_2 \cos \theta_2, \dot{y} - \frac{1}{2} a \dot{\theta}_2 \sin \theta_2 \right).$$

Each rod has a moment of inertia about its center of mass of $ma^2/12$. Hence the
total kinetic energy is

$$
\begin{aligned}
T &= \frac{1}{2} m \left[\dot{x}^2 + \dot{y}^2 + \frac{1}{4} a^2 \dot{\theta}_1^2 + a \dot{\theta}_1 (\dot{x} \cos \theta_1 - \dot{y} \sin \theta_1) \right] + \frac{1}{24} ma^2 \dot{\theta}_1^2 \\
&\quad + \frac{1}{2} m \left[\dot{x}^2 + \dot{y}^2 + \frac{1}{4} a^2 \dot{\theta}_1^2 + a \dot{\theta}_2 (\dot{x} \cos \theta_2 - \dot{y} \sin \theta_2) \right] + \frac{1}{24} ma^2 \dot{\theta}_2^2 \\
&= \frac{1}{2} m \left[2 (\dot{x}^2 + \dot{y}^2) + a \dot{x} (\dot{\theta}_1 \cos \theta_1 + \dot{\theta}_2 \cos \theta_2) - a \dot{y} (\dot{\theta}_1 \sin \theta_1 + \dot{\theta}_2 \sin \theta_2) \right] \\
&\quad + \frac{1}{6} ma^2 (\dot{\theta}_1^2 + \dot{\theta}_2^2) .
\end{aligned}
$$

The impulse $\overline{P}$ is applied at A in a direction perpendicular to the line ABC. Thus
the virtual moment of the impulse is $\overline{P}\delta(y + a \cos \theta_2)$ and the generalized components of the impulse are

$$Q_x = 0, \quad Q_y = \overline{P}, \quad Q_{\theta_1} = 0, \quad Q_{\theta_2} = -a\overline{P} \sin \theta_2.$$

Lagrange's equations for impulsive motion are

$$\left(\frac{\partial T}{\partial \dot{q}_j}\right)_f - \left(\frac{\partial T}{\partial \dot{q}_j}\right)_i = Q_j,$$

where i, f refer to the initial and final states of the system relative to the application of impulse. Note that at $t = 0$ when the impulse is applied, $\theta_1 = -\pi/2$, $\theta_2 = \pi/2$. Furthermore, for the initial state, $\dot{\theta}_1 = \dot{\theta}_2 = \dot{x} = \dot{y} = 0$. As

$$\frac{\partial T}{\partial \dot{x}} = 2m\dot{x} + \frac{1}{2}ma(\dot{\theta}_1 \cos\theta_1 + \dot{\theta}_2 \cos\theta_2),$$

$$\frac{\partial T}{\partial \dot{y}} = 2m\dot{y} - \frac{1}{2}ma(\dot{\theta}_1 \sin\theta_1 + \dot{\theta}_2 \sin\theta_2),$$

$$\frac{\partial T}{\partial \dot{\theta}_1} = \frac{1}{2}ma\dot{x}\cos\theta_1 - \frac{1}{2}ma\dot{y}\sin\theta_1 + \frac{1}{3}ma^2\dot{\theta}_1,$$

$$\frac{\partial T}{\partial \dot{\theta}_2} = \frac{1}{2}ma\dot{x}\cos\theta_2 - \frac{1}{2}ma\dot{y}\sin\theta_2 + \frac{1}{3}ma^2\dot{\theta}_2,$$

Lagrange's equations give

$$2m\dot{x} = 0,$$

$$2m\dot{y} + \frac{1}{2}ma(\dot{\theta}_1 - \dot{\theta}_2) = \overline{P},$$

$$\frac{1}{2}ma\dot{y} + \frac{1}{3}ma^2\dot{\theta}_1 = 0,$$

$$-\frac{1}{2}ma\dot{y} + \frac{1}{3}ma^2\dot{\theta}_2 = -a\overline{P}.$$

The solution is

$$\dot{x} = 0, \quad \dot{y} = -\frac{\overline{P}}{m}, \quad \dot{\theta}_1 = \frac{3\overline{P}}{2ma}, \quad \dot{\theta}_2 = -\frac{9\overline{P}}{2ma}.$$

Hence immediately after the application of impulse, the center of mass of BC has velocity

$$\left(\dot{x}, \dot{y} + \frac{1}{2}a\dot{\theta}_1\right) = \left(0, -\frac{\overline{P}}{4m}\right),$$

and that of AB has velocity

$$\left(\dot{x}, \dot{y} - \frac{1}{2}a\dot{\theta}_2\right) = \left(0, \frac{5\overline{P}}{4m}\right).$$

2025

Consider a particle of mass m moving in a plane under a central force

$$F(r) = -\frac{k}{r^2} + \frac{k'}{r^3}$$

(assume $k > 0$).

a. What is the Lagrangian for this system in terms of the polar coordinate r, θ and their velocities?

b. Write down the equations of motion for r and θ, and show that the orbital angular momentum l is a constant of the motion.

c. Assume that $l^2 > -mk'$. Find the equation for the orbit, i.e. r as a function of θ.

(Columbia)

Sol:

a. As

$$F(r) = -\frac{k}{r^2} + \frac{k'}{r^3},$$

$$V(r) = -\int_\infty^r F(r)\,dr = -\frac{k}{r} + \frac{k'}{2r^2}.$$

The Lagrangian is then

$$L = T - V = \frac{1}{2}m(\dot{r}^2 + r^2\dot{\theta}^2) + \frac{k}{r} - \frac{k'}{2r^2}.$$

b. Lagrange's equations give the equations of motion

$$m(\ddot{r} - r\dot{\theta}^2) + \frac{k}{r^2} - \frac{k'}{r^3} = 0, \tag{1}$$

$$m(r\ddot{\theta} + 2\dot{r}\dot{\theta}) = 0.$$

The second equation has first integral $mr^2\dot{\theta} = $ constant. This quantity is the angular momentum of the mass about the origin $l = r \cdot mr\dot{\theta}$.

c. Let $u = r^{-1}$. As $r = u^{-1}$,

$$\dot{r} = -u^{-2}\frac{du}{d\theta}\dot{\theta} = -u^{-2}\frac{du}{d\theta}\frac{l}{mr^2} = -\frac{l}{m}\frac{du}{d\theta},$$

$$\ddot{r} = -\frac{l}{m}\frac{d^2u}{d\theta^2}\dot{\theta} = -\frac{l^2}{m^2}u^2\frac{d^2u}{d\theta^2},$$

$$mr\dot{\theta}^2 = \frac{l^2}{mr^3} = \frac{l^2u^3}{m},$$

Eq. (1) becomes

$$\frac{d^2u}{d\theta^2} + \left(1 + \frac{mk'}{l^2}\right)u - \frac{mk}{l^2} = 0.$$

A special solution is

$$u = \frac{mk}{l^2}\left(\frac{l^2}{l^2 + mk'}\right) = \frac{mk}{l^2 + mk'},$$

As $l^2 > -mk'$, i.e. $\frac{mk'}{l^2} > -1$,

$$1 + \frac{mk'}{l^2} > 0$$

and the general solution is

$$u = A\cos\left(\sqrt{1 + \frac{mk'}{l^2}}\theta + \alpha\right) + \frac{mk}{l^2 + mk'},$$

where A, α are constants. By a suitable choice of coordinates, α can be put to zero. Hence the equation of the trajectory can be written as

$$r = \left[A\cos\left(\sqrt{1 + \frac{mk'}{l^2}}\theta\right) + \frac{mk}{l^2 + mk'}\right]^{-1}.$$

2026

A point particle of mass m is constrained to move frictionlessly on the inside surface of a circular wire hoop of radius r, uniform density and mass M. The hoop is in the xy-plane, can roll on a fixed line (the x-axis), but does not slide, nor can it lose contact with the x-axis. The point particle is acted on by gravity exerting a force along the negative y-axis. At $t = 0$ suppose the hoop is at rest. At this time the particle is at the top of the hoop and is given a velocity v_0 along the x-axis. What is the velocity v_f, with respect to the fixed axis, when the particle comes to the bottom of the hoop? Simplify your answer in the limits $m/M \to 0$ and $M/m \to 0$.

(Columbia)

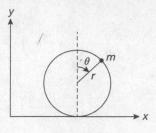

Fig. 2.20

Sol: Use a fixed coordinate frame as shown in Fig. 2.20 and let the coordinates of the center of the hoop be (x, y). Then the mass m has coordinates

$$(x + r\sin\theta, r + r\cos\theta)$$

and velocity

$$(\dot{x} + r\dot{\theta}\cos\theta, -r\dot{\theta}\sin\theta).$$

As the hoop has moment of inertia Mr^2, the system has kinetic energy

$$T = \frac{1}{2}m(\dot{x}^2 + r^2\dot{\theta}^2 + 2r\dot{x}\dot{\theta}\cos\theta) + \frac{1}{2}M\dot{x}^2 + \frac{1}{2}Mr^2\left(\frac{\dot{x}}{r}\right)^2$$

and potential energy

$$V = mg(r + r\cos\theta).$$

Hence the Lagrangian is

$$L = T - V = M\dot{x}^2 + \frac{1}{2}m(\dot{x}^2 + r^2\dot{\theta}^2 + 2r\dot{x}\dot{\theta}\cos\theta) - mgr(1 + \cos\theta).$$

As $\partial L/\partial x = 0$, Lagrange's equation gives

$$(2M + m)\dot{x} + mr\dot{\theta}\cos\theta = \text{constant.} \tag{1}$$

At $t = 0$, m is at the top of the hoop, $\dot{x} = 0$, $\theta = 0$, $r\dot{\theta} = v_0$, giving the value of the constant as mv_0. When m is at the bottom of the hoop, $\theta = \pi$, the velocity of the mass is

$$v_f = \dot{x} + r\dot{\theta}\cos\pi = \dot{x} - r\dot{\theta},$$

and Eq. (1) becomes

$$2M\dot{x} + mv_f = mv_0.$$

The total energy is conserved so that between these two points we have

$$M\dot{x}^2 + \frac{1}{2}mv_f^2 = \frac{1}{2}mv_0^2 + 2mgr.$$

Eliminating $\dot{x}$ between the last two equations gives

$$(2M + m)v_f^2 - 2mv_0v_f - [(2M - m)v_0^2 + 8Mgr] = 0.$$

The solutions are

$$v_f = \frac{mv_0 \pm 2\sqrt{M^2v_0^2 + 2(2M + m)Mgr}}{2M + m}.$$

In the limit $m/M \to 0$, $v_f \to \pm\sqrt{v_0^2 + 4gr}$. The negative sign is to be chosen as for $M \gg m$, $\dot{x}$ is small and $v_f \simeq -r\dot{\theta}$. In the limit $M/m \to 0$, $v_f \to v_0$.

2027

a. A particle slides on the inside of a smooth vertical paraboloid of revolution $r^2 = az$. Show that the constraint force has a magnitude = constant . $\left(1 + \dfrac{4r^2}{a^2}\right)^{-\frac{3}{2}}$. What is its direction?

b. A particle of mass m is acted on by a force whose potential is $V(r)$.

  1. Set up the Lagrangian function in a spherical coordinate system which is rotating with angular velocity ω about the z-axis.

  2. Show that your Lagrangian has the same form as in a fixed coordinate system with the addition of a velocity-dependent potential U (which gives the centrifugal and Coriolis forces).

3. Calculate from U the components of the centrifugal and Coriolis forces in the radial (r) and azimuthal (ϕ) directions.

(Wisconsin)

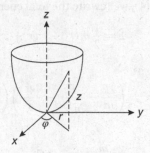

Fig. 2.21

Sol:

a. Use cylindrical coordinates (r, φ, z) as shown in Fig. 2.21. In Cartesian coordinates the particle, mass m, has coordinates

$$(r\cos\varphi, r\sin\varphi, z),$$

velocity

$$(\dot{r}\cos\varphi - r\dot{\varphi}\sin\varphi, \dot{r}\sin\varphi + r\dot{\varphi}\cos\varphi, \dot{z}),$$

and hence Lagrangian

$$L = T - V = \frac{1}{2}m(\dot{r}^2 + r^2\dot{\varphi}^2 + \dot{z}^2) - mgz.$$

The constraint equation is

$$f(r, \varphi, z) = -r^2 + az = 0,$$

or

$$-2rdr + adz = 0.$$

Lagrange's equations

$$\frac{d}{dt}\frac{\partial L}{\partial \dot{q}_i} - \frac{\partial L}{\partial q_i} = Q_i,$$

where Q_i are the generalized forces of constraint, then give, making use of Lagrange's undetermined multiplier λ,

$$m\ddot{r} - mr\dot{\varphi}^2 = -2r\lambda, \tag{1}$$
$$m\ddot{z} + mg = a\lambda, \tag{2}$$
$$mr^2\dot{\varphi} = \text{constant} = J, \quad \text{say.} \tag{3}$$

The equation of constraint $z = \dfrac{r^2}{a}$ gives

$$\dot{z} = \frac{2r\dot{r}}{a}, \qquad \ddot{z} = \frac{2r\ddot{r}}{a} + \frac{2\dot{r}^2}{a}. \tag{4}$$

Using Eqs. (3) and (4), we rewrite the total energy

$$E = \frac{1}{2}m(\dot{r}^2 + r^2\dot{\varphi}^2 + \dot{z}^2) + mgz,$$

which is conserved, as

$$\dot{r}^2 = \left(\frac{2E}{m} - \frac{J^2}{m^2r^2} - \frac{2gr^2}{a}\right)\left(1 + \frac{4r^2}{a^2}\right)^{-1}, \tag{5}$$

and Eq. (2) as

$$\frac{m}{a}(2r\ddot{r} + 2\dot{r}^2) + mg = a\lambda.$$

Making use of Eqs. (1) and (3), this becomes

$$a\lambda\left(1 + \frac{4r^2}{a^2}\right) = \frac{2m\dot{r}^2}{a} + mg + \frac{2J^2}{mar^2}.$$

Expression (5) then reduces it to

$$\lambda = \left(\frac{4E}{a^2} + \frac{8J^2}{ma^4} + \frac{mg}{a}\right)\left(1 + \frac{4r^2}{a^2}\right)^{-2} = \text{constant} \cdot \left(1 + \frac{4r^2}{a^2}\right)^{-2}.$$

The force of constraint is thus

$$\mathbf{f} = -2r\lambda\mathbf{e}_r + a\lambda\mathbf{e}_z,$$

of magnitude

$$f = a\lambda\sqrt{1 + \frac{4r^2}{a^2}}$$

$$= \text{constant} \cdot \left(1 + \frac{4r^2}{a^2}\right)^{-\frac{3}{2}}.$$

This force is in the *rz*-plane and is perpendicular to the inside surface of the paraboloid. (It makes an angle arctan $(-a/2r)$ with the *r*-axis while the slope of the parabola is $2r/a$).

b. As shown in Fig. 2.22, in spherical coordinates (r, θ, φ) an infinitesimal displacement of the particle can be resolved as

$$\delta\mathbf{r} = (\delta r, r\delta\theta, r\delta\varphi \sin\theta),$$

and its velocity as

$$\dot{\mathbf{r}} = (\dot{r}, r\dot{\theta}, r\dot{\varphi} \sin\theta).$$

1. Suppose the coordinate frame rotates with angular velocity ω about the *z*-axis. Then the velocity of the particle with respect to a fixed frame is

$$\mathbf{v}' = \dot{\mathbf{r}} + \boldsymbol{\omega} \times \mathbf{r},$$

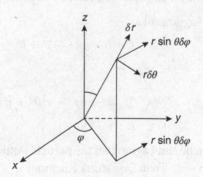

Fig. 2.22

so the kinetic energy of the particle is

$$T = \frac{1}{2}m[\dot{r}^2 + 2\dot{\mathbf{r}} \cdot \boldsymbol{\omega} \times \mathbf{r} + (\boldsymbol{\omega} \times \mathbf{r})^2].$$

Referring to the rotating frame and using spherical coordinates we have

$$\mathbf{r} = (r, 0, 0),$$
$$\boldsymbol{\omega} = (\omega \cos\theta, -\omega \sin\theta, 0),$$
$$\boldsymbol{\omega} \times \mathbf{r} = (0, 0, \omega r \sin\theta),$$
$$2\dot{\mathbf{r}} \cdot \boldsymbol{\omega} \times \mathbf{r} = 2\omega r^2 \dot{\varphi} \sin^2\theta,$$
$$(\boldsymbol{\omega} \times \mathbf{r})^2 = \omega^2 r^2 \sin^2\theta,$$
$$\dot{\mathbf{r}}^2 = \dot{r}^2 + r^2\dot{\varphi}^2 + r^2\dot{\varphi}^2 \sin^2\theta.$$

Hence

$$L = T - V$$
$$= \frac{1}{2} m(\dot{r}^2 + r^2\dot{\theta}^2 + r^2\dot{\varphi}^2 \sin^2\theta + 2\omega r^2\dot{\varphi} \sin^2\theta + \omega^2 r^2 \sin^2\theta) - V(r).$$

Note that this is the Lagrangian of the particle with respect to a fixed frame which is to be used in Lagrange's equations, using coordinates referring to the rotating frame.

2. The Lagrangian can be written as

$$L = \frac{1}{2} m(\dot{r}^2 + r^2\dot{\theta}^2 + r^2\dot{\varphi}^2 \sin^2\theta) - U - V$$

with

$$U = -\frac{1}{2} m(2\omega r^2\dot{\varphi} \sin^2\theta + \omega^2 r^2 \sin^2\theta).$$

Thus L has the form of the Lagrangian the particle would have if the coordinate frame referred to were fixed and the particle were under a potential $U + V$, i.e. with an additional velocity-dependent potential U.

3. Write the Lagrangian as

$$L = T' - U - V = L' - U,$$

where

$$T' = \frac{1}{2} m(\dot{r}^2 + r^2\dot{\theta}^2 + r^2\dot{\varphi}^2 \sin^2\theta),$$
$$L' = T' - V$$

are the kinetic and Lagrangian the particle would have if the coordinate frame referred to were fixed. Lagrange's equations

$$\frac{d}{dt}\left(\frac{\partial L}{\partial \dot{q}_i}\right) - \frac{\partial L}{\partial q_i} = 0$$

can be written as

$$\frac{d}{dt}\left(\frac{\partial L'}{\partial \dot{q}_i}\right) - \frac{\partial L'}{\partial q_i} = \frac{d}{dt}\left(\frac{\partial U}{\partial \dot{q}_i}\right) - \frac{\partial U}{\partial q_i} \equiv Q_i'.$$

Q_i' are the generalized forces that have to be introduced because of the fact that the frame referred to is rotating. Differentiating U we find

$$Q_r' = 2m\omega r\dot{\varphi}\sin^2\theta + m\omega^2 r\sin^2\theta,$$
$$Q_\theta' = 2m\omega r^2\dot{\varphi}\sin\theta\cos\theta + m\omega^2 r^2\sin\theta\cos\theta,$$
$$Q_\varphi' = -2m\omega r\dot{r}\sin^2\theta - 2m\omega r^2\dot{\theta}\sin\theta\cos\theta.$$

The generalized components Q_j' of a force $\mathbf{F}'$ are defined by

$$\mathbf{F}' \cdot \delta\mathbf{r} = \sum_j Q_j'\delta q_j,$$

i.e.

$$F_r\delta r + F_\theta r\delta\theta + F_\varphi r\sin\theta\delta\varphi = Q_r\delta r + Q_\theta\delta\theta + Q_\varphi\delta\varphi.$$

Hence

$$F_r = Q_r' = 2m\omega r\dot{\varphi}\sin^2\theta + m\omega^2 r\sin^2\theta,$$
$$F_\theta = \frac{Q_\theta'}{r} = 2m\omega r\dot{\varphi}\sin\theta\cos\theta + m\omega^2 r\sin\theta\cos\theta,$$
$$F_\varphi = \frac{Q_\varphi'}{r\sin\theta} = -2m\omega\dot{r}\sin\theta - 2m\omega r\dot{\theta}\cos\theta$$

are the components of the centrifugal and Coriolis forces in the directions of $\mathbf{e}_r, \mathbf{e}_\theta, \mathbf{e}_\varphi$. Note that the velocity-dependent terms are due to the Coriolis force while the remaining terms are due the centrifugal force.

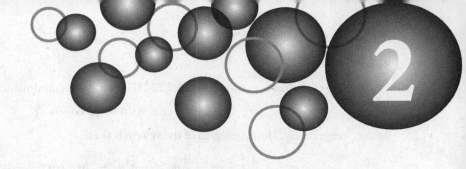

SMALL OSCILLATIONS (2028–2067)

2028

A mass M is constrained to slide without friction on the track AB as shown in Fig. 2.23. A mass m is connected to M by a massless inextensible string. (Make small angle approximation.)

   a. Write a Lagrangian for this system.

   b. Find the normal coordinates (and describe them).

   c. Find expressions for the normal coordinates as functions of time.

(Wisconsin)

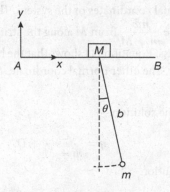

Fig. 2.23

Sol:

a. Use coordinates as shown in Fig. 2.23. M and m have coordinates

$$(x, 0), \quad (x + b \sin \theta, -b \cos \theta)$$

respectively. The Lagrangian of the system is then

$$L = T - V = \frac{1}{2}M\dot{x}^2 + \frac{1}{2}m(\dot{x}^2 + b^2\dot{\theta}^2 + 2b\dot{x}\dot{\theta} \cos \theta) + mgb \cos \theta.$$

b. For small oscillations, θ and $\dot{\theta}$ are small quantities and we have the approximat[e] Lagrangian

$$L = \frac{1}{2}M\dot{x}^2 + \frac{1}{2}m(\dot{x}^2 + b^2\dot{\theta}^2 + 2b\dot{x}\dot{\theta}) + mgb\left(1 - \frac{1}{2}\theta^2\right).$$

Lagrange's equations

$$\frac{d}{dt}\left(\frac{\partial L}{\partial \dot{q}_i}\right) - \frac{\partial L}{\partial q_i} = 0$$

then give $(m + M)\dot{x} + mb\dot{\theta} = C$, a constant, $\ddot{x} + b\ddot{\theta} + g\theta = 0$.
In the above, the first equation can be written as

$$(m + M)\dot{\eta} = C \tag{1}$$

by setting

$$\eta = x + \frac{mb\theta}{m + M}.$$

As $(m + M)\ddot{x} + mb\ddot{\theta} = 0$, the second equation can be written as

$$\frac{Mb\ddot{\theta}}{m + M} + g\theta = 0. \tag{2}$$

The two new equations of motion are now independent of each other. Hence η and θ are the normal coordinates of the system. The center of mass of the system occurs at a distance $\dfrac{mb}{m + M}$ from M along the string. Hence η is the x-coordinate of the center of mass. Equation (1) shows that the horizontal motion of the center of mass is uniform. The other normal coordinate, θ, is the angle the string makes with the vertical.

c. Equation (1) has the solution

$$\eta = \frac{Ct}{m + M} + D,$$

and Eq. (2) has solution

$$\theta = A \cos(\omega t + B),$$

where

$$\omega = \sqrt{\frac{(m + M)g}{Mb}}$$

is the angular frequency of small oscillations of the string and A, B, C, D are constants.

2029

A simple pendulum is attached to a support which is driven horizontally with time as shown in

a. Set up the Lagrangian for the system in terms of the generalized coordinates θ and y, where θ is the angular displacement from equilibrium and $y(t)$ is the horizontal position of the pendulum support.

b. Find the equation of motion for θ.

c. For small angular displacements and a sinusoidal motion of the support

$$y = y_0\cos(\omega t).$$

Find the steady-state solution to the equation of motion.

(Wisconsin)

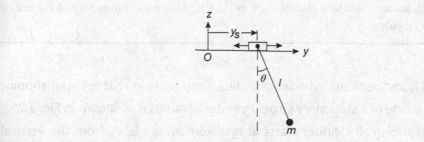

Fig. 2.24

Sol:

a. The mass m has coordinates

$$(y_s + l\sin\theta, -l\cos\theta)$$

and velocity

$$(\dot{y}_s + l\dot{\theta}\cos\theta, l\dot{\theta}\sin\theta).$$

Hence the Lagrangian is

$$L = T - V = \frac{m}{2}(\dot{y}_s^2 + l^2\dot{\theta}^2 + 2l\dot{y}_s\dot{\theta}\cos\theta) + mgl\cos\theta.$$

b. Lagrange's equation

$$\frac{d}{dt}\left(\frac{\partial L}{\partial \dot{\theta}}\right) - \frac{\partial L}{\partial \theta} = 0$$

gives

$$l\ddot{\theta} + \ddot{y}_s \cos\theta + g\sin\theta = 0.$$

c. For $y_s = y_0\cos(\omega t)$ and small θ, the above reduces to

$$\ddot{\theta} + \omega_0^2\theta = \frac{y_0}{l}\omega^2\cos(\omega t)$$

with $\omega_0 = \sqrt{\frac{g}{l}}$. A particular solution is obtained by putting $\theta = A\cos(\omega t)$. Substitution gives

$$A = \frac{y_0\omega^2}{l(\omega_0^2 - \omega^2)}.$$

The general solution is then

$$\theta = \frac{y_0\omega^2\cos(\omega t)}{l(\omega_0^2 - \omega^2)} + A\cos(\omega_0 t) + B\sin(\omega_0 t).$$

Resonance will take place if $\omega_0 \approx \omega$. As long as $\omega \neq \omega_0$, the motion of the system is steady.

2030

A solid homogeneous cylinder of radius r and mass m rolls without slipping on the inside of a stationary larger cylinder of radius R as shown in Fig. 2.25.

a. If the small cylinder starts at rest from an angle θ_0 from the vertical, what is the total downward force it exerts on the outer cylinder as it passes through the lowest point?

b. Determine the equation of motion of the inside cylinder using Lagrangian techniques.

c. Find the period of small oscillations about the stable equilibrium position.

(Wisconsin)

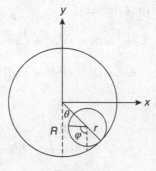

Fig. 2.25

Sol: Take coordinates as shown in Fig. 2.25. The center of mass of the rolling cylinder has coordinates

$$((R - r) \sin \theta, -(R - r) \cos \theta)$$

and velocity

$$((R - r)\dot{\theta} \cos \theta, (R - r)\dot{\theta} \sin \theta).$$

The cylinder has moment of inertia $\frac{1}{2}mr^2$ and the condition of rolling without slipping means

$$(R - r)\theta = r\varphi.$$

a. Initially $\dot{\theta} = 0$ at $\theta = \theta_0$. Suppose the cylinder has velocity v when it passes through the lowest point $\theta = 0$. Conservation of the total energy $T+V$ gives

$$\frac{1}{2}mv^2 + \frac{1}{4}mr^2\dot{\varphi}^2 - mg(R - r) = -mg(R - r)\cos \theta_0,$$

or, with $r\dot{\varphi} = (R - r)\dot{\theta}, v = (R - r)\dot{\theta}$,

$$mv^2 = \frac{4}{3}(R - r)(1 - \cos \theta_0)mg.$$

The force exerted by the cylinder on the outer cylinder as it passes through the lowest point is vertically downward and has magnitude

$$mg + m(R - r)\dot{\theta}^2 = mg + \frac{mv^2}{(R - r)}$$

$$= mg + \frac{4}{3}(1 - \cos \theta_0)mg$$

$$= \frac{1}{3}mg(7 - 4\cos \theta_0).$$

b. The Lagrangian of the cylinder is

$$L = T - V = \frac{1}{2}m(R - r)^2\dot{\theta}^2 + \frac{1}{4}mr^2\dot{\varphi}^2 + mg(R - r)\cos\theta$$

$$= \frac{3}{4}m(R - r)^2\dot{\theta}^2 + mg(R - r)\cos\theta.$$

Lagrange's equation

$$\frac{d}{dt}\left(\frac{\partial L}{\partial\dot{\theta}}\right) - \frac{\partial L}{\partial\theta} = 0$$

gives

$$\ddot{\theta} + \frac{2}{3}\left(\frac{g}{R - r}\right)\sin\theta = 0.$$

c. For small oscillations about the equilibrium position $\theta = 0$, the equation of motion reduces to

$$\ddot{\theta} + \frac{2}{3}\left(\frac{g}{R - r}\right)\theta = 0.$$

This has the form of the equation for simple harmonic motion. Hence the equilibrium is stable and has period

$$T = \frac{2\pi}{\omega} = \pi\sqrt{\frac{6(R - r)}{g}}.$$

2031

A bead of mass m is constrained to move on a hoop of radius b. The hoop rotates with constant angular velocity ω around a vertical axis which coincides with a diameter of the hoop.

a. Set up the Lagrangian and obtain equations of motion of the bead.

b. Find the critical angular velocity Ω below which the bottom of the hoop provides a stable equilibrium position for the bead.

c. Find the stable equilibrium position for $\omega > \Omega$.

(Wisconsin)

Sol:

a. Use a rotating frame attached to the hoop as shown in Fig. 2.26. The mass m has coordinates $(b\sin\theta, b\cos\theta)$ and velocity $(b\dot{\theta}\cos\theta, -b\dot{\theta}\sin\theta)$ referring to the rotating frame. In addition to the potential $mgb \cos \theta$ due to gravity, a potential due to a fictitious centrifugal force $mx\omega^2$ has to be introduced. As

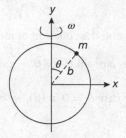

Fig. 2.26

$$mx\omega^2 = -\frac{\partial U}{\partial x},$$

we can take

$$U = -\frac{1}{2}m\omega^2 x^2 = -\frac{1}{2}m\omega^2 b^2 \sin^2 \theta.$$

Hence

$$L = T - U - V = \frac{1}{2}mb^2(\dot{\theta}^2 + \omega^2 \sin^2 \theta) - mgb \cos \theta.$$

Lagrange's equation

$$\frac{d}{dt}\left(\frac{\partial L}{\partial \dot{\theta}}\right) - \frac{\partial L}{\partial \theta} = 0$$

then gives

$$b\ddot{\theta} - b\omega^2 \sin \theta \cos\theta - g \sin \theta = 0.$$

b. At the bottom of the hoop, $\theta = \pi$. Let $\theta = \pi + \alpha$, where α is a small quantity. As

$$\sin \theta = \sin(\pi + \alpha) = -\sin \alpha \approx -\alpha,$$
$$\cos \theta = \cos(\pi + \alpha) = -\cos \alpha \approx -1,$$

the equation of motion becomes

$$\ddot{\alpha} + \left(\frac{g}{b} - \omega^2\right)\alpha = 0.$$

For α to oscillate about the equilibrium position, i.e. for the equilibrium to be stable, we require

$$\frac{g}{b} - \omega^2 > 0, \quad \text{i.e.} \quad \omega < \sqrt{\frac{g}{b}}.$$

Hence for stable equilibrium, ω must be smaller than a critical angular frequenc

$$\Omega = \sqrt{\frac{g}{b}}.$$

c.   At equilibrium, $\ddot{\theta} = 0$ and the equation of motion becomes

$$b\omega^2 \sin\theta \cos\theta + g\sin\theta = 0.$$

Having considered the case $\theta = 0$ in (b), we can take $\sin\theta \neq 0$ and so the abov
gives

$$\cos\theta_0 = -\frac{g}{b\omega^2}$$

for the other equilibrium position.

To test the stability of this equilibrium, let $\beta = \theta - \theta_0$, where β is a small quan
tity. As

$$\sin\theta = \sin(\theta_0 + \beta) \approx \sin\theta_0 + \beta\cos\theta_0,$$
$$\cos\theta = \cos(\theta_0 + \beta) \approx \cos\theta_0 - \beta\sin\theta_0,$$

the equation of motion becomes

$$b\ddot{\beta} - b\omega^2 \sin\theta_0 \cos\theta_0 - b\omega^2(\cos^2\theta_0 - \sin^2\theta_0)\beta - g\sin\theta_0 - g\beta\cos\theta_0 = 0,$$

i.e.

$$\ddot{\beta} - \omega^2(2\cos^2\theta_0 - 1)\beta - \frac{g}{b}\beta\cos\theta_0 = 0,$$

or, using the value of $\cos\theta_0$,

$$\ddot{\beta} + \left(1 - \frac{g^2}{b^2\omega^2}\right)\beta = 0.$$

Hence the equilibrium is stable since as $\omega > \Omega$, $1 - \dfrac{g^2}{b^2\omega^2} > 0$.

2032

Consider the longitudinal motion of the system of masses and springs illus-
trated in Fig. 2.27, with $M > m$.

a.   What are the normal-mode frequencies of the system?

b.   If the left-hand mass receives an impulse P_0 at $t = 0$, find the motion of
the left-hand mass as a function of time.

c. If, alternatively, the middle mass is driven harmonically at a frequency $\omega_0 = 2\sqrt{\dfrac{k}{m}}$, will it move in or out of phase with the driving motion? Explain.

(*Wisconsin*)

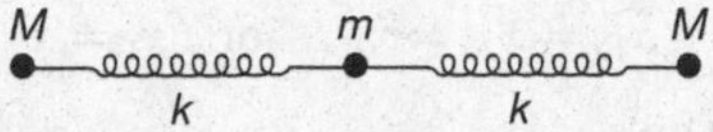

Fig. 2.27

Sol:

a. Let x_1, x_2, x_3 be the displacements of the three masses, counting from the left, from their equilibrium positions. The Lagrangian of the system is

$$L = T - V = \frac{1}{2}M\dot{x}_1^2 + \frac{1}{2}m\dot{x}_2^2 + \frac{1}{2}M\dot{x}_3^2 - \frac{1}{2}k(x_2 - x_1)^2 - \frac{1}{2}k(x_3 - x_2)^2.$$

Lagrange's equations

$$\frac{d}{dt}\left(\frac{\partial L}{\partial \dot{q}_i}\right) - \frac{\partial L}{\partial q_i} = 0$$

then give

$$\begin{aligned}
M\ddot{x}_1 + k(x_1 - x_2) &= 0, \\
m\ddot{x}_2 + k(x_2 - x_1) + k(x_2 - x_3) &= 0, \qquad (1) \\
M\ddot{x}_3 + k(x_3 - x_2) &= 0.
\end{aligned}$$

Try a solution of the type

$$x_i = x_{i0}e^{i\omega t}.$$

Substitution gives

$$\begin{aligned}
(k - \omega^2 M)x_{10} - kx_{20} &= 0, \\
-kx_{10} + (2k - \omega^2 m)x_{20} - kx_{30} &= 0, \qquad (2) \\
-kx_{20} + (k - \omega^2 M)x_{30} &= 0.
\end{aligned}$$

For a solution where not all amplitudes vanish, we require

$$\begin{vmatrix}
k - \omega^2 M & -k & 0 \\
-k & 2k - \omega^2 m & -k \\
0 & -k & k - \omega^2 M
\end{vmatrix} = 0,$$

which has solutions

$$\omega = 0, \quad \pm\sqrt{\frac{k}{M}}, \quad \pm\sqrt{\frac{k}{m}\left(1 + \frac{2M}{m}\right)}.$$

Hence the system has three normal-mode (angular) frequencies

$$\omega_1 = 0, \quad \omega_2 = \sqrt{\frac{k}{M}}, \quad \omega_3 = \sqrt{\frac{k}{m}\left(1 + \frac{2M}{m}\right)}.$$

b. For $\omega = \omega_1$, Eqs. (2) give

$$x_{10} = x_{20} = x_{30}, \quad \text{or} \quad x_1 = x_2 = x_3.$$

Equations (1) then give

$$x_1 = x_2 = x_3 = at + b,$$

where a, b are constants, showing that in this mode the three masses undergo translation as a rigid body without oscillation.

For $\omega = \omega_2$, Eqs. (2) give

$$x_2 = 0, \quad x_3 = -x_1,$$

and Eqs. (1) give

$$\ddot{x}_1 + \omega_2^2 x_1 = 0, \quad \ddot{x}_3 + \omega_2^2 x_3 = 0.$$

The solutions are then

$$x_1 = A\sin(\omega_2 t) + B\cos(\omega_2 t),$$
$$x_2 = 0,$$
$$x_3 = -x_1.$$

In this mode the middle mass stays stationary while the two end masses oscillate harmonically exactly out of phase with each other.

For $\omega = \omega_3$, we have, similarly,

$$x_1 = C\sin(\omega_3 t) + D\cos(\omega_3 t),$$
$$x_2 = -\frac{2Mx_1}{m},$$
$$x_3 = x_1.$$

Here the two outer masses oscillate with the same amplitude and phase, while the inner one oscillates out of phase and with a different amplitude.

The general longitudinal motion of the system is some linear combination of the normal-modes:

$$x_1 = at + b + A\sin(\omega_2 t) + B\cos(\omega_2 t) + C\sin(\omega_3 t) + D\cos(\omega_3 t),$$
$$x_2 = at + b - \frac{2M}{m}[C\sin(\omega_3 t) + D\cos(\omega_3 t)],$$
$$x_3 = at + b - A\sin(\omega_2 t) - B\cos(\omega_2 t) + C\sin(\omega_3 t) + D\cos(\omega_3 t),$$

The initial conditions that at $t = 0$,

$$x_1 = x_2 = x_3 = 0, \quad \dot{x}_1 = \frac{P_0}{m}, \quad \dot{x}_2 = \dot{x}_3 = 0$$

then give

$$a = \frac{P_0}{m + 2M},$$

$$A = \frac{P_0}{2M\omega_2},$$

$$C = \frac{P_0 m}{2M(m + 2M)\omega_3},$$

$$b = B = D = 0.$$

Hence the motion of the left-hand mass is given by

$$x_1 = P_0 \left[\frac{t}{m + 2M} + \frac{\sin(\omega_2 t)}{2M\omega_2} + \frac{m \sin(\omega_3 t)}{2M(m + 2M)\omega_3} \right].$$

c. Suppose the middle mass has motion given by

$$x_2 = x_{20} \sin(\omega_0 t).$$

The first equation of (1) now becomes

$$\ddot{x}_1 + \omega_2^2 x_1 = \omega_2^2 x_{20} \sin(\omega_0 t).$$

In steady state x_1 moves with the same frequency as the driving motion:

$$x_1 = x_{10} \sin(\omega_0 t).$$

Substitution in the above gives

$$x_1 = \left(\frac{\omega_2^2}{\omega_2^2 - \omega_0^2} \right) x_{20} \sin(\omega_0 t) = \left(\frac{m}{m - 4M} \right) x_{20} \sin(\omega_0 t)$$

As $m - 4M < 0$, the left-hand mass will move out of phase with the driving motion.

2033

Two pendulums of equal length *l* and equal mass *m* are coupled by a massless spring of constant *k* as shown in Fig. 2.28. The unstretched length of the spring is equal to the distance between the supports.

a. Set up the exact Lagrangian in terms of appropriate generalized coordinates and velocities.

b. Find the normal coordinates and frequencies of small vibrations about equilibrium.

c. Suppose that initially the two masses are at rest. An impulsive force gives a horizontal velocity v toward the right to the mass on the left. What is the motion of the system in terms of the normal coordinates?

(*Wisconsin*)

Sol:

a. Assume the masses are constrained to move in a vertical plane. Let the distance between the two supports be d, which is also the unstretched length of the spring and use Cartesian coordinates as shown in Fig. 2.28. The masses have coordinates

$$(l \sin \theta_1, -l \cos \theta_1), \quad (d + l \sin \theta_2, -l \cos \theta_2)$$

and velocities

$$(l \dot\theta_1 \cos \theta_1, l \dot\theta_1 \sin \theta_1), \quad (l \dot\theta_2 \cos \theta_2, l \dot\theta_2 \sin \theta_2),$$

respectively. The length of the spring is the distance between the two masses:

$$\sqrt{(d + l \sin \theta_2 - l \sin \theta_1)^2 + (l \cos \theta_2 - l \cos \theta_1)^2}.$$

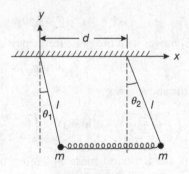

Fig. 2.28

Hence the Lagrangian of the system is

$$L = T - V = \frac{1}{2}ml^2(\dot\theta_1^2 + \dot\theta_2^2) + mgl(\cos \theta_1 + \cos \theta_2)$$

$$- \frac{1}{2}k(\sqrt{d^2 + 2dl(\sin \theta_2 - \sin \theta_1) + 2l^2 - 2l^2 \cos(\theta_2 - \theta_1)} - d)^2.$$

b. As

$$-\frac{\partial L}{\partial \theta_1}$$

$$= mgl \sin \theta_1 - k(\sqrt{d^2 + 2dl(\sin \theta_2 - \sin \theta_1) + 2l^2 - 2l^2 \cos(\theta_2 - \theta_1)} - d)$$

$$\times \frac{dl \cos \theta_1 + l^2 \sin(\theta_2 - \theta_1)}{\sqrt{d^2 + 2dl(\sin \theta_2 - \sin \theta_1) + 2l^2 - 2l^2 \cos(\theta_2 - \theta_1)}}$$

$$\approx mgl\theta_1 - kl\left[\sqrt{d^2 + 2dl(\theta_2 - \theta_1)} - d\right] \times \frac{d + l(\theta_2 - \theta_1)}{\sqrt{d^2 + 2dl(\theta_2 - \theta_1)}}$$

$$\approx mgl\theta_1 - kl\left(1 - \frac{1}{\sqrt{1 + \dfrac{2l(\theta_2 - \theta_1)}{d}}}\right)[d + l(\theta_2 - \theta_1)]$$

$$\approx mgl\theta_1 - kl\left[\frac{l(\theta_2 - \theta_1)}{d}\right][d + l(\theta_2 - \theta_1)]$$

$$\approx mgl\theta_1 - kl^2(\theta_2 - \theta_1),$$

neglecting second and higher order terms in θ_1, θ_2 which are small quantities. Similarly,

$$-\frac{\partial L}{\partial \theta_2} = mgl\theta_2 + kl^2(\theta_2 - \theta_1).$$

Thus the equations of motion for small oscillations are

$$\ddot{\theta}_1 + \frac{g\theta_1}{l} - \frac{k(\theta_2 - \theta_1)}{m} = 0,$$

$$\ddot{\theta}_2 + \frac{g\theta_2}{l} + \frac{k(\theta_2 - \theta_1)}{m} = 0.$$

Let

$$\eta = \frac{1}{2}(\theta_1 + \theta_2), \quad \xi = \frac{1}{2}(\theta_1 - \theta_2)$$

and the above give

$$\ddot{\eta} + \frac{g\eta}{l} = 0,$$

$$\ddot{\xi} + \left(\frac{g}{l} + \frac{2k}{m}\right)\xi = 0.$$

These show that η and ξ are the normal coordinates with the corresponding normal (angular) frequencies

$$\omega_1 = \sqrt{\frac{g}{l}}, \quad \omega_2 = \sqrt{\frac{g}{l} + \frac{2k}{m}}.$$

c. The solutions of the equations of motion in the normal coordinates are

$$\eta = A\cos(\omega_1 t) + B\sin(\omega_1 t),$$
$$\xi = C\cos(\omega_2 t) + D\sin(\omega_2 t).$$

At $t = 0$, $\theta_1 = \theta_2 = 0$, giving $\eta = \xi = 0$; and $\dot{\theta}_1 = \dfrac{v}{l}$, $\dot{\theta}_2 = 0$, giving $\dot{\eta} = \dot{\xi} = \dfrac{v}{2}$. Thus

$$A = C = 0, \quad B = \frac{v}{2l\omega_1}, \quad D = \frac{v}{2l\omega_2},$$

and

$$\eta = \frac{v \sin(\omega_1 t)}{2l\omega_1}, \quad \xi = \frac{v \sin(\omega_2 t)}{2l\omega_2},$$

giving the motion of the system in terms of the normal coordinates.

2034

Four identical masses are connected by four identical springs and constrained to move on a frictionless circle of radius b as shown in Fig. 2.29.

a. How many normal-modes of small oscillations are there?

b. What are the frequencies of small oscillations?

(Wisconsin)

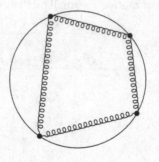

Fig. 2.29

Sol:

a. Take the lengths of arc s_1, s_2, s_3, and s_4 of the four masses from their initial equilibrium positions as the generalized coordinates. The kinetic energy of the system is

$$T = \frac{1}{2}m(\dot{s}_1^2 + \dot{s}_2^2 + \dot{s}_3^2 + \dot{s}_4^2).$$

As the springs are identical, at equilibrium the four masses are positioned symmetrically on the circle, i.e. the arc between two neighboring masses, the nth and the $(n+1)$th, subtends an angle $\dfrac{\pi}{2}$ at the center. When the neighboring masses are displaced from the equilibrium positions, the spring connecting them will extend by

$$2b \sin\left[\frac{1}{2}\left(\frac{s_{n+1}}{b} - \frac{s_n}{b} + \frac{\pi}{2}\right)\right] - 2b \sin\frac{\pi}{4} \approx \frac{1}{\sqrt{2}}(s_{n+1} - s_n),$$

for small oscillations for which s_n are small.

Thus the potential energy is

$$V = \frac{k}{2}(s_1^2 + s_2^2 + s_3^2 + s_4^2 - s_1 s_2 - s_2 s_3 - s_3 s_4 - s_4 s_1).$$

This system has four degrees of freedom and hence four normal-modes.

b. The T and V matrices are

$$T = \begin{pmatrix} m & 0 & 0 & 0 \\ 0 & m & 0 & 0 \\ 0 & 0 & m & 0 \\ 0 & 0 & 0 & m \end{pmatrix}, \quad V = \begin{pmatrix} k & -\dfrac{k}{2} & 0 & -\dfrac{k}{2} \\ -\dfrac{k}{2} & k & -\dfrac{k}{2} & 0 \\ 0 & -\dfrac{k}{2} & k & -\dfrac{k}{2} \\ -\dfrac{k}{2} & 0 & -\dfrac{k}{2} & k \end{pmatrix},$$

so the secular equation is

$$|V - \omega^2 T| = \begin{vmatrix} k - m\omega^2 & -\dfrac{k}{2} & 0 & -\dfrac{k}{2} \\ -\dfrac{k}{2} & k - m\omega^2 & -\dfrac{k}{2} & 0 \\ 0 & -\dfrac{k}{2} & k - m\omega^2 & -\dfrac{k}{2} \\ -\dfrac{k}{2} & 0 & -\dfrac{k}{2} & k - m\omega^2 \end{vmatrix} = 0,$$

which has four roots $0, 0, \sqrt{\dfrac{k}{m}}, \sqrt{\dfrac{2k}{m}}$. Hence the angular frequencies of small oscillations are $\sqrt{\dfrac{k}{m}}$ and $\sqrt{\dfrac{2k}{m}}$.

2035

A simple pendulum of length $4l$ and mass m is hung from another simple pendulum of length $3l$ and mass m. It is possible for this system to perform small oscillations about equilibrium such that a point on the lower pendulum undergoes no horizontal displacement. Locate that point.

(Wisconsin)

Sol: Use Cartesian coordinates as shown in Fig. 2.30. The upper and lower masses have, respectively, coordinates

$$(3l \sin \theta_1, -3l \cos \theta_1),$$
$$(3l \sin \theta_1 + 4l \sin \theta_2, -3l \cos \theta_1 - 4l \cos \theta_2)$$

and velocities

$$(3l\dot{\theta}_1 \cos \theta_1, 3l\dot{\theta}_1 \sin \theta_1),$$
$$(3l\dot{\theta}_1 \cos \theta_1 + 4l\dot{\theta}_2 \cos \theta_2, 3l\dot{\theta}_1 \sin \theta_1 + 4l\dot{\theta}_2 \sin \theta_2).$$

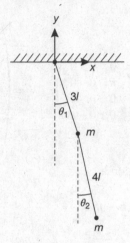

Fig. 2.30

The Lagrangian of the system is then

$$L = T - V = \frac{1}{2}m[18l^2\dot{\theta}_1^2 + 16l^2\dot{\theta}_2^2 + 24l^2\dot{\theta}_1\dot{\theta}_2\cos(\theta_1 - \theta_2)]$$
$$+ mg(6l\cos\theta_1 + 4l\cos\theta_2).$$

Lagrange's equations

$$\frac{d}{dt}\left(\frac{\partial L}{\partial \dot{q}_i}\right) - \frac{\partial L}{\partial q_i} = 0$$

give

$$3\ddot{\theta}_1 + 2\ddot{\theta}_2\cos(\theta_1 - \theta_2) + 2\dot{\theta}_2^2\sin(\theta_1 - \theta_2) + \frac{g\sin\theta_1}{l} = 0,$$

or, retaining only first order terms for small oscillations,

$$3\ddot{\theta}_1 + 2\ddot{\theta}_2 + \frac{g\theta_1}{l} = 0,$$

and, similarly,

$$3\ddot{\theta}_1 + 4\ddot{\theta}_2 + \frac{g\theta_2}{l} = 0.$$

Try $\theta_1 = \theta_{10}e^{i\omega t}$, $\theta_2 = \theta_{20}e^{i\omega t}$. The above equations give

$$\left(\frac{g}{l} - 3\omega^2\right)\theta_{10} - 2\omega^2\theta_{20} = 0,$$

$$-3\omega^2\theta_{10} + \left(\frac{g}{l} - 4\omega^2\right)\theta_{20} = 0.$$

The secular equation

$$\begin{vmatrix} \frac{g}{l} - 3\omega^2 & -2\omega^2 \\ -3\omega^2 & \frac{g}{l} - 4\omega^2 \end{vmatrix} = \left(\frac{g}{l} - \omega^2\right)\left(\frac{g}{l} - 6\omega^2\right) = 0$$

has roots

$$\omega = \pm\sqrt{\frac{g}{l}}, \quad \pm\sqrt{\frac{g}{6l}}.$$

Hence there are two normal-mode frequencies. For

$$\omega_1 = \sqrt{\frac{g}{l}}, \quad \theta_{20} = -\theta_{10} \quad \text{or} \quad \theta_2 = -\theta_1;$$

for

$$\omega_2 = \sqrt{\frac{g}{6l}}, \quad \theta_{20} = \frac{3}{2}\theta_{10} \quad \text{or} \quad \theta_2 = \frac{3}{2}\theta_1.$$

The general small oscillations are a linear combination of the two normal-modes. A point on the lower pendulum at distance ξ from the upper mass has x-coordinate $3l\sin\theta_1 + \xi\sin\theta_2$ and thus x-component velocity

$$\dot{x} = 3l\dot{\theta}_1\cos\theta_1 + \xi\dot{\theta}_2\cos\theta_2 \approx 3l\dot{\theta}_1 + \xi\dot{\theta}_2.$$

For it to have no horizontal displacement, $\dot{x} = 0$. For the ω_1 mode, $\dot{\theta}_2 = -\dot{\theta}_1$, this requires

$$(3l - \xi)\dot{\theta}_1 = 0, \quad \text{or} \quad \xi = 3l.$$

For the ω_2 mode, $\dot{\theta}_2 = \frac{3}{2}\dot{\theta}_1$, $\dot{x} = 0$ would require

$$3\left(l + \frac{\xi}{2}\right)\dot{\theta}_1 = 0.$$

As ξ is positive this is not possible unless $\dot\theta_1 = 0$, i.e. there is no motion. Therefor when the system undergoes small oscillations with angular frequency $\sqrt{\frac{g}{l}}$, a point o the lower pendulum at distance $3l$ from the upper mass has no horizontal displacemen

---

2036

a. Find the Lagrangian equations of motion for the coplanar double oscillator shown in Fig. 2.31 in the vibration limit, assuming massles strings or connecting rods. From them find the normal frequencies o the system.

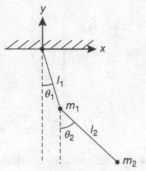

Fig. 2.31

b. Now consider a simple pendulum of mass *m*, again in the small-vibration limit. Suppose the string of length *l* is shortened very slowly (by being pulled up through a frictionless hole in the support as shown in Fig. 2.32), so that the fractional change in *l* over one period is small. How does the amplitude of vibration of *m* vary with *l*?

(Wisconsin)

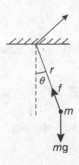

Fig. 2.32

Sol:

a. The coordinates of m_1, m_2 are

$$(l_1 \sin \theta_1, -l_1 \cos \theta_1),$$
$$(l_1 \sin \theta_1 + l_2 \sin \theta_2, -l_1 \cos \theta_1 - l_2 \cos \theta_2)$$

and their velocities are

$$(l_1 \dot{\theta}_1 \cos \theta_1, l_1 \dot{\theta}_1 \sin \theta_1),$$
$$(l_1 \dot{\theta}_1 \cos \theta_1 + l_2 \dot{\theta}_2 \cos \theta_2, l_1 \dot{\theta}_1 \sin \theta_1 + l_2 \dot{\theta}_2 \sin \theta_2)$$

respectively. The Lagrangian of the system is then

$$
\begin{aligned}
L = T - V &= \frac{1}{2} m_1 [l_1^2 \dot{\theta}_1^2 + l_2^2 \dot{\theta}_2^2 + 2 l_1 l_2 \dot{\theta}_1 \dot{\theta}_2 \cos(\theta_1 - \theta_2)] \\
&\quad + m_1 g l_1 \cos \theta_1 + m_2 g (l_1 \cos \theta_1 + l_2 \cos \theta_2) \\
&\approx \frac{1}{2}(m_1 + m_2) l_1^2 \dot{\theta}_1^2 + \frac{1}{2} m_2 (l_2^2 \dot{\theta}_2^2 + 2 l_1 l_2 \dot{\theta}_1 \dot{\theta}_2) \\
&\quad + (m_1 + m_2) g l_1 \left(1 - \frac{\theta_1^2}{2} \right) + m_2 g l_2 \left(1 - \frac{\theta_2^2}{2} \right),
\end{aligned}
$$

neglecting terms higher than second order in the small quantities θ_1, θ_2 in the small vibration limit. Lagrange's equations

$$\frac{d}{dt}\left(\frac{\partial L}{\partial \dot{q}_i} \right) - \frac{\partial L}{\partial q_i} = 0$$

then give

$$\ddot{\theta}_1 + \left(\frac{m_2}{m_1 + m_2} \right) \frac{l_2}{l_1} \ddot{\theta}_2 + \frac{g}{l_1} \theta_1 = 0,$$

$$\ddot{\theta}_1 + \frac{l_2}{l_1} \ddot{\theta}_2 + \frac{g}{l_1} \theta_2 = 0.$$

Let $\theta_1 = \theta_{10} e^{i\omega t}$, $\theta_2 = \theta_{20} e^{i\omega t}$ and obtain the secular equation

$$
\begin{vmatrix}
\dfrac{g}{l_1} - \omega^2 & -\dfrac{m_2 l_2 \omega^2}{(m_1 + m_2) l_1} \\[2ex]
-\omega^2 & \dfrac{g - l_2 \omega^2}{l_1}
\end{vmatrix} = 0,
$$

or

$$\left(\frac{m_1}{m_1 + m_2} \right) l_1 l_2 \omega^4 - g(l_1 + l_2)\omega^2 + g^2 = 0.$$

The normal frequencies ω_1, ω_2 are given by the solutions of this equation:

$$
\begin{aligned}
\left.\begin{array}{c} \omega_1^2 \\ \omega_2^2 \end{array}\right\} &= \frac{g}{2 m_1 l_1 l_2} \\
&\times \left[(m_1 + m_2)(l_2 + l_2) \pm \sqrt{(m_1 + m_2)^2 (l_1 + l_2)^2 - 4(m_1 + m_2) m_1 l_1 l_2} \right].
\end{aligned}
$$

b. As shown in Fig. 2.32. The forces on m are the tension f in the string and the grav‐ity mg. These provide for the centripetal force:

$$f - mg \cos \theta = mr\dot{\theta}^2.$$

When the string is shortened by dr, the work done by f is

$$dW = \mathbf{f} \cdot d\mathbf{r} = -fdr$$
$$\approx -mgdr + \left(\frac{1}{2}mg\theta^2 - mr\dot{\theta}^2 \right) dr$$
$$= -mgdr + dE,$$

where dE is the part relating to the oscillations, for small angle oscillations. As the change in r, the length of the string, is small over a period, we can take average

$$d\overline{E} = \left(\frac{1}{2}mg\overline{\theta^2} - mr\overline{\dot{\theta}^2} \right) dr.$$

Also, the vibration can be considered simple harmonic, i.e.

$$\theta = \theta_0 \cos (\omega t + \varphi),$$

where $\omega = \sqrt{\dfrac{g}{r}}$. Then if $T = \dfrac{2\pi}{\omega}$ is the period we have

$$\overline{\theta^2} = \frac{1}{T} \int_0^T \theta^2 dt = \frac{1}{2}\theta_0^2,$$

$$\overline{\dot{\theta}^2} = \frac{1}{T} \int_0^T \dot{\theta}^2 dt = \frac{\omega^2}{T} \int_0^T \theta^2 dt = \omega^2\overline{\theta^2},$$

i.e.

$$mr\overline{\dot{\theta}^2} = mg\overline{\theta^2}.$$

The energy of the pendulum is

$$\frac{1}{2}mr^2\dot{\theta}^2 - mgr \cos \theta \approx -mgr + \frac{1}{2}mr^2\dot{\theta}^2 + \frac{1}{2}mgr\theta^2,$$

so that

$$\overline{E} = \frac{1}{2}mr^2\overline{\dot{\theta}^2} + \frac{1}{2}mgr\overline{\theta^2} = mgr\overline{\theta^2}.$$

Hence

$$d\overline{E} = \left(\frac{1}{2}\overline{E} - \overline{E} \right)\frac{dr}{r},$$

or

$$\frac{d\overline{E}}{\overline{E}} = -\frac{dr}{2r}.$$

Integrating we have

$$\overline{E}r^{\frac{1}{2}} = \text{constant},$$

or

$$\theta_0^4 r^3 = \text{constant}.$$

Let the amplitudes at string lengths r, l be θ_r, θ_l respectively, then

$$\theta_r^4 = \frac{\theta_l^4 l^3}{r^3}.$$

2037

A particle in an isotropic three-dimensional harmonic oscillator potential has a natural angular frequency ω_0. Find its vibration frequencies if it is charged and is simultaneously acted on by uniform magnetic and electric fields. Discuss your result in the weak and strong field limits.

(Wisconsin)

Sol: Assume that the uniform magnetic and electric fields, B and E, are mutually perpendicular and take their directions as along the z- and x-axes respectively. Then as

$$B\mathbf{k} = \triangledown \times \mathbf{A}, \quad E\mathbf{i} = \triangledown\Phi,$$

we can take the vector and scalar potentials as

$$\mathbf{A} = \frac{1}{2}(-By\mathbf{i} + Bx\mathbf{j}), \quad \Phi = -Ex.$$

As the particle is an isotropic harmonic oscillator of natural angular frequency ω_0 and has charge e, say, its potential energy is

$$V = \frac{1}{2}m\omega_0^2 r^2 + e\Phi - e\dot{\mathbf{r}} \cdot \mathbf{A},$$

where $\mathbf{r} = (x, y, z)$ is the displacement of the particle from the origin, in SI units. Hence the Lagrangian is

$$L = T - V = \frac{1}{2}m(\dot{x}^2 + \dot{y}^2 + \dot{z}^2) - \frac{1}{2}m\omega_0^2(x^2 + y^2 + z^2)$$
$$+ eEx + \frac{1}{2}eB(-\dot{x}y + x\dot{y}).$$

Lagrange's equations

$$\frac{d}{dt}\left(\frac{\partial L}{\partial \dot{q}_i}\right) - \frac{\partial L}{\partial q_i} = 0.$$

then give

$$\ddot{x} + \omega_0^2 x - \frac{eB\dot{y}}{m} - \frac{eE}{m} = 0,$$

$$\ddot{y} + \omega_0^2 y + \frac{eB\dot{x}}{m} = 0,$$

$$\ddot{z} + \omega_0^2 z = 0.$$

The last equation shows that the vibration in the z-direction takes place with the natural angular frequency ω_0. Letting $x = x' + \dfrac{eE}{m\omega_0^2}$, the first two equations become

$$\ddot{x}' + \omega_0^2 x' - \frac{eB\dot{y}}{m} = 0,$$

$$\ddot{y} + \omega_0^2 y + \frac{eB\dot{x}'}{m} = 0.$$

Try a solution of the type

$$x' = A'e^{i\omega t}, \quad y = B'e^{i\omega t}$$

and we obtain the matrix equation

$$\begin{pmatrix} \omega_0^2 - \omega^2 & -\dfrac{ieB\omega}{m} \\ \dfrac{ieB\omega}{m} & \omega_0^2 - \omega^2 \end{pmatrix} \begin{pmatrix} A' \\ B' \end{pmatrix} = 0.$$

The secular equation

$$\begin{vmatrix} \omega_0^2 - \omega^2 & -\dfrac{ieB\omega}{m} \\ \dfrac{ieB\omega}{m} & \omega_0^2 - \omega^2 \end{vmatrix} = (\omega_0^2 - \omega^2)^2 - \left(\frac{eB\omega}{m}\right)^2 = 0$$

then gives

$$\omega^2 \pm \frac{eB\omega}{m} - \omega_0^2 = 0,$$

which has two positive roots

$$\omega_+ = \frac{1}{2}\left(\frac{eB}{m} + \sqrt{\left(\frac{eB}{m}\right)^2 + 4\omega_0^2}\right),$$

$$\omega_- = \frac{1}{2}\left(-\frac{eB}{m} + \sqrt{\left(\frac{eB}{m}\right)^2 + 4\omega_0^2}\right).$$

Hence the three normal-mode angular frequencies are ω_0, ω_+ and ω_-. Note that the last two modes of oscillations are caused by the magnetic field alone, whereas the electric field only causes a displacement $\dfrac{eE}{m\omega_0^2}$ along its direction.

For weak fields, $\dfrac{eB}{m} \ll \omega_0$, we have

$$\omega_+ = \omega_0 + \frac{eB}{2m}, \quad \omega_- = \omega_0 - \frac{eB}{2m}.$$

For strong fields, $\dfrac{eB}{m} \gg \omega_0$, we have

$$\omega_+ \approx \frac{1}{2}\left[\frac{eB}{m} + \frac{eB}{m}\left(1 + \frac{2m^2\omega_0^2}{e^2B^2}\right)\right]$$

$$= \frac{eB}{m} + \frac{m\omega_0^2}{eB},$$

$$\omega_- \approx \frac{1}{2}\left[-\frac{eB}{m} + \frac{eB}{m}\left(1 + \frac{2m^2\omega_0^2}{e^2B^2}\right)\right]$$

$$= \frac{m\omega_0^2}{eB}.$$

2038

Three particles of equal mass m move without friction in one dimension. Two of the particles are each connected to the third by a massless spring of spring constant k. Find normal-modes of oscillation and their corresponding frequencies.

(CUSPEA)

Sol: Number the masses from the left as shown in Fig. 2.33 and let x_1, x_2, x_3 be the displacements of the respective masses from their equilibrium positions. The Lagrangian of the system is

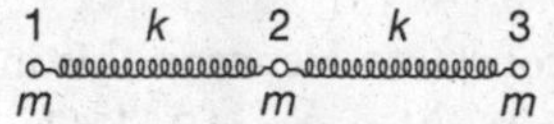

Fig. 2.33

$$L = T - V = \frac{1}{2}m(\dot{x}_1^2 + \dot{x}_2^2 + \dot{x}_3^2) - \frac{1}{2}k[(x_2 - x_1)^2 + (x_3 - x_2)^2].$$

Lagrange's equations

$$\frac{d}{dt}\left(\frac{\partial L}{\partial \dot{q}_i}\right) - \frac{\partial L}{\partial q_i} = 0$$

give

$$
\begin{aligned}
m_1 \ddot{x}_1 + k(x_1 - x_2) &= 0, \\
m\ddot{x}_2 + k(x_2 - x_1) + k(x_2 - x_3) &= 0, \\
m\ddot{x}_3 + k(x_3 - x_2) &= 0.
\end{aligned}
\tag{1}
$$

Trying a solution of the type

$$x_1 = Ae^{i\omega t}, \quad x_2 = Be^{i\omega t}, \quad x_3 = Ce^{i\omega t},$$

we can write the above as a matrix equation

$$
\begin{pmatrix}
k - m\omega^2 & -k & 0 \\
-k & 2k - m\omega^2 & -k \\
0 & -k & k - m\omega^2
\end{pmatrix}
\begin{pmatrix}
A \\ B \\ C
\end{pmatrix} = 0.
\tag{2}
$$

The secular equation

$$
\begin{vmatrix}
k - m\omega^2 & -k & 0 \\
-k & 2k - m\omega^2 & -k \\
0 & -k & k - m\omega^2
\end{vmatrix}
= m\omega^2 (k - m\omega^2)(m\omega^2 - 3k) = 0
$$

has three non-negative roots

$$\omega_1 = 0, \quad \omega_2 = \sqrt{\frac{k}{m}}, \quad \omega_3 = \sqrt{\frac{3k}{m}}.$$

These are the normal-mode angular frequencies of the system. The corresponding normal-modes are as follows.

i. $\omega_1 = 0$

Equation (2) gives $A = B = C$ and thus $x_1 = x_2 = x_3$. The first of Eqs. (1) then gives

$$\ddot{x}_1 = 0, \quad \text{or} \quad x_1 = at + b,$$

where a, b are constants. Hence in this mode the three masses undergo uniform translation together as a rigid body and no vibration occurs.

ii. $\omega_2 = \sqrt{\frac{k}{m}}$

Equation (2) gives $B = 0$, $A = -C$. In this mode the middle mass remains stationary while the outer masses oscillate symmetrically with respect to it. The displacements are

$$x_1 = A \cos (\omega_2 t + \varphi),$$
$$x_2 = 0,$$
$$x_3 = -A \cos (\omega_2 t + \varphi),$$

φ being a constant.

iii. $\omega_3 = \sqrt{\dfrac{3k}{m}}$

Equation (2) gives $B = -2A$, $C = A$. In this mode the two outer masses oscillate with the same amplitude and phase while the middle mass oscillates exactly out of phase with twice the amplitude with respect to the other two masses. The displacements are

$$x_1 = A \cos (\omega_3 t + \varphi),$$
$$x_2 = -2A \cos (\omega_3 t + \varphi),$$
$$x_3 = A \cos (\omega_3 t + \varphi).$$

The three normal-modes are shown in Fig. 2.34.

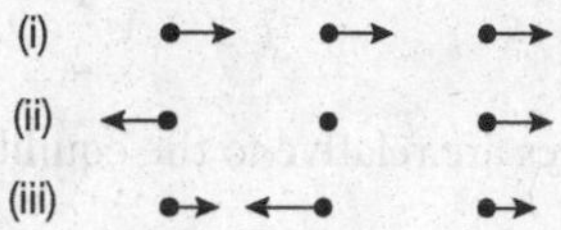

Fig. 2.34

2039

A rectangular plate of mass *M*, length *a* and width *b* is supported at each of its corners by a spring with spring constant *k* as shown in Fig. 2.35. The springs are confined so that they can move only in the vertical direction. For small amplitudes, find the normal-modes of vibration and their frequencies. Describe each of the modes.

(UC, Berkeley)

Sol: Use Cartesian coordinates with origin at the center of mass C of the plate when the plate is in equilibrium, the z-axis vertically upwards, the x- and y-axes along the axes of symmetry in the plane of the plate, and let the angles of rotation about the x- and y-axes be φ, θ respectively, as shown in Fig. 2.36. If z is the vertical coordinate of C, the vertical coordinates of the four corners are for small angle oscillations.

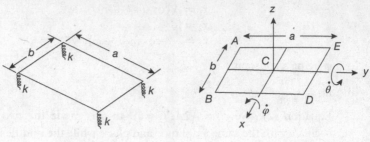

Fig. 2.35   Fig. 2.36

$$z_A = z - \frac{1}{2}a\varphi + \frac{1}{2}b\theta,$$

$$z_B = z - \frac{1}{2}a\varphi - \frac{1}{2}b\theta,$$

$$z_D = z + \frac{1}{2}a\varphi - \frac{1}{2}b\theta,$$

$$z_E = z + \frac{1}{2}a\varphi + \frac{1}{2}b\theta,$$

As the coordinates are relative to the equilibrium positions, the Lagrangian is

$$L = T - V$$
$$= \frac{1}{2}M\dot{z}^2 + \frac{1}{24}Ma^2\dot{\varphi}^2 + \frac{1}{24}Mb^2\dot{\theta}^2 - \frac{1}{2}k(z_A^2 + z_B^2 + z_D^2 + z_E^2) - Mgz$$
$$= \frac{1}{2}M\dot{z}^2 + \frac{1}{24}Ma^2\dot{\varphi}^2 + \frac{1}{24}Mb^2\dot{\theta}^2 - \frac{1}{2}k(4z^2 + a^2\varphi^2 + b^2\theta^2) - Mgz.$$

Lagrange's equations

$$\frac{d}{dt}\left(\frac{\partial L}{\partial \dot{q}_i}\right) - \frac{\partial L}{\partial q_i} = 0$$

then give

$$M\ddot{z} + 4kz + Mg = 0,$$

$$\frac{1}{12}M\ddot{\varphi} + k\varphi = 0,$$

$$\frac{1}{12}M\ddot{\theta} + k\theta = 0.$$

By putting $z = z' - \dfrac{Mg}{4k}$, the first equation can be written as

$$M\ddot{z}' + 4kz' = 0.$$

The equations show that the normal-mode angular frequencies are

$$\omega_1 = 2\sqrt{\frac{k}{M}}, \quad \omega_2 = \omega_3 = 2\sqrt{\frac{3k}{M}}.$$

If we define

$$\xi_1 = \sqrt{M}z', \quad \xi_2 = \sqrt{\frac{M}{12}}a\varphi, \quad \xi_3 = \sqrt{\frac{M}{12}}b\theta,$$

we can, neglecting a constant term in the potential energy, write

$$T = \frac{1}{2}(\dot{\xi}_1^2 + \dot{\xi}_2^2 + \dot{\xi}_3^2),$$

$$V = \frac{1}{2}(\omega_1^2\xi_1^2 + \omega_2^2\xi_2^2 + \omega_3^2\xi_3^2).$$

These are both in quadratic form, slowing that ξ_1, ξ_2, ξ_3 are the normal-mode coordinates.

Denoting the amplitudes of z', φ, θ by $z_0', \varphi_0, \theta_0$ respectively, we obtain from the equations of motion

$$(4k - M\omega^2)z_0 = 0,$$

$$\left(k - \frac{1}{12}M\omega^2\right)\varphi_0 = 0,$$

$$\left(k - \frac{1}{12}M\omega^2\right)\theta_0 = 0.$$

It can be seen that if $\omega = \omega_1$ then $z_0 \neq 0$, $\varphi_0 = \theta_0 = 0$. If $\omega = \omega_2$ or ω_3, then $z_0 = 0$, and one or both of φ_0, θ_0 are not zero.

2040

A particle moves without friction on the inside wall of an axially symmetric vessel given by

$$z = \frac{1}{2}b(x^2 + y^2)$$

where b is a constant and z is in the vertical direction, as shown in Fig. 2.37.

a. The particle is moving in a circular orbit at height $z = z_0$. Obtain its energy and angular momentum in terms of z_0, b, g (gravitational acceleration), and the mass m of the particle.

b. The particle in the horizontal circular orbit is poked downwards slightly. Obtain the frequency of oscillation about the unperturbed orbit for very small oscillation amplitude.

(UC, Berkeley)

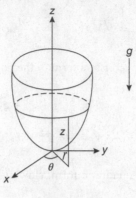

Fig. 2.37

Sol:

a. Use coordinates as shown in Fig. 2.37. As $x = r\cos\theta$, $y = r\sin\theta$, the vessel can be represented by

$$z = \frac{1}{2}b(x^2 + y^2) = \frac{1}{2}br^2.$$

The Lagrangian of the particle is

$$L = T - V = \frac{1}{2}m(\dot{r}^2 + r^2\dot{\theta}^2 + \dot{z}^2) - mgz$$

$$= \frac{1}{2}m(\dot{r}^2 + r^2\dot{\theta}^2 + b^2r^2\dot{r}^2) - \frac{1}{2}mgbr^2.$$

Lagrange's equation for r then gives

$$(1 + b^2r^2)\ddot{r} - b^2r\dot{r}^2 - r\dot{\theta}^2 + gbr = 0. \tag{1}$$

As the particle motion is confined to a circle of height z_0 and radius r_0, say, we have

$$r = r_0, \quad \dot{r} = \ddot{r} = 0, \quad z_0 = \frac{1}{2}br_0^2,$$

$$\dot{\theta}^2 = gb = \Omega^2, \quad \text{say.}$$

The total energy of the particle is then

$$T + V = \frac{1}{2}m(r_0^2\Omega^2 + gbr_0^2) = mgbr_0^2 = 2mgz_0,$$

and the angular momentum about the center of the circle is

$$J = mr \cdot r\dot\theta = mr_0^2\Omega = 2mz_0\sqrt{\frac{g}{b}}.$$

b. For the perturbed motion, let $r = r_0 + \rho$ where $\rho \ll r_0$, Lagrange's equation for θ shows that the angular momentum $mr^2\dot\theta$ is conserved. Hence.

$$r\dot\theta^2 = \frac{r^4\dot\theta^2}{r^3} = \frac{r_0^4\Omega^2}{r^3} = \frac{r_0^4gb}{r^3},$$

and Eq. (1) becomes

$$(1 + b^2r_0^2)\ddot\rho + 4gb\rho = 0$$

by neglecting terms of order higher than the first in the small quantities $\rho, \dot\rho, \ddot\rho$. The angular frequency of small amplitude oscillations about r_0 is therefore

$$\omega = \sqrt{\frac{4gb}{1 + b^2r_0^2}} = 2\sqrt{\frac{gb}{1 + 2bz_0}}.$$

2041

A spring mass system is set up on an inclined plane as shown in the figure. The mass of the body is m and the spring is compressed by an amount d when the mass is momentarily at rest. Assuming a frictionless surface and a massless spring, determine the frequency of oscillation if the subsequent motion of the mass is simple harmonic. If the same arrangement is made on a rough surface where the coefficient of friction between the block and the surface is μ, determine the compression of the spring given that the velocity with which the body hits the spring is v_0.

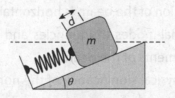

Fig. 2.38

Sol:

$$mg \sin \theta - kd = 0 \quad \text{or} \quad k = \frac{mg \sin \theta}{d}$$

$$\omega = \sqrt{\frac{g \sin \theta}{d}}.$$

$$f = \frac{1}{2\pi}\sqrt{\frac{g \sin \theta}{d}}$$

For a rough surface, $\frac{mv_0^2}{2} = \frac{kx_f^2}{2} + Fx_f$, where F is the friction force and x_f is the final distance.

Substituting $F = \mu mg \cos \theta$

$$-\mu mg \cos \theta x_f = \frac{kx_f^2}{2} - \frac{mv_0^2}{2}$$

$$kx_f^2 + 2k'x_f - mv_0^2 = 0,$$

where $k' = \mu mg \cos \theta$.

$$x_f = \frac{-2\mu mg \cos \theta \pm \sqrt{(2\mu mg \cos \theta)^2 + 4kmv_0^2}}{2k}$$

$$= \frac{-\mu mg \cos \theta - \sqrt{(\mu mg \cos \theta)^2 + kmv_0^2}}{k}$$

Since the positive sign will give unreasonable results, the negative sign alone is considered.

2042

An uniform log with length L, cross-sectional area A and mass M is floating vertically in water ($\rho = 1.0$) and is attached by a spring with spring constant K to a uniform beam which is pivoted at the center as shown in Fig. 2.39(a). The beam has the same mass and is twice the length of the log. The log is constrained to move vertically and the natural length of the spring is such that the equilibrium position of the beam is horizontal.

a. Find the normal-modes (frequencies and ratio of displacements) for small displacements of the beam.

b. Discuss the physical significance of the normal-modes in the limit of a very strong spring.

(UC, Berkeley)

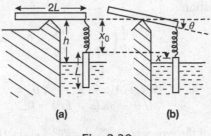

Fig. 2.39

Sol:

a. Use coordinates as shown in Fig. 2.39(b) with x denoting the displacement of the top of the vertical rod from its equilibrium position (the downward direction being taken as positive), and θ the angle of rotation of the beam. At equilibrium (Fig. 2.39(a)), the spring is in its natural length x_0 and does not exert a force on the rod. With $\rho = 1$ we have

$$Mg = [L - (h - x_0)]Ag.$$

When the beam has rotated an angle θ (Fig. 2.39(b)) the spring is extended by $x - L\theta$ and the upward thrust of the water is

$$-[L - (h - x_0 - x)]Ag = -\frac{\partial V_t}{\partial x},$$

giving

$$V_t = Ag \int_0^x \{[L - (h - x_0)] + x'\} dx'$$

$$= [L - (h - x_0)]Agx + \frac{1}{2}Agx^2$$

$$= Mgx + \frac{1}{2}Agx^2.$$

Hence the total potential energy is

$$V = -Mgx + Mgx + \frac{1}{2}Agx^2 + \frac{1}{2}K(x - L\theta)^2$$

$$= \frac{1}{2}Agx^2 + \frac{1}{2}K(x - L\theta)^2.$$

The beam has moment of inertia $\frac{1}{3}ML^2$, so the total kinetic energy is

$$T = \frac{1}{2}M\dot{x}^2 + \frac{1}{6}L^2\dot{\theta}^2.$$

Thus the Lagrangian is

$$L = T - V = \frac{1}{2}M\dot{x}^2 + \frac{1}{6}ML^2\dot{\theta}^2 - \frac{1}{2}Agx^2 - \frac{1}{2}K(x - L\theta)^2.$$

Lagrange's equations

$$\frac{d}{dt}\left(\frac{\partial L}{\partial \dot{q}_i}\right) - \frac{\partial L}{\partial q_i} = 0$$

give

$$M\ddot{x} + Agx + K(x - L\theta) = 0,$$
$$ML\ddot{\theta} - 3K(x - L\theta) = 0.$$

Try a solution of the type $x = De^{i\omega t}$, $\theta = Be^{i\omega t}$ and write the above as

$$(K + Ag - M\omega^2)D - KLB = 0,$$
$$-3KD + (3KL - ML\omega^2)B = 0.$$

The secular equation is then

$$\begin{vmatrix} K + Ag - M\omega^2 & -KL \\ -3K & 3KL - ML\omega^2 \end{vmatrix} = 0,$$

or

$$M^2\omega^4 - M(4K + Ag)\omega^2 + 3KAg = 0.$$

The two positive roots

$$\omega \pm = \sqrt{\frac{4K + Ag \pm \sqrt{(4K + Ag)^2 - 12KAg}}{2M}}$$

are the two normal-mode angular frequencies of the system for small oscillations of the beam.

The ratios of the displacements are

$$\frac{x}{L\theta} = \frac{D}{BL} = \frac{3K - M\omega^2}{3K} = \frac{2K - Ag \mp \sqrt{(4K + Ag)^2 - 12KAg}}{6K}$$

with the top sign for ω_+ and the bottom sign for ω_-.

b. In the limit of a very strong spring, $K \to \infty$. As $M\ddot{x}$, Agx, $ML\ddot{\theta}$ are all finite, this requires that $x - L\theta \to 0$, i.e. $x \to L\theta$. Eliminating the $K(x - L\theta)$ terms from the equations of motion and making use of $L\ddot{\theta} \approx \ddot{x}$ we find

$$4M\ddot{x} + 3Agx = 0$$

and hence the angular frequency of oscillation

$$\omega = \sqrt{\frac{3Ag}{4M}}.$$

The ratio of the displacements is

$$\frac{x}{L\theta} \approx 1,$$

and they are in the same phase. Note that these results cannot be obtained from the previous ones by putting $K \to \infty$ because the constraint relations are different. Physically, the constraint $x \approx L\theta$ means that the system oscillates with the spring keeping its length constant, which is expected for a very strong spring.

2043

Two unequal masses M and $m(M > m)$ hang from a support by strings of equal lengths l. The masses are coupled by a spring of spring constant K and of unstretched length equal to the distance between the support points as shown in Fig. 2.40. Find the normal-mode frequencies for the small oscillations along the line between the two masses. Give the relation between the motion of M and that of m in each mode. Write down the most general solution.

Now specialize for the case where at $t = 0$, m is at rest at its equilibrium position, and M is released from rest with an initial positive displacement. If the total energy of the system is E_0 and the spring is very weak, find the maximum energy acquired by m during the subsequent motion for the case $\frac{M}{m} = 2$. (Where did you use the assumption that the spring is weak?)

(UC, Berkeley)

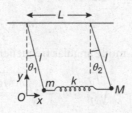

Fig. 2.40

Sol: Use coordinates as shown in Fig. 2.40 with origin O at the equilibrium position of the mass m and the x- and y-axes along the horizontal and vertical directions respectively and let the distance between the two supports be L. The masses m and M then have coordinates

$$(l \sin \theta_1, l(1 - \cos \theta_1)), \quad (L + l \sin \theta_2, l(1 - \cos \theta_2))$$

and velocities

$$(l\dot{\theta}_1 \cos \theta_1, l\dot{\theta}_1 \sin \theta_1), \quad (l\dot{\theta}_2 \cos \theta_2, l\dot{\theta}_2 \sin \theta_2)$$

respectively. The Lagrangian of the system is

$$L = T - V$$
$$= \frac{1}{2}ml^2\dot{\theta}_1^2 + \frac{1}{2}Ml^2\dot{\theta}_2^2 - \frac{1}{2}Kl^2(\sin\theta_2 - \sin\theta_1)^2$$
$$- mgl(1 - \cos\theta_1) - Mgl(1 - \cos\theta_2)$$
$$\approx \frac{1}{2}ml^2\dot{\theta}_1^2 + \frac{1}{2}Ml^2\dot{\theta}_2^2 - \frac{1}{2}Kl^2(\theta_2 - \theta_1)^2 - \frac{1}{2}gl(m\theta_1^2 + M\theta_2^2)$$

for small oscillations in the horizontal direction.

Lagrange's equations

$$\frac{d}{dt}\left(\frac{\partial L}{\partial \dot{q}_i}\right) - \frac{\partial L}{\partial q_i} = 0$$

then give

$$ml\ddot{\theta}_1 + (mg + Kl)\theta_1 - Kl\theta_2 = 0,$$
$$Ml\ddot{\theta}_2 + (Mg + Kl)\theta_2 - Kl\theta_1 = 0.$$

Try a solution of the type $\theta_1 = Ae^{i\omega t}$, $\theta_2 = Be^{i\omega t}$ and write the above as

$$\begin{pmatrix} mg + Kl - ml\omega^2 & -Kl \\ -Kl & Mg + Kl - Ml\omega^2 \end{pmatrix}\begin{pmatrix} A \\ B \end{pmatrix} = 0.$$

The secular equation is then

$$\begin{vmatrix} mg + Kl - ml\omega^2 & -Kl \\ -Kl & Mg + Kl - Ml\omega^2 \end{vmatrix} = 0,$$

yielding the normal-mode angular frequencies

$$\omega_1 = \sqrt{\frac{g}{l}}, \quad \omega_2 = \sqrt{\frac{mMg + Kl(m + M)}{mMl}}.$$

As

$$\frac{A}{B} = \frac{Mg + Kl - Ml\omega^2}{Kl},$$

we have

$$\frac{A}{B} = 1 \quad \text{for} \quad \omega = \omega_1,$$

$$\frac{A'}{B'} = -\frac{M}{m} \quad \text{for} \quad \omega = \omega_2.$$

Hence, for $\omega = \omega_1$,

$$\theta_1 = A' \cos(\omega_2 t + \varphi_2), \quad \theta_2 = -\frac{m}{M} A' \cos(\omega_2 t + \varphi_2),$$

for $\omega = \omega_2$,

$$\theta_1 = A' \cos(\omega_2 t + \varphi_2), \quad \theta_2 = -\frac{m}{M} A' \cos(\omega_2 t + \varphi_2),$$

and the most general solution is

$$\theta_1 = A \cos(\omega_1 t + \varphi_1) + A' \cos(\omega_2 t + \varphi_2),$$
$$\theta_2 = A \cos(\omega_1 t + \varphi_1) - \frac{m}{M} A' \cos(\omega_2 t + \varphi_2).$$

Initially at $t = 0$, $\dot{\theta}_1 = \dot{\theta}_2 = 0$, giving $\dot{\varphi}_1 = \dot{\varphi}_2 = 0$, and $\theta_1 = 0$, $\theta_2 = \theta_0$, giving

$$A = \frac{M}{m + M} \theta_0, \quad A' = -A.$$

If the initial total energy is E_0, then as

$$E_0 = \frac{1}{2} K l^2 \theta_0^2 + \frac{1}{2} M g l \theta_0^2,$$

we have, as θ_0 is positive,

$$\theta_0 = \sqrt{\frac{2E_0}{(Kl + Mg)l}}.$$

If in addition, $M = 2m$, the general solution reduces to

$$\theta_1 = \frac{2}{3} \theta_0 [\cos(\omega_1 t) - \cos(\omega_2 t)],$$
$$\theta_2 = \frac{2}{3} \theta_0 [\cos(\omega_1 t) + \frac{1}{2} \cos(\omega_2 t)],$$

with

$$\omega_1 = \sqrt{\frac{g}{l}}, \quad \omega_2 = \sqrt{\frac{2mg + 3Kl}{2ml}}, \quad \theta_0 = \sqrt{\frac{2E_0}{(2mg + Kl)l}}.$$

The energy of m is

$$E_1 = \frac{1}{2} m l^2 \dot{\theta}_1^2 + \frac{1}{2} m g l \theta_1^2.$$

If the spring is very weak, we can take $Kl \ll mg$ so that

$$\omega_2 = \sqrt{\frac{g}{l}\left(1 + \frac{3Kl}{2mg}\right)} \approx \sqrt{\frac{g}{l}\left(1 + \frac{3Kl}{4mg}\right)} = \omega_1(1 + \delta),$$

$$\theta_0 = \sqrt{\frac{E_0}{mgl}}\left(1 - \frac{\delta}{3}\right),$$

where

$$\delta = \frac{3Kl}{4mg} \ll 1.$$

We then have

$$E = \frac{2}{9}mgl\theta_0^2[1 + (1 + \delta)^2 \sin^2(\omega_2 t) + \cos^2(\omega_2 t)$$
$$-2(1 + \delta)\sin(\omega_1 t)\sin(\omega_2 t) - 2\cos(\omega_1 t)\cos(\omega_2 t)]$$
$$\approx \frac{4}{9}E_0[1 - \cos(\omega_1 \delta t)],$$

neglecting δ as compared with unity. Hence the maximum energy of m is $\frac{8}{9}E_0$.

2044

Two small spheres of mass M are suspended between two rigid supports as shown in Fig. 2.41. We assume that both particles can move in the plane of the figure, sideways and up and down. The three springs are equal, of spring constant K. The springs are under tension: in its unstretched condition each spring would be of length $\frac{a}{2}$. The springs are assumed massless and perfectly elastic. Assuming small oscillations about the equilibrium configuration shown above, find the frequencies for the four normal-modes of the system.

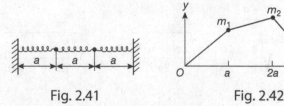

Fig. 2.41   Fig. 2.42

(UC, Berkeley)

Sol: Since the motion is confined to the plane of the diagram of Fig. 2.41, the sideway motion is to be interpreted as longitudinal along the springs.

Let (x_1, y_1) and (x_2, y_2) be the horizontal and vertical displacements of the spheres, numbered from the left, from their respective positions of equilibrium. Using coordinates as shown in Fig. 2.42, m_1, m_2 have coordinates $(a + x_1, y_1)$, $(2a + x_2, y_2)$ respectively. Taking the equilibrium configuration (Fig. 2.41) as the state of zero potential, we have for the system the potential energy

$$V = \frac{1}{2}K\left[\sqrt{(a + x_1)^2 + y_1^2} - \frac{a}{2}\right]^2 - \frac{1}{2}K\left(\frac{a}{2}\right)^2$$

$$+ \frac{1}{2}K\left[\sqrt{(a + x_2 - x_1)^2 + (y_2 - y_1)^2} - \frac{a}{2}\right]^2$$

$$- \frac{1}{2}K\left(\frac{a}{2}\right)^2 + \frac{1}{2}K\left[\sqrt{(a - x_2)^2 + y_2^2} - \frac{a}{2}\right]^2$$

$$- \frac{1}{2}K\left(\frac{a}{2}\right)^2 + Mg(y_1 + y_2).$$

Consider

$$\frac{1}{2}K\left[\sqrt{(a + x_1)^2 + y_1^2} - \frac{a}{2}\right]^2 - \frac{1}{2}K\left(\frac{a}{2}\right)^2.$$

$$= \frac{1}{2}K\left[a^2 + x_1^2 + 2ax_1 + y_1^2 + \frac{a^2}{4} - a\sqrt{a^2 + x_1^2 + 2ax_1 + y_1^2}\right] - \frac{1}{8}Ka^2.$$

As the term involving the square-root sign can be written as

$$a^2\left(1 + \frac{x_1^2 + 2ax_1 + y_1^2}{a^2}\right)^{\frac{1}{2}}$$

$$\approx a^2\left[1 + \frac{1}{2}\left(\frac{x_1^2 + 2ax_1 + y_1^2}{a^2}\right) + \frac{1}{2!}\left(\frac{1}{2}\right)\left(-\frac{1}{2}\right)\frac{4x_1^2}{a^2}\right]$$

$$= a^2 + \frac{1}{2}(x_1^2 + 2ax_1 + y_1^2) - \frac{1}{2}x_1^2$$

retaining only terms of orders up to the second in the small quantities x_1, y_1, the above becomes

$$\frac{1}{2}K\left(x_1^2 + ax_1 + \frac{1}{2}y_1^2\right).$$

The same approximation is taken over the other terms. Hence

$$
\begin{aligned}
V \approx \frac{1}{2}K \Bigg[& x_1^2 + ax_1 + \frac{y_1^2}{2} + (x_2 - x_1)^2 + a(x_2 - x_1) \\
& + \frac{1}{2}(y_2 - y_1)^2 + x_2^2 - ax_2 + \frac{1}{2}y_2^2 \Bigg] + Mg(y_1 + y_2) \\
= & \frac{1}{2}K(2x_1^2 + 2x_2^2 + y_1^2 + y_2^2 - 2x_1x_2 - y_1y_2) + Mg(y_1 + y_2).
\end{aligned}
$$

The Lagrangian is then

$$
\begin{aligned}
L & = T - V \\
& = \frac{1}{2}M(\dot{x}_1^2 + \dot{y}_1^2 + \dot{x}_2^2 + \dot{y}_2^2) \\
& \quad - \frac{1}{2}K(2x_1^2 + 2x_2^2 + y_1^2 + y_2^2 - 2x_1x_2 - y_1y_2) - Mg(y_1 + y_2).
\end{aligned}
$$

Lagrange's equations

$$
\frac{d}{dt}\left(\frac{\partial L}{\partial \dot{q}_i} \right) - \frac{\partial L}{\partial q_i} = 0
$$

give

$$
\begin{aligned}
M\ddot{x}_1 + 2Kx_1 - Kx_2 &= 0, \\
M\ddot{x}_2 + 2Kx_2 - Kx_1 &= 0, \\
M\ddot{y}_1 + Ky_1 - \frac{1}{2}Ky_2 + Mg &= 0, \\
M\ddot{y}_2 + Ky_2 - \frac{1}{2}Ky_1 + Mg &= 0.
\end{aligned}
$$

It is seen that the equations naturally separate into two groups, those in x_1, x_2 and those in y_1, y_2. Let

$$
x_i = A_i e^{i\omega t}.
$$

Then the first two equations give the secular equation

$$
\begin{vmatrix} 2K - M\omega^2 & -K \\ -K & 2K - M\omega^2 \end{vmatrix} = (3K - M\omega^2)(K - M\omega^2) = 0,
$$

yielding two normal-mode angular frequencies

$$
\omega_1 = \sqrt{\frac{K}{M}}, \quad \omega_2 = \sqrt{\frac{3K}{M}}
$$

for longitudinal oscillations.

For the second group of two equations, let

$$y_1' = y_1 + \frac{2Mg}{K}, \qquad y_2' = y_2 + \frac{2Mg}{K}.$$

They can then be written as

$$M\ddot{y}_1' + Ky_1' - \frac{1}{2}Ky_2' = 0,$$

$$M\ddot{y}_2' + Ky_2' - \frac{1}{2}Ky_1' = 0.$$

Trying a solution of the type

$$y_i' = B_i e^{i\omega t},$$

we obtain the secular equation

$$\begin{vmatrix} K - M\omega^2 & -\dfrac{K}{2} \\ -\dfrac{K}{2} & K - M\omega^2 \end{vmatrix} = \left(\frac{3}{2}K - M\omega^2\right)\left(\frac{1}{2}K - M\omega^2\right) = 0,$$

which yields the normal-mode angular frequencies

$$\omega_3 = \sqrt{\frac{K}{2M}}, \qquad \omega_4 = \sqrt{\frac{3K}{2M}}$$

for vertical oscillations.

2045

A simple pendulum of length L is suspended at the rim of a wheel of radius b which rotates within the vertical plane with constant angular velocity Ω (Fig. 2.43). We consider only the motion in which the bob of the pendulum swings in the plane of the wheel.

a. Write an exact differential equation of motion for the angular displacement θ of the bob. Also write a simplified form valid when the oscillation amplitude is very small.

b. Assume that both the radius b and the oscillation amplitude of the bo▮
are very small. Give an approximate steady-state solution of the equa▮
tion of motion valid under the assumptions.

(You may ignore transients which will die out if there is a sligh▮
dissipation.)

(UC, Berkeley

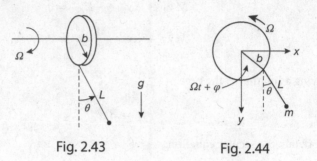

Fig. 2.43   Fig. 2.44

Sol:

a. Use coordinates as shown in Fig. 2.44. The mass m has coordinates

$$(b \sin(\Omega t + \varphi) + L \sin \theta, \, b \cos(\Omega t + \varphi) + L \cos \theta)$$

and velocity

$$(b\Omega \cos(\Omega t + \varphi) + L\dot{\theta} \cos \theta, -b\Omega \sin(\Omega t + \varphi) - L\dot{\theta} \sin \theta),$$

where φ is a constant.

The Lagrangian of m is then

$$L = T - V$$
$$= \frac{1}{2}m[b^2\Omega^2 + L^2\dot{\theta}^2 + 2bL\Omega\dot{\theta} \cos(\theta - \Omega t - \varphi)]$$
$$+ mg[b \cos(\Omega t + \varphi) + L \cos \theta]$$

Lagrange's equation

$$\frac{d}{dt}\left(\frac{\partial L}{\partial \dot{\theta}}\right) - \frac{\partial L}{\partial \theta} = 0$$

gives

$$L\ddot{\theta} + b\Omega^2 \sin(\theta - \Omega t - \varphi) + g \sin \theta = 0.$$

For small-amplitude oscillations, $\sin\theta \approx \theta$, $\cos\theta \approx 1$,

$$\sin(\theta - \Omega t - \varphi) \approx \theta\cos(\Omega t + \varphi) - \sin(\Omega t + \varphi),$$

and the equation of motion becomes

$$L\ddot{\theta} + [b\Omega^2\cos(\Omega t + \varphi) + g]\theta - b\Omega^2\sin(\Omega t + \varphi) = 0.$$

b. For b and θ small, we have, retaining terms of only up to the first order of b, $\theta, \dot{\theta}$,

$$L\ddot{\theta} + g\theta - b\Omega^2\sin(\Omega t + \varphi) = 0.$$

In the steady state, the pendulum will swing with the same frequency as the rotation of the wheel, so we can assume

$$\theta = \alpha\cos(\Omega t + \varphi) + \beta\sin(\Omega t + \varphi),$$

where α, β are constants. Substitution in the equation of motion gives

$$(-L\Omega^2 + g)[\alpha\cos(\Omega t + \varphi) + \beta\sin(\Omega t + \varphi)] - b\Omega^2\sin(\Omega t + \varphi) = 0.$$

As this equation must be true for any arbitrary time, the coefficients of $\cos(\Omega t + \varphi)$ and $\sin(\Omega t + \varphi)$ must separately vanish:

$$-\alpha L\Omega^2 + g\alpha = 0,$$
$$g\beta - \beta L\Omega^2 - b\Omega^2 = 0.$$

As Ω is given, we must have $\alpha = 0$ in the first equation. The second equation gives

$$\beta = \frac{b\Omega^2}{g - L\Omega^2}.$$

Hence the steady-state solution is

$$\theta = \frac{b\Omega^2\sin(\Omega t + \varphi)}{g - L\Omega^2}.$$

2046

Three equal point masses m move on a circle of radius b under forces derivable from the potential energy

$$V(\alpha, \beta, \gamma) = V_0(e^{-\alpha} + e^{-\beta} + e^{-\gamma}).$$

where α, β, γ are their angular separations in radians, as shown in Fig. 2.45. When $\alpha = \beta = \gamma = \dfrac{2\pi}{3}$, the system is in equilibrium. Find the normal-mode frequencies for small departure from equilibrium.

(Note that α, β, γ are not independent since $\alpha + \beta + \gamma = 2\pi$.)

(UC, Berkeley)

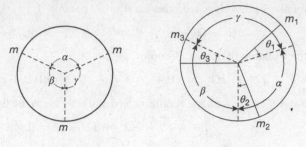

Fig. 2.45   Fig. 2.46

Sol: Let $\theta_1, \theta_2, \theta_3$ be the angular displacements of the three masses from their equilib rium positions as shown in Fig. 2.46. We have

$$\alpha = \frac{2\pi}{3} + \theta_2 - \theta_1,$$

$$\beta = \frac{2\pi}{3} + \theta_3 - \theta_2,$$

$$\gamma = \frac{2\pi}{3} + \theta_1 - \theta_3.$$

As

$$e^{-x} \approx 1 - \frac{x}{1!} + \frac{x^2}{2!} - \cdots,$$

we can write the potential energy as

$$V = V_0 e^{-\frac{2\pi}{3}} \left[e^{-(\theta_2 - \theta_1)} + e^{-(\theta_3 - \theta_2)} + e^{-(\theta_1 - \theta_3)} \right]$$

$$\approx V_0 e^{-\frac{2\pi}{3}} \left[3 - (\theta_2 - \theta_1) - (\theta_3 - \theta_2) - (\theta_1 - \theta_3) \right.$$
$$\left. + \frac{1}{2}(\theta_2 - \theta_1)^2 + \frac{1}{2}(\theta_3 - \theta_2)^2 + \frac{1}{2}(\theta_1 - \theta_3)^2 \right]$$

$$= A(3 + \theta_1^2 + \theta_2^2 + \theta_3^2 - \theta_1 \theta_2 - \theta_2 \theta_3 - \theta_3 \theta_1)$$

with $A = V_0 \exp\left(-\frac{2\pi}{3}\right)$, retaining terms of orders up to the second in the small quantities $\theta_1, \theta_2, \theta_3$.

As the velocities are $b\dot{\theta}_1, b\dot{\theta}_2, b\dot{\theta}_3$, the kinetic energy is

$$T = \frac{1}{2} B(\dot{\theta}_1^2 + \dot{\theta}_2^2 + \dot{\theta}_3^2)$$

with $B = mb^2$.

The Lagrangian is therefore

$$L = T - V$$
$$= B(\dot{\theta}_1^2 + \dot{\theta}_2^2 + \dot{\theta}_3^2) - A(3 + \theta_1^2 + \theta_2^2 + \theta_3^2 - \theta_1\theta_2 - \theta_2\theta_3 - \theta_3\theta_1).$$

Lagrange's equations

$$\frac{d}{dt}\left(\frac{\partial L}{\partial \dot{q}_i}\right) - \frac{\partial L}{\partial q_i} = 0$$

then give

$$B\ddot{\theta}_1 + A(2\theta_1 - \theta_2 - \theta_3) = 0,$$
$$B\ddot{\theta}_2 + A(2\theta_2 - \theta_3 - \theta_1) = 0,$$
$$B\ddot{\theta}_3 + A(2\theta_3 - \theta_1 - \theta_2) = 0.$$

Trying a solution of the type $\theta_i = C_i e^{i\omega t}$, we find the secular equation

$$\begin{vmatrix} 2A - B\omega^2 & -A & -A \\ -A & 2A - B\omega^2 & -A \\ -A & -A & 2A - B\omega^2 \end{vmatrix} = 0,$$

or, after some arithmetic manipulations,

$$\begin{vmatrix} 0 & -2A & A - B\omega^2 \\ 0 & 2A - B\omega^2 & -A \\ -3A + B\omega^2 & -A & 2A - B\omega^2 \end{vmatrix} = B\omega^2(-3A + B\omega^2)^2 = 0.$$

Hence the normal-mode angular frequencies are

$$\omega_1 = 0, \quad \omega_2 = \omega_3 = \sqrt{\frac{3A}{B}} = \sqrt{\frac{3V_0 \exp\left(\frac{-2\pi}{3}\right)}{mb^2}}.$$

Note that ω_1 does not give rise to oscillations, for in this case the equations of motion give $\theta_1 = \theta_2 = \theta_3$ and the system as a whole rotates with a constant angular velocity. The other two normal-modes are degenerate and there is only one normal-mode frequency

$$\frac{1}{b}\sqrt{\frac{3V_0}{m}} \exp\left(-\frac{\pi}{3}\right).$$

2047

Three point particles, two of mass m and one of mass M, are constrained to lie on a horizontal circle of radius r. They are mutually connected by springs, each of constant K, that follow the arc of the circle and that are of equal length when the system is at rest as shown in Fig. 2.47. Assuming motion that stretches the springs only by a small amount from the equilibrium length ($2\pi r/3$),

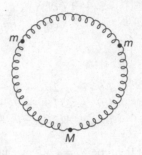

Fig. 2.47

a. describe qualitatively the modes of motion that are simple harmonic in time (the normal-modes);

b. find a precise set of normal coordinates, one corresponding to each mode;

c. find the frequency of each mode.

(UC, Berkeley)

Sol:

a. As the system is not acted upon by external torque, its angular momentum is conserved. This means that there is a normal-mode in which the system rotates as a whole. Consequently there are only two vibrational degrees of freedom. Let $\theta_1, \theta_2, \theta_3$ be respectively the angular displacements of m, M, m from their equilibrium positions and let their amplitudes be c_1, c_2, c_3. When considering the vibration of the masses relative to their equilibrium positions, we can take the total angular momentum of the system to be zero. Then the two vibrational normal-modes correspond to

$$c_2 = 0, \quad c_1 = -c_3 \quad \text{and} \quad c_1 = c_3 = -\frac{Mc_2}{2m}.$$

b. Let the natural length of each spring be a and denote the equilibrium length by b, i.e.

$$b = \frac{2\pi r}{3}.$$

The Lagrangian of the system is

$$L = T - V$$
$$= \frac{1}{2}mr^2(\dot{\theta}_1^2 + \dot{\theta}_3^2) + \frac{1}{2}Mr^2\dot{\theta}_2^2$$
$$- \frac{1}{2}K[(b + r\theta_2 - r\theta_1 - a)^2 + (b + r\theta_3 - r\theta_2 - a)^2 + (b + r\theta_1 - r\theta_3 - a)^2].$$

Lagrange's equations

$$\frac{d}{dt}\left(\frac{\partial L}{\partial \dot{q}_i}\right) - \frac{\partial L}{\partial q_i} = 0$$

then give the differential equations of motion

$$m\ddot{\theta}_1 + K(2\theta_1 - \theta_2 - \theta_3) = 0,$$
$$M\ddot{\theta}_2 + K(2\theta_2 - \theta_3 - \theta_1) = 0,$$
$$m\ddot{\theta}_3 + K(2\theta_3 - \theta_1 - \theta_2) = 0.$$

The above sum up to

$$m\ddot{\theta}_1 + M\ddot{\theta}_2 + m\ddot{\theta}_3 = 0,$$

and the first and third equations give

$$m(\ddot{\theta}_1 - \ddot{\theta}_3) + 3K(\theta_1 - \theta_3) = 0.$$

These can be written as

$$m\ddot{\xi} = 0, \tag{1}$$
$$m\ddot{\eta} + 3K\eta = 0 \tag{2}$$

if we set

$$\xi = \theta_1 + \frac{M\theta_2}{m} + \theta_3,$$
$$\eta = \theta_1 - \theta_3.$$

Hence ξ and η are normal-mode coordinates of the system.

Equation (1) shows that $\omega_1 = 0$. Thus corresponding to this mode in which the system rotates as a whole and there is no oscillation.

Equation (2) shows that

$$\omega_2 = \sqrt{\frac{3K}{m}}.$$

To find the third normal coordinate, we choose the coordinate transformation

$$q_1 = \theta_1, \quad q_2 = \theta_2\sqrt{\frac{M}{m}}, \quad q_3 = \theta_3$$

to make the kinetic energy a sum of squares:

$$T = \frac{1}{2}mr^2(\dot{q}_1^2 + \dot{q}_2^2 + \dot{q}_3^2).$$

q_1, q_2, q_3 are just like Cartesian coordinates. The transformation between the three normal coordinates and the three "Cartesian" coordinates q_1, q_2, q_3 must be linear. We already have

$$\xi = q_1 + q_2\sqrt{\frac{M}{m}} + q_3, \quad \eta = q_1 - q_3.$$

Assume the third normal coordinate to be

$$\zeta = Aq_1 + Bq_2 + Cq_3.$$

It should be orthogonal to the ξ-, η-axes. Resolving along the q_i-axes we have

$$\xi = \left(1, \sqrt{\frac{M}{m}}, 1\right), \quad \eta = (1, 0, -1), \quad \zeta = (A, B, C).$$

Orthogonality means that

$$\zeta \cdot \xi = A + B\sqrt{\frac{M}{m}} + C = 0,$$

$$\zeta \cdot \eta = A - C = 0,$$

which yield $A = C$, $B = -2A\sqrt{\frac{m}{M}}$. Since a normal coordinate remains so after multiplying it with a nonzero constant, we can set $A = 1$, then

$$\zeta = q_1 - 2q_2\sqrt{\frac{m}{M}} + q_3 = \theta_1 - 2\theta_2 + \theta_3.$$

The equations of motion then give

$$\ddot{\zeta} + \left(\frac{(2m + M)K}{mM}\right)\zeta = 0,$$

yielding

$$\omega_3 = \sqrt{\frac{(2m + M)K}{mM}}.$$

c. $\omega_1, \omega_2, \omega_3$ are the normal-mode angular frequencies corresponding to the three normal coordinates ξ, η, ζ respectively.

2048

A ring of mass M and radius R is supported from a pivot located at one point of the ring, about which it is free to rotate in its own vertical plane. A bead of mass m slides without friction about the ring (Fig. 2.48).

a. Write the Lagrangian for this system.

b. Write the equations of motion.

c. Describe the normal-modes for small oscillations in the limits $m \gg M$ and $m \ll M$.

d. Find the frequencies of the normal-modes of small oscillations for general m and M.

(UC, Berkeley)

Sol:

a. Use coordinates as shown in Fig. 2.48. The mass m and the center of mass of the ring have coordinates

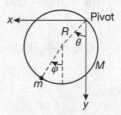

Fig. 2.48

$$(R \sin \theta + R \sin \varphi, \; R \cos \theta + R \cos \varphi), \quad (R \sin \theta, \; R \cos \theta)$$

and velocities

$$(R \dot{\theta} \cos \theta + R \dot{\varphi} \cos \varphi, \; -R \dot{\theta} \sin \theta - R \dot{\varphi} \sin \varphi), \quad (R \dot{\theta} \cos \theta, \; -R \dot{\theta} \sin \theta)$$

respectively. The ring has moment of inertial $2MR^2$ about the pivot. The Lagrangian of the system is then

$$L = T - V$$
$$= MR^2 \dot{\theta}^2 + \frac{1}{2} mR^2 [\dot{\theta}^2 + \dot{\varphi}^2 + 2 \dot{\theta} \dot{\varphi} \cos(\theta - \varphi)]$$
$$+ (M + m)gR \cos \theta + mgR \cos \varphi,$$

taking the pivot as the reference level of potential energy.

b. Lagrange's equations

$$\frac{d}{dt}\left(\frac{\partial L}{\partial \dot{q}_i}\right) - \frac{\partial L}{\partial q_i} = 0$$

give the equations of motion

$$(2M + m)R\ddot{\theta} + mR\ddot{\varphi}\cos(\theta - \varphi)$$
$$+ mR\dot{\varphi}^2\sin(\theta - \varphi) + (m + M)g\sin\theta = 0,$$
$$R\ddot{\varphi} + R\ddot{\theta}\cos(\theta - \varphi) - R\dot{\theta}^2\sin(\theta - \varphi) + g\sin\varphi = 0\cdot$$

(c),(d) For small oscillations, $\theta, \varphi, \dot{\theta}, \dot{\varphi}$ are small and the above reduce to

$$(2M + m)R\ddot{\theta} + mR\ddot{\varphi} + (M + m)g\theta = 0,$$

$$R\ddot{\varphi} + R\ddot{\theta} + g\varphi = 0.$$

Try a solution of the type $\theta = Ae^{i\omega t}$, $\varphi = Be^{i\omega t}$ and write these equations as matrix equation

$$\begin{pmatrix} (M + m)g - (2M + m)R\omega^2 & -mR\omega^2 \\ -R\omega^2 & g - R\omega^2 \end{pmatrix} \begin{pmatrix} A \\ B \end{pmatrix} = 0.$$

The secular equation

$$\begin{vmatrix} (m + M)g - (2M + m)R\omega^2 & -mR\omega^2 \\ -R\omega^2 & g - R\omega^2 \end{vmatrix}$$

$$= (2R\omega^2 - g)[MR\omega^2 - (m + M)g] = 0$$

has positive roots

$$\omega_1 = \sqrt{\frac{g}{2R}}, \quad \omega_2 = \sqrt{\frac{(m + M)g}{MR}},$$

which are the normal-mode angular frequencies of the system. The ratio of the amplitudes is

$$\frac{A}{B} = \frac{g - R\omega^2}{R\omega^2} = \begin{cases} 1 & \text{for } \omega = \omega_1, \\ -\dfrac{m}{M + m} & \text{for } \omega = \omega_2. \end{cases}$$

If $m \gg M$,

$$\omega_1 = \sqrt{\frac{g}{2R}}, \quad \frac{A}{B} = 1,$$

i.e. θ and φ have the same amplitude and phase;

$$\omega_2 = \sqrt{\frac{mg}{MR}}, \quad \frac{A}{B} = -1,$$

i.e. θ and φ have the same amplitude but opposite phases.

If $m \ll M$,

$$\omega_1 = \sqrt{\frac{g}{2R}}, \quad \frac{A}{B} = 1,$$

i.e. θ and φ have the same amplitude and phase as in the above;

$$\omega_2 = \sqrt{\frac{g}{R}}, \quad \frac{A}{B} = -\frac{m}{M},$$

i.e. θ has a much smaller amplitude than φ and the two oscillations are opposite in phase.

2049

A body of mass m is constrained to move in vertical parabola with $y = kx^2$. Describe the motion of the particle near the vertex, assuming no friction.

Sol: Resolving the forces along and perpendicular to the tangent,

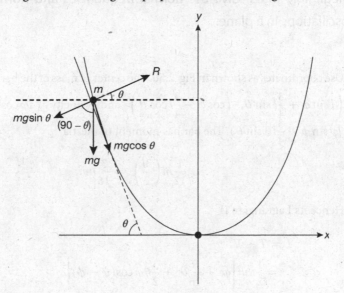

Fig. 2.49

$$mg \sin \theta = R$$
$$mg \cos \theta = m\ddot{s}$$

From the first two equations

$$R \cot \theta = m\ddot{s}$$

Equation of the parabola

$$y = kx^2$$

Hence

$$\frac{dy}{dx} = 2kx = -\cot\theta$$

and

$$m\ddot{s} = -2Rkx$$

Now $s = r\theta$ and $\sin\theta = \dfrac{x}{r}$

Since θ is small near the vertex $\sin\theta \approx \theta$ and $s \approx x$

For motion near the vertex, $m\ddot{x} = -2Rkx$ representing simple harmonic motion (SHM).

2050

A thin uniform bar of mass m and length $\dfrac{3l}{2}$ is suspended by a string of length l and negligible mass. Give the normal frequencies and normal-modes for small oscillations in a plane.

(Columbia)

Sol: Use coordinates as shown in Fig. 2.50. The center of mass of the bar has coordinates $\left(l\sin\varphi + \dfrac{3}{4}l\sin\theta, -l\cos\varphi - \dfrac{3}{4}l\cos\theta\right)$ and velocity $\left(l\dot{\varphi}\cos\varphi + \dfrac{3}{4}l\dot{\theta}\cos\theta, \right.$ $\left. l\dot{\varphi}\sin\phi + \dfrac{3}{4}l\dot{\theta}\sin\theta\right)$. The bar has moment of inertia

$$\frac{1}{12}m\left(\frac{3l}{2}\right)^2 = \frac{3}{16}ml^2.$$

Hence its Lagrangian is

$$
\begin{aligned}
L &= T - V \\
&= \frac{1}{2}ml^2\left[\dot{\varphi}^2 + \frac{9}{16}\dot{\theta}^2 + \frac{3}{2}\dot{\theta}\dot{\varphi}\cos(\theta - \phi)\right] \\
&\quad + \frac{3}{32}ml^2\dot{\theta}^2 + mgl\left(\cos\varphi + \frac{3}{4}\cos\theta\right) \\
&\approx \frac{1}{2}ml^2\left(\frac{3}{4}\dot{\theta}^2 + \dot{\varphi}^2 + \frac{3}{2}\dot{\theta}\dot{\varphi}\right) + \frac{7}{4}mgl - \frac{1}{2}mgl\left(\varphi^2 + \frac{3}{4}\theta^2\right)
\end{aligned}
$$

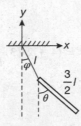

Fig. 2.50

for small oscillations, retaining only terms of up to the second order of the small quantities $\theta, \varphi, \dot{\theta}, \dot{\varphi}$.

Lagrange's equation

$$\frac{d}{dt}\left(\frac{\partial L}{\partial \dot{q}_i}\right) - \frac{\partial L}{\partial q_i} = 0$$

give

$$\frac{3}{4}l\ddot{\theta} + l\ddot{\varphi} + g\varphi = 0,$$
$$l\ddot{\theta} + l\ddot{\varphi} + g\theta = 0.$$

With a solution of the form

$$\theta = Ae^{i\omega t}, \quad \varphi = Be^{i\omega t},$$

the above give

$$\begin{pmatrix} -\frac{3}{4}l\omega^2 & g - l\omega^2 \\ g - l\omega^2 & -l\omega^2 \end{pmatrix}\begin{pmatrix} A \\ B \end{pmatrix} = 0.$$

The secular equation

$$\begin{vmatrix} -\frac{3}{4}l\omega^2 & g - l\omega^2 \\ g - l\omega^2 & -l\omega^2 \end{vmatrix} = 0,$$

i.e.

$$l^2\omega^4 - 8lg\omega^2 + 4g^2 = 0,$$

has solutions

$$\omega^2 = (4 \pm 2\sqrt{3})\frac{g}{l} = (1 \pm \sqrt{3})^2\frac{g}{l},$$

or

$$\omega = (\sqrt{3} \pm 1)\sqrt{\frac{g}{l}},$$

since ω has to be positive. Hence the normal-mode angular frequencies are

$$\omega_1 = (\sqrt{3} + 1)\sqrt{\frac{g}{l}}, \qquad \omega_2 = (\sqrt{3} - 1)\sqrt{\frac{g}{l}}.$$

The ratio of amplitudes is

$$\frac{B}{A} = \frac{g - l\omega^2}{l\omega^2} = \begin{cases} \sqrt{-\dfrac{3}{2}} & \text{for } \omega = \omega_1, \\ \sqrt{\dfrac{3}{2}} & \text{for } \omega = \omega_2. \end{cases}$$

Thus in the normal-mode given by ω_1, θ and φ are opposite in phase, while in that given by ω_2, θ and φ are in phase. In both cases the ratio of the amplitude of φ to that of θ is

$$\sqrt{3} : 2.$$

2051

A simple pendulum consisting of a mass m and weightless string of length l is mounted on a support of mass M which is attached to a horizontal spring with force constant k as shown in Fig. 2.51.

a. Set up Lagrange's equations.

b. Find the frequencies for small oscillations.

(Columbia)

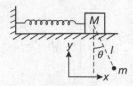

Fig. 2.51

Sol:

a. Use coordinates with origin at the position of m when the system is in equilibrium, and the x- and y- axes along the horizontal and vertical directions respectively as shown in Fig. 2.51. Then M and m have coordinates and velocities

$$(x, l), \quad (x + l\sin\theta, l - l\cos\theta)$$
$$(\dot{x}, 0), \quad (\dot{x} + l\dot{\theta}\cos\theta, l\dot{\theta}\sin\theta)$$

respectively. The Lagrangian of the system is

$$
L = T - V
$$
$$
= \frac{1}{2}M\dot{x}^2 + \frac{1}{2}m(\dot{x}^2 + l^2\dot{\theta}^2 + 2l\dot{x}\dot{\theta}\cos\theta) - Mgl - mgl(1 - \cos\theta) - \frac{1}{2}kx^2.
$$

Lagrange's equations

$$
\frac{d}{dt}\left(\frac{\partial L}{\partial \dot{q}_i}\right) - \frac{\partial L}{\partial q_i} = 0
$$

then give

$$
(M + m)\ddot{x} - ml\dot{\theta}^2\sin\theta + ml\ddot{\theta}\cos\theta + kx = 0,
$$
$$
l\ddot{\theta} + \ddot{x}\cos\theta + g\sin\theta = 0.
$$

b. For small oscillations, $x, \theta, \dot{x}, \dot{\theta}$ are small quantities. Neglecting terms of orders higher than two, the equations of motion become

$$
(M + m)\ddot{x} + ml\ddot{\theta} + kx = 0,
$$
$$
l\ddot{\theta} + \ddot{x} + g\theta = 0.
$$

Set

$$
x = A\exp(i\omega t), \quad \theta = B\exp(i\omega t).
$$

These equations become

$$
\begin{pmatrix} k - (M + m)\omega^2 & -ml\omega^2 \\ -\omega^2 & g - l\omega^2 \end{pmatrix}\begin{pmatrix} A \\ B \end{pmatrix} = 0.
$$

The secular equation

$$\begin{vmatrix} k - (M + m)\omega^2 & -ml\omega^2 \\ -\omega^2 & g - l\omega^2 \end{vmatrix} = Ml\omega^2 - [g(M + m) + kl]\omega^2 + gk = 0$$

has two positive roots

$$\omega_1 = \left[\frac{g(M + m) + kl + \sqrt{[g(M + m) + kl]^2 - 4Mlgk}}{2Ml} \right]^{\frac{1}{2}},$$

$$\omega_2 = \left[\frac{g(M + m) + kl - \sqrt{[g(M + m) + kl]^2 - 4Mlgk}}{2Ml} \right]^{\frac{1}{2}},$$

which are the normal-mode angular frequencies of the system.

2052

Two masses, $2m$ and m, are suspended from a fixed frame by elastic springs as shown in Fig. 2.52. The elastic constant (force/unit length) of each spring is k. Consider only vertical motion.

a. Calculate the frequencies of the normal-modes of oscillations of this system.

b. The upper mass $2m$ is slowly displaced downwards from the equilibrium position by a distance l and then let go, so that the system performs free oscillations. Calculate the subsequent motion of the lower mass m.

(Columbia)

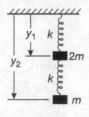

Fig. 2.52

Sol: Let the natural lengths of the upper and lower springs be l_1, l_2, and denote the positions of the upper and lower masses by y_1, y_2 as shown in Fig. 2.52, respectively. The Lagrangian is then

$$L = T - V$$
$$= m\dot{y}_1^2 + \frac{1}{2}m\dot{y}_2^2 + 2mgy_1 + mgy_2 - \frac{1}{2}k(y_1 - l_1)^2 - \frac{1}{2}k(y_2 - y_1 - l_2)^2$$
$$= \frac{1}{2}m(2\dot{y}_1^2 + \dot{y}_2^2) + mg(2y_1 + y_2) - \frac{1}{2}k[(y_1 - l_1)^2 + (y_2 - y_1 - l_2)^2].$$

Lagrange's equations

$$\frac{d}{dt}\left(\frac{\partial L}{\partial \dot{q}_i}\right) - \frac{\partial L}{\partial q_i} = 0$$

give

$$2m\ddot{y}_1 + 2ky_1 - ky_2 = 2mg + kl_1 - kl_2,$$
$$m\ddot{y}_2 + ky_2 - ky_1 = mg + kl_2.$$

Let $y_1 = y_1' + \eta_1, y_2 = y_2' + \eta_2$.

The above can be written as

$$2m\ddot{y}_1' + 2ky_1' - ky_2' = 0,$$
$$m\ddot{y}_2' + ky_2' - ky_1' = 0,$$

if we set

$$\eta_1 = \frac{3mg + kl_1}{k}, \quad \eta_2 = \frac{4mg + kl_1 + kl_2}{k}.$$

Note that $y_1 = \eta_1, y_2 = \eta_2$ are the equilibrium positions of the masses $2m$ and m respectively, as can be seen from the force equations

$$3mg = k(y_1 - l_1),$$
$$mg = k(y_2 - y_1 - l_2).$$

With a solution of the type

$$y_1' = Ae^{i\omega t}, y_2' = Be^{i\omega t},$$

we have

$$\begin{pmatrix} 2k - 2m\omega^2 & -k \\ -k & k - m\omega^2 \end{pmatrix}\begin{pmatrix} A \\ B \end{pmatrix} = 0.$$

The secular equation

$$\begin{vmatrix} 2k - 2m\omega^2 & -k \\ -k & k - m\omega^2 \end{vmatrix} = 0$$

has two positive roots

$$\omega_\pm = \sqrt{\frac{k}{m}\left(1 \pm \frac{1}{\sqrt{2}}\right)},$$

which are the angular frequencies of the normal-modes of oscillation. As

$$\frac{B}{A} = \frac{2k - 2m\omega^2}{k} = \mp\sqrt{2},$$

the corresponding normal-modes are $\begin{pmatrix} 1 \\ -\sqrt{2} \end{pmatrix}$ and $\begin{pmatrix} 1 \\ \sqrt{2} \end{pmatrix}$.

(b) The general motion of the system is given by

$$y_1' = A\cos(\omega_+ t + \varphi_1) + A'\cos(\omega_- t + \varphi_2),$$
$$y_2' = -\sqrt{2}A\cos(\omega_+ t + \varphi_1) + \sqrt{2}A'\cos(\omega_- t + \varphi_2).$$

The initial condition is that at $t = 0$,

$$y_1' = y_2' = l, \quad \dot{y}_1' = \dot{y}_2' = 0.$$

This gives

$$\varphi_1 = \varphi_2 = 0,$$

$$A = \frac{l}{2}\left(1 - \frac{1}{\sqrt{2}}\right), \quad A' = \frac{l}{2}\left(1 + \frac{1}{\sqrt{2}}\right).$$

Hence the motion of the mass $2m$ is described by

$$y_2 = \left(\frac{1}{2} - \frac{1}{\sqrt{2}}\right)l\cos\left[\sqrt{\frac{k}{m}\left(1 + \frac{1}{\sqrt{2}}\right)}\,t\right]$$

$$+ \left(\frac{1}{2} + \frac{1}{\sqrt{2}}\right)l\cos\left[\sqrt{\frac{k}{m}\left(1 - \frac{1}{\sqrt{2}}\right)}\,t\right] + l_1 + l_2 + \frac{4mg}{k}.$$

2053

Three massless springs of natural length $\sqrt{2}$ and spring constant K are attached to a point particle of mass m and to the fixed points $(-1, 1)$, $(1, 1)$ and $(-1, -1)$ as shown in Fig. 2.53. The point mass m is allowed to move in the (x, y)-plane only.

   a. Write the Lagrangian for the system.

   b. Is there a stable equilibrium for the point mass? Where is it?

   c. Give the Lagrangian appropriate for small oscillations.

   d. Introduce normal coordinates and solve for the motion of the particle in the small oscillation approximation.

   e. Sketch the normal-modes of vibration.

(Columbia)

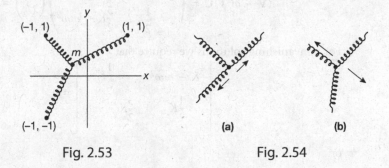

Fig. 2.53   Fig. 2.54

Sol:

   a. Let the coordinates of the mass m be (x, y). Its Lagrangian is then

$$L = T - V$$

$$= \frac{1}{2}m\dot{x}^2 + \frac{1}{2}m\dot{y}^2$$

$$- \frac{1}{2}[K\sqrt{(x-1)^2 + (y-1)^2} - \sqrt{2}]^2.$$

$$- \frac{1}{2}K[\sqrt{(x+1)^2 + (y-1)^2} - \sqrt{2}]^2$$

$$- \frac{1}{2}K[\sqrt{(x+1)^2 + (y+1)^2} - \sqrt{2}]^2$$

b. From the conditions of a stable equilibrium

$$\frac{\partial V}{\partial x} = 0, \quad \frac{\partial V}{\partial y} = 0, \quad \frac{\partial^2 V}{\partial x^2} + \frac{2\partial^2 V}{\partial x \partial y} + \frac{\partial^2 V}{\partial y^2} > 0,$$

we find one stable equilibrium position $(0, 0)$.

c. For small oscillations, $x, y, \dot{x}, \dot{y}$ are small quantities. Expanding L and retaining only the lowest-order terms in these small quantities we have

$$L = \frac{1}{2}m\dot{x}^2 + \frac{1}{2}m\dot{y}^2 - \frac{1}{4}K(3x^2 + 2xy + 3y^2).$$

d. The kinetic and potential energies can respectively be represented by matrices

$$\mathbf{T} = \begin{pmatrix} m & 0 \\ 0 & m \end{pmatrix}, \quad \mathbf{V} = \begin{pmatrix} \frac{3}{2}K & \frac{1}{2}K \\ \frac{1}{2}K & \frac{3}{2}K \end{pmatrix}.$$

We have the matrix equation

$$(\mathbf{V} - \omega^2\mathbf{T})\mathbf{U} = \begin{pmatrix} \frac{3}{2}K - m\omega^2 & \frac{1}{2}K \\ \frac{1}{2}K & \frac{3}{2}K - m\omega^2 \end{pmatrix}\begin{pmatrix} U_1 \\ U_2 \end{pmatrix} = 0.$$

For nonvanishing solutions we require that

$$\begin{vmatrix} \frac{3}{2}K - m\omega^2 & \frac{1}{2}K \\ \frac{1}{2}K & \frac{3}{2}K - m\omega^2 \end{vmatrix} = 0,$$

or

$$(2K - m\omega^2)(K - m\omega^2) = 0.$$

Its two positive roots give the normal frequencies and the corresponding normal-modes of vibrations

$$\omega_1 = \sqrt{\frac{2K}{m}}, \quad U_1 = \begin{pmatrix} 1 \\ 1 \end{pmatrix}$$

$$\omega_2 = \sqrt{\frac{K}{m}}, \quad U_2 = \begin{pmatrix} 1 \\ -1 \end{pmatrix}.$$

The general motion of the particle for small oscillations is then

$$\begin{pmatrix} x \\ y \end{pmatrix} = A\begin{pmatrix} 1 \\ 1 \end{pmatrix}\cos(\omega_1 t + \varphi_1) + B\begin{pmatrix} 1 \\ -1 \end{pmatrix}\cos(\omega_2 t + \varphi_2),$$

where A, B, φ_1, φ_2 are constants to be determined from the initial conditions. The normal coordinates are given by

$$\bar{q}_s = \sum_{i,j} U_i a_{ij} q_j,$$

where a_{ij} are the elements of the matrix $\mathbf{T}$. Thus for the ω_1 mode, the normal coordinate is

$$\xi = U_1 m x + U_2 m y = U_1 m (x + y).$$

The constant factor $U_1 m$ is immaterial and we can take

$$\xi = x + y.$$

Similarly for the ω_2 mode

$$\eta = U_1 m (x - y)$$

and we can take

$$\eta = x - y.$$

ξ, η are the normal coordinates of the system.

e.　For $\omega_1 = \sqrt{\dfrac{2K}{m}}$,

$$U_1 = \begin{pmatrix} 1 \\ 1 \end{pmatrix}$$

so the point mass oscillates along the line $y = -x$ as shown in Fig. 2.54 (a).

For $\omega_2 = \sqrt{\dfrac{K}{m}}$,

$$U_2 = \begin{pmatrix} 1 \\ -1 \end{pmatrix},$$

and the point mass oscillates along the line $y = -x$ as shown in Fig. 2.54 (b).

2054

One simple pendulum is hung from another; that is, the string of the lower pendulum is tied to the bob of the upper one. Using arbitrary lengths for the strings and arbitrary masses for the bobs, set up the Lagrangian of the system. Use the angles each string makes with the vertical as generalized coordinates. Discuss small oscillations of this system. What are the normal-modes? What are the corresponding frequencies? Show that in the special case of equal masses and equal lengths the frequencies are given by $\sqrt{\dfrac{g(2 \pm \sqrt{2})}{l}}$. Under what conditions will the system move as a single piece?

(*Columbia*)

Sol:

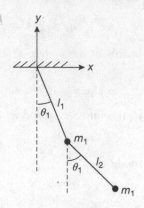

Fig. 2.55

Let m_1, m_2 be the masses of the bobs and l_1, l_2 the lengths of the two strings, as shown in Fig. 2.55. The two bobs m_1, m_2 have coordinates

$$(l_1 \sin \theta_1, -l_1 \cos \theta_1), \quad (l_1 \sin \theta_1 + l_2 \sin \theta_2, -l_1 \cos \theta_1 - l_2 \cos \theta_2)$$

and velocities

$$(l_1 \dot\theta_1 \cos \theta_1, l_1 \dot\theta_1 \sin \theta_1), (l_1 \dot\theta_1 \cos \theta_1 + l_2 \dot\theta_2 \cos \theta_2, l_1 \dot\theta_1 \sin \theta_1 + l_2 \dot\theta_2 \sin \theta_2)$$

respectively. Then the kinetic energy T of the system is given by

$$2T = m_1 l_1^2 \dot\theta_1^2 + m_2 [l_1^2 \dot\theta_2^2 + l_2^2 \dot\theta_2^2 + 2 l_1 l_2 \dot\theta_1 \dot\theta_2 \cos(\theta_2 - \theta_1)]$$

$$= (m_1 + m_2) l_1^2 \dot\theta_1^2 + m_2 l_2^2 \dot\theta_2^2 + 2 m_2 l_1 l_2 \dot\theta_1 \dot\theta_2 \cos(\theta_2 - \theta_1)$$

$$\approx (m_1 + m_2) l_1^2 \dot\theta_1^2 + m_2 l_2^2 \dot\theta_2^2 + 2 m_2 l_1 l_2 \dot\theta_1 \dot\theta_2,$$

and the potential energy V is given by

$$2V = -2 m_1 g l_1 \cos \theta_1 - 2 m_2 g (l_1 \cos \theta_1 + l_2 \cos \theta_2)$$

$$\approx -2(m_1 + m_2) g l_1 \left(1 - \frac{1}{2}\theta_1^2\right) - m_2 g l_2 \left(1 - \frac{1}{2}\theta_2^2\right)$$

$$= V_0 + (m_1 + m_2) g l_1 \theta_1^2 + m_2 g l_2 \theta_2^2.$$

For small oscillations, we have retained only terms of up to the order two of the small quantities $\theta_1, \theta_2, \dot{\theta}_1, \dot{\theta}_2$. The Lagrangian of the system is given by $L = T - V$. To find the normal-modes we write these in matrix form:

$$2T = \sum_{i,j=1}^{2} M_{ij}\dot{\theta}_i\dot{\theta}_j = \dot{\Theta}'\mathbf{M}\dot{\Theta},$$

$$2V = V_0 + \sum_{i,j=1}^{2} K_{ij}\theta_i\theta_j = V_0 + \Theta'\mathbf{K}\Theta,$$

with

$$\mathbf{M} = \begin{pmatrix} (m_1 + m_2)l_1^2 & m_2l_1l_2 \\ m_2l_1l_2 & m_2l_2^2 \end{pmatrix},$$

$$\mathbf{K} = \begin{pmatrix} (m_1 + m_2)gl_1 & 0 \\ 0 & m_2gl_2 \end{pmatrix},$$

$$\Theta = \begin{pmatrix} \theta_1 \\ \theta_2 \end{pmatrix}, \quad \dot{\Theta} = \begin{pmatrix} \dot{\theta}_1 \\ \dot{\theta}_2 \end{pmatrix},$$

and $\Theta', \dot{\Theta}'$ being the transpose matrices of $\Theta, \dot{\Theta}$ respectively. Considering a solution of the type

$$\begin{pmatrix} \theta_2 \\ \theta_2 \end{pmatrix} = \begin{pmatrix} A_1 \\ A_2 \end{pmatrix} \cos(\omega t + \varepsilon),$$

we have

$$(\mathbf{K} - \omega^2\mathbf{M})\mathbf{A} = 0,$$

i.e.

$$\begin{pmatrix} (m_1 + m_2)l_1(g - l_1\omega^2) & -m_2l_1l_2\omega^2 \\ -m_2l_1l_2\omega^2 & m_2l_2(g - l_2\omega^2) \end{pmatrix} \begin{pmatrix} A_1 \\ A_2 \end{pmatrix} = 0.$$

For A_1, A_2 not to be zero identically we require

$$\begin{vmatrix} (m_1 + m_2)l_1(g - l_1\omega^2) & -m_2l_1l_2\omega^2 \\ -m_2l_1l_2\omega^2 & m_2l_2(g - l_2\omega^2) \end{vmatrix} = 0,$$

or

$$m_1l_1l_2\omega^4 - (l_2 + l_2)(m_1 + m_2)g\omega^2 + (m_1 + m_2)g^2 = 0.$$

Its positive roots

$$
\omega_{\pm} = \left[\frac{g}{2l_1 l_2 m_1} \left\{ (m_1 + m_2)(l_1 + l_2) \right. \right.
$$

$$
\left. \left. \pm \sqrt{(m_1 + m_2)[m_2(l_1 + l_2)^2 + m_1(l_1 - l_2)^2]} \right\} \right]^{\frac{1}{2}}
$$

are the normal-mode angular frequencies of the system. As

$$
\frac{A_1}{A_2} = \frac{l_2}{l_1} \left(\frac{g}{l_2 \omega^2} - 1 \right)
$$

$$
= \frac{1}{2l_1} \left\{ (l_1 - l_2) \mp \sqrt{\frac{m_2(l_1 + l_2)^2 + m_1(l_1 - l_2)^2}{m_1 + m_2}} \right\},
$$

the normal-modes are given by

$$
\begin{pmatrix} \theta_1 \\ \theta_2 \end{pmatrix} = \begin{pmatrix} \dfrac{l_1 - l_2}{2l_1} \mp \dfrac{1}{2l_1}\sqrt{[m_2(l_1 + l_2)^2 + m_1(l_1 - l_2)^2]/(m_1 + m_2)} \\ 1 \end{pmatrix}
$$

$$
\times A \pm \cos(\omega \pm t + \varepsilon \pm),
$$

where the top and bottom signs correspond to ω_+, and ω_- respectively.

The general solution is

$$
\theta_1 = \left\{ \frac{l_1 - l_2}{2l_1} - \frac{1}{2l_1}\sqrt{\frac{m_2(l_1 + l_2)^2 + m_1(l_1 - l_2)^2}{m_1 + m_2}} \right\} A_+ \cos(\omega_+ t + \varepsilon_+)
$$

$$
+ \left\{ \frac{l_1 - l_2}{2l_1} + \frac{1}{2l_1}\sqrt{\frac{m_2(l_1 + l_2)^2 + m_1(l_1 - l_2)^2}{m_1 + m_2}} \right\} A_- \cos(\omega_- t + \varepsilon_-),
$$

$$
\theta_2 = A_+ \cos(\omega_+ t + \varepsilon_+) + A_- \cos(\omega_- t + \varepsilon_-),
$$

where A_+, A_-, ε_+ and ε_- are constants to be determined from the initial conditions.

In the special case of equal masses and equal lengths, $m_1 = m_2 = m, l_1 = l_2 = l$, the normal frequencies are

$$
\omega_{\pm} = \sqrt{\frac{g}{l}(2 \pm \sqrt{2})}.
$$

For the system to move as a single piece, we require $\theta_1 = \theta_2$, i.e.

$$\frac{1}{2l_1}\left[(l_1 - l_2) \mp \sqrt{\frac{m_2(l_1 + l_2)^2 + m_1(l_1 - l_2)^2}{m_1 + m_2}}\right] = 1,$$

or

$$(m_1 + m_2)(l_1 + l_2) = \mp\sqrt{(m_1 + m_2)^2(l_1 + l_2)^2 - 4m_1(m_1 + m_2)l_1l_2}.$$

As the left-hand side is positive, the bottom sign of the right-hand side has to be used. Furthermore, squaring both sides gives

$$l_1l_2m_1(m_1 + m_2) = 0.$$

This requires either $l_1 = 0$, or $l_2 = 0$, or $m_1 = 0$. Each of these cases will reduce the two-pendulum system into a one-pendulum one. Hence the two-pendulum system cannot move as a single piece.

2055

a. Consider two simple pendulums each of mass m and length l joined by a massless spring with spring constant k as shown in Fig. 2.56(a). The distance between the pivots is chosen so that the spring is unstreched when the pendulums are vertical. Find the frequencies and normal-modes for the small oscillations of this system about equilibrium.

b. Now consider an infinite row of pendulums with each pendulum connected to its neighbors just as the pair in part (a) is connected, as shown in Fig. 2.56(b). Find the normal-modes and the corresponding frequencies for this new system.

(Columbia)

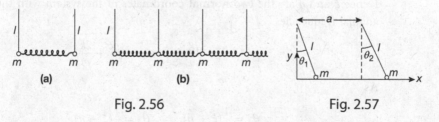

Fig. 2.56   Fig. 2.57

Sol:

a. Let a be the natural length of each spring. Number the pendulums from the left and use coordinates with the origin at the equilibrium position of the bob of pendulum 1 and the x-, y-axes along the horizontal and vertical directions, as shown in Fig. 2.57. Then the two bobs have coordinates

$$(l \sin \theta_1, l(1 - \cos \theta_1)), \quad (a + l \sin \theta_2, l(1 - \cos \theta_2))$$

and velocities

$$(l\dot{\theta}_1 \cos \theta_1, l\dot{\theta}_1 \sin \theta_1), (l\dot{\theta}_2 \cos \theta_2, l\dot{\theta}_2 \sin \theta_2)$$

respectively. The Lagrangian of the system is

$$
\begin{aligned}
L &= T - V \\
&= \frac{1}{2}m(l^2\dot{\theta}_1^2 + l^2\dot{\theta}_2^2) - mgl(2 - \cos \theta_1 - \cos \theta_2) \\
&\quad - \frac{1}{2}k(a + l \sin \theta_2 - l \sin \theta_1 - a)^2 \\
&\approx \frac{1}{2}l^2(\dot{\theta}_1^2 + \dot{\theta}_2^2) - \frac{1}{2}mgl(\theta_1^2 + \theta_2^2) - \frac{1}{2}kl^2(\theta_2 - \theta_1)^2
\end{aligned}
$$

for small oscillations.

Lagrange's equations

$$\frac{d}{dt}\left(\frac{\partial L}{\partial \dot{q}_i}\right) - \frac{\partial L}{\partial q_i} = 0$$

give

$$
\begin{aligned}
ml^2\ddot{\theta}_1 + mgl\theta_1 - kl^2(\theta_2 - \theta_1) &= 0, \\
ml^2\ddot{\theta}_2 + mgl\theta_2 + kl^2(\theta_2 - \theta_1) &= 0.
\end{aligned}
$$

Let $\xi = \theta_1 + \theta_2, \eta = \theta_1 - \theta_2$. The sum and difference of the above two equations give

$$
\begin{aligned}
l\ddot{\xi} + g\xi &= 0, \\
ml\ddot{\eta} + (mg + 2kl)\eta &= 0.
\end{aligned}
$$

Hence ξ and η are the two normal coordinates of the system with the normal angular frequencies

$$\omega_1 = \sqrt{\frac{g}{l}}, \qquad \omega_2 = \sqrt{\frac{g}{l} + \frac{2k}{m}}.$$

As

$$\theta_1 = \frac{1}{2}(\xi + \eta), \qquad \theta_2 = \frac{1}{2}(\xi - \eta),$$

their amplitudes u_1, u_2 have the ratio

$$u_1 : u_2 = 1 : 1$$

for the ω_1 mode, for which $\eta = 0$, and

$$u_1 : u_2 = 1 : -1$$

for the ω_2 mode, for which $\xi = 0$.

b. The same treatment gives

$$L = T - V$$
$$= \frac{1}{2}ml^2(\dot{\theta}_1^2 + \dot{\theta}_2^2 + \cdots + \dot{\theta}_n^2 + \cdots) - \frac{1}{2}mgl(\theta_1^2 + \theta_2^2 + \cdots + \theta_n^2 + \ldots)$$
$$- \frac{1}{2}kl^2[(\theta_2 - \theta_1)^2 + (\theta_3 - \theta_2)^2 + \ldots$$
$$+ (\theta_n - \theta_{n-1})^2 + (\theta_{n+1} - \theta_n)^2 + \cdots].$$

Lagrange's equations then give

$$ml^2\ddot{\theta}_n + mgl\theta_n + kl^2[(\theta_n - \theta_{n-1}) - (\theta_{n+1} - \theta_n)] = 0,$$

i.e.

$$ml\ddot{\theta}_n + mg\theta_n + kl(2\theta_n - \theta_{n+1} - \theta_{n-1}) = 0.$$

Since θ_n remains finite as $n \to \infty$, assume the amplitude varies periodically along the x-axis and try

$$\theta_n = Ae^{i(\kappa na - \omega t)},$$

where the "wave number" $\kappa = \dfrac{2\pi}{\lambda}$, with the "wavelength" λ being integral multiples of a, i.e. $\lambda = pa$, $p = 1, 2, 3, \ldots$. Substitution gives

$$\omega = \sqrt{\frac{g}{l} + \frac{2k}{m}[1 - \cos(\kappa a)]}.$$

The first few normal angular frequencies are for

$$p = 1, \qquad \omega_1 = \sqrt{\frac{g}{l}},$$

$$p = 2, \qquad \omega_2 = \sqrt{\frac{g}{l} + \frac{4k}{m}},$$

$$p = 3, \qquad \omega_3 = \sqrt{\frac{g}{l} + \frac{3k}{m}},$$

$$p = 4, \qquad \omega_4 = \sqrt{\frac{g}{l} + \frac{2k}{m}},$$

....

The corresponding normal-modes (for $p = 1, 2, 3, 4, \ldots$) are

$$
\begin{pmatrix} \theta_1 \\ \theta_2 \\ \vdots \end{pmatrix} = \begin{pmatrix} 1 \\ 1 \\ \vdots \end{pmatrix} Ae^{-i\omega t}, \quad \begin{pmatrix} -1 \\ 1 \\ -1 \\ 1 \\ \vdots \end{pmatrix} Ae^{-i\omega t},
$$

$$
\begin{pmatrix} e^{i\frac{2}{3}\pi} \\ e^{i\frac{4}{3}\pi} \\ 1 \\ e^{i\frac{2}{3}\pi} \\ e^{i\frac{4}{3}\pi} \\ 1 \\ \vdots \end{pmatrix} Ae^{-i\omega t}, \quad \begin{pmatrix} e^{i\frac{1}{2}\pi} \\ e^{i\pi} \\ e^{i\frac{3}{2}\pi} \\ 1 \\ e^{i\frac{1}{2}\pi} \\ e^{i\pi} \\ e^{i\frac{3}{2}\pi} \\ 1 \\ \vdots \end{pmatrix} Ae^{-i\omega t}, \ldots
$$

2056

Consider a particle of mass m moving in two dimensions in a potential

$$
V(x, y) = -\frac{1}{2}kx^2 + \frac{1}{2}\lambda_0 x^2 y^2 + \frac{1}{4}\lambda_1 x^4, \quad k, \lambda_0, \lambda_1 > 0.
$$

a. At what point (x_0, y_0) is the particle in stable equilibrium?
b. Give the Lagrangian appropriate for small oscillations about this equilibrium position.
c. What are the normal frequencies of vibration in (b)?

(Columbia)

Sol:

a. A point where $\partial V/\partial x = 0$, $\partial V/\partial y = 0$, $\partial^2 V/\partial x^2 > 0$, $\partial^2 V/\partial y^2 > 0$ and

$$
\frac{\partial^2 V}{\partial x^2}(dx)^2 + 2\frac{\partial^2 V}{\partial x \partial y}dxdy + \frac{\partial^2 V}{\partial y^2}(dy)^2 > 0
$$

is a point of stable equilibrium. For the given potential we find two such points, $(\sqrt{k/\lambda_1}, 0)$ and $(-\sqrt{k/\lambda_1}, 0)$.

b. V is a minimum at a point of stable equilibrium (x_0, y_0). At a neighboring point (x, y), we have, to second order of the small qualities $x - x_0, y - y_0$,

$$V(x, y) = V(x_0, y_0) + \frac{1}{2}\left[\left(\frac{\partial^2 V}{\partial x^2}\right)_{x_0, y_0} (x - x_0)^2\right.$$

$$\left. + 2\left(\frac{\partial^2 V}{\partial x \partial y}\right)_{x_0, y_0} (x - x_0)(y - y_0) + \left(\frac{\partial^2 V}{\partial y^2}\right)_{x_0, y_0} (y - y_0)^2\right] + \cdots$$

$$\approx -\frac{k^2}{4\lambda_1} + \frac{1}{2}\left[2k\left(x - \sqrt{\frac{k}{\lambda_1}}\right)^2 + \frac{k\lambda_0 y^2}{\lambda_1}\right]$$

for the equilibrium point $\left(\sqrt{\frac{k}{\lambda_1}}, 0\right)$.

Translate the coordinate system to the new origin $\left(\sqrt{\frac{k}{\lambda_1}}, 0\right)$:

$$x' = x - \sqrt{\frac{k}{\lambda_1}}, \quad y' = y,$$

and take the new origin as the reference level for potential energy. Then

$$V(x', y') = \frac{1}{2}k\left(x'^2 + \frac{\lambda_0}{\lambda_1}y'^2\right)$$

and the Lagrangian is

$$L = T - V = \frac{1}{2}m(\dot{x}'^2 + \dot{y}'^2) - \frac{1}{2}k\left(x'^2 + \frac{\lambda_0}{\lambda_1}y'^2\right).$$

Similarly for the other point of equilibrium, we set

$$x'' = x + \sqrt{\frac{k}{\lambda_1}}, \quad y'' = y$$

and obtain the same Lagrangian, but with x'', y'' replacing x', y'.

b. The secular equation

$$|V - \omega^2 T| = 0,$$

or

$$\begin{vmatrix} 2k - m\omega^2 & 0 \\ 0 & \dfrac{k\lambda_0}{\lambda_1} - m\omega^2 \end{vmatrix} = 0$$

has positive roots

$$\omega_1 = \sqrt{\frac{2k}{m}}, \quad \omega_2 = \sqrt{\frac{k\lambda_0}{m\lambda_1}}.$$

These are the normal angular frequencies for small oscillations of the system, about either of the points of equilibrium.

2057

A negligibly thin piece of metal of mass m in the shape of a square hangs from two identical springs at two corners as shown in Fig. 2.58. The springs can move only in the vertical plane. Calculate the frequencies of vibration of the normal-modes of small amplitude oscillations.

(UC, Berkeley)

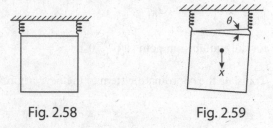

Fig. 2.58   Fig. 2.59

Sol: Let x be the vertical displacement of the center of mass of the square from its equilibrium position and θ the angle of rotation of the square in the vertical plane containing the springs as shown in Fig. 2.59. The square has moment of inertial $\frac{1}{6}ms^2$, s being the length of each side of the square. For small θ, the extensions of the springs are $x + \frac{1}{2}s\theta$ and $x - \frac{1}{2}s\theta$. Hence the kinetic and potential energies are

$$T = \frac{1}{2}m\dot{x}^2 + \frac{1}{12}ms^2\dot{\theta}^2,$$

$$V = -mgx + \frac{1}{2}k\left[\left(x + \frac{1}{2}s\theta\right)^2 + \left(x - \frac{1}{2}s\theta\right)^2\right],$$

where k is the spring constant, taking the potential reference level at the equilibrium position, and the Lagrangian is

$$L = T - V = \frac{1}{2}m\dot{x}^2 + \frac{1}{12}ms^2\dot{\theta}^2 + mgx - k\left(x^2 + \frac{1}{4}s^2\theta^2\right).$$

Lagrange's equations

$$\frac{d}{dt}\left(\frac{\partial L}{\partial \dot{q}_i}\right) - \frac{\partial L}{\partial q_i} = 0$$

give

$$m\ddot{x} + 2kx - mg = 0,$$
$$\frac{1}{6}ms^2\ddot{\theta} + \frac{1}{2}ks^2\theta = 0.$$

Let $x' = x - \dfrac{mg}{2k}$ and we can write the first equation as

$$m\ddot{x}' + 2kx' = 0.$$

Thus x' and θ are the normal coordinates of the system with the corresponding normal angular frequencies

$$\omega_1 = \sqrt{\frac{2k}{m}}, \qquad \omega_2 = \sqrt{\frac{3k}{m}}.$$

2058

A small sphere, mass m and radius r, hangs like a pendulum between two plates of a capacitor, as shown in Fig. 2.60, from an insulating rod of length l. The plates are grounded and the potential of the sphere is V.

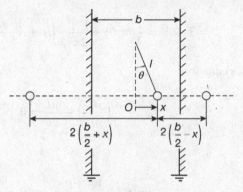

Fig. 2.60

The position of the sphere is displaced by an amount Δx. Calculate the frequency of small oscillations and specify for what conditions of the voltage V such oscillations occur. Make reasonable approximations to simplify the calculation.

(UC, Berkeley)

Sol: We assume that the mass of the insulating rod and the radius of the sphere are very small and can be neglected. The charge on the sphere is

$$q = 4\pi\varepsilon_0 rV,$$

ε_0 being the permittivity of free space. According to the method of images, th[e] forces between the sphere and the plates of the capacitor are the same as thos[e] between the charges on the sphere and its images symmetrically located a[t] positions as shown in Fig. 2.60. Take x-axis along the horizontal with origin a[t] the equilibrium position. The kinetic and potential energies of the system ar[e] respectively

$$T = \frac{1}{2}m\dot{x}^2,$$

$$V = -\frac{q^2}{4\pi\varepsilon_0}\left[\frac{1}{b+2x} + \frac{1}{b-2x}\right] + mgl(1 - \cos\theta)$$

$$= -\frac{q^2}{4\pi\varepsilon_0}\frac{2b}{b^2 - 4x^2} + mgl(1 - \cos\theta).$$

For small x, $x \approx l\theta$,

$$\frac{1}{b^2 - 4x^2} \approx \frac{1}{b^2}\left(1 + \frac{4l^2\theta^2}{b^2}\right);$$

and the Lagrangian is

$$L = T - V = \frac{1}{2}ml^2\dot{\theta}^2 + \frac{q^2}{2\pi\varepsilon_0 b}\left(1 + \frac{4l^2\theta^2}{b^2}\right) - \frac{1}{2}mgl\theta^2.$$

Lagrange's equation

$$\frac{d}{dt}\left(\frac{\partial L}{\partial\dot{\theta}}\right) - \frac{\partial L}{\partial\theta} = 0$$

gives

$$ml^2\ddot{\theta} - \frac{4q^2}{\pi\varepsilon_0}\frac{l^2\theta}{b^3} + mgl\theta = 0.$$

Hence the angular frequency of small oscillations is

$$\omega = \sqrt{\frac{g}{l} - \frac{4q^2}{\pi\varepsilon_0}\frac{1}{mb^3}} = \sqrt{\frac{g}{l} - \frac{64\pi\varepsilon_0 r^2 V^2}{mb^3}}.$$

The condition for such oscillations to take place is that ω be real, i.e.

$$V < \sqrt{\frac{gmb^3}{64\pi\varepsilon_0 r^2 l}}.$$

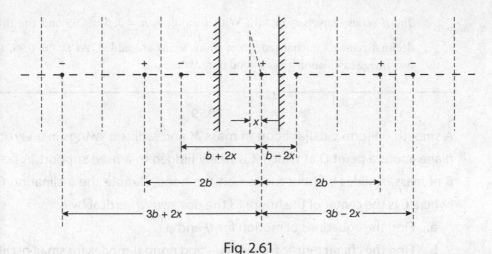

Fig. 2.61

Note that the above solution is only approximate since the images themselves will produce more images, some of which are shown in Fig. 2.61, which also have to be taken into account. Thus the potential due to electrostatic interactions is

$$U = -\frac{q^2}{4\pi\varepsilon_0}\sum_{n=1}^{\infty}\left[\frac{1}{(2n-1)b-2x}+\frac{1}{(2n-1)b+2x}-\frac{2}{2nb}\right]$$

$$\approx -\frac{q^2}{2\pi\varepsilon_0}\sum_{n=1}^{\infty}\left\{\frac{1}{(2n-1)b}\left[1+\frac{4x^2}{(2n-1)^2b^2}\right]-\frac{1}{2nb}\right\}$$

$$= -\frac{q^2}{2\pi\varepsilon_0 b}\sum_{n=1}^{\infty}\left[\left(\frac{1}{2n-1}-\frac{1}{2n}\right)+\frac{4x^2}{(2n-1)^3b^2}\right]$$

$$= -\frac{q^2}{2\pi\varepsilon_0 b}\left(\alpha+\frac{4l^2\theta^2\beta}{b^2}\right)$$

with

$$\alpha = \sum_{n=1}^{\infty}\frac{1}{2n(2n-1)}, \quad \beta = \sum_{n=1}^{\infty}\frac{1}{(2n-1)^3}.$$

This would give

$$\omega = \sqrt{\frac{g}{l}-\frac{64\pi\varepsilon_0 r^2 V^2 \beta}{mb^3}}$$

and the condition for oscillations

$$V < \sqrt{\frac{gmb^3}{64\pi\varepsilon_0 r^2 l\beta}}.$$

The β series converges rapidly. With maximum $n = 3, \beta = 1.05$ and the thir
decimal remains unchanged when more terms are added. As $\beta^{-\frac{1}{2}} = 0.98$, th
two-image calculation gives a good approximation.

2059

A smooth uniform circular hoop of mass M and radius a swings in a vertica
plane about a point O at which it is freely hinged to a fixed support. A bead
B of mass m slides without friction on the hoop. Denote the inclination OC
(where C is the center of the hoop) to the downward vertical by φ.

a. Find the equations of motion for θ and φ.

b. Find the characteristic frequencies and normal-modes for small oscilla
tions about the position of stable equilibrium.

(Chicago)

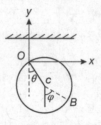

Fig. 2.62

Sol:

a. The moment of inertia of the hoop about O is

$$I = Ma^2 + Ma^2 = 2Ma^2.$$

Use coordinates as shown in Fig. 2.62. The coordinates and velocity of the bead
are respectively

$(a \sin \theta + a \sin \varphi, -a \cos \theta - a \cos \varphi)$, $(a\dot{\theta} \cos \theta + a\dot{\varphi} \cos \varphi, a\dot{\theta} \sin \theta$
$+ a\dot{\varphi} \sin \varphi)$.

The Lagrangian of the system is

$$L = T - V = Ma^2\dot{\theta}^2 + \frac{1}{2}ma^2[\dot{\theta}^2 + \dot{\varphi}^2 + 2\dot{\theta}\dot{\varphi}\cos(\theta - \varphi)]$$
$$+ Mga\cos\theta + mga(\cos\theta + \cos\varphi)$$
$$= \frac{1}{2}(2M + m)a^2\dot{\theta}^2 + \frac{1}{2}ma^2\dot{\varphi}^2 + ma^2\dot{\theta}\dot{\varphi}\cos(\theta - \varphi)$$
$$+ (M + m)ga\cos\theta + mga\cos\varphi.$$

Lagrange's equations give

$$(2M + m)a\ddot{\theta} + ma\ddot{\varphi}\cos(\theta - \varphi) + ma\dot{\varphi}^2\sin(\theta - \varphi) + (M + m)g\sin\theta = 0,$$
$$a\ddot{\theta}\cos(\theta - \varphi) + a\ddot{\varphi} - a\dot{\theta}^2\sin(\theta - \varphi) + g\sin\varphi = 0$$

b. For small oscillations, retaining terms up to second order in the small quantities $\theta, \varphi, \dot{\theta}, \dot{\varphi}$, we have from the above

$$\ddot{\theta} + \left(\frac{M + m}{2M + m}\right)\frac{g\theta}{a} + \frac{m}{2M + m}\ddot{\varphi} = 0,$$

$$\ddot{\theta} + \frac{g}{a}\varphi + \ddot{\varphi} = 0.$$

For a solution of the type $\theta = A\exp(i\omega t)$, $\phi = B\exp(i\omega t)$, the above become

$$\left[\left(\frac{M + m}{2M + m}\right)\frac{g}{a} - \omega^2\right]A - \frac{m\omega^2}{2M + m}B = 0,$$

$$- \omega^2 A + \left(\frac{g}{a} - \omega^2\right)B = 0.$$

For nonzero solutions the determinant of the coefficients must vanish. Thus

$$\left[M\omega^2 - \frac{(M + m)g}{a}\right]\left(2\omega^2 - \frac{g}{a}\right) = 0,$$

whose two positive roots

$$\omega_1 = \sqrt{\frac{g}{2a}}, \quad \omega_2 = \sqrt{\left(\frac{M + m}{M}\right)\frac{g}{a}}$$

are the characteristic angular frequencies of the system for small oscillations. As $\frac{A}{B} = \frac{g}{a\omega^2} - 1$, we have for $\omega = \omega_1$, $\frac{A}{B} = 1$ and the normal-mode $\begin{pmatrix} 1 \\ 1 \end{pmatrix}$, for $\omega = \omega_2$, $\frac{A}{B} = -\frac{m}{M + m}$ and the normal-mode $\begin{pmatrix} 1 \\ -\frac{M + m}{m} \end{pmatrix}$.

2060

A small body of mass m and charge q is constrained to move without friction on the interior of a cone of opening angle 2α. A charge $-q$ is fixed at the apex of the cone as shown in Fig. 2.63. There is no gravity. Find the frequency of small oscillations about equilibrium trajectories of the moving body in terms of $\dot{\varphi}_0$, the equilibrium angular velocity of the body around the inside of the cone. Assume $v \ll c$ so that radiation is negligible.

(UC, Berkeley)

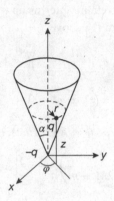

Fig. 2.63

Sol: Use coordinates as shown in Fig. 2.63. In the Cartesian system, m has coordinates $(r\cos\varphi, r\sin\varphi, z)$, or, as $z = r\cot\alpha$, $(r\cot\varphi, r\sin\varphi, r\cot\alpha)$, and velocity

$$(\dot{r}\cos\varphi - r\dot{\varphi}\sin\varphi, \ \dot r\sin\varphi + r\dot{\varphi}\cos\varphi, \ \dot{r}\cot\alpha).$$

The Lagrangian is then

$$L = T - V = \frac{1}{2}m(\dot{r}^2 + r^2\dot{\varphi}^2 + \dot{r}^2\cot^2\alpha) + \frac{q^2}{4\pi\varepsilon_0}\cdot\frac{\sin\alpha}{r}$$

$$= \frac{1}{2}m(\dot{r}^2\csc^2\alpha + r^2\dot{\varphi}^2) + \frac{q^2\sin\alpha}{4\pi\varepsilon_0 r}.$$

Lagrange's equations

$$\frac{d}{dt}\left(\frac{\partial L}{\partial \dot{q}_i}\right) - \frac{\partial L}{\partial q_i} = 0$$

give

$$m\ddot{r}\csc^2\alpha - mr\dot{\varphi}^2 + \frac{q^2\sin\alpha}{4\pi\varepsilon_0 r^2} = 0,$$

$$mr^2\dot{\varphi} = J \text{ (constant)},$$

or, combining the above,

$$m\ddot{r}\csc^2\alpha - \frac{J^2}{mr^3} + \frac{q^2\sin\alpha}{4\pi\varepsilon_0 r^2} = 0. \tag{1}$$

For the equilibrium trajectory,

$$\ddot{r} = 0, \quad r = r_0, \quad \dot{\varphi} = \dot{\varphi}_0,$$

the above becomes

$$\frac{J^2}{mr_0^3} = \frac{q^2\sin\alpha}{4\pi\varepsilon_0 r_0^2}.$$

For small oscillations about equilibrium, let $r = r_0 + \xi$, where $\xi \ll r_0$. Then

$$\frac{1}{r^2} \approx \frac{1}{r_0^2}\left(1 - \frac{2\xi}{r_0}\right), \quad \frac{1}{r^3} = \frac{1}{r_0^3}\left(1 - \frac{3\xi}{r_0}\right),$$

and Eq. (1) becomes

$$m\ddot{\xi} + \frac{q^2\sin^3\alpha}{4\pi\varepsilon_0 r_0^3}\xi = 0.$$

Hence the angular frequency for small oscillations is

$$\omega = \sqrt{\frac{q^2\sin^3\alpha}{4\pi\varepsilon_0 mr_0^3}} = \dot{\varphi}_0\sin\alpha,$$

as

$$\dot{\varphi}_0 = \sqrt{\frac{J^2}{m^2 r_0^4}} = \sqrt{\frac{q^2\sin\alpha}{4\pi\varepsilon_0 mr_0^3}}.$$

2061

A flywheel of moment of inertia I rotates about its center in a horizontal plane. A mass m can slide freely along one of the spokes and is attached to the center of the wheel by a spring of natural length l and force constant k as shown in Fig. 2.64.

a. Find an expression for the energy of this system in terms of $r, \dot{r}$, and the angular momentum J.

b. Suppose the flywheel initially has a constant angular velocity Ω_0 and the spring has a steady extension $r = r_0$. Use the result of part (a) to determine the relation between Ω_0 and r_0 and the frequency of small oscillations about this initial configuration.

(MIT

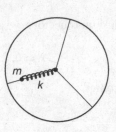

Fig. 2.64

Sol:

a. Let r be the distance of m from the center and $\dot{\theta}$ the angular velocity of the flywheel at time t. The system has angular momentum

$$J = I\dot{\theta} + mr^2\dot{\theta}$$

and energy

$$T + V = \frac{1}{2}I\dot{\theta}^2 + \frac{1}{2}m(\dot{r}^2 + r^2\dot{\theta}^2) + \frac{1}{2}(r - l)^2$$

$$= \frac{J^2}{2(I + mr^2)} + \frac{1}{2}m\dot{r}^2 + \frac{1}{2}k(r - l)^2.$$

b. The Lagrangian of the system is

$$L = T - V = \frac{1}{2}I\dot{\theta}^2 + \frac{1}{2}m\dot{r}^2 + \frac{1}{2}mr^2\dot{\theta}^2 - \frac{1}{2}k(r - l)^2.$$

Lagrange's equations

$$\frac{d}{dt}\left(\frac{\partial L}{\partial \dot{q_i}}\right) - \frac{\partial L}{\partial q_i} = 0$$

give

$$m\ddot{r} - mr\dot{\theta}^2 + k(r - l) = 0, \qquad (1)$$

$$(I + mr^2)\dot{\theta} = \text{constant} = J,$$

or, combining the two,

$$m\ddot{r} - \frac{mrJ^2}{(I + mr^2)^2} + k(r - l) = 0. \tag{2}$$

Initially, $\ddot{r} = 0$, $r = r_0$, $\dot{\theta} = \Omega_0$, $J = (I + mr_0^2)\Omega_0$. For small oscillations about this equilibrium configuration, let $r = r_0 + \rho$, where $\rho \ll r_0$. As

$$\frac{mrJ^2}{(I + mr^2)^2} \approx \frac{m(r_0 + \rho)J^2}{(I + mr_0^2 + 2mr_0\rho)^2}$$

$$\approx \frac{mr_0J^2}{(I + mr_0^2)^2}\left(1 + \frac{\rho}{r_0} - \frac{4mr_0\rho}{I + mr_0^2}\right)$$

$$\approx \frac{mr_0J^2}{(I + mr_0^2)^2}\left[1 - \left(\frac{3mr_0^2 - I}{I + mr_0^2}\right)\frac{\rho}{r_0}\right]$$

$$\approx \frac{mr_0J^2}{(I + mr_0^2)^2} - \left(\frac{3mr_0^2 - I}{I + mr_0^2}\right)m\Omega_0^2\rho,$$

$$\frac{mr_0J^2}{(I + mr_0^2)^2} = mr_0\Omega_0^2 = k(r_0 - l),$$

Eq. (2) becomes

$$\ddot{\rho} + \left[\frac{k}{m} + \left(\frac{3mr_0^2 - I}{I + mr_0^2}\right)\Omega_0^2\right]\rho = 0.$$

Therefore, provided that I is such that

$$\left(\frac{I - 3mr_0^2}{I + mr_0^2}\right)\Omega_0^2 < \frac{k}{m},$$

the system will oscillate about the initial configuration with angular frequency

$$\omega = \sqrt{\frac{k}{m} + \left(\frac{3mr_0^2 - I}{I + mr_0^2}\right)\Omega_0^2}$$

after a small perturbation. Note that Eq. (1) implies

$$r_0 = \frac{kl}{k - m\Omega_0^2},$$

i.e. r_0 itself is related to Ω_0.

2062

Three point-like masses (two of them equal) and the massless springs (constant K) connecting them are constrained to move in a frictionless tube of radius R.

This system is in gravitational field (**g**) as shown in Fig. 2.65. The springs are c
zero length at equilibrium and the masses may move through one anothe
Using Lagrangian methods, find the normal-modes of small vibration abou
the position of equilibrium of this system and describe each of the modes.

(UC, Berkeley

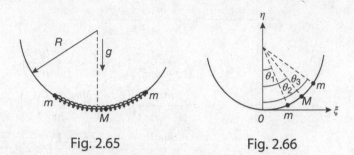

Fig. 2.65   Fig. 2.66

Sol: Use Cartesian coordinates (ξ, η) as shown in Fig. 2.66. The ith mass has coordi
nates $(R \sin \theta_i, R(1 - \cos \theta_i))$. For small oscillations these can be approximated
as $\left(R\theta_i, \frac{1}{2}R\theta_i^2 \right)$, or $\left(x_i, \frac{1}{2}x_i^2 \right)$ with $x_i = R\theta_i$. Then, neglecting terms of orders
greater than two of the small quantities x_i, $\dot{x}_i$, we have for the kinetic and poten-
tial energies

$$T = \frac{1}{2}m\dot{x}_1^2 + \frac{1}{2}\dot{x}_2^2 + \frac{1}{2}m\dot{x}_3^2,$$

$$V = \frac{1}{2}K(x_1 - x_2)^2 + \frac{1}{2}K(x_2 - x_3)^2 + \frac{1}{2}m(x_1^2 + x_3^2) + \frac{1}{2}Mx_2^2,$$

and the Lagrangian

$$L = \frac{1}{2}m\dot{x}_1^2 + \frac{1}{2}M\dot{x}_2^2 + \frac{1}{2}m\dot{x}_3^2$$
$$- \left[\frac{1}{2}\left(K + \frac{mg}{R} \right)(x_1^2 + x_3^2) + \frac{1}{2}\left(2K + \frac{Mg}{R} \right)x_2^2 - K(x_1x_2 + x_2x_3) \right].$$

Lagrange's equations give

$$m\ddot{x}_1 + \left(K + \frac{mg}{R} \right)x_1 - Kx_2 = 0,$$

$$M\ddot{x}_2 + \left(2K + \frac{Mg}{R} \right)x_2 - K(x_1 + x_3) = 0,$$

$$m\ddot{x}_3 + \left(K + \frac{mg}{R} \right)x_3 - Kx_2 = 0.$$

Letting

$$x_j = A_j e^{i\omega t}$$

in the above we obtain the matrix equation

$$
\begin{pmatrix}
K + \dfrac{mg}{R} - m\omega^2 & -K & 0 \\[2ex]
-K & 2K + \dfrac{Mg}{R} - M\omega^2 & -K \\[2ex]
0 & -K & K + \dfrac{mg}{R} - m\omega^2
\end{pmatrix}
\begin{pmatrix}
A_1 \\ A_2 \\ A_3
\end{pmatrix} = 0.
\tag{1}
$$

For solutions in which not all A_i are zero, we require

$$
\begin{vmatrix}
K + \dfrac{mg}{R} - m\omega^2 & -K & 0 \\[2ex]
-K & 2K + \dfrac{Mg}{R} - M\omega^2 & -K \\[2ex]
0 & -K & K + \dfrac{mg}{R} - m\omega^2
\end{vmatrix} = 0
$$

whose three non-negative roots are the angular frequencies of the normal-modes of the system:

$$\omega_1 = \sqrt{\frac{g}{R} + \frac{K}{m}},$$

$$\left.\begin{matrix}\omega_2 \\ \omega_3\end{matrix}\right\} = \sqrt{\frac{g}{R} + \frac{K}{2m} + \frac{K}{M} \pm K\left(\frac{1}{4m^2} + \frac{1}{mM} + \frac{1}{M^2}\right)^{\frac{1}{2}}}.$$

Equation (1) gives

$$\frac{A_2}{A_1} = \frac{A_2}{A_3} = 1 + \frac{mg}{RK} - \frac{m\omega^2}{K},$$

$$KA_1 - \left(2K + \frac{Mg}{R} - M\omega^2\right)A_2 + KA_3 = 0.$$

These equations give for ω_1: $A_3 = -A_1$, $A_2 = 0$;

for ω_2: $B_3 = B_1$, $\dfrac{B_2}{B_1}$ = negative;

for ω_3: $C_3 = C_1$, $\dfrac{C_2}{C_1}$ = positive.

Hence the three corresponding normal-modes are

$$
\begin{pmatrix} A_1 \\ 0 \\ -A_1 \end{pmatrix}, \quad
\begin{pmatrix} B_1 \\ B_2 \\ B_1 \end{pmatrix}, \quad
\begin{pmatrix} C_1 \\ C_2 \\ C_1 \end{pmatrix}
$$

for ω_1, ω_2, ω_3 respectively, where

$$B_2 = \left[\frac{1}{2} - \frac{m}{M} - m\sqrt{\frac{1}{4m^2} + \frac{1}{mM} + \frac{1}{M^2}}\right]B_1,$$

$$C_2 = \left[\frac{1}{2} - \frac{m}{M} + m\sqrt{\frac{1}{4m^2} + \frac{1}{mM} + \frac{1}{M^2}}\right]C_1.$$

The three normal modes are depicted in Fig. 2.67.

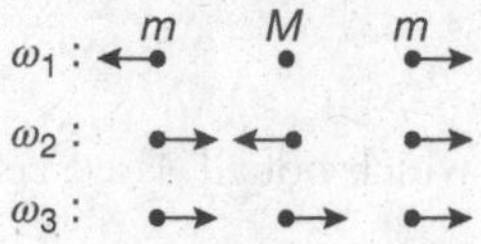

Fig. 2.67

2063

In the theory of small oscillations one frequently encounters Lagrangian of the form $L = T - V$, where

$$T = \sum_{i,j=1}^{N} \dot{q}_i a_{ij} \dot{q}_j, \quad V = \sum_{i,j=1}^{N} q_i b_{ij} q_j.$$

The matrices $\mathbf{A} = (a_{ij})$ and $\mathbf{B} = (b_{ij})$ are real and symmetric.

a. Prove that $\mathbf{A}$ is positive definite, i.e.

$$\mathbf{x}^+\mathbf{A}\mathbf{x} \geq 0$$

for an arbitrary column matrix $\mathbf{x}$. Prove that in general the eigenvalues of such a matrix are greater than or equal to zero. Show that we need not be concerned with zero eigenvalues.

b. Prove the existence of the matrices $\mathbf{A}^{\pm\frac{1}{2}}$.

c. Introduce new coordinates θ_j by

$$q_i = \sum_{j=1}^{N} (\mathbf{A}^{-\frac{1}{2}}\mathbf{S})_{ij}\theta_j,$$

where S is an $N \times N$ matrix. Show that S can be chosen so that $\mathbf{A}$ and $\mathbf{B}$ are diagonalized. Interpret the diagonal elements of the transformed $\mathbf{B}$.

(SUNY, Buffalo)

Sol:

a. By definition,

$$T = \frac{1}{2} \sum_k m_k \dot{x}_k^2 \geq 0$$

in Cartesian coordinates. After a linear transformation

$$x_k = x_k(q_1, q_2, ..., q_N),$$

it becomes

$$T = \sum_{i,j}^{n} \dot{q}_i a_{ij} \dot{q}_j,$$

but is still ≥ 0. In matrix form,

$$\mathbf{T} = \dot{\mathbf{q}}^\dagger \mathbf{A} \dot{\mathbf{q}},$$

where

$$\dot{\mathbf{q}} = \begin{pmatrix} \dot{q}_1 \\ \dot{q}_2 \\ \vdots \\ \dot{q}_N \end{pmatrix}$$

and the dagger denotes its transpose matrix. As the velocities $\dot{x}_1, \dot{x}_2, ...$ and hence the generalized velocities $\dot{q}_1, \dot{q}_2, ...$ are arbitrary, we have

$$\mathbf{T} = \mathbf{x}^\dagger \mathbf{A} \mathbf{x} \geq 0$$

for an arbitrary column matrix $\mathbf{x}$. That is, $\mathbf{A}$ is positive definite.

Suppose $\mathbf{x}_g$ is an eigenvector of $\mathbf{A}$ with eigenvalue λ_g. By definition,

$$\mathbf{A}\mathbf{x}_g = \lambda_g \mathbf{x}_g,$$

where λ_g is a real number as $\mathbf{A}$ is symmetrical and real. Then

$$\mathbf{x}_g^\dagger \mathbf{A}\mathbf{x}_g = \mathbf{x}_g^\dagger \lambda_g \mathbf{x}_g = \lambda_g \mathbf{x}_g^\dagger \mathbf{x}_g = \lambda_g \sum_{i=1}^{N} x_{gi}^2.$$

As this is greater or equal to zero as shown above, the eigenvalues $\lambda_g \geq 0$.

If $\lambda_g = 0$, there is no oscillation for the corresponding mode, which then does not concern us. The vibrational degrees of freedom are simply reduced by one.

b. For the matrices $\mathbf{A}^{\pm \frac{1}{2}}$ to exist we require that

$$\det|\mathbf{A}| > 0.$$

A real symmetrize matrix can be diagonalized by an orthogonal matrix $\mathbf{S}$, i.e. one for which $\mathbf{S}^\dagger \mathbf{S} = I$, the unit matrix:

$$\mathbf{S}^\dagger \mathbf{A} \mathbf{S} = \lambda,$$

where λ is a diagonal matrix elements $\lambda_{ij} = \lambda_i \delta_{ij}$. Writing $|\mathbf{A}|$ for $\det|\mathbf{A}|$, we have

$$|\mathbf{A}| = |\mathbf{A}||\mathbf{S}^\dagger||\mathbf{S}| = |\mathbf{S}^\dagger \mathbf{A} \mathbf{S}| = |\lambda| = \prod_{i=1}^{N} \lambda_i > 0$$

by the result of (a) (any zero λ has been eliminated). Hence $\mathbf{A}^{\pm\frac{1}{2}}$ exists.

b. Introduce new coordinates θ_j by

$$q_i = \sum_{j=1}^{N} (\mathbf{A}^{-\frac{1}{2}}\mathbf{S})_{ij}\theta_j,$$

where $\mathbf{S}$ which diagonalizes $\mathbf{A}$ is orthogonal. Consider

$$\begin{aligned}
T = \dot{\mathbf{q}}^\dagger \mathbf{A} \dot{\mathbf{q}} &= (\mathbf{A}^{-\frac{1}{2}}\mathbf{S}\dot{\theta})^\dagger \mathbf{A} \mathbf{A}^{-\frac{1}{2}}\mathbf{S}\dot{\theta} \\
&= \dot{\theta}^\dagger \mathbf{S}^\dagger (\mathbf{A}^{-\frac{1}{2}})^\dagger \mathbf{A} \mathbf{A}^{-\frac{1}{2}}\mathbf{S}\dot{\theta}.
\end{aligned}$$

As $\mathbf{A}$ is real symmetric, $\mathbf{A}^\dagger = \mathbf{A}$ and

$$(\mathbf{A}^{-\frac{1}{2}})^\dagger = (\mathbf{A}^\dagger)^{-\frac{1}{2}} = \mathbf{A}^{-\frac{1}{2}},$$

the above becomes

$$T = \dot{\theta}^\dagger \mathbf{S}^\dagger \mathbf{S}\dot{\theta} = \dot{\theta}^\dagger \mathbf{I} \dot{\theta}.$$

Similarly

$$V = \mathbf{q}^\dagger \mathbf{B} \mathbf{q} = \theta^\dagger \mathbf{S}^\dagger \mathbf{A}^{-\frac{1}{2}} \mathbf{B} \mathbf{A}^{-\frac{1}{2}} \mathbf{S}\theta.$$

As $\mathbf{A}, \mathbf{B}$ are real symmetric,

$$(\mathbf{A}^{-\frac{1}{2}} \mathbf{B} \mathbf{A}^{-\frac{1}{2}})^\dagger = (\mathbf{A}^{-\frac{1}{2}})^\dagger \mathbf{B}^\dagger (\mathbf{A}^{-\frac{1}{2}})^\dagger = \mathbf{A}^{-\frac{1}{2}} \mathbf{B} \mathbf{A}^{-\frac{1}{2}}.$$

$\mathbf{A}^{-\frac{1}{2}} \mathbf{B} \mathbf{A}^{-\frac{1}{2}}$ is real symmetric and can be diagonalized by the orthogonal matrix $\mathbf{S}$. We therefore have

$$T = \sum_{j=1}^{N} \dot{\theta}_j^2, \quad V = \sum_{j=1}^{N} B_j \theta_j^2,$$

where B_j are the diagonal elements of the diagonalized matrix of $\mathbf{A}^{-\frac{1}{2}} \mathbf{B} \mathbf{A}^{-\frac{1}{2}}$, i.e.

$$(\mathbf{S}^\dagger \mathbf{A}^{-\frac{1}{2}} \mathbf{B} \mathbf{A}^{-\frac{1}{2}} \mathbf{S})_{ij} = B_j \delta_{ij}.$$

The Lagrangian is

$$L = T - V = \sum_{j=1}^{N} (\dot{\theta}_j^2 - B_j \theta_j^2)$$

and Lagrange's equations

$$\frac{d}{dt}\left(\frac{\partial L}{\partial \dot{\theta}_i}\right) - \frac{\partial L}{\partial \theta_i} = 0$$

give

$$\ddot{\theta}_i + B_i \theta_i = 0, \quad i = 1, 2, \ldots, N.$$

Hence B_i are the squares of the normal angular frequencies ω_i of the system.

2064

A flyball governor consists of two masses m connected to arms of length l and a mass M as shown in Fig. 2.68. The assembly is constrained to rotate around a shaft on which the mass M can slide up and down without friction. Neglect the mass of the arms, air friction, and assume that the diameter of the mass M is small. Suppose first that the shaft is constrained to rotate at an angular velocity ω_0.

a. Calculate the equilibrium height of the mass M.

b. Calculate the frequency of small oscillations around this value.

Suppose the shaft is now allowed to rotate freely.

c. Does the frequency of small oscillation change? If so, calculate the new value.

(*Princeton*)

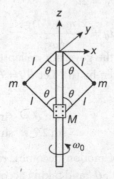

Fig. 2.68

Sol:

a. Use a rotating coordinate frame with the x-axis in the plane of the governor arm as shown in Fig. 2.68. In this frame the masses m, m and M have coordinate $(-l\sin\theta, 0, -l\cos\theta)$, $(l\sin\theta, 0, -l\cos\theta)$, $(0, 0, -2l\cos\theta)$ respectively. In fixed coordinate frame with the same origin and z-axis, the velocity is given by $\dot{\mathbf{r}}' = \dot{\mathbf{r}} + \boldsymbol{\omega}_0 \times \mathbf{r}$, where $\boldsymbol{\omega}_0 = (0, 0, \omega_0)$. Hence the corresponding velocities are $(-l\dot\theta\cos\theta, l\omega_0\sin\theta, l\dot\theta\sin\theta)$, $(l\dot\theta\cos\theta, -l\omega_0\sin\theta, l\dot\theta\sin\theta)$, $(0, 0, -2l\dot\theta\sin\theta)$. Thus the kinetic energy, potential energy and Lagrangian of the system are respectively

$$T = ml^2\omega_0^2\sin^2\theta + ml^2\dot\theta^2 + 2Ml^2\dot\theta^2\sin^2\theta,$$
$$V = -2mgl\cos\theta - 2Mgl\cos\theta,$$
$$L = T - V = ml^2\omega_0^2\sin^2\theta + ml^2\dot\theta^2 + 2Ml^2\dot\theta^2\sin^2\theta + 2(M+m)gl\cos\theta.$$

Lagrange's equation

$$\frac{d}{dt}\left(\frac{\partial L}{\partial\dot\theta}\right) - \frac{\partial L}{\partial\theta} = 0$$

then gives

$$2(m + 2M\sin^2\theta)l\ddot\theta + 2Ml\dot\theta^2\sin 2\theta - ml\omega_0^2\sin 2\theta + 2(m+M)g\sin\theta = 0.$$

At equilibrium, $\ddot\theta = 0$, $\dot\theta = 0$, $\theta = \theta_0$ and the above becomes

$$ml\omega_0^2\sin 2\theta_0 = 2(m+M)g\sin\theta_0. \tag{1}$$

Solving for θ_0 we obtain two equilibrium positions:

    i. $\theta_0 = 0$,

    ii. $\cos\theta_0 = \dfrac{(m+M)g}{ml\omega_0^2}$.

The distances of the mass M at the two equilibrium positions from the top of the shaft are respectively

    i. $2l\cos\theta_0 = 2l$,

    ii. $2l\cos\theta_0 = \dfrac{2(m+M)g}{m\omega_0^2}$.

b. When $\theta_0 = 0$, the governor collapses and there is no oscillation. Consider the equilibrium given by (ii). Let $\theta' = \theta - \theta_0$, then $\dot\theta = \dot\theta'$. For small oscillations, $\theta' \ll \theta_0$,

$$\sin\theta \approx \sin\theta_0 + \theta'\cos\theta_0,$$
$$\sin 2\theta \approx \sin 2\theta_0 + 2\theta'\cos 2\theta_0.$$

The equation of motion becomes, retaining only first order terms of the small quantities θ', $\dot\theta'$, $\ddot\theta'$ and taking account of (1),

$$(m + 2M\sin^2\theta_0)l\ddot\theta' + [(m+M)g\cos\theta_0 - ml\omega_0^2\cos 2\theta_0]\theta' = 0.$$

Hence the oscillation frequency is

$$f = \frac{1}{2\pi} \sqrt{\frac{(m + M)g \cos \theta_0 - ml\omega_0^2 \cos 2\theta_0}{(m + 2M \sin^2 \theta_0)l}}.$$

b. One would expect the oscillation frequency to be different since the angular velocity ω_0 in the above is arbitrary. Let φ be the angle of rotation about the shaft. Putting $\omega = \dot{\varphi}$ in the Lagrangian we have

$$L = ml^2\dot{\varphi}^2 \sin^2 \theta + ml^2\dot{\theta}^2 + 2Ml^2\dot{\theta}^2 \sin^2 \theta + 2(m + M)gl \cos \theta.$$

Lagrange's equations give

$$\dot{\varphi} \sin^2 \theta = c \quad \text{(a constant)},$$

$$2(m + 2M \sin^2 \theta)l\ddot{\theta} + 2Ml\dot{\theta}^2 \sin 2\theta - ml\dot{\varphi}^2 \sin 2\theta + 2(m + M)g \sin \theta = 0,$$

which combine to give

$$(m + 2M \sin^2 \theta)l\ddot{\theta} + Ml\dot{\theta}^2 \sin 2\theta - mlc^2 \frac{\cos \theta}{\sin^3 \theta} + (m + M)g \sin \theta = 0. \qquad (2)$$

At equilibrium, $\ddot{\theta} = 0$, $\dot{\theta} = 0$ and $\theta = \theta_0$, which is given by

$$mlc^2 \frac{\cos \theta_0}{\sin^3 \theta_0} = (m + M)g \sin \theta_0.$$

For small oscillations about θ_0, let $\theta = \theta_0 + \theta'$, where $\theta' \ll \theta_0$. As

$$mlc^2 \frac{\cos \theta}{\sin^3 \theta} \approx mlc^2 \frac{\cos \theta_0 - \theta' \sin \theta_0}{(\sin \theta_0 + \theta' \cos \theta_0)^3}$$

$$\approx mlc^2 \frac{\cos \theta_0}{\sin^3 \theta_0}(1 - \theta' \tan \theta_0 - 3\theta' \cot\theta_0)$$

$$= (m + M)g \sin \theta_0 \left[1 - \left(\frac{1 + 2\cos^2 \theta_0}{\sin \theta_0 \cos \theta_0}\right)\theta'\right],$$

Eq. (2) becomes

$$(m + 2M \sin^2 \theta_0)l\ddot{\theta}' + (m + M)g \frac{(1 + 3\cos^2 \theta_0)}{\cos \theta_0}\theta' = 0.$$

Hence the frequency of small oscillations is

$$f = \frac{1}{2\pi} \sqrt{\frac{(m + M)g(1 + 3\cos^2 \theta_0)}{(m + 2M \sin^2 \theta_0)l \cos \theta_0}}.$$

2065

A particle of mass M moves along the x-axis under the influence of the potential energy $V(x) = -Kx \exp(-ax)$, where K and a are positive constants. Find the equilibrium position and the period of small oscillations about this equilibrium position. Consider also the cases where K and/or a are negative.

(*Princeton*)

Sol: Expand the potential near a point x_0:

$$V(x) = V(x_0) + \left(\frac{\partial V}{\partial x}\right)_{x_0}(x - x_0) + \frac{1}{2}\left(\frac{\partial^2 V}{\partial x^2}\right)_{x_0}(x - x_0)^2 + \ldots.$$

For x_0 to be an equilibrium position,

$$\left(\frac{\partial V}{\partial x}\right)_{x_0} = K(ax_0 - 1)e^{-ax_0} = 0,$$

giving

$$x_0 = \frac{1}{a}.$$

As

$$\left(\frac{\partial^2 V}{\partial x^2}\right)_{V_0} = aK(2 - ax_0)e^{-ax_0} = \frac{ak}{e} > 0,$$

the equilibrium is stable.

Let

$$\xi = x - x_0 = x - \frac{1}{a}$$

and take x_0 as the reference level of potential energy. Then the potential at ξ is

$$V(\xi) = \frac{1}{2}\left(\frac{\partial^2 V}{\partial x^2}\right)_{x_0}\xi^2 = \frac{aK}{2e}\xi^2.$$

The Lagrangian is then

$$L = T - V = \frac{1}{2}M\dot{\xi}^2 - \frac{aK}{2e}\xi^2.$$

Lagrange's equation

$$\frac{d}{dt}\left(\frac{\partial L}{\partial \dot{\xi}}\right) - \frac{\partial L}{\partial \xi} = 0$$

yields

$$M\ddot{\xi} + \frac{aK}{e}\xi = 0.$$

This shows that the angular frequency of small oscillations about the equilibrium position is

$$\omega = \sqrt{\frac{aK}{Me}},$$

and the period is

$$T = \frac{2\pi}{\omega} = 2\pi\sqrt{\frac{Me}{aK}}.$$

If both a and K are negative, then aK is positive and the above results still hold.

If only one of a, K is negative then

$$\left(\frac{\partial^2 V}{\partial x^2}\right)_{x_0} < 0,$$

which means that the potential at equilibrium is a maximum and the equilibrium is unstable. Hence no oscillation occurs. This can also be seen from the equation of motion, which would give an imaginary ω.

2066

A particle of mass m moves under gravity on a smooth surface the equation of which is $z = x^2 + y^2 - xy$, the z-axis being vertical, pointing upwards.

a. Find the equations of motion of the particle.

b. Find the frequencies of the normal-modes for small oscillations about the position of stable equilibrium.

c. If the particle is displaced from equilibrium slightly and then released, what must be the ratio of the x and y displacements to guarantee that only the higher frequency normal-mode is excited?

(Wisconsin)

Sol:

a. As

$$z = x^2 + y^2 - xy,$$
$$\dot{z} = 2x\dot{x} + 2y\dot{y} - \dot{x}y - x\dot{y} = \dot{x}(2x - y) + \dot{y}(2y - x).$$

The Lagrangian is

$$L = T - V$$
$$= \frac{1}{2}m[\dot{x}^2 + \dot{y}^2 + \dot{x}^2(2x - y)^2 + \dot{y}^2(2y - x)^2 + 2\dot{x}\dot{y}(2x - y)(2y - x)]$$
$$- mg(x^2 + y^2 - xy).$$

Lagrange's equations

$$\frac{d}{dt}\left(\frac{\partial L}{\partial \dot{q}_i}\right) - \frac{\partial L}{\partial q_i} = 0$$

give

$$\frac{d}{dt}[\dot{x} + \dot{x}(2x - y)^2 + \dot{y}(2x - y)(2y - x)]$$
$$= 2\dot{x}^2(2x - y) - \dot{y}^2(2y - x) + 2\dot{x}\dot{y}(2y - x) - \dot{x}\dot{y}(2x - y) - 2gx + gy,$$
$$\frac{d}{dt}[\dot{y} + \dot{y}(2y - x)^2 + \dot{x}(2x - y)(2y - x)]$$
$$= 2\dot{y}^2(2y - x) - \dot{x}^2(2x - y) + 2\dot{x}\dot{y}(2x - y) - \dot{x}\dot{y}(2y - x) - 2gy + gx.$$

b. As

$$\frac{\partial V}{\partial x} = mg(2x - y), \quad \frac{\partial V}{\partial y} = mg(2y - x),$$

equilibrium occurs at the origin $(0, 0)$. For small oscillations about the origin, $x, y, \dot{x}, \dot{y}$ are small quantities and the equations of motion reduce to

$$\ddot{x} + 2gx - gy = 0,$$
$$\ddot{y} + 2gy - gx = 0.$$

Considering a solution of the type

$$x = x_0 e^{i\omega t}, \quad y = y_0 e^{i\omega t},$$

we find the secular equation

$$\begin{vmatrix} 2g - \omega^2 & -g \\ -g & 2g - \omega^2 \end{vmatrix} = (g - \omega^2)(3g - \omega^2) = 0.$$

Its position roots

$$\omega_1 = \sqrt{g}, \quad \omega_2 = \sqrt{3g}$$

are the angular frequencies of the normal-modes of the system. Note that as ω_1, ω_2 are real the equilibrium is stable.

b. As

$$\frac{y_0}{x_0} = \frac{2g - \omega^2}{g},$$

for the higher frequency mode to be excited we require $\dfrac{y_0}{x_0} = -1$. Hence the initial displacements of x and y must be equal in magnitude and opposite in sign. Note that under this condition the lower frequency mode, which requires $y_0/x_0 = 1$, is not excited.

2067

A rigid structure consists of three massless rods joined at a point attached to two point masses (each of mass m) as shown in Fig. 2.69, with $AB = BC = L$, $BD = l$, the angle $ABD = DBC = \theta$. The rigid system is supported at the point D and rocks back and forth with a small amplitude of oscillation. What is the oscillation frequency? What is the limit on l for stable oscillations?

(CUSPEA)

Sol: The structure oscillates in a vertical plane. Take it as the xy-plane as shown in Fig. 2.70 with the origin at the point of support D and the y-axis vertically upwards. We have

$$\overline{AD} = \overline{CD} = b = \sqrt{L^2 + l^2 - 2Ll \cos \theta},$$

and the angles between $\overline{AD}$ and $\overline{CD}$ with the vertical are $\alpha + \varphi, \alpha - \varphi$ respectively, where $\alpha = \theta + \psi$, ψ being given by

$$\frac{b}{\sin \theta} = \frac{l}{\sin \psi}.$$

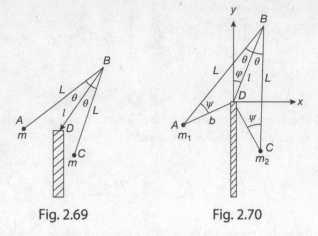

Fig. 2.69   Fig. 2.70

The masses m_1, m_2 have coordinates

$$(-b\sin(\alpha + \varphi), -b\cos(\alpha + \varphi)), \ (b\sin(\alpha - \varphi), -b\cos(\alpha - \varphi))$$

and velocities

$$(-b\dot{\varphi}\cos(\alpha + \varphi), b\dot{\varphi}\sin(\alpha + \varphi)), \ (-b\dot{\varphi}\cos(\alpha - \varphi), -b\dot{\varphi}\sin(\alpha - \varphi))$$

respectively. Thus the Lagrangian is

$$L = T - V = mb^2\dot{\varphi}^2 + mgb[\cos(\alpha + \varphi) + \cos(\alpha - \varphi)].$$

Lagrange's equation

$$\frac{d}{dt}\left(\frac{\partial L}{\partial \dot{\varphi}}\right) - \frac{\partial L}{\partial \varphi} = 0$$

then gives

$$2mb^2\ddot{\varphi} + mgb[\sin(\alpha + \varphi) - \sin(\alpha - \varphi)] = 0.$$

For small oscillations, $\varphi \ll \alpha$ and

$$\sin(\alpha \pm \varphi) \approx \sin\alpha \pm \varphi\cos\alpha,$$

so the equation of motion reduces to

$$b\ddot{\varphi} + \varphi g\cos\alpha = 0,$$

giving the angular frequency as

$$\omega = \sqrt{\frac{g \cos \alpha}{b}}.$$

As

$$\cos \alpha = \cos (\theta + \psi) = \cos \theta \cos \psi - \sin \theta \sin \psi$$
$$= \frac{1}{b}(\sqrt{b^2 - l^2 \sin^2 \theta} \cos \theta - l \sin^2 \theta)$$
$$= \frac{1}{b}(L \cos \theta - l),$$

we have

$$\omega = \sqrt{\frac{g(L \cos \theta - l)}{L^2 + l^2 - 2Ll \cos \theta}}.$$

Since

$$\frac{\partial^2 V}{\partial \varphi^2} = mgb[\cos(\alpha + \varphi) + \cos(\alpha - \varphi)]$$
$$= 2mgb \cos \alpha$$

at the equilibrium position $\varphi = 0$, oscillations are stable if $\cos \alpha > 0$. This requires that

$$L \cos \theta - l > 0,$$

or

$$l < L \cos \theta.$$

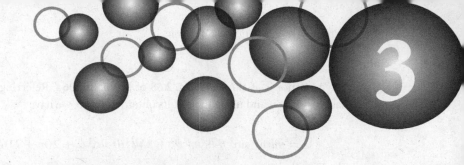

HAMILTON'S CANONICAL EQUATIONS (2068–2084)

2068

A flyball governor for a steam engine consists of two balls, each of mass m, attached by means of four hinged arms, each of length l, to sleaves located on a vertical rod. The lower sleeve has mass M and negligible moment of inertial, and is free to slide up and down the rod without friction. The upper sleeve is fastened to the rod. The system is constrained to rotate with constant angular velocity ω.

a. Choose suitable coordinates and write the Lagrangian and Hamiltonian functions for the system. Neglect weights of arms and rod, and neglect friction.

b. Discuss the motion.

c. Determine the height z of the lower sleeve above its lowest position, as a function of ω, for steady motion. Find the frequency of small oscillations about this steady motion.

(Wisconsin)

Sol:

a. The governor is as shown in Fig. 2.68 of Problem **2064**. Referring to the coordinates as shown and using the results obtained there, we have

$$L = T - V$$
$$= ml^2\omega^2 \sin^2\theta + ml^2\dot{\theta}^2 + 2Ml^2\dot{\theta}^2 \sin^2\theta + 2(m + M)gl\cos\theta.$$

The Hamiltonian is

$$H = \dot{\theta}p_\theta - L$$

with the generalized momentum p_θ defined as

$$p_\theta = \frac{\partial L}{\partial \dot{\theta}} = 2(m + 2M\sin^2\theta)l^2\dot{\theta}.$$

Thus

$$H = \dot{\theta}p_\theta - ml^2\omega^2 \sin^2\theta - (m + 2M\sin^2\theta)l^2\dot{\theta}^2 - 2(m + M)gl\cos\theta$$

$$= \frac{p_\theta^2}{4(m + 2M\sin^2\theta)l^2} - ml^2\omega^2\sin^2\theta - 2(m + M)gl\cos\theta.$$

b. Lagrange's equation

$$\frac{d}{dt}\left(\frac{\partial L}{\partial \dot{\theta}}\right) - \frac{\partial L}{\partial \theta} = 0$$

gives

$$2(m + 2M\sin^2\theta)l\ddot{\theta} + 2Ml\dot{\theta}^2\sin 2\theta - ml\omega^2\sin 2\theta + 2(m + M)g\sin\theta = 0.$$

The motion is discussed in Problem **2064**. Briefly, M will oscillate up and down the vertical rod about an equilibrium position given by

$$\cos\theta_0 = \frac{(m + M)g}{ml\omega^2}.$$

c. At equilibrium, M has z coordinate $-2l\cos\theta_0$. Hence its height above the lowest point is

$$2l - 2l\cos\theta_0 = 2l\left[1 - \frac{(m + M)g}{ml\omega^2}\right].$$

The angular frequency of small oscillations about the equilibrium position is (Problem **2064**)

$$\Omega = \sqrt{\frac{(m + M)g\cos\theta_0 - ml\omega^2\cos 2\theta_0}{(m + 2M\sin^2\theta_0)l}}$$

$$= \omega\sqrt{\frac{m\sin^2\theta_0}{m + 2M\sin^2\theta_0}}$$

with

$$\sin^2 \theta_0 = 1 - \left[\frac{(m+M)g}{ml\omega^2} \right]^2.$$

2069

Consider the two-body system consisting of (1) a point particle of mass m and (2) a rotator of finite size and mass M (see Fig. 2.71). This rotator is a rigid body which has uniform density, has an axis of symmetry, and, like the particle of mass m, is free to move. Discuss the motion of this system if the particle is attracted to every element of the rotator by a Coulomb or gravitational force. Include in your discussion answers to the following questions.

a. How many degrees of freedom does this system have?

b. What would be a suitable set of coordinates?

c. What is the Lagrangian (or Hamiltonian)? (Write it down or say how you would try.)

d. On what coordinates does the interaction between the particle and the rotator depend?

e. How many constants of motion can you infer, and what are they physically?

f. What orbits of this system are closely similar to orbits of two point masses? Describe the nature of their (small) difference. What is the nature of the motion of the rotator relative to its center of mass?

(Wisconsin)

Fig. 2.71

Sol:

a. The system has 9 degrees of freedom, of which 3 belong to the mass m and 6 belong to the rigid rotator.

b. One may take generalized coordinates as follows: 3 coordinates x, y, z describing the position of the mass m, 3 coordinates X, Y, Z describing the position of the center of mass of the rigid rotator, 3 Euler's angles φ, θ, ψ describing rotation relative to the center of mass of the rotator, the axis of symmetry of the rotator being taken as the Z'-axis of the rest coordinate system of the rotator.

c. The kinetic energy of the system consists of three parts: kinetic energy of the point mass m and the translational and rotational kinetic energies of the rotator, namely,

$$T = T_1 + T_2 + T_3,$$

with

$$T_1 = \frac{1}{2}m(\dot{x}^2 + \dot{y}^2 + \dot{z}^2),$$

$$T_2 = \frac{1}{2}M(\dot{X}^2 + \dot{Y}^2 + \dot{Z}^2),$$

$$T_3 = \frac{1}{2}(\omega_1, \omega_2, \omega_3) \begin{pmatrix} I_{11} & I_{12} & 0 \\ I_{21} & I_{22} & 0 \\ 0 & 0 & I_{33} \end{pmatrix} \begin{pmatrix} \omega_1 \\ \omega_2 \\ \omega_3 \end{pmatrix},$$

where $\omega_1, \omega_2, \omega_3$ are related to Euler's angles (Problem **1212**) by

$$\omega_1 = \dot{\theta} \cos \psi + \dot{\varphi} \sin \theta \sin \psi,$$
$$\omega_2 = -\dot{\theta} \sin \psi + \dot{\varphi} \sin \theta \cos \psi,$$
$$\omega_3 = \dot{\varphi} \cos \theta + \dot{\psi},$$

and the inertia tensor is with respect to the center of mass of the rotator with the Z'-axis in the direction of the axis of symmetry. The calculation of the potential energy is more complex. Imagine a series of spherical shells centered at the mass m and consider a shell of inner and outer radii r and $r + dr$ respectively. The potential due to Coulomb interaction between the element dM of the rotator in the shell and the particle is

$$dV = -\frac{GmdM}{r},$$

where G is the gravitational constant. Then the total potential of the system is

$$V = -Gm \int \frac{dM}{r}.$$

The Lagrangian of the system, $L = T - V$, can then be obtained.

d. The interaction between the particle and the rotator depends on $X - x$, $Y - y$, $Z - z$, φ and θ.

e. As the interaction is conservative and the space is uniform and isotropic, the constants of motion are the energy $T + V$, total angular momentum (each of the three components) and total momentum (each of the three components) of the system.

f. When the mass and the rotator are far removed from each other, their orbits are closely similar to those of two point masses. The difference stems from the fact that for the rotator the center of mass and the center of gravitational force do not coincide, so the torque of the gravitational force about the center of mass makes the rotator revolve around its center of mass.

2070

A motor turns a vertical shaft to which is attached a simple pendulum of length *l* and mass *m* as shown in Fig. 2.72. The pendulum is constrained to move in a plane. This plane is rotated at constant angular speed ω by the motor.

a. Find the equations of motion of the mass *m*.

b. Solve the equations of motion, obtaining the position of the mass as a function of time for all possible motions of this system. For this part use small angle approximations.

c. Find the angular hequencies of any oscillatory motions.

d. Find an expression for the torque that the motor must supply.

e. Is the total energy of this system constant in time? Is the Hamiltonian function constant in time? Explain briefly.

(UC, Berkeley)

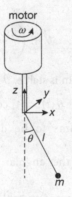

Fig. 2.72

Sol:

a. Use rotating coordinates as shown in Fig. 2.72 with the *x*- and *z*-axes in the plane of oscillation of the pendulum. In this frame the mass *m* has coordinates

$$(l \sin \theta, \ 0, \ -l \cos \theta)$$

and velocity

$$(l\dot\theta \cos\theta, 0, l\dot\theta \sin\theta).$$

In the fixed frame m has an additional velocity

$$\begin{aligned} \boldsymbol{\omega} \times \mathbf{r} &= (0,0,\omega) \times (l\sin\theta, 0, -l\cos\theta) \\ &= (0, \omega l \sin\theta, 0). \end{aligned}$$

Hence the Lagrangian of the system is

$$L = T - V = \frac{1}{2}ml^2\dot\theta^2 + \frac{1}{2}ml^2\omega^2 \sin^2\theta + mgl\cos\theta.$$

Lagrange's equation

$$\frac{d}{dt}\left(\frac{\partial L}{\partial \dot\theta}\right) - \frac{\partial L}{\partial \theta} = 0$$

then gives

$$\ddot\theta + \left(\frac{g}{l} - \omega^2 \cos\theta\right)\sin\theta = 0.$$

b.   For equilibrium, $\ddot\theta = 0$. The equation of motion gives the equilibrium positions as

$$\theta_1 = 0, \quad \theta_2 = \arccos\left(\frac{g}{l\omega^2}\right).$$

For oscillation near $\theta_1 = 0$, in the small angle approximation the equation of motion reduces to

$$\ddot\theta + \left(\frac{g}{l} - \omega^2\right)\theta = 0.$$

If $\omega < \sqrt{\frac{g}{l}}$, the equilibrium is stable. θ is harmonic and can be represented by

$$\theta(t) = A_1 \cos(\Omega_1 t + \varphi_1),$$

where A_1, φ_1 are constants to be determined from the initial conditions, and $\Omega_1 = \sqrt{\frac{g}{l} - \omega^2}$ is the angular frequency of small angle oscillations. If $\omega > \sqrt{\frac{g}{l}}$, the equilibrium is unstable.

For oscillations near θ_2, let $\theta = \theta_2 + \alpha$, where $\alpha \ll \theta_2$. The equation of motion is then, in the first approximation,

$$\ddot\alpha + \left(\frac{g}{l} - \omega^2\cos\theta_2 + \omega^2\alpha\sin\theta_2\right)(\sin\theta_2 + \alpha\cos\theta_2) = 0,$$

or

$$\ddot{\alpha} + \alpha\omega^2 \sin^2 \theta_2 = 0.$$

The Solution is

$$\alpha(t) = A_2 \cos(\Omega_2 t + \varphi_2),$$

where $\Omega_2 = \omega \sin \theta_2 = \frac{1}{l\omega}\sqrt{l^2\omega^4 - g^2}$ is the angular frequency of small oscillations about θ_2, A_2, φ_2 are constants to be determined from the initial conditions. Hence

$$\theta(t) = A_2 \cos(\Omega_2 t + \varphi_2) + \theta_2.$$

c. For small angle oscillations about θ_1, the angular frequency is $\Omega_1 = \sqrt{\frac{g}{l} - \omega^2}$; and about θ_2, $\Omega_2 = \frac{1}{l\omega}\sqrt{l^2\omega^4 - g^2}$.

d. The angular momentum about the z-axis is

$$J = ml \sin \theta \cdot l \sin \theta \cdot \omega = ml^2\omega \sin^2 \theta.$$

The torque the motor must supply is therefore

$$M = \frac{dJ}{dt} = ml^2\omega \sin(2\theta)\frac{d\theta}{dt},$$

where for θ the expressions obtained in (b) are to be used.

e. The kinetic energy in the fixed frame, T, is not a homogeneous quadratic function of the generalized velocity, so the mechanical energy is not conservative. Physically, the pendulum is constrained to oscillate in a plane which is rotating. So the constraint is not a stable one and the mechanical energy is not conserved. On the other hand, not being an explicit function of t, the Hamiltonian H is conserved.

Note that while in the fixed frame the mechanical energy is not conserved, as the system is an unstable holomorphic one and all the external forces are conservative, the generalized energy H is conserved. We have

$$H = \dot{\theta}\frac{\partial L}{\partial \dot{\theta}} - L$$

$$= \frac{1}{2}ml^2\dot{\theta}^2 - \frac{1}{2}ml^2\omega^2 \sin^2 \theta - mgl \cos \theta = \text{constant}.$$

In the rotating frame fixed to the motor, because of the fictitious centrifugal force

$$ml\omega^2 \sin \theta = -\frac{\partial V}{\partial(l \sin \theta)},$$

the potential energy is

$$V = -\frac{1}{2}ml^2\omega^2 \sin^2 \theta - mgl \cos \theta,$$

so that the total energy is

$$\frac{1}{2}ml^2\dot{\theta}^2 + V = H = \text{constant}.$$

Therefore, whether the mechanical energy is conserved or not depends on the choice of reference frame.

2071

The classical interaction between two inert gas atoms, each of mass m, is given by the potential

$$V(r) = -\frac{2A}{r^6} + \frac{B}{r^{12}}, \qquad A, B > 0, \qquad r = |\mathbf{r}_1 - \mathbf{r}_2|.$$

a. Give the Hamiltonian for the system of the two atoms.

b. Describe completely the lowest energy classical state(s) of this system.

c. If the energy is slightly higher than the lowest [part (b)], what are the possible frequencies of the motion of the system?

(Wisconsin)

Sol:

a. The center of mass of the system is given by $\mathbf{R} = \frac{1}{2}(\mathbf{r}_1 + \mathbf{r}_2) = (x, y, z)$, the reduced mass is $\mu = \frac{m^2}{m + m} = \frac{m}{2}$, and the total mass is $M = 2m$. Let $\mathbf{r} = \mathbf{r}_1 - \mathbf{r}_2$. Then the kinetic energy of the system is

$$T = \frac{1}{2}M\dot{R}^2 + \frac{1}{2}\mu\dot{\mathbf{r}}$$

$$= \frac{1}{2}M\dot{\mathbf{R}}^2 + \frac{1}{2}\mu(\dot{r}^2 + r^2\dot{\theta}^2 + r^2\dot{\varphi}^2 \sin^2 \theta)$$

and the Lagrangian is

$$L = T - V$$

$$= \frac{1}{2}M(\dot{x}^2 + \dot{y}^2 + \dot{z}^2) + \frac{1}{2}\mu(\dot{r}^2 + r^2\dot{\theta}^2 + r^2\dot{\varphi}^2 \sin^2 \theta) + \frac{2A}{r^6} - \frac{B}{r^{12}},$$

where r, θ, φ are the spherical coordinates of a frame fixed at the center of mass. The generalized momenta are

$$p_x = \frac{\partial L}{\partial \dot{x}} = M\dot{x}, \quad p_y = \frac{\partial L}{\partial \dot{y}} = M\dot{y}, \quad p_z = \frac{\partial L}{\partial \dot{z}} = M\dot{z},$$

$$p_r = \frac{\partial L}{\partial \dot{r}} = \mu\dot{r}, \quad p_\theta = \frac{\partial L}{\partial \dot{\theta}} = \mu r^2 \dot{\theta}, \quad p_\varphi = \frac{\partial L}{\partial \dot{\varphi}} = \mu r^2 \dot{\varphi} \sin^2\theta.$$

The Hamiltonian is

$$H = \sum_i p_i \dot{q}_i - L$$

$$= \frac{1}{2M}(p_x^2 + p_y^2 + p_z^2) + \frac{1}{2\mu}\left(p_r^2 + \frac{1}{r^2}p_\theta^2 + \frac{1}{r^2 \sin^2\theta}p_\varphi^2\right) - \frac{2A}{r^6} + \frac{B}{r^{12}}.$$

b. The lowest energy state corresponds to $p_x = p_y = p_z = p_r = p_\theta = p_\varphi = 0$ and an $\mathbf{r}_0$ which minimizes

$$-\frac{2A}{r^6} + \frac{B}{r^{12}}.$$

Letting

$$\frac{d}{dt}\left(-\frac{2A}{r^6} + \frac{B}{r^{12}}\right) = 0,$$

we obtain $r_0 = (B/A)^{\frac{1}{6}}$ as the distance between the two atoms for the lowest energy classical state. For this state the energy of the system is

$$H = \frac{-A^2}{B}.$$

c. If the energy is only slightly higher than the lowest and the degrees of freedom corresponding to x, y, z, θ, φ are not excited yet ($p_x = p_y = p_z = p_\theta = p_\phi = 0$), we have

$$H = \frac{p_r^2}{2\mu} - \frac{2A}{r^6} + \frac{B}{r^{12}}.$$

As

$$\left(\frac{d^2V}{dr^2}\right)_{r_0} = 72A\left(\frac{A}{B}\right)^{\frac{4}{3}},$$

the Lagrangian is

$$L = T - V = \frac{1}{2}\mu\dot{r}^2 - 36A\left(\frac{A}{B}\right)^{\frac{4}{3}}(r - r_0)^2 = \frac{1}{2}\mu\dot{\rho}^2 - 36A\left(\frac{A}{B}\right)^{\frac{4}{3}}\rho^2,$$

where $\rho = r - r_0 \ll r_0$. Lagrange's equation gives

$$\mu\ddot{\rho} + 72A\left(\frac{A}{B}\right)^{\frac{4}{3}}\rho = 0.$$

Hence

$$\omega = \sqrt{\frac{72A}{\mu}\left(\frac{A}{B}\right)^{\frac{4}{3}}} = 12\left(\frac{A}{m}\right)^{\frac{1}{2}}\left(\frac{A}{B}\right)^{\frac{2}{3}}.$$

2072

Consider a particle of mass m which is constrained to move on the surface of a sphere of radius R. There are no external forces of any kind on the particle.

a. What is the number of generalized coordinates necessary to describe the problem?

b. Choose a set of generalized coordinates and write the Lagrangian of the system.

c. What is the Hamiltonian of the system? Is it conserved?

d. Prove that the motion of the particle is along a great circle of the sphere.

(*Columbia*)

Sol:

a. As the particle is constrained to move on the surface of a sphere, there are two degrees of freedom and hence two generalized coordinates are needed.

b. Choose (θ, φ) of spherical coordinates as the generalized coordinates. As there are no external forces, $V = 0$. The Lagrangian of the system is

$$L = T = \frac{1}{2}mv^2 = \frac{1}{2}mR^2(\dot{\theta}^2 + \dot{\varphi}^2 \sin^2 \theta).$$

c. As $p_i = \dfrac{\partial L}{\partial \dot{q}_i}$, we have

$$p_\theta = mR^2\dot{\theta}, \quad p_\varphi = mR^2\dot{\varphi}\sin^2 \theta,$$

and

$$H = p_\theta\dot{\theta} + p_\varphi\dot{\varphi} - L = \frac{1}{2mR^2}\left(p_\theta^2 + \frac{p_\varphi^2}{\sin^2 \theta}\right).$$

Since the Hamiltonian H is not an explicit function of time, it is a constant of the motion, or, in other words, conserved.

d. Hamilton's equation

$$\dot{p}_\varphi = -\frac{\partial H}{\partial \varphi}$$

gives

$$p_\varphi = \dot\varphi \sin^2\theta = \text{constant}.$$

We can choose the set of coordinates (θ, φ) so that the initial condition is $\dot\varphi = 0$ at $t = 0$. Then the above constant is zero at all time: $\dot\varphi \sin^2\theta = 0$. As θ cannot be zero at all time, $\dot\varphi = 0$, or $\varphi = $ constant, the motion of the particle is along a great circle of the sphere.

2073

A light, uniform U-shaped tube is partially filled with mercury (total mass M, mass per unit length ρ) as shown in Fig. 2.73. The tube is mounted so that it can rotate about one of the vertical legs. Neglect friction, the mass and moment of inertia of the glass tube, and the moment of inertia of the mercury column on the axis of rotation.

a. Calculate the potential energy of the mercury column and describe its possible motion when the tube is not spinning.

b. The tube is set in rotation with an initial angular velocity ω_0 with the mercury column at rest vertically with a displacement z_0 from equilibrium.

  1. Give the Lagrangian for the system.

  2. Give the equation of motion.

  3. What quantities are conserved in the motion? Give expressions for these quantities.

  4. Describe the motion qualitatively as completely as you can.

(Wisconsin)

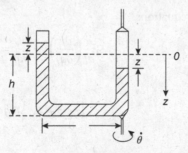

Fig. 2.73

Sol:

a. Let z be the distance of the top of the mercury column from its equilibrium position. Suppose an external force F acting on the descending top causes it to descend slowly a distance dz. Then $F = 2\rho zg$ and its work done is

$$dW = Fdz = 2z\rho gdz.$$

This work is stored as potential energy dV. Hence the potential energy of the mercury column is $V = \rho gz^2$. If the tube is not spinning, the mercury column will oscillate about the equilibrium position and the Lagrangian of the system is

$$L = \frac{1}{2}\rho s\dot{z}^2 - \rho gz^2,$$

where $s = l + 2h$. Lagrange's equation gives

$$\ddot{z} + \frac{2g}{s}z = 0.$$

Hence the mercury column will oscillate with an angular frequency

$$\omega = \sqrt{\frac{2g}{s}} = \sqrt{\frac{2g}{l + 2h}}.$$

b. The system has 2 degrees of freedom when the U-shaped tube is spinning z and the angle of rotation θ are taken as the generalized coordinates.

1. We have

$$T = \frac{1}{2}\rho(h - z)\dot{z}^2 + \frac{1}{2}\rho\int_0^l (\dot{z}^2 + x^2\dot{\theta}^2)\,dx + \frac{1}{2}\rho(h + z)(\dot{z}^2 + l^2\dot{\theta}^2),$$
$$V = \rho gz^2,$$

so the Lagrangian is

$$L = T - V = \frac{1}{2}\rho s\dot{z}^2 + \frac{1}{2}\rho\left(\frac{l}{3} + h + z\right)l^2\dot{\theta}^2 - \rho gz^2.$$

2. Lagrange's equations

$$\frac{d}{dt}\left(\frac{\partial L}{\partial \dot{q}_i}\right) - \frac{\partial L}{\partial q_i} = 0$$

give

$$s\ddot{z} + 2gz - \frac{1}{2}l^2\dot{\theta}^2 = 0,$$

$$\rho\left(\frac{l}{3} + h + z\right)l^2\dot{\theta} = \text{constant}.$$

As $p_\theta = \dfrac{\partial L}{\dot{\theta}}$, the last equation can be written as $p_\theta = $ constant. With the initial conditions $\dot{\theta} = \omega_0$, $\dot{z} = 0$, $z = z_0$ at $t = 0$, we have

$$p_\theta = \rho\left(\frac{l}{3} + h + z_0\right)l^2\omega_0.$$

3. The Hamiltonian of the system is

$$H = p_z\dot{z} + p_\theta\dot{\theta} - L$$
$$= \frac{1}{2}\rho s\dot{z}^2 + \frac{1}{2}\rho\left(\frac{l}{3} + h + z\right)l^2\dot{\theta}^2 + \rho g z^2 = T + V,$$

where $p_z = \dfrac{\partial L}{\partial \dot{z}} = \rho s\dot{z}$. Thus H is equal to the total energy of the system. In terms of the canonical variable we have

$$H = \frac{p_z^2}{2\rho s} + \frac{p_\theta^2}{2\left(\dfrac{l}{3} + h + z\right)\rho l^2} + \rho g z^2.$$

As H does not depend on t explicitly, it is a constant of the motion, in addition to the constant

$$p_\theta = \rho\left(\frac{l}{3} + h + z\right)l^2\dot{\theta}.$$

Using the initial conditions given we obtain

$$H = \frac{1}{2}\left(\frac{l}{3} + h + z_0\right)\rho l^2\omega_0^2 + \rho g z_0^2.$$

4. The motion of the mercury column consists of two components. One is the rotation together with the tube. The angular velocity of rotation changes in connection with the up-and-down motion of the column. When z increases the rotation slows down, and vice verse, to keep the angular momentum about the vertical axis constant. The other component is the motion of the column in the tube. The equation of motion in z is

$$s\ddot{z} + 2gz = \frac{A}{\left(\dfrac{l}{3} + h + z\right)^2},$$

where $A = \dfrac{p_\theta^2}{2\rho^2 l^2}$ is a constant. Generally speaking, there are three equilibrium positions corresponding to the three roots of the equation

$$2gz = \frac{A}{\left(\dfrac{l}{3} + h + z\right)^2}.$$

Near each equilibrium position, the column undergoes small oscillation. Suppose z_1 is one of the equilibrium positions, i.e.

$$2gz_1 = \frac{A}{\left(\frac{1}{3} + h + z_1\right)^2}.$$

For small oscillations let $z = z_1 + z'$ where z' is a small quantity. The equation of motion becomes

$$s\ddot{z}' + 2gz' = \frac{-2Az'}{\left(\frac{1}{3} + h + z_1\right)^3},$$

giving the angular frequency of oscillations

$$\Omega = \sqrt{\frac{2g}{s} + \frac{2A}{s\left(\frac{1}{3} + h + z_1\right)^3}}.$$

As Ω is real, the equilibrium positions are all stable.

2074

A particle under the action of gravity slides on the inside of a smooth paraboloid of revolution whose axis is vertical. Using the distance from the axis, r, and the azimuthal angle φ as generalized coordinates, find

a. The Lagrangian of the system.

b. The generalized momenta and the corresponding Hamiltonian.

c. The equation of motion for the coordinate r as a function of time.

d. If $\dfrac{d\varphi}{dt} = 0$, show that the particle can execute small oscillations about the lowest point of the paraboloid, and find the frequency of these oscillations.

(Columbia)

Sol: Suppose the paraboloid of revolution is generated by a parabola which in cylindrical coordinates (r, φ, z) is represented by

$$z = Ar^2,$$

where A is a positive constant.

a. The Lagrangian of the system is

$$L = T - V = \frac{1}{2}m(\dot{r}^2 + r^2\dot{\varphi}^2 + \dot{z}^2) - mgz$$

$$= \frac{1}{2}m(1 + 4A^2r^2)\dot{r}^2 + \frac{1}{2}mr^2\dot{\varphi}^2 - Amgr^2.$$

b. The generalized momenta are

$$p_r = \frac{\partial L}{\partial \dot{r}} = m(1 + 4A^2r^2)\dot{r},$$

$$p_\varphi = \frac{\partial L}{\partial \dot{\varphi}} = mr^2\dot{\varphi},$$

and the Hamiltonian is

$$H = p_r\dot{r} + p_\varphi\dot{\varphi} - L$$

$$= \frac{1}{2}m(1 + 4A^2r^2)\dot{r}^2 + \frac{1}{2}mr^2\dot{\varphi}^2 + Amgr^2$$

$$= \frac{P_r^2}{2m(1 + 4A^2r^2)} + \frac{P_\varphi^2}{2mr^2} + Amgr^2.$$

c. Lagrange's equations

$$\frac{d}{dt}\left(\frac{\partial L}{\partial \dot{q}_i}\right) - \frac{\partial L}{\partial q_i} = 0$$

give

$$m(1 + 4A^2r^2)\ddot{r} + 4mA^2r\dot{r}^2 - mr\dot{\varphi} + 2Amgr = 0,$$

$$mr^2\dot{\varphi} = \text{constant}.$$

Letting the constant be mh and eliminating $\dot{\varphi}$ from the first equation, we obtain the equation for r:

$$(1 + 4A^2r^2)r^3\ddot{r} + 4A^2r^4\dot{r}^2 + 2Agr^4 = h^2.$$

d. If $\dot{\varphi} = 0$, the first equation of (c) becomes

$$(1 + 4A^2r\theta^2)\ddot{r} + 4A^2r\dot{r}^2 + 2Agr = 0.$$

The lowest point of the paraboloid is given by $r = 0$. For small oscillations in its vicinity, $r, \dot{r}, \ddot{r}$ are small quantities. Then to first approximation the above becomes

$$\ddot{r} + 2Agr = 0.$$

As the coefficient of r is positive, the particle executes simple harmonic motion about $r = 0$ with angular frequency

$$\omega = \sqrt{2Ag}.$$

2075

A nonrelativistic electron of mass m, charge $-e$ in a cylindrical magnetor
moves between a wire of radius a at a negative electric potential $-\phi_0$ and
concentric cylindrical conductor of radius R at zero potential. There is a uni
form constant magnetic field B parallel to the axis. Use cylindrical coordinate
r, θ, z. The electric and magnetic vector potentials can be written as

$$\phi = -\phi_0 \frac{\ln(r/R)}{\ln(a/R)}, \quad \mathbf{A} = \frac{1}{2}Br\mathbf{e}_\theta$$

($\mathbf{e}_\theta$ is a unit vector in the direction of increasing θ).

  a. Write the Lagrangian and Hamiltonian functions.
  b. Show that there are three constants of the motion. Write them down
    and discuss the kinds of motion which can occur.
  c. Assuming that an electron leaves the inner wire with zero initia
    velocity, there is a value of the magnetic field B_c such that for $B \leq B_c$
    the electron can reach the outer cylinder, and for $B > B_c$ the electron
    cannot reach the outer cylinder. Find B_c and make a sketch of the
    electron's trajectory for this case.
    You may assume that $R \gg a$.

(Wisconsin)

Sol:

  a. In SI units, the Lagrangian is

$$L = T - V = \frac{1}{2}m\dot{\mathbf{r}}^2 + e\phi - e\dot{\mathbf{r}} \cdot \mathbf{A}.$$

As

$$\dot{\mathbf{r}} = (\dot{r}, r\dot{\theta}, \dot{z}), \quad \mathbf{A} = \left(0, \frac{1}{2}Br, 0\right),$$

the above becomes

$$L = \frac{1}{2}m(\dot{r}^2 + r^2\dot{\theta}^2 + \dot{z}^2) + e\phi - \frac{1}{2}eBr^2\dot{\theta}.$$

The generalized momenta are

$$p_r = \frac{\partial L}{\partial \dot{r}} = m\dot{r}, \quad p_\theta = \frac{\partial L}{\partial \dot{\theta}} = mr^2\dot{\theta} - \frac{1}{2}eBr^2, \quad p_z = \frac{\partial L}{\partial \dot{z}} = m\dot{z},$$

and the Hamiltonian is

$$H = p_r \dot{r} + p_\theta \dot{\theta} + p_z \dot{z} - L$$

$$= \frac{1}{2} m (\dot{r}^2 + r^2 \dot{\theta}^2 + \dot{z}^2) - e\phi$$

$$= \frac{p_r^2}{2m} + \frac{1}{2mr^2}\left(p_\theta + \frac{1}{2}eBr^2\right)^2 + \frac{p_z^2}{2m} - e\phi$$

$$= \frac{1}{2m}\left[p_r^2 + \left(\frac{p_\theta}{r} + \frac{1}{2}eBr\right)^2 + p_z^2\right] - e\phi.$$

b. As H is not an explicit function of time, it is a constant of the motion. Also, as

$$\dot{p}_i = -\frac{\partial H}{\partial q_i},$$

if H does not contain q_i explicitly, p_i is a constant of the motion. Hence p_θ, p_z are constants of the motion. Explicitly,

$$H = \frac{1}{2m}\left[p_r^2 + \left(\frac{p_\theta}{r} + \frac{1}{2}eBr\right)^2 + p_z^2\right] - e\phi = E,$$

$$p_\theta = mr^2\dot{\theta} - \frac{1}{2}eBr^2 = C_1,$$

$$p_z = m\dot{z} = C_2,$$

where E, C_1, C_2 are constants.

c. The initial conditions $r = a, \dot{r} = \dot{\theta} = \dot{z} = 0$ at $t = 0$ give

$$E = -e\phi = e\phi_0, \quad C_1 = -\frac{1}{2}eBa^2, \quad C_2 = 0.$$

$p_z = C_2 = 0$ means that there is no motion along the z-direction. $H = E$ gives

$$\frac{1}{2m}\left[p_r^2 + \left(\frac{1}{2}eB\right)^2\left(r - \frac{a^2}{r}\right)^2\right] + e\phi_0\frac{\left[\ln\left(\frac{r}{R}\right)\right]}{\ln\left(\frac{a}{R}\right)} = e\phi_0.$$

Suppose a value B_c of the magnetic field will just make the electron reach the outer cylinder. As then $p_r = 0$ at $r = R$, the above gives

$$\frac{1}{2m}\left(\frac{1}{2}eB_c\right)^2\left(R - \frac{a^2}{R}\right)^2 = e\phi_0.$$

If we assume that $a \ll R$, this reduces to

$$B_c = \frac{2}{R}\sqrt{\frac{2m\phi_0}{e}}.$$

At $r = R$, p_r is given by[3]

$$\frac{1}{2m}\left[p_r^2 + \left(\frac{1}{2}eBR\right)^2\right] = e\phi_0,$$

or

$$p_r^2 = 2me\phi_0 - \left(\frac{1}{2}eBR\right)^2 = (B_c^2 - B^2)\left(\frac{1}{2}eR\right)^2.$$

p_r is real at $r = R$ if $B \le B_c$. Hence under this condition the electron can reach th
outer cylinder. If $B > B_c$, p_r is imaginary at $r = R$ and the electron cannot reac
the outer cylinder. For the latter case the trajectory of the electron is as sketche
in Fig. 2.74.

Fig. 2.74

2076

Consider the Lagrangian

$$L = \frac{1}{2}m(\dot{x}^2 - \omega^2 x^2)e^{\gamma t}$$

for the motion of a particle of mass m in one dimension (x). The constants m, γ
and ω are real and positive.

a. Find the equation of motion.

b. Interpret the equation of motion by stating the kinds of force to which
the particle is subject.

c. Find the canonical momentum, and from this construct the Hamiltonian function.

d. Is the Hamiltonian a constant of the motion? Is the energy conserved?
Explain.

e. For the initial conditions $x(0) = 0$ and $\dot{x}(0) = v_0$, what is $x(t)$ asymptotically as $t \to \infty$?

(Wisconsin)

Sol:

a. Lagrange's equation

$$\frac{d}{dt}\left(\frac{\partial L}{\partial \dot{x}}\right) - \frac{\partial L}{\partial x} = 0$$

gives the equation of motion

$$\ddot{x} + \omega^2 x = -\gamma \dot{x}.$$

b. The particle moves as a damped harmonic oscillator. It is subject to an restoring force $-m\omega^2 x$ and a damping force $-m\gamma\dot{x}$ proportional to its speed.

c. The canonical momentum is

$$p = \frac{\partial L}{\partial \dot{x}} = me^{\gamma t}\dot{x}$$

and the Hamiltonian is

$$H = p\dot{x} - L$$
$$= me^{\gamma t}\dot{x}^2 - \frac{1}{2}me^{\gamma t}\dot{x}^2 + \frac{1}{2}me^{\gamma t}\omega^2 x^2.$$
$$= \frac{p^2 e^{-\gamma t}}{2m} + \frac{1}{2}m\omega^2 x^2 e^{\gamma t}$$

d. Since H depends explicitly on time, it is not a constant of motion. It follows that energy is not conserved also. Physically, in the course of the motion, the damping force continually does negative work, causing dissipation of energy.

e. Try a Sol of the type $x \sim e^{i\Omega t}$. Substitution in the equation of motion gives

$$\Omega^2 - i\gamma\Omega - \omega^2 = 0,$$

which has Solutions

$$\Omega = \frac{i}{2}(\gamma \pm \sqrt{\gamma^2 - 4\omega^2}).$$

Hence

$$x = A\exp\left[-\frac{1}{2}(\gamma + \sqrt{\gamma^2 - 4\omega^2})t\right] + B\exp\left[-\frac{1}{2}(\gamma - \sqrt{\gamma^2 - 4\omega^2})t\right].$$

The initial conditions $x = 0$, $\dot{x} = v_0$ at $t = 0$ give

$$B = -A, \quad A = -\frac{v_0}{\sqrt{\gamma^2 - 4\omega^2}}.$$

If $\gamma < 2\omega$, let $\frac{1}{2}\sqrt{\gamma^2 - 4\omega^2} = i\omega_1$. Then

$$x = -\frac{v_0}{2i\omega_1}e^{-\frac{\gamma t}{2}}(e^{-i\omega_1 t} - e^{i\omega_1 t}) = \frac{v_0}{\omega_1}e^{-\frac{\gamma t}{2}}\sin(\omega_1 t),$$

so that $x \to 0$ as $t \to \infty$.

If $\gamma > 2\omega$, both $\gamma \pm \sqrt{\gamma^2 - 4\omega^2}$ are real and positive so that there will be no oscil lation and x will decrease monotonically to zero as $t \to \infty$.

2077

A particle is confined inside a box and can move only along the *x*-axis. The ends of the box move toward the center with a speed small compared with the particle's speed (Fig. 2.75).

a. If the momentum of the particle is p_0 when the walls of the box are at a distance x_0 apart, find the momentum of the particle at any later time. Collisions with the walls are perfectly elastic. Assume that at all times the speed of the particle is much less than the speed of light.

b. When the walls are a distance *x* apart what average external force must be applied to each wall in order to move it at constant speed *V*?

(*UC, Berkeley*)

Sol:

a. Consider a collision of the particle with one of the walls. As the collision is per- fectly elastic, the relative speeds before and after the collision are equal. If the particle is incident with speed v and reflected back with speed v' and the wall has speed V towards the particle, we have

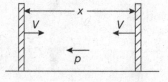

Fig. 2.75

$$v + V = v' - V,$$

i.e.

$$v' = v + 2V.$$

Thus after each colhsion, the magnitude of the particle momentum gains an amount $2mV$, m being the mass of the particle. When the walls are at a distance x apart, as V is much smaller than the speed of the particle, the interval between two consecutive collisions is

$$T = \frac{x}{\left(\dfrac{p}{m}\right)} = \frac{xm}{p},$$

p being the particle momentum. The change of momentum in time dt is

$$dp = 2mV\frac{dt}{T} = \frac{2Vpdt}{x}.$$

As the walls move toward each other with speed V,

$$x = x_0 - 2Vt,$$

measuring time from the moment $x = x_0$. Then

$$dp = -\frac{pdx}{x}.$$

As $p = p_0$ when $x = x_0$, its integration gives

$$p = \frac{p_0 x_0}{x} = \frac{p_0 x_0}{x_0 - 2Vt}.$$

b. Consider a collision of the particle with *one* wall. The momentum acquired by the particle is

$$\mathbf{p} + 2m\mathbf{V} - (-\mathbf{p}) = 2\mathbf{p} + 2m\mathbf{V}.$$

The interval between two consecutive collisions with *the* wall is

$$T' = \frac{2x}{\left(\dfrac{p}{m}\right)} = \frac{2xm}{p},$$

so that the change of momentum due to collisions with the wall in a time dt is

$$dp = 2(p + mV)\frac{dt}{T'}.$$

Hence

$$\frac{dp}{dt} = \frac{(p + mV)p}{xm} \approx \frac{p^2}{xm} = \frac{p_0^2 x_0^2}{mx^3}$$

as $\frac{p}{m} \gg V$. This is the force exerted by the wall on the particle. To keep the walls moving at constant speed, a force of the same magnitude must apply to each wall. The problem can also be solved using the Hamiltonian formalism. Use a reference frame attached to one of the walls, say the left-hand wall. As shown in Fig. 2.75, the particle has velocity $-\frac{p}{m} - V$. The Hamiltonian is

$$H = \frac{1}{2}m\left(\frac{p}{m} + V\right)^2$$

$$= \frac{1}{2m}(p + mV)^2 \approx \frac{p^2}{2m} = \frac{p_0^2 x_0^2}{2mx^2}.$$

The force on the particle is $\dot{p}$ which is given by Hamilton's equation:

$$\dot{p} = -\frac{\partial H}{\partial x} = \frac{p_0^2 x_0^2}{mx^3}.$$

2078

The Poisson bracket is defined by

$$[a, b] = \sum_k \left(\frac{\partial a}{\partial q_k} \frac{\partial b}{\partial p_k} - \frac{\partial a}{\partial p_k} \frac{\partial b}{\partial q_k} \right).$$

a. Show that for a dynamical quantity $a(q, p, t)$

$$\frac{da}{dt} = [a, H] + \frac{\partial a}{\partial t}.$$

A two-dimensional oscillator has energies

$$T(\dot{x}, \dot{y}) = \frac{1}{2}m(\dot{x}^2 + \dot{y}^2),$$

$$V(x, y) = \frac{1}{2}K(x^2 + y^2) + Cxy,$$

where C and K are constants.

b. Show by a coordinate transformation that this oscillator is equivalent to a nonisotropic harmonic oscillator.

c. Find two independent constants of the motion and verify using part (a).

d. If $C = 0$ find a third constant of the motion.

e. Show that for the isotropic oscillator the symmetric matrix

$$A_{ij} = \frac{p_i p_j}{2m} + \frac{1}{2}Kx_i x_j$$

is a constant of the motion by expressing each element in terms of the known constants of motion.

(Wisconsin)

Sol:

a. Using Hamilton's canonical equations

$$\dot{q}_k = \frac{\partial H}{\partial p_k}, \quad \dot{p}_k = -\frac{\partial H}{\partial q_k},$$

we find

$$\frac{da}{dt} = \sum_k \frac{\partial a}{\partial q_k} \dot{q}_k + \sum_k \frac{\partial a}{\partial p_k} \dot{p}_k + \frac{\partial a}{\partial t}$$

$$= \sum_k \frac{\partial a}{\partial q_k} \frac{\partial H}{\partial p_k} - \sum_k \frac{\partial a}{\partial p_k} \frac{\partial H}{\partial q_k} + \frac{\partial a}{\partial t}$$

$$= [a, H] + \frac{\partial a}{\partial t}.$$

b. Introducing the new variables

$$\eta = \frac{1}{\sqrt{2}}(x + y), \qquad \xi = \frac{1}{\sqrt{2}}(x - y),$$

we have

$$x = \frac{1}{\sqrt{2}}(\eta + \xi), \qquad y = \frac{1}{\sqrt{2}}(\eta - \xi).$$

Then

$$T = \frac{1}{4}m[(\dot{\eta} + \dot{\xi})^2 + (\dot{\eta} - \dot{\xi})^2]$$

$$= \frac{1}{2}m(\dot{\eta}^2 + \dot{\xi}^2),$$

$$V = \frac{1}{4}K[(\eta + \xi)^2 + (\eta - \xi)^2] + \frac{1}{2}C(\eta^2 - \xi^2)$$

$$= \frac{1}{2}K(\eta^2 + \xi^2) + \frac{1}{2}C(\eta^2 - \xi^2)$$

$$= \frac{1}{2}(K + C)\eta^2 + \frac{1}{2}(K - C)\xi^2,$$

$$L = L_1 + L_2,$$

with

$$L_1 = \frac{1}{2}m\dot{\eta}^2 - \frac{1}{2}(K + C)\eta^2,$$

$$L_2 = \frac{1}{2}m\dot{\xi}^2 - \frac{1}{2}(K - C)\xi^2.$$

Note that the form of L_1 and L_2 indicates that η and ξ are normal coordinates. Hence the system is equivalent to two harmonic oscillators with angular frequencies

$$\omega_1 = \sqrt{\frac{K + C}{m}}, \qquad \omega_2 = \sqrt{\frac{K - C}{m}}$$

respectively. As the frequencies are different the system acts as a non-isotropic harmonic oscillator.

c. As the canonical momenta are by definition

$$p_\eta = \frac{\partial L}{\partial \dot\eta} = m\dot\eta, \qquad p_\xi = \frac{\partial L}{\partial \dot\xi} = m\dot\xi,$$

$$H = p_\eta \dot\eta + p_\xi \dot\xi - L$$

$$= \frac{1}{2m}p_\eta^2 + \frac{1}{2m}p_\xi^2 + \frac{1}{2}(K + C)\eta^2 + \frac{1}{2}(K - C)\xi^2.$$

This can also be written as

$$H = H_1 + H_2$$

with H_1, H_2 corresponding to L_1, L_2 respectively, i.e.

$$H_1 = \frac{p_\eta^2}{2m} + \frac{1}{2}(K + C)\eta^2, \qquad H_2 = \frac{p_\xi^2}{2m} + \frac{1}{2}(K - C)\xi^2.$$

As H_1, H_2 do not contain t explicitly, we have

$$\frac{dH_1}{dt} = [H_1, H] = [H_1, H_1 + H_2] = [H_1, H_1] + [H_1, H_2] = [H_1, H_2]$$

$$= \frac{\partial H_1}{\partial \eta}\frac{\partial H_2}{\partial p_\eta} + \frac{\partial H_1}{\partial \xi}\frac{\partial H_2}{\partial p_\xi} - \frac{\partial H_1}{\partial p_\eta}\frac{\partial H_2}{\partial \eta} - \frac{\partial H_1}{\partial p_\xi}\frac{\partial H_2}{\partial \xi} = 0$$

and, similarly,

$$\frac{dH_2}{dt} = 0.$$

Hence H_1, H_2 are two independent constants of the motion.

d. If $C = 0$, $\omega_1 = \omega_2$ and the oscillator becomes an isotropic one. The Hamiltonian is

$$H = \frac{1}{2m}p_\eta^2 + \frac{1}{2m}p_\xi^2 + \frac{1}{2}K\eta^2 + \frac{1}{2}K\xi^2.$$

Let $J = m(\eta p_\xi - \xi p_\eta)$. Then as

$$\frac{dJ}{dt} = [J, H]$$

$$= \frac{\partial J}{\partial \eta}\frac{\partial H}{\partial p_\eta} - \frac{\partial J}{\partial p_\eta}\frac{\partial H}{\partial \eta} + \frac{\partial J}{\partial \xi}\frac{\partial H}{\partial p_\xi} - \frac{\partial J}{\partial p_\xi}\frac{\partial H}{\partial \xi}$$

$$= p_\xi p_\eta + mK\xi\eta - p_\eta p_\xi - mK\eta\xi = 0,$$

J is also a constant of the motion.

e. For the isotropic oscillator, $C = 0$ and x, y already are normal coordinates. As shown above, there are three known constants of the motion:

$$E_1 = \frac{p_x^2}{2m} + \frac{1}{2}Kx^2, \qquad E_2 = \frac{p_y^2}{2m} + \frac{1}{2}Ky^2, \qquad J = m(xp_y - yp_x).$$

As $A_{11} = E_1$, $A_{22} = E_2$, the diagonal elements of the matrix A_{ij} are constants of the motion. Consider

$$E_1 E_2 - \frac{KJ^2}{4m^3} = \left(\frac{p_x^2}{2m} + \frac{1}{2}Kx^2\right)\left(\frac{p_y^2}{2m} + \frac{1}{2}Ky^2\right) - \frac{K}{4m}(xp_y - yp_x)^2$$

$$= \left(\frac{p_x p_y}{2m} + \frac{1}{2}Kxy\right)^2.$$

Since the left-hand side is a constant, $A_{12} = A_{21} =$ constant. Hence **A** is a matrix of constant elements.

2079

Consider the system of particles $m_1 = m_2$ connected by a rope of length l with m_2 constrained to stay on the surface of an upright cone of half-angle α and m_1 hanging freely inside the cone, the rope passing through a hole at the top of the cone as shown in Fig. 2.76. Neglect friction.

a. Give an appropriate generalized coordinate system for the problem.

b. Write the Lagrangian of the system and the equation of motion for each generalized coordinate.

c. Write the Hamiltonian for the system.

d. Express the angular frequency for m_2 moving in a circular orbit in terms of the variables of the problem.

(Wisconsin)

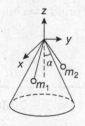

Fig. 2.76

Sol:

a. Use spherical coordinates with origin at the top of the cone as shown in Fig. 2.76. The coordinates of m_1, m_2 are respectively $(r, 0, \varphi)$, $(l - r, \pi - \alpha, \beta)$. The variables $r, \theta, \varphi, \beta$ are taken to be the generalized coordinates.

b. The velocities of m_1, m_2 are respectively $(\dot{r}, r\dot{\theta}, r\dot{\varphi}\sin\theta)$, $(-\dot{r}, 0, (l - r)$
$\dot{\beta}\sin(\pi - \alpha))$. The Lagrangian of the system is then

$$L = T - V$$
$$= \frac{1}{2}m[2\dot{r}^2 + r^2\dot{\theta}^2 + r^2\dot{\varphi}^2\sin^2\theta + (l - r)^2\dot{\beta}^2\sin^2(\pi - \alpha)]$$
$$- mgr\cos\theta + mn(l - r)\cos\alpha.$$

Lagrange's equations

$$\frac{d}{dt}\left(\frac{\partial L}{\partial \dot{q}_i}\right) - \frac{\partial L}{\partial q_i} = 0$$

give

$$mr^2\dot{\varphi}\sin^2\theta = p_\varphi, \quad \text{a constant,}$$
$$m(l - r)^2\dot{\beta}\sin^2(\pi - \alpha) = p_\beta, \quad \text{a constant,}$$
$$2\ddot{r} - r(\dot{\theta}^2 + \dot{\varphi}^2\sin^2\theta) + (l - r)\dot{\beta}^2\sin^2\alpha + g(\cos\theta + \cos\alpha) = 0,$$
$$r\ddot{\theta} + 2\dot{r}\dot{\theta} - r\dot{\varphi}^2\sin\theta\cos\theta - g\sin\theta = 0.$$

c. Two other canonical momenta are

$$p_r = \frac{\partial L}{\partial \dot{r}} = 2m\dot{r}, \quad p_\theta = \frac{\partial L}{\partial \dot{\theta}} = mr^2\dot{\theta}.$$

The Hamiltonian is

$$H = p_r\dot{r} + p_\theta\dot{\theta} + p_\varphi\dot{\varphi} + p_\beta\dot{\beta} - L$$
$$= \frac{p_r^2}{4m} + \frac{p_\theta^2}{2mr^2} + \frac{p_\varphi^2}{2mr^2\sin^2\theta} + \frac{p_\beta^2}{2m(l - r)^2\sin^2\alpha}$$
$$+ mgr\cos\theta - mg(L - r)\cos\alpha.$$

d. If m_2 moves in a circular orbit, $r =$ constant and the angular frequency of revolution is

$$\dot{\beta} = \frac{p_\beta}{m(l - r)^2\sin^2\alpha}.$$

2080

The transformation equations between two sets of coordinates are

$$Q = \ln(1 + q^{\frac{1}{2}}\cos p)$$

$$P = 2(1 + q^{\frac{1}{2}}\cos p)q^{\frac{1}{2}}\sin p.$$

a. Show directly from these transformation equations that Q, P are canonical variables if q and p are.

b. Show that the function that generates this transformation between the two sets of canonical variables is

$$F_3 = -[\exp(Q) - 1]^2 \tan p.$$

(*SUNY, Buffalo*)

Sol:

a. As $[Q, Q] = 0$, $[P, P] = 0$,

$$[Q, P] = \frac{\partial Q}{\partial q}\frac{\partial P}{\partial p} - \frac{\partial P}{\partial q}\frac{\partial Q}{\partial p}$$

$$= \frac{q^{-\frac{1}{2}}\cos p}{1 + q^{\frac{1}{2}}\cos p}\left[-q\sin^2 p + (1 + q^{\frac{1}{2}}\cos p)q^{\frac{1}{2}}\cos p\right]$$

$$+ \frac{q^{\frac{1}{2}}\sin^2 p}{1 + q^{\frac{1}{2}}\cos p}\left[\cos p + (1 + q^{\frac{1}{2}}\cos p)q^{-\frac{1}{2}}\right] = 1,$$

the transformation is canonical. Then if q, p are canonical variables, so are Q, P.

b. Solving the transformation equations for q and p we obtain

$$q = (e^Q - 1)^2 \sec^2 p,$$
$$P = 2e^Q(e^Q - 1)\tan p.$$

Since the transformation is canonical, there exists a generating function $F_3(Q, p)$ such that

$$q = -\frac{\partial F_3}{\partial p}, \qquad P = -\frac{\partial F_3}{\partial Q}.$$

give the transformation equations. As

$$dF_3 = \frac{\partial F_3}{\partial Q}dQ + \frac{\partial F_3}{\partial p}dp = -PdQ - qdp$$

$$= -d[(e^Q - 1)^2]\tan p - (e^Q - 1)^2 d\tan p$$

$$= -d[(e^Q - 1)^2 \tan p],$$

we obtain

$$F_3 = -(e^Q - 1)^2 \tan p.$$

2081

A particle of mass m moves in one dimension q in a potential energy field $V(q$
and is retarded by a damping force $-2m\gamma\dot{q}$ proportional to its velocity.

a.   Show that the equation of motion can be obtained from the Lagrangian

$$L = \exp(2\gamma t)\left[\frac{1}{2}m\dot{q}^2 - V(q)\right]$$

and that the Hamiltonian is

$$H = \frac{p^2\exp(-2\gamma t)}{2m} + V(q)\exp(2\gamma t),$$

where $p = m\dot{q}\exp(2\gamma t)$ is the momentum conjugate to q.

b.   For the generating function

$$F_2(q, P, t) = \exp(\gamma t)qP$$

find the transformed Hamiltonian $K(Q, P, t)$. For an oscillator potential

$$V(q) = \frac{1}{2}m\omega^2 q^2$$

show that the transformed Hamiltonian yields a constant of the motion

$$K = \frac{p^2}{2m} + \frac{1}{2}m\omega^2 Q + \gamma QP.$$

c.   Obtain the Solution $q(t)$ for the damped oscillator from the constant of
the motion in (b) in the underdamped case $\gamma < \omega$. You may need the
integral

$$\int \frac{dx}{\sqrt{1-x^2}} = \sin^{-1}x.$$

(Wisconsin)

Sol:

a.   Lagrange's equation

$$\frac{d}{dt}\left(\frac{\partial L}{\partial \dot{q}}\right) - \frac{\partial L}{\partial q} = 0$$

gives

$$m\ddot{q} = -\frac{\partial V}{\partial q} - 2m\gamma\dot{q}.$$

The particle is seen to be subject to a potential force $-\dfrac{\partial V}{\partial q}$ and a damping force $-2m\gamma\dot{q}$ proportional to its speed. Hence the given Lagrangian is appropriate. The Hamiltonian is given by

$$H = p\dot{q} - L$$

with

$$p = \frac{\partial L}{\partial \dot{q}} = m\dot{q}e^{2\gamma t}.$$

Thus

$$H = \left(\frac{1}{2}m\dot{q}^2 + V\right)e^{2\gamma t} = \frac{p^2 e^{-2\gamma t}}{2m} + V(q)e^{2\gamma t}.$$

b. For the generating function $F_2(q, P, t)$ we have

$$p = \frac{\partial F_2}{\partial q}, \quad Q = \frac{\partial F_2}{\partial P}, \quad K = H + \frac{\partial F_2}{\partial t}.$$

As $F_2 = qPe^{\gamma t}$,

$$p = Pe^{\gamma t}, \quad Q = qe^{\gamma t},$$

$$K = \frac{p^2 e^{-2\gamma t}}{2m} + V(q)e^{2\gamma t} + \gamma qPe^{\gamma t}.$$

For an oscillator of potential

$$V = \frac{1}{2}m\omega^2 q^2 = \frac{1}{2}m\omega^2 Q^2 e^{-2\gamma t},$$

the transformed Hamiltonian is

$$K = \frac{P^2}{2m} + \frac{1}{2}m\omega^2 Q^2 + \gamma QP.$$

As it does not depend on time explicitly, K is a constant of the motion.

c. Hamilton's canonical equations are

$$\dot{P} = -\frac{\partial K}{\partial Q} = -m\omega^2 Q - \gamma P,$$

$$\dot{Q} = \frac{\partial K}{\partial P} = \frac{P}{m} + \gamma Q.$$

Differentiating the second equation and making use of the original equations we have

$$\ddot{Q} + (\omega^2 - \gamma^2)Q = 0.$$

In the underdamped case, $\omega > \gamma$ and we may set

$$\omega_1 = \sqrt{\omega^2 - \gamma^2},$$

where ω_1 is real positive. The Solution is then

$$Q = A \sin(\omega_1 t + \varphi),$$

where A, φ are constants. As

$$P = m(\dot{Q} - \gamma Q),$$

we have

$$K = \frac{1}{2}m[(\dot{Q} - \gamma Q)^2 + \omega^2 Q^2 + 2\gamma Q(\dot{Q} - \gamma Q)] = \frac{1}{2}m(\dot{Q}^2 + \omega_1^2 Q^2) = \frac{1}{2}m\omega_1^2 A^2,$$

giving

$$A = \frac{1}{\omega_1}\sqrt{\frac{2K}{m}}.$$

Hence the solution is

$$q = Qe^{-\gamma t} = \frac{1}{\omega_1}\sqrt{\frac{2K}{m}}e^{-\gamma t}\sin(\omega_1 t + \varphi).$$

2082

Suppose that a system with time-independent Hamiltonian $H_0(q, p)$ has imposed on it an external oscillating field, so that the Hamiltonian becomes $H = H_0(q, p) - \varepsilon q \sin \omega t$, where ε and ω are given constants.

a. What is the physical interpretation of $\varepsilon \sin \omega t$?

b. How are the canonical equations of motion modified?

c. Find a canonical transformation which restores the canonical form of the equations of motion. What is the "new" Hamiltonian?

(Wisconsin)

Sol:

a. A possible interpretation is shown in the following example. A particle of charge e moves in an electric field uniform in space but oscillating in time, namely an electrical field whose strength is represented by $(\varepsilon/e) \sin \omega t$. Then $\varepsilon \sin \omega t$ is the force exerted on the particle by the electric field.

b. Hamilton's canonical equations of motion are now

$$\dot{q} = \frac{\partial H}{\partial p} = \frac{\partial H_0}{\partial p},$$

$$\dot{p} = -\frac{\partial H}{\partial q} = -\frac{\partial H_0}{\partial q} + \varepsilon \sin(\omega t).$$

c. We have

$$H(q, p, t) = H_0(q, p) - \varepsilon q \sin(\omega t)$$

and wish to find a new Hamiltonian

$$K(Q, P) = H_0(q, p)$$

by a canonical transformation. Let the generating function be $F_2(q, P, t)$. As

$$\frac{\partial F_2}{\partial t} = K - H = \varepsilon q \sin(\omega t),$$

we take

$$F_2 = qP - \frac{\varepsilon q}{\omega} \cos(\omega t).$$

The transformation equations are

$$p = \frac{\partial F_2}{\partial q} = P - \frac{\varepsilon}{\omega} \cos(\omega t),$$

or

$$P = p + \frac{\varepsilon}{\omega} \cos(\omega t),$$

and

$$Q = \frac{\partial F_2}{\partial P} = q.$$

Then

$$K(Q, P) = H + \frac{\partial F_2}{\partial t}$$

$$= H_0(q, p) - \varepsilon q \sin(\omega t) + \varepsilon q \sin(\omega t)$$

$$= H_0(Q, P - \frac{\varepsilon}{\omega} \cos(\omega t)),$$

$$\frac{\partial K}{\partial P} = \frac{\partial H_0}{\partial p} \frac{\partial p}{\partial P} = \frac{\partial H_0}{\partial p} = \dot{q} = \dot{Q},$$

$$-\frac{\partial K}{\partial Q} = -\frac{\partial H_0}{\partial q} \frac{\partial q}{\partial Q} = -\frac{\partial H_0}{\partial q} = \dot{p} - \varepsilon \sin(\omega t) = \dot{P},$$

use having been of the results in (b). Hence the transformation restores the canonical form of the equations of motion with the Hamiltonian H_0.

2083

a. Solve the Hamilton-Jacobi equation for the generating functio $S(q, \alpha, t)$ in the case of a single particle moving under the Hamiltonia $H = \frac{1}{2}p^2$. Find the canonical transformation $q = q(\beta, \alpha)$, and $p = p(\beta, \alpha)$ where β and α are the transformed coordinate and momentum respectively. Interpret your result.

b. If there is a perturbing Hamiltonian $H' = \frac{1}{2}q^2$, then α will no longe be constant. Express the transformed Hamiltonian K (using the same transformation found in part (a)) in terms of α, β and t. Solve for $\beta(t)$ and $\alpha(t)$ and show that the perturbed Solution

$$q[\beta(t), \alpha(t)], \quad p[\beta(t), \alpha(t)]$$

c. is simple harmonic. You may need the integrals

$$\int \frac{dx}{x^2 + 1} = \tan^{-1} x,$$

$$-\int \tan x \, dx = \ln(\cos x).$$

(Wisconsin)

Sol:

a. The Hamilton-Jacobi equation

$$\frac{\partial S}{\partial t} + H(q, p, t) = 0$$

with $p = \dfrac{\partial S}{\partial q}$ becomes, for this case,

$$\frac{\partial S}{\partial t} + \frac{1}{2}\left(\frac{\partial S}{\partial q}\right)^2 = 0.$$

As H does not depend on q, t explicitly, we can take the two terms on the left-hand side as equal to $-\gamma, \gamma$ respectively, where γ is at most a function of p. Then

$$S = \sqrt{2\gamma}q - \gamma t.$$

Setting $\alpha = \sqrt{2\gamma}$, we have the generating function

$$S = \alpha q - \frac{1}{2}\alpha^2 t.$$

The constant α can be taken to be the new momentum P. The transformation equations are thus

$$p = \frac{\partial S}{\partial q} = \alpha,$$

$$Q = \frac{\partial S}{\partial P} = \frac{\partial S}{\partial \alpha} = q - \alpha t = \beta, \qquad \text{say.}$$

As $g = \beta + \alpha t$, the particle moves with uniform velocity β in the q, p system.

b. The perturbed Hamiltonian is

$$H = \frac{p^2}{2} + \frac{q^2}{2}.$$

It is transformed to

$$K = H + \frac{\partial S}{\partial t} = \frac{p^2}{2} + \frac{q^2}{2} - \frac{\alpha^2}{2} = \frac{1}{2}(\beta + \alpha t)^2$$

by the transformation equations in (a). Hamilton's equations

$$\dot{Q} = \frac{\partial K}{\partial P}, \qquad \dot{P} = -\frac{\partial K}{\partial Q}$$

give

$$\dot{\beta} = (\beta + \alpha t)t, \qquad \dot{\alpha} = -(\beta + \alpha t).$$

Note that $\alpha \equiv P$, $\beta \equiv Q$ can no longer be considered constant as H has been changed. The last equations combine to give

$$\ddot{\alpha} + \alpha = 0,$$

showing that α is harmonic:

$$\alpha = \alpha_0 \sin(t + \varphi),$$

where α_0, φ are constants, and thus

$$\beta = -\dot{\alpha} - \alpha t = -\alpha_0[\cos(t + \varphi) + t\sin(t + \varphi)].$$

The transformation equations then give

$$p = \alpha = \alpha_0 \sin(t + \varphi),$$
$$q = \beta + \alpha t = -\dot{\alpha} = -\alpha_0 \cos(t + \varphi).$$

Hence the Sol for the perturbed system is harmonic.

2084

a. Let us apply a shearing force on a rectangular solid block as shown i[n]
Fig. 2.77. Find the relation between the displacement u and the applied
force within elastic limits.

b. The elastic properties of a solid support elastic waves. Assume a transverse plane wave which proceeds in the x-direction and whose oscillations are in the y-direction. Derive the equations of motion for the
displacement.

c. Find the expression for the speed of the transverse elastic wave.

(SUNY, Buffalo)

Sol:

a. Hooke's law for shearing

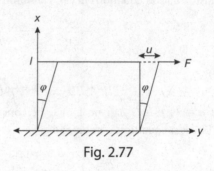

Fig. 2.77

$$\frac{F}{A} = n\varphi,$$

where F is the shearing force, n the shear modulus of the material of the block, φ
the shear angle, and A the cross sectional area of the block parallel to F, gives the
resulting displacement as

$$u = l\varphi = \frac{lF}{An},$$

as φ is a small angle.

b. The potential energy of a unit volume of the block due to shear strain is

$$\frac{1}{lA}\int_0^u F du' = \frac{1}{lA}\int_0^\varphi An\varphi' l d\varphi' = \frac{1}{2}n\varphi^2 = \frac{1}{2}n\left(\frac{\partial u}{\partial x}\right)^2.$$

The kinetic energy of the block during shearing is

$$\int_0^l \frac{1}{2}\rho A\left(\frac{\partial u}{\partial t}\right)^2 dx,$$

ρ being the density of the block. Within the elastic limits, energy is considered and Hamilton's principle

$$\delta \int_{t_1}^{t_2} L\,dt = 0$$

applies. Thus

$$\delta \int_{t_1}^{t_2} (T - V)\,dt = \delta \int_{t_1}^{t_2}\int_0^l \left[\frac{n}{2}\left(\frac{\partial u}{\partial x}\right)^2 - \frac{\rho}{2}\left(\frac{\partial u}{\partial t}\right)^2\right] A\,dx\,dt = 0.$$

As, integrating by parts, we have

$$\int_0^l \delta\left(\frac{\partial u}{\partial x}\right)^2 dx = 2\int_0^l \left(\frac{\partial u}{\partial x}\right)\frac{\partial}{\partial x}(\delta u)\,dx$$

$$= 2\left(\frac{\partial u}{\partial x}\right)\delta u\Big|_0^l - 2\int_0^l \frac{\partial^2 u}{\partial x^2}\delta u\,dx,$$

$$\int_0^l \delta\left(\frac{\partial u}{\partial t}\right)^2 dt = 2\left(\frac{\partial u}{\partial t}\right)\delta u\Big|_{t_1}^{t_2} - 2\int_0^l \left(\frac{\partial^2 u}{\partial t^2}\right)\delta u\,dt,$$

and as $\delta u = 0$ at $y = 0$, l and $t = t_1, t_2$, the above becomes

$$-\int_{t_1}^{t_2}\int_0^l \left(n\frac{\partial^2 u}{\partial x^2} - \rho\frac{\partial^2 u}{\partial t^2}\right)\delta u A\,dx\,dt = 0,$$

giving

$$\frac{\partial^2 u}{\partial x^2} - \frac{\rho}{n}\frac{\partial^2 u}{\partial t^2} = 0$$

as the equation of motion for the displacement u.

c. The equation shows that u, which is in the y-direction, propagates along the x-direction as a transverse wave with speed

$$v = \sqrt{\frac{n}{\rho}}.$$

Part III
Special Relativity

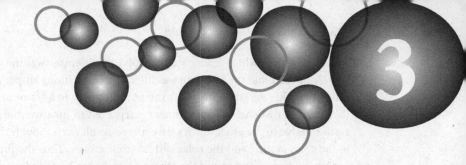

SPECIAL
RELATIVITY
(3001–3054)

a. Briefly describe the dilemma which necessitated the development of the special theory of relativity.

b. Describe an earlier theory which could have eliminated the need for special relativity and name an experiment which proved this theory to be wrong.

c. Describe one modern experiment which lends credence to the special theory of relativity.

(Wisconsin)

Sol:

a. According to Maxwell's electromagnetic theory, the velocity of propagation of electromagnetic waves in free space, c, is a constant independent of the velocity of the source of the electromagnetic radiation. This is contrary to the Galilean transformation which was known to apply between inertial frames. If Maxwell's theory holds in one inertial frame, it would not hold in another inertial frame that has relative motion with respect to the first. The dilemma was that either Maxwell's electromagnetic theory or Newtonian mechanics holds but not both, even though both appeared to be well established.

b.   An earlier theory which attempted to resolve the dilemma was the theory of ether. It presupposed that the universe was filled with a fictitious all-pervasive medium called the ether and that Maxwell's theory holds only in a frame at rest relative to the ether. But Michelson's experiment purported to measure the velocity of the earth relative to the ether always gave a zero result even though the earth moves in the solar system and the solar system itself moves. Thus the presence of ether cannot be demonstrated and the ether theory has to be abandoned.

c.   Take as example Herter's experiment measuring the time of arrival of two photons emitted in the annihilation of a positron in flight. The detectors were at different locations which had the same distance from the place where the annihilation took place. It was found that the two photons arrived at the detectors simultaneously. This indicates that light emitted in different directions from a rapidly moving source has a constant speed.

3002

A space traveler with velocity v synchronizes his clock ($t' = 0$) with his earth friend ($t = 0$). The earthman then observes both clocks simultaneously, t directly and t' through a telescope. What does t read when t' reads one hour?

(UC, Berkeley)

Sol: Let Σ, Σ' be inertial frames attached to the earth and spaceship respectively with the x-axes along the direction of relative velocity, and set $t_1 = t_1' = 0$, $x_1 = x_1' = 0$ when the clocks are synchronized. Consider the event that the spaceship clock reads t_2'. The transformation equations are

$$ct_2 = \gamma(ct_2' + \beta x_2') = \gamma ct_2',$$
$$x_2 = \gamma(x_2' + \beta ct_2') = \gamma \beta ct_2',$$

where $\beta = \dfrac{v}{c}$, $\gamma = \dfrac{1}{\sqrt{1 - \beta^2}}$, as $x_2' = x_1' = 0$. Light signal takes a time

$$\Delta t = \frac{x_2}{c} = \gamma \beta t_2'$$

to reach the earthman. Hence his clock will read

$$t_2 + \Delta t = \gamma(1 + \beta) t_2' = \sqrt{\frac{1 + \beta}{1 - \beta}}\, t_2' = \sqrt{\frac{c + v}{c - v}}$$

when he sees $t_2' = 1$ hour through a telescope.

3003

A light source at rest at position $x = 0$ in reference frame S emits two pulses (called P_1 and P_2) of light, P_1 at $t = 0$ and P_2 at $t = \tau$. A frame S' moves with velocity $v\hat{x}$ with respect to S. An observer in frame S' receives the initial pulse P_1 at time $t' = 0$ at $x' = 0$.

a. Calculate the time τ' between the reception of the pulses at $x' = 0$ as a function of τ and $\beta = \dfrac{v}{c}$.

b. From (a) determine an exact expression for the longitudinal Doppler effect, that is, calculate λ' in terms of λ and β, where λ and λ' are the vacuum wavelengths of light as measured in S and S' respectively.

c. Calculate to first and second order in $\dfrac{v}{c}$ the Doppler shift of the H_β emission ($\lambda = 4861.33$ Å) from the neutralization to H atoms of protons accelerated through a potential of 20 kV. Assuming that emission occurs after acceleration and while the protons drift with constant velocity. Also, assume that the optical axis of the spectrometer is parallel to the motion of the protons.

(Chicago)

Sol:

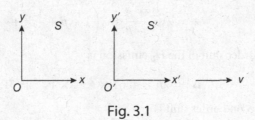

Fig. 3.1

a. The inertial frames S, S' are as shown in Fig. 3.1. Assume that the origins O and O' coincide at $t = t' = 0$ so that the emission of P_1 and its arrival at the observer both occur at $x = 0$, $t = 0$, $x' = 0$, $t' = 0$ as given. The emission of P_2 is at $x = 0$, $t = \tau$ in S and

$$x' = \gamma(x - \beta ct) = -\gamma\beta c\tau,$$

$$t' = \gamma\left(t - \frac{\beta x}{c}\right) = \gamma\tau$$

in S. The signal take a time

$$\Delta t' = \frac{|x'|}{c} = \gamma\beta\tau$$

to arrive at the observer. Hence the observer records the time of arrival as

$$t' + \Delta t' = \gamma(1 + \beta)\tau,$$

or

$$\tau' = \tau\sqrt{\frac{1 + \beta}{1 - \beta}}.$$

b. As τ, τ' are the intervals between two consecutive pulses in S, S' respectively, we have the frequencies

$$\nu = \frac{1}{\tau}, \nu' = \frac{1}{\tau'}$$

and wavelengths

$$\lambda = \frac{c}{\nu} = c\tau, \qquad \lambda' = c\tau' = \lambda\sqrt{\frac{1 + \beta}{1 - \beta}}$$

in S and S' respectively.

c. The protons have energy 20 keV each, much smaller that the rest mass energy of 936 MeV, so that their velocity can be obtained nonrelativistically. Thus

$$\beta = \frac{v}{c} = \sqrt{\frac{2E}{mc^2}} = \sqrt{\frac{40 \times 10^{-3}}{936}} = 0.00654.$$

As β is small, we can expand $\lambda'(\beta)$ as a power series

$$\lambda' = \lambda\sqrt{\frac{1 + \beta}{1 - \beta}} \approx \lambda\left(1 + \beta + \frac{1}{2}\beta^2 + \cdots\right).$$

The first order shift of the H_β emission is

$$\Delta\lambda = \lambda\beta = 4861 \times 6.54 \times 10^{-3} = 31.8 \text{ Å},$$

and the second order shift is

$$\Delta\lambda = \frac{1}{2}\lambda\beta^2 = 0.10 \text{ Å}.$$

3004

a. Consider Lorentz transformations (LT) between the hames S, S' and S'' indicated in Fig. 3.2, where the x-axes are all parallel, and S and S'' are moving in the positive x-direction. Prove that for this type of transformation the inverse of an LT is an LT, and that the resultant of two LT's is another LT.

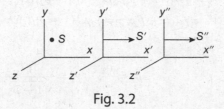

Fig. 3.2

If the velocity of S' relative to S is v_1, and the velocity of S'' relative to S' is v_2, derive the expression for the velocity of S'' relative to S.

b. In particle physics, the interaction between particles is thought of as arising from the exchange of a particle as shown in Fig. 3.3. Prove that the particle exchanged is not real but virtual.

(SUNY, Buffalo)

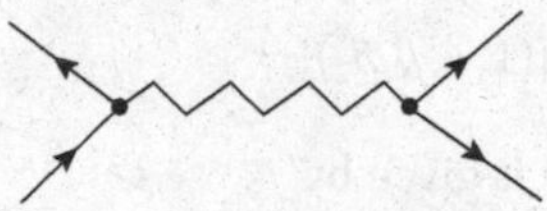

Fig. 3.3 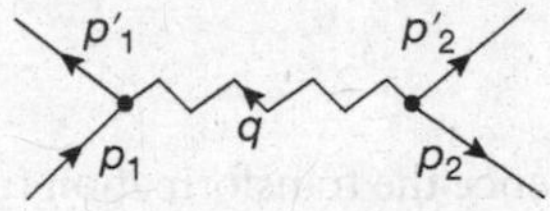   Fig. 4

Sol: The Lorentz transformation between the frames S, S' is given by

$$x' = \gamma_1(x - \beta_1 ct), \quad y' = y, \quad z' = z, \quad ct' = \gamma_1(ct - \beta_1 x),$$

where

$$\beta_1 = \frac{v_1}{c}, \quad \gamma_1 = \frac{1}{\sqrt{1 - \beta_1^2}}.$$

According to the principle of relativity, all inertial frames are equivalent, so the transformation from S to S' should have the same form as the transformation from S' to S. However, as the velocity of S' relative to S is v_1 the velocity of S relative to S' is $-v_1$. The transformation from S' to S, i.e. the inverse transformation, is therefore

$$x = \gamma_1(x' + \beta_1 ct'), \quad y = y', \quad z = z', \quad ct = \gamma_1(ct' + \beta_1 x'),$$

which is seen to be also a Lorentz transformation.

Consider

$$x'' = \gamma_2(x' - \beta_2 ct') = \gamma_2\gamma_1[(x - \beta_1 ct) - \beta_2(ct - \beta_1 x)]$$
$$= \gamma_2\gamma_1[(1 + \beta_1\beta_2)x - (\beta_1 + \beta_2)ct],$$
$$ct'' = \gamma_2(ct' - \beta_2 x') = \gamma_2\gamma_1[(1 + \beta_1\beta_2)ct - (\beta_1 + \beta_2)x],$$

where

$$\beta_2 = \frac{v_2}{c}, \quad \gamma_2 = \frac{1}{\sqrt{1 - \beta_2^2}}.$$

Writing

$$\beta = \frac{\beta_1 + \beta_2}{1 + \beta_1\beta_2}, \quad \gamma = \frac{1}{\sqrt{1 - \beta^2}},$$

we have

$$\frac{1}{\gamma^2} = 1 - \left(\frac{\beta_1 + \beta_2}{1 + \beta_1\beta_2}\right)^2 = \frac{(1 - \beta_1^2)(1 - \beta_2^2)}{(1 + \beta_1\beta_2)^2} = \frac{1}{\gamma_1^2\gamma_2^2(1 + \beta_1\beta_2)^2},$$

or

$$\gamma = \gamma_1\gamma_2(1 + \beta_1\beta_2).$$

Hence the transformation from S to S'' is given by

$$x'' = \gamma(x - \beta ct), \quad y'' = y, \quad z'' = z, \quad ct'' = \gamma(ct - \beta x),$$

showing that it is also a Lorentz transformation. Thus the resultant of two LT's is also an LT.

Note that βc is the velocity of S'' relative to S. This can be shown directly as follows. Consider the transformation between S and S'. Differentiating we have

$$dx = \gamma_1(dx' + \beta_1 cdt'),$$
$$cdt = \gamma_1(cdt' + \beta_1 dx'),$$

and

$$v = \frac{dx}{dt} = \frac{v' + v_1}{1 + \dfrac{v_1 v'}{c^2}}.$$

Thus with v', the velocity of a point relative to S', and v_1, the velocity of S' relative to S, the velocity of the point relative to S is given by the above relation. If the point is at rest in S'', then $v' = v_2$ and the relation gives

$$\beta = \frac{\beta_1 + \beta_2}{1 + \beta_1\beta_2}$$

as expected.

b. As shown in Fig. 3.4, by exchanging a particle of 4-momentum q in the interaction, the 4-momenta of particles 1 and 2, P_1 and P_2 change to p'_1 and p'_2, respectively. The conservation of 4-momentum requires

$$p'_1 = p_1 + q, \quad p'_2 = p_2 - q.$$

Let the mass of particle 1 be m_1 and that of the exchange particle of 4-momentum q be m and consider the first 4-momentum equation. The momentum part gives

$$\mathbf{q} = \mathbf{p}'_1 - \mathbf{p}_1,$$

or

$$q^2 = p'^2_1 + p^2_1 - 2\mathbf{p}_1 \cdot \mathbf{p}'_1,$$

i.e.

$$m^2\gamma^2\beta^2 = m^2_1\gamma'^2_1\beta'^2_1 + m^2\gamma^2_1\beta^2_1 - 2m^2\gamma_1\gamma'_1\beta_1\beta'_1 \cos^2\theta, \tag{1}$$

θ being the angle between $\mathbf{p}_1$ and $\mathbf{p}'_1$, and

$$m\gamma = m_1\gamma'_1 - m_1\gamma_1,$$

or

$$m^2\gamma^2 = m^2_1\gamma'^2_1 + m^2_1\gamma^2_1 - 2m^2_1\gamma_1\gamma'_1. \tag{2}$$

Equations (1) and (2) combine to give

$$m^2 = 2m^2_1(1 - \gamma_1\gamma'_1 + \gamma_1\gamma'_1\beta_1\beta'_1 \cos\theta).$$

We have to show that $m^2 < 0$ so that the interaction cannot be real, but has to be virtual. As $\gamma^2\beta^2 = \gamma^2 - 1$, we have to show

$$\sqrt{\gamma^2_1 - 1}\sqrt{\gamma'^2_1 - 1} \cos\theta < \gamma_1\gamma'_1 - 1,$$

or

$$(\gamma^2_1 - 1)(\gamma'^2_1 - 1)\cos^2\theta < (\gamma_1\gamma'_1 - 1)^2.$$

This would be true if the following is true:

$$(\gamma^2_1 - 1)(\gamma'^2_1 - 1) < (\gamma_1\gamma'_1 - 1)^2,$$

i.e.

$$-\gamma^2_1 - \gamma'^2_1 < -2\gamma_1\gamma'_1,$$

or

$$-(\gamma_1 - \gamma'_1)^2 < 0.$$

Since this always holds, the interaction has to be virtual.

3005

a. Given that $(\mathbf{r}, ct)$ is a relativistic 4-vector, justify the statement that $(c\mathbf{k}, \omega)$ is a relativistic 4-vector.

b. Given that an atom at rest emits light of angular frequency ω_0 and that this atom is traveling at velocity v either directly towards or away from an observer, use the Lorentz transformation to derive a formula for the frequency observed by the observer for the two cases (towards or away from the observer).

(UC, Berkeley)

Sol:

a. Consider a plane electromagnetic wave

$$\mathbf{E} = \mathbf{E}_0 e^{i(\mathbf{k}\cdot\mathbf{r}-\omega t)}, \qquad \mathbf{H} = \mathbf{H}_0 e^{i(\mathbf{k}\cdot\mathbf{r}-\omega t)}$$

in an inertial frame Σ. In another inertial frame Σ' moving with relative velocity v along the x-direction, the field vectors $\mathbf{E}'$, $\mathbf{H}'$ are given by

$$\mathbf{E}'_{\parallel} = \mathbf{E}_{\parallel}, \qquad \mathbf{H}'_{\parallel} = \mathbf{H}_{\parallel},$$
$$\mathbf{E}'_{\perp} = \gamma(\mathbf{E}_{\perp} + \mu_0 \mathbf{v} \times \mathbf{H}_{\perp}),$$
$$\mathbf{H}'_{\perp} = \gamma(\mathbf{H}_{\perp} - \varepsilon_0 \mathbf{v} \times \mathbf{E}_{\perp}).$$

These relations require that the exponential function in $\mathbf{E}$ and $\mathbf{H}$ be invariant:

$$\mathbf{k}' \cdot \mathbf{r}' - \omega' t' = \mathbf{k} \cdot \mathbf{r} - \omega t.$$

Since $(\mathbf{r}, ct)$ is a 4-vector, its components transform according to

$$x' = \gamma(x - \beta ct), \quad y' = y, \quad z' = z, \quad ct' = \gamma(ct - \beta x).$$

Letting $\mathbf{k} = (k_1, k_2, k_3)$, $\mathbf{k}' = (k'_1, k'_2, k'_3)$ we have

$$\mathbf{k}' \cdot \mathbf{r}' - \omega' t' = k'_1 \gamma(x - \beta ct) + k'_2 y + k'_3 z - \omega' \gamma\left(t - \frac{\beta x}{c}\right)$$

$$= \gamma\left(k'_1 + \frac{\beta \omega'}{c}\right)x + k'_2 y + k'_3 z - \gamma(\omega' + \beta c k'_1)t$$

$$= k_1 x + k_2 y + k_3 z - \omega t.$$

Comparing the coefficients of the independent variables x, y, z, t on the two sides of the equation, we find

$$ck_1 = \gamma(ck'_1 + \beta\omega'), \quad ck_2 = ck'_2, \quad ck_3 = ck'_3, \quad \omega = \gamma(\omega' + \beta c k'_1).$$

These relations are exactly the same as those for $(\mathbf{r}, ct)$:

$$x = \gamma(x' + \beta ct'), \quad y = y', \quad z = z', \quad ct = \gamma(ct' + \beta x').$$

Hence $(c\mathbf{k}, \omega)$ is a relativistic 4-vector.

b. Let the observer be at the origin of Σ and the atom be at the origin of Σ' (the atom is moving away from the observer with velocity βc). The angular frequency as measured by the observer is ω. As light from the atom that reaches the observer must have been emitted in the $-x$ direction, we have

$$\mathbf{k}' = (-k', 0, 0) = \left(-\frac{\omega'}{c}, 0, 0\right)$$

by definition. The transformation relation then give

$$\omega = \gamma(\omega' - \beta\omega') = \gamma(1 - \beta)\omega_0 = \omega_0 \sqrt{\frac{1 - \beta}{1 + \beta}} = \omega_0 \sqrt{\frac{c - v}{c + v}}.$$

If the atom is moving towards the observer, βc in the above is to be replaced by $-\beta c$ and we have

$$\omega = \omega_0 \sqrt{\frac{1 + \beta}{1 - \beta}} = \omega_0 \sqrt{\frac{c + v}{c - v}}.$$

3006

An observer in space observes that two events on the earth occur at a time interval of t seconds. The spacecraft is travelling at a speed of 0.8 c. If the spacecraft moves at a speed of 0.9 c, what would the observers measure?

Sol: Let Δt_0 be the time between the events as seen from the earth.

$$\Delta t = \frac{\Delta t_0}{\sqrt{1 - \dfrac{v^2}{c^2}}}$$

$$\Delta t_0 = \Delta t \sqrt{1 - \frac{v^2}{c^2}} = t\sqrt{1 - 0.64} = 0.6t$$

$$\Delta t = \frac{\Delta t_0}{\sqrt{1 - \dfrac{v^2}{c^2}}} = \frac{0.6t}{\sqrt{1 - 0.18}} = \frac{0.6t}{\sqrt{0.82}} = \frac{0.6t}{0.424} = 1.415t.$$

3007

Two rockets of lengths L_1 and $L_2 = 0.75L_1$ approach each other with speed of 0.5 c and 0.75 c, respectively. What is the length of each from the other' perspective?

Sol: From the point of view of L_1,

$$u'_2 = \frac{u_1 - u_2}{1 - \dfrac{u_1 u_2}{c^2}} = \frac{1.25\,c}{1 + 0.375} = 0.9\,c \quad .$$

and

$$L'_2 = L_2 \sqrt{1 - \frac{v^2}{c^2}} = L_2 \sqrt{1 - 0.81} = 0.44 L_2 = 0.44 \times 0.75 L_1 = 0.33 L_1.$$

From the point of view of L_1,

$$u'_1 = \frac{u_2 - u_1}{1 - \dfrac{u_1 u_2}{c^2}} = \frac{-1.25\,c}{1 + 0.375} = -0.9\,c$$

and

$$L'_1 = L_1 \sqrt{1 - \frac{v^2}{c^2}} = L_1 \sqrt{1 - 0.81} = 0.44 L_1.$$

3008

An atomic clock is carried once around the world by a jet plane and then compared with a previously synchronized and similar clock that did not travel. Approximately how large a discrepancy does special relativity predict?

(Columbia)

Sol: Suppose the jet plane moves with velocity βc. Let its rest frame be S' and the earth's frame be S. The two frames can be considered as approximately inertial. Lorentz transformation $ct = \gamma(ct' + \beta x')$ give for $\Delta x' = 0$, since the clock is fixed in Σ', $\Delta t = \gamma \Delta t'$, where $\gamma = \dfrac{1}{\sqrt{1 - \beta^2}}$. Then for $\beta \ll 1$, we have

$$\Delta t = \frac{1}{\sqrt{1 - \beta^2}} \Delta t' \simeq \left(1 + \frac{1}{2}\beta^2\right)\Delta t',$$

or

$$\Delta t - \Delta t' \simeq \frac{1}{2}\beta^2 \Delta t'.$$

Take for example a jet fighter flying at 1000 m/s, about three times the speed of sound. The earth has radius 6400 km, so the fighter takes

$$\frac{2\pi \times 6400 \times 10^3}{1000} = 4.02 \times 10^4 \text{ s}$$

to fly once around the earth. The clock carried by the fighter will be slower by

$$\Delta t - \Delta t' \approx \left(\frac{1000}{3 \times 10^8}\right)^2 \times \frac{4.02 \times 10^4}{2} = 2.2 \times 10^{-7} \text{ s}.$$

3009

a. Write down the Lorentz transformation for the position 4-vector and derive the transformation for the momentum 4-vector.

b. Show that the Doppler effect on light frequency can be expressed as

i. $\nu = \nu_0 \sqrt{\dfrac{1+\beta}{1-\beta}}$ when the source and observer are approaching;

ii. $\nu = \nu_0 \sqrt{\dfrac{1+\beta}{1-\beta}}$ when the source and observer are receding;

iii. $\nu = \dfrac{\nu_0}{\sqrt{1-\beta^2}}$ when the source and observer are in perpendicular directions passing each other.

(SUNY, Buffalo)

Sol:

a. Consider two inertial frames Σ, Σ' with the corresponding axes parallel to each other such that Σ' moves with a velocity $v = \beta c$ along the x direction and that the origins coincide at $t = t' = 0$. The Lorentz transformation for the position 4-vector $x^\alpha \equiv (\mathbf{r}, ct) \equiv (x, y, z, ct)$ is

$$x'^\alpha = Q^\alpha_\beta x^\beta,$$

where

$$Q^\alpha_\beta = \begin{pmatrix} \gamma & 0 & 0 & -\beta\gamma \\ 0 & 1 & 0 & 0 \\ 0 & 0 & 1 & 0 \\ -\beta\gamma & 0 & 0 & \gamma \end{pmatrix}$$

with $\gamma = (1 - \beta^2)^{-\frac{1}{2}}$.

The momentum 4-vector is defined as

$$p^\alpha \equiv (\mathbf{p}c, E),$$

where $E = mc^2$ is the total energy. As all 4-vectors transform in the same way, it transformation is given by

$$\begin{pmatrix} P'_x c \\ P'_y c \\ P'_z c \\ E \end{pmatrix} = \begin{pmatrix} \gamma & 0 & 0 & -\beta\gamma \\ 0 & 1 & 0 & 0 \\ 0 & 0 & 1 & 0 \\ -\beta\gamma & 0 & 0 & \gamma \end{pmatrix} \begin{pmatrix} p_x c \\ p_y c \\ p_z c \\ E \end{pmatrix} = \begin{pmatrix} \gamma(p_x c - \beta E) \\ p_y c \\ p_z c \\ \gamma(E - \beta p_x c) \end{pmatrix}.$$

b. The wave 4-vector is defined by

$$k^\alpha \equiv (\mathbf{k}c, \omega).$$

Its transformation

$$k'^\alpha = Q^\alpha_\beta k^\beta$$

can be written as

$$k'_x = \gamma\left(k_x - \frac{\beta\omega}{c}\right), \quad k'_y = k_y, \quad k'_z = k_z, \quad \omega' = \gamma(\omega - \beta k_x c).$$

To obtain the Doppler effect, let the frames of the light source and observer be Σ', Σ respectively.

i. When the source and observer approach each other let $\beta_0 c$ be the relative velocity of the former relative to the latter. Then $\beta = -\beta_0$. The inverse transformation is

$$\omega = \gamma(\omega' + \beta k'_x c) = \gamma(\omega' - \beta_0 k'_x c).$$

As $k'_y = k'_z = 0$, $k'_x = -k' = -\dfrac{\omega'}{c}$, we have

$$\omega = \gamma(1 + \beta_0)\omega' = \omega' \sqrt{\frac{1 + \beta_0}{1 + \beta_0}},$$

or

$$\nu = \nu_0 \sqrt{\frac{1 + \beta_0}{1 - \beta_0}},$$

where $\omega' = 2\pi\nu_0$ is the proper angular frequency of the light and ω is the angular frequency as measured by the observer. Note that $k_x = -k$ as the light has to be emitted backwards to reach the observer.

ii. When the source and observer are receding from each other, we have $\beta = \beta_0$, $\beta_0 c$ being the velocity of the former relative to the latter. Thus

$$\omega = \gamma(1 - \beta_0)\omega' = \omega\sqrt{\frac{1 + \beta_0}{1 + \beta_0}},$$

or

$$\nu = \nu_0\sqrt{\frac{1 - \beta_0}{1 + \beta_0}}.$$

iii. When the source and observer are in perpendicular directions passing each other, let the source be at $(0, y', 0)$ in Σ' and the observer be at $(0, 0, 0)$ in Σ. They pass each other at $t = t' = 0$, when $k'_x = 0$, $k'_y = -k$, $k'_z = 0$. The transformation equation for ω then gives

$$\omega = \gamma(\omega' + \beta k'_x c) = \gamma\omega' = \frac{\omega'}{\sqrt{1 - \beta^2}},$$

or

$$\nu = \frac{\nu_0}{\sqrt{1 - \beta^2}}.$$

3010

A monochromatic transverse wave with frequency υ propagates in a direction which makes an angle of 60° with the x-axis in the reference frame K of its source. The source moves in the x-direction at speed $\upsilon = \frac{4}{5}c$ towards an observer at rest in the K' frame (where his x'-axis is parallel to the x -axis). The observer measures the frequency of the wave.

a. Determine the measured frequency ν' in terms of the proper frequency υ of the wave.

b. What is the angle of observation in the K' frame?

(SUNY, Buffalo)

Sol: The frame K of the light source moves with velocity βc relative to K', the observer's frame. The (inverse) transformation of the components of the wave 4-vector is given by

$$k'_x c = \gamma(k_x c + \beta\omega), \quad k'_y c = k_y c, \quad k'_z c = k_z c, \quad \omega' = \gamma(\omega + \beta k_x c),$$

where $\gamma = (1 - \beta^2)^{-\frac{1}{2}}$. The angular frequency of the wave in K is $\omega = 2\pi\nu$. If the angle between the light and the x-axis is θ, then

$$k_x = k \cos\theta, \quad k_y = k \sin\theta, \quad k_z = 0, \quad \omega = kc.$$

Thus

$$\omega' = \gamma(\omega + \beta\omega \cos\theta) = \gamma(1 + \beta \cos\theta)\omega,$$

or

$$\nu' = \frac{(1 + \beta \cos \theta)\nu}{\sqrt{1 - \beta^2}}.$$

The above can also be written as

$$k' = \gamma(1 + \beta \cos \theta)k.$$

As

$$k'_x = \gamma\left(k_x + \frac{\beta\omega}{c}\right) = \gamma k(\cos \theta + \beta),$$

the angle $\mathbf{k}'$ makes with the x'-axis is given by

$$\cos \theta' = \frac{k'_x}{k'} = \frac{\cos \theta + \beta}{1 + \beta \cos \theta}.$$

With $\beta = 0.8$, $\theta = 60°$, we have

a. $$\nu' = \left(\frac{1 + 0.8 \cos 60°}{\sqrt{1 - 0.8^2}}\right)\nu = \frac{1.4}{0.6}\nu = \frac{7}{3}\nu,$$

b. $$\cos \theta' = \frac{0.5 + 0.8}{1 + 0.8 \times 0.5} = \frac{13}{14},$$

giving $\theta' = 21.8°$.

3011

Consider two twins. Each twin's heart beats once per second, and each twin broadcasts a radio pulse at each heartbeat. The stay-at-home twin remains at rest in an inertial frame. The traveler starts at rest at time zero, very rapidly accelerates up to velocity v (within less than a heartbeat, and without perturbing his heart!). The traveler travels for time t_1 by his clock, all the while sending out pulses and receiving pulses from home. Then at time t_1 he suddenly reverses his velocity and arrives back home at time $2t_1$. How many pulses did he send out altogether? How many pulses did he receive during the outgoing trip? How many did he receive on the ingoing half of his trip? What is the ratio of total pulses received and sent? Next consider the twin who stays at home. He sends out pulses during the entire trip of the traveler. He receives pulses from the traveler. From time zero to t_2 (by his clock) he receives Doppler-lowered-frequency pulses. At time t_2 he starts receiving Doppler-raised-frequency pulses. Let t_3 be the time interval from time t_2 till the end of the trip. How many pulses does he receive during interval t_2? During t_3? What is the ratio between these? What is the ratio of the total number of pulses he sends to the total he receives? Compare this result with the analogous result for the traveler.

(UC, Berkeley)

Sol: Consider inertial frames Σ, Σ' with Σ' moving with velocity v relative to Σ in the direction of the x-axis. The transformation relations for space-time and angular frequency four-vectors are

$$x' = \gamma(x - vt), \qquad x = \gamma(x' - vt'),$$

$$y' = y, \qquad z' = z,$$

$$t' = \gamma\left(t - \frac{vx}{c^2}\right), \qquad t = \gamma\left(t' - \frac{vx'}{c^2}\right),$$

$$\omega' = \gamma(\omega - vk_x), \qquad \omega = \gamma(\omega' - vk'_x),$$

$$k'_y = k_y, \qquad k'_z = k_z,$$

where

$$\gamma = \frac{1}{\sqrt{1 - \beta^2}} \quad \text{with} \quad \beta = \frac{|v|}{c}, \quad |k| = \frac{\omega}{c}, \quad \omega = 2\pi\nu,$$

ν being the frequency.

Let Σ, Σ' be the rest frames of the twin A who stays at home and the twin B who travels, respectively, with A,B located at the respective origins. As the times of acceleration and deceleration of B are small compared with the time of the trips, Σ' can still be considered inertial. Measure time in seconds so that ν has numerical value one in the rest frame. At the start of the journey of B, $t = t' = 0$.

Consider from the point of view of B.

i. The total trip takes time $\Delta t' = 2t_1$. Thus B sends out $2t_1$ pulses for the entire trip.

ii. For the outgoing trip, $\beta = \dfrac{v}{c}$, $k_x = \dfrac{\omega}{c}$, and the pulses received by B have frequency

$$\nu' = \gamma\left(\nu - \frac{1}{2\pi}\beta c k_x\right) = \gamma(1 - \beta)\nu = \gamma(1 - \beta)$$

since $\nu = 1$ as Σ is the rest frame of A. Hence B receives

$$\nu' t_1 = \gamma(1 - \beta)t_1 = t_1 \sqrt{\frac{1 - \beta}{1 + \beta}}$$

pulses during the outgoing trip.

iii.   For the ingoing trip, $\beta = -\dfrac{v}{c}$, $k_x = \dfrac{\omega}{c}$, and

$$\nu' = \gamma(1 + \beta)\nu = \gamma(1 + \beta).$$

Hence B receives

$$\nu' t_1 = \gamma(1 + \beta)t_1 = t_1 \sqrt{\frac{1 + \beta}{1 - \beta}}$$

pulses during the ingoing trip.

iv. $\dfrac{\text{total pulses received by } B}{\text{total pulses sent by } B} = \dfrac{\gamma(1 - \beta)t_1 + \gamma(1 + \beta)t_1}{2t_1} = \gamma = \dfrac{1}{\sqrt{1 - \beta^2}}.$

Consider from the point of view of A

i.   In the interval $t = 0$ to $t = t_2$, A receives Doppler-lowered-frequency pulses indicating that B is moving away during the interval, i.e. $\beta = \dfrac{v}{c}$. As the pulses have to be emitted in $-x'$ direction to reach $A, k'_x = -\dfrac{\omega}{c}$. Thus

$$\nu = \gamma(\nu' - \beta \nu') = \gamma(1 - \beta)\nu' = \gamma(1 - \beta),$$

since $\nu' = 1$ as Σ' is the rest frame of B, and the number of pulses received is $\gamma(1 - \beta)t_2$. The interval of time during which B, starting at $t = t' = 0$, moves away from A is transformed by

$$\Delta t = \gamma \left(\Delta t' + \frac{\beta \Delta x'}{c} \right) = \gamma \Delta t' = \gamma t_1$$

since $\Delta x' = 0$, B being stationary in Σ'. However, A and B communicate by light pulses, whose time of travel

$$\frac{x}{c} = \frac{\gamma(x' + \beta c t')}{c} = \gamma \beta t' = \gamma \beta t_1,$$

where x is the coordinate of B in Σ, must be taken into account. Hence

$$t_2 = \Delta t + \frac{x}{c} = \gamma(1 + \beta)t_1 = t_1 \sqrt{\frac{1 + \beta}{1 - \beta}},$$

i.e. the number of pulses received is

$$\gamma(1 - \beta)t_2 = t_2 \sqrt{\frac{1 - \beta}{1 + \beta}} = t_1.$$

ii. In the time interval t_3 from $t = t_2$ to the end of journey, A receives Doppler-raised-frequency pulses, indicating that Σ' moves toward Σ, i.e. $\beta = -\dfrac{v}{c}$. As $k'_x = -\dfrac{\omega}{c}$,

$$\nu = \gamma(\nu' + \beta\nu') = \gamma(1 + \beta)\nu' = \gamma(1 + \beta).$$

By a similar argument as that in (i) we have

$$t_3 = \gamma t_1 - \gamma\beta t_1 = \gamma(1 - \beta)t_1.$$

Hence the number of pulses received is

$$\gamma(1 + \beta)t_3 = t_1.$$

iii. $\dfrac{\text{lowered-frequency pulses received by } A}{\text{raised-frequency pulses received by } A} = \dfrac{t_1}{t_1} = 1.$

iv. $\dfrac{\text{total number of pulses sent by } A}{\text{total number of pulses received by } A}$

$$= \frac{t_2 + t_3}{2t_1} = \frac{\gamma(1 + \beta)t_1 + \gamma(1 - \beta)t_1}{2t_1} = \gamma = \frac{1}{\sqrt{1 - \beta^2}}.$$

This is the same as the ratio of the number of pulses received by B to that sent by B during the entire trip, as expected since counting of numbers is invariant under Lorentz transformation.

3012

A spaceship has a transmitter and a receiver. The ship, which is proceeding at constant velocity directly away from the mother earth, sends back a signal pulse which is reflected from the earth. Forty seconds later on the ship's clock the signal is picked up and the frequency received is one half the transmitter frequency.

   a. At the time when the radar pulse bounces off the earth what is the position of the earth as measured in the spaceship frame?

   b. What is the velocity of the spaceship relative to the earth?

   c. At the time when the radar pulse is received by the spaceship where is the ship in the earth frame?

(UC, Berkeley)

Sol: Let the spaceship and the earth be at the origins of inertial frames Σ' and Σ respectively with Σ' moving with velocity βc relative to Σ in the x direction such that $x' = x = 0$ at $t = t' = 0$.

a. The velocity of the radar pulse is c in all directions. So in Σ' the pulse takes time $\dfrac{40}{2} = 20$ s to reach the earth. Hence the position of the earth when the pulse bounces off the earth is $x' = -20\,c = -6 \times 10^9$ m as measured in the ship's frame.

b. In Σ the angular frequency ω of the signal is observed to be

$$\omega = \gamma(\omega' + \beta c k'_x)$$

with $\omega' = \omega_0$, the proper angular frequency of the signal, $k'_x = -\dfrac{\omega_0}{c}$ as the signal has to go in the $-x'$ direction to reach the earth, and $\gamma = \sqrt{\dfrac{1}{1-\beta^2}}$. Thus

$$\omega = \gamma(1 - \beta)\omega_0.$$

After reflection from the earth's surface, the angular frequency will be observed in Σ' as

$$\omega'' = \gamma(\omega - \beta c k_x)$$

with $k_x = \dfrac{\omega}{c}$, $\omega = \gamma(1 - \beta)\omega_0$. Thus

$$\omega'' = \gamma(1 - \beta)\omega = \gamma^2(1 - \beta)^2\omega_0 = \left(\frac{1-\beta}{1+\beta}\right)\omega_0 = \frac{1}{2}\omega_0,$$

yielding

$$\beta = \frac{1}{3}.$$

Hence the velocity of the spaceship relative to the earth is

$$v = \frac{1}{3} \times 3 \times 10^8 = 10^8 \text{ m/s}.$$

c. In Σ', when the signal bounces off the earth the time is

$$t' = \frac{-20c}{-\beta c} = 60 \text{ s}$$

as the earth moves with relative velocity $-\beta c$. When the signal is received by the ship, the time is $t' = 60 + 20 = 80$ s. As the ship is stationary at the origin of Σ', $x' = 0$. This instant is perceived in Σ as a time

$$t = \gamma\left(t' + \frac{\beta x'}{c}\right) = \gamma t' = 80\gamma.$$

As the ship moves away from the earth at a velocity $\beta c = \dfrac{1}{3}c$, its position in Σ at this instant is

$$x = \beta c t = \frac{80}{3}\gamma c = 8.5 \times 10^9 \text{ m}.$$

3013

A point source S of monochromatic light emits radiation of frequency f. An observer A moves at constant speed v along a straight line that passes at a distance d from the source as shown in Fig. 3.5.

a. Derive an expression for the observed frequency as function of the distance x from the point of closest approach O.

b. Sketch an approximate graph of your answer to (a) for the case of $\dfrac{v}{c} = 0.80$.

(UC, Berkeley)

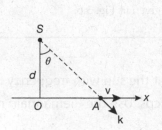

Fig. 3.5

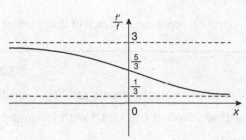

Fig. 3.6

Sol:

a. Let the rest frames of the light source S and the observer A be Σ and Σ' respectively, taking the direction of the relative velocity v as along the x, x'-axes. The transformation of the wave 4-vector components is

$$ck'_x = \gamma(ck_x - \beta\omega), \quad k_y = k'_y, \quad k'_z = k_z, \quad \omega' = \gamma(\omega - \beta ck_x),$$

where $|\mathbf{k}| = \dfrac{\omega}{c}, \beta = \dfrac{v}{c}, \gamma = \dfrac{1}{\sqrt{1 - \beta^2}}$. As $k_x = k \sin\theta, k = \dfrac{\omega}{c}$, we have

$$\omega' = \gamma(\omega - \beta\omega \sin\theta) = \gamma\omega(1 - \beta \sin\theta).$$

With $\sin\theta = \dfrac{x}{\sqrt{d^2 + x^2}}, \omega = 2\pi f', \omega = 2\pi f$, the above gives the observed frequency as

$$f' = \left(1 - \frac{\beta x}{\sqrt{d^2 + x^2}}\right)\frac{f}{\sqrt{1 - \beta^2}}.$$

b. If $\beta = 0.8, \gamma = \dfrac{1}{\sqrt{1 - 0.64}} = \dfrac{10}{6}$ and

$$f' = \frac{1}{3}\left(5 - \frac{4x}{\sqrt{d^2 + x^2}}\right)f.$$

To find the shape of $\frac{f'}{f}$, consider the following:

$$x \to -\infty, \quad \frac{f'}{f} = \frac{1}{3}\left(5 + \frac{4}{\sqrt{\left(\frac{d}{x}\right)^2 + 1}}\right) \to 3,$$

$$x \to \infty, \quad \frac{f'}{f} = \frac{1}{3}\left(5 - \frac{4}{\sqrt{\left(\frac{d}{x}\right)^2 + 1}}\right) \to \frac{1}{3},$$

$$x = 0 \qquad \frac{f'}{f} = \frac{5}{3}.$$

An approximate sketch of the graph of $\frac{f'}{f}$ is given in Fig. 3.6.

3014

Consider monochromatic radiation emitted at the sun with frequency ν_s cps, and received at the earth with frequency ν_e cps. Use the Riemannain matrix form

$$g_{00} = \left(1 + \frac{2\Phi}{c^2}\right), \quad g_{11} = g_{22} = g_{33} = -1, \quad g_{\mu \neq \nu} = 0,$$

where Φ is the gravitational potential energy per unit mass, to derive the "gravitational red shift" $\frac{(\nu_e - \nu_s)}{\nu_s}$ as a function of the difference of gravitational potentials at the sun and earth.

(SUNY, Buffalo)

Sol: In a gravitational field it is always possible to define a frame relative to which the field vanishes over a limited region and which behaves as an inertial frame. A frame freely falling in the gravitational field is such a frame. A standard clock at rest in such a frame measures the local proper time interval.

Consider the emission of monochromatic radiation by an atom at rest at point P_1 in a gravitational field and use a coordinate frame in which the atom is at rest. If the period is t in the coordinate time, the period τ in the local proper time is

$$\tau = t\sqrt{g_{00}(P_1)}.$$

Suppose successive crests of the radiation emitted from P_1 at coordinate times $t_0, t_0 + t$ are received at another fixed point P_2 at coordinate times $t_0 + T$ and

$t_0 + T + t$, where T is the difference between the coordinate times of emission at P_1 and reception at P_2. If the gravitational field is static, T is a constant and the period measured in the coordinate time is

$$(t_0 + T + t) - (t_0 + T) = t.$$

However, a standard clock measuring the local proper time at P_2 will give the period as

$$\tau' = t\sqrt{g_{00}(P_2)}.$$

Hence the frequency ν of the line emitted at P_1 and the frequency ν' observed at P_2, as measured by identical standard clocks, are related by

$$\frac{\nu'}{\nu} = \frac{\tau}{\tau'} = \sqrt{\frac{g_{00}(P_1)}{g_{00}(P_2)}}.$$

If P_1, P_2 are on the sun and the earth respectively, this gives the gravitational red shift as

$$\frac{\nu_e - \nu_s}{\nu_s} = \sqrt{\frac{g_{00}(\mathbf{r}_s)}{g_{00}(\mathbf{r}_e)}} - 1$$

$$= \sqrt{\frac{1 + \dfrac{2\Phi(\mathbf{r}_s)}{c^2}}{1 + \dfrac{2\Phi(\mathbf{r}_e)}{c^2}}} - 1$$

$$\approx \frac{\Phi(\mathbf{r}_s) - \Phi(\mathbf{r}_e)}{c^2}.$$

3015

A mirror is moving through vacuum with relativistic speed v in the x direction. A beam of light with frequency ω_i is normally incident (from $x = +\infty$) on the mirror, as shown in Fig. 3.7.

  a. What is the frequency of the reflected light expressed in terms of ω_i, c and v?

  b. What is the energy of each reflected photon?

  c. The average energy flux of the incident beam is P_i (watts/m²). What is the average reflected energy flux?

(MIT)

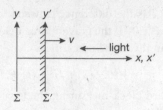

Fig. 3.7

Sol:

a. Let Σ, Σ' be the rest frames of the light source and observer, and of the mirror respectively. The transformation for angular frequency is given by

$$\omega' = \gamma(\omega - \beta ck_x), \quad \omega = \gamma(\omega' + \beta ck_x'),$$

where $\beta = \dfrac{v}{c}, \gamma = \dfrac{1}{\sqrt{1 - \beta^2}}$. For the incident light, $\omega = \omega_i$, $k_x = -\dfrac{\omega c}{c}$, the mirror perceives

$$\omega_i' = \gamma(\omega_i + \beta\omega_i) = \gamma(1 + \beta)\omega_i.$$

On reflection, $\omega_r' = \omega_i'$. The observer in Σ will perceive

$$\omega_r = \gamma(\omega_r' + \beta ck_x')$$

with $k_x' = \omega_r'/c$, or

$$\omega_r = \gamma(1 + \beta)\omega_r' = \gamma^2(1 + \beta)^2\omega_i$$

$$= \left(\frac{1 + \beta}{1 - \beta}\right)\omega_i = \left(\frac{c + v}{c - v}\right)\omega_i$$

as the angular frequency of the reflected light.

b. The energy of each reflected photon is

$$\hbar w_r = \left(\frac{c + v}{c - v}\right)\hbar\omega_i.$$

c. If n is the number of photons per unit volume of the beam, its average energy flux is $nc\hbar\omega$. The average energy flux of the reflected beam is therefore

$$P_r = nc\hbar w_r = \left(\frac{c + v}{c - v}\right)nc\hbar\omega_i = \left(\frac{c + v}{c - v}\right)P_i.$$

3016

As seen by an inertial observer O, photons of frequency ν are incident, at an angle θ_i to the normal, on a plane mirror. These photons are reflected back at an angle θ_r to the normal and at a frequency ν' as shown in Fig. 3.8. Find θ_r

and ν' in terms of θ_i and ν if the mirror is moving in the x direction with velocity v relative to O. What is the result if the mirror were moving with a velocity v in the y direction?

(*Princeton*)

Sol: Let Σ, Σ' be the rest frames of the observer and the mirror, as shown in Figs. 3.8 and 3.9 respectively and use the transformation relations

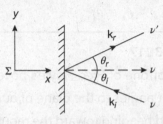

Fig. 3.8   Fig. 3.9

$$ck'_x = \gamma(ck_x - \beta\omega), \qquad ck_x = \gamma(ck'_x + \beta\omega'),$$
$$ck'_y = ck_y, \qquad\qquad ck'_z = ck_z,$$
$$\omega' = \gamma(\omega - \beta ck_x), \qquad \omega = \gamma(\omega' + \beta ck'_x).$$

With $k_i = \dfrac{\omega_i}{c} = \dfrac{2\pi\nu}{c}$, $k_r = \dfrac{\omega_r}{c} = \dfrac{2\pi\nu'}{c}$, we have for the incident light

$$-k'_i \cos \theta'_i = \gamma.\left(-k_i \cos \theta_i - \frac{\beta\omega_i}{c}\right) = -\gamma k_i(\cos \theta_i + \beta),$$

$$\omega'_i = \gamma(\omega_i + \beta ck_i \cos \theta_i) = \gamma\omega_i(1 + \beta \cos \theta_i),$$

or

$$k'_i \cos \theta'_i = \gamma k_i(\cos \theta_i + \beta), \qquad k'_i = \gamma k_i(1 + \beta \cos \theta_i).$$

On reflection, $\omega'_r = \omega'_i$, $\theta'_r = \theta'_i$, so for the reflected light we have

$$\omega_r = \gamma(\omega'_r + \beta ck'_r \cos \theta'_r) = \gamma(\omega'_i + \beta\omega'_i \cos \theta'_i),$$
$$k_r \cos \theta_r = \gamma k'_r(\cos \theta'_r + \beta) = \gamma k'_i(\cos \theta'_i + \beta),$$

or

$$k_r = \gamma k'_i(1 + \beta \cos \theta'_i)$$
$$= \gamma^2 k_i(1 + \beta \cos \theta_i) + \gamma^2\beta k_i(\cos \theta_i + \beta)$$
$$= \gamma^2 k_i(1 + 2\beta \cos \theta_i + \beta^2),$$
$$k_r \cos \theta_r = \gamma^2 k_i[(1 + \beta^2) \cos \theta_i + 2\beta],$$

i.e.

$$v' = \frac{v(1 + 2\beta \cos \theta_i + \beta^2)}{1 - \beta^2},$$

$$\cos \theta_r = \frac{(1 + \beta^2) \cos \theta_i + 2\beta}{1 + 2\beta \cos \theta_i + \beta^2}.$$

If the mirror moves in the y-direction, the motion will have no effect on the reflection process and we still have

$$v' = v, \quad \theta_r = \theta_i.$$

3017

In a simplified version of the ending of one of Fred Hoyle's novels, the hero traveling at high Lorentz factor at right angles to the plane of our galaxy (Fig. 3.10), said he appeared to be inside and heading toward the mouth of a "gold-fish bowl" with a blue rim and a red body (Fig. 3.11). Feynman betted 25 cents that the light from the galaxy would not look that way. We want to see who was right. Take the relative speed to be $\beta = 0.99$ and the angle φ in the frame of the galaxy to be 45° (Fig. 3.10).

a. Derive (or recall) an expression for the relativistic aberration and use it to calculate (Fig. 3.11) the direction from which light from the edge of the galaxy appears to come when viewed in the spacecraft.

b. Derive (or recall) the relativistic Doppler effect and use it to calculate the frequency ratio v'/v for light from the edge.

c. Calculate φ' and v'/v at enough angles φ to decide who won the bet.

(UC, Berkeley)

Sol:

a. Let Σ', Σ be inertial frames attached to the spaceship and the galaxy respectively with Σ' moving with velocity v along the x-direction which is perpendicular to the galactic plane as shown in Fig. 3.10. The velocities of a point, $\mathbf{u}$ and $\mathbf{u}'$, in Σ and Σ' are related by the transformation for velocity

$$u'_x = \frac{u_x - v}{1 - \frac{vu_x}{c^2}}, \quad u'_y = \frac{u_y}{\gamma\left(1 - \frac{vu_x}{c^2}\right)}, \quad u'_z = \frac{u_z}{\gamma\left(1 - \frac{vu_x}{c^2}\right)},$$

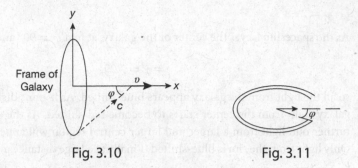

Fig. 3.10 Fig. 3.11

where $\gamma = \dfrac{1}{\sqrt{1-\beta^2}}$ with $\beta = \dfrac{v}{c}$. Consider light coming from a point at the rim of the galactic circle as shown in the figure, for which

$$u_x = c\cos\varphi, \quad u_y = c\sin\varphi, \quad u_z = 0.$$

Then

$$u_x' = c\cos\varphi' = \frac{c\cos\varphi - v}{1 - \beta\cos\varphi},$$

or

$$\cos\varphi' = \frac{\cos\varphi - \beta}{1 - \beta\cos\varphi} = \frac{0.707 - 0.99}{1 - 0.99 \times 0.707} = -0.943,$$

giving $\varphi' = 160.6°$. This is the angle the direction of the light makes with the direction of motion of the spaceship as seen by the traveler. This angle is supplementary to the angle φ' shown in Fig. 3.11.

b. The transformation for angular frequency,

$$\begin{aligned}
\omega' &= \gamma(\omega - \beta c k_x) \\
&= \gamma(\omega - \beta c k\cos\varphi) \\
&= \gamma\omega(1 - \beta\cos\varphi),
\end{aligned}$$

gives

$$\frac{\nu'}{\nu} = \frac{\omega'}{\omega} = \gamma(1 - \beta\cos\varphi) = \frac{1 - 0.99 \times 0.707}{\sqrt{1 - 0.99^2}} = 2.13.$$

c. The above result shows that the light from the rim is blue-shifted. For light from the center, $\varphi = 0$ and

$$\frac{\nu'}{\nu} = \gamma(1 - \beta) = 0.071,$$

showing that it is red-shifted. The critical direction between blue shift and red shift is given by $\nu' = \nu$, or

$$\cos\varphi = \frac{1}{\beta}\left(1 - \frac{1}{\gamma}\right) = 0.868,$$

i.e. $\varphi = 29.8°$.

As the spaceship leaves the center of the galaxy, at first $\varphi = 90°$ and

$$\frac{\nu'}{\nu} = \gamma = 7.09,$$

so all the light from the galaxy appears blue-shifted. As it gains distance from th galaxy, light from the center starts to become red-shifted. As the spaceship goe further out, light from a larger and larger central region will appear red-shifted Only light from the rim is blue-shifted. Finally at a large distance away, $\varphi = 0$ and $\frac{\nu'}{\nu} = 0.071$, so all the light from the galaxy is red-shifted. Thus the statement o Red Hoyle's hero is correct and Feynman loses the bet.

3018

Two identical particles are moving toward each other with a speed of 0.8 c What is the velocity of one particle from the point of view of the other par- ticle? Do the same calculation if the particles were moving with a velocity of 0.008 c and hence prove the principle of correspondence.

Sol: Let $u_1 = 0.8\,c$ and $u_2 = -0.8\,c$ in the stationary frame of reference.

From the frame of reference of u_1,

$$u_2' = \frac{u_1 - u_2}{1 - \dfrac{u_1 u_2}{c^2}} = \frac{1.6\,c}{1 + .64} = 0.98\,c.$$

If $u_1 = 0.008\,c$ and $u_2 = -0.008\,c$ in the stationary frame of reference, according to Newtonian mechanics $u_2' = 0.08\,c + 0.08\,c = 0.16\,c$.

Using the relativistic formula, in the frame of reference of u_2',

$$u_1 = \frac{u_1 - u_2}{1 - \dfrac{u_1 u_2}{c^2}} = \frac{0.16\,c}{1 + 0.0064} = 0.16\,c,$$

which is the same as that obtained by Newtonian physics.

3019

Two particles with the same mass m are emitted in the same direction, with momenta $5mc$ and $10mc$ respectively. As seen from the slower one, what is the velocity of the faster particle, and vice verse? (c = speed of light).

(Wisconsin)

Sol: In the laboratory frame K_0, the slower particle has momentum

$$m\gamma_1 v_1 = m\gamma_1 \beta_1 c = 5mc,$$

giving

$$\gamma_1 \beta_1 = \sqrt{\gamma_1^2 - 1} = 5,$$

or

$$\gamma_1^2 = 26.$$

Hence

$$\beta_1^2 = 1 - \frac{1}{26} = \frac{25}{26}, \quad \text{or} \quad v_1 = \sqrt{\frac{25}{26}} c.$$

Similarly for the faster particle, the velocity is

$$v_2 = \sqrt{\frac{100}{101}} c.$$

Let K_1, K_2 be the rest frames of the slower and faster particles respectively. The transformation for velocity between K_0 and K, which moves with velocity v in the x direction relative to K_0, is

$$u'_x = \frac{u_x - v}{1 - \frac{u_x v}{c^2}}.$$

Thus in K_1, the velocity of the faster particle is

$$v'_2 = \frac{v_2 - v_1}{1 - \frac{u_2 v_1}{c^2}} = \left(\frac{\sqrt{\frac{100}{101}} - \sqrt{\frac{25}{26}}}{1 - \sqrt{\frac{100}{101} \cdot \frac{25}{26}}} \right) c = 0.595c.$$

In K_2, the velocity of the slower particle is

$$v'_1 = \frac{v_1 - v_2}{1 - \frac{v_1 v_2}{c^2}} = -0.595c.$$

3020

Observer 1 sees a particle moving with velocity v on a straight-line trajectory inclined at an angle φ to his z-axis. Observer 2 is moving with velocity u relative to observer 1 along the z-direction. Derive formulas for the velocity and

direction of motion of the particle as by seen observer 2. Check that you ge
the proper result in the limit $v \to c$.

(UC, Berkeley)

Sol: Let K, K' be the rest frames of the observers 1 and 2 respectively with parallel axe
such that the x-axis is in the plane of v and u as shown in Fig. 3.12. The transfor
mation for velocity gives

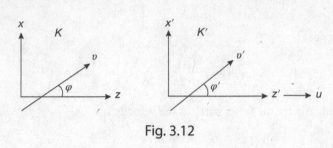

Fig. 3.12

$$v'_z = \frac{v_z - u}{1 - \dfrac{uv_z}{c^2}} = \frac{v \cos \varphi - u}{1 - \dfrac{uv \cos \varphi}{c^2}},$$

$$v'_x = \frac{v_x}{\gamma\left(1 - \dfrac{uv_z}{c^2}\right)} = \frac{v \sin \varphi \sqrt{1 - \dfrac{u^2}{c^2}}}{1 - \dfrac{uv \cos \varphi}{c^2}}, \qquad v'_y = \frac{v_y}{\gamma\left(1 - \dfrac{uv_z}{c^2}\right)} = 0,$$

with $\gamma = \dfrac{1}{\sqrt{1 - \left(\dfrac{u}{c}\right)^2}}$. Hence

$$v' = \sqrt{v'^2_z + v'^2_x}$$

$$= \frac{\sqrt{v^2 + u^2 - 2vu \cos \varphi - \dfrac{u^2 v^2 \sin^2 \varphi}{c^2}}}{1 - \dfrac{uv \cos \varphi}{c^2}},$$

$$\tan \varphi' = \frac{v'_x}{v'_z} = \frac{v \sin \varphi \sqrt{1 - \dfrac{u^2}{c^2}}}{v \cos \varphi - u}.$$

Thus observer 2 sees a particle moving with velocity v' on a straight-line trajectory inclined at an angle φ' to the z'-axis.

In the limit $v \to c$,

$$v' \to \frac{\sqrt{c^2 + u^2 - 2cu\cos\varphi - u^2\sin^2\varphi}}{1 - \dfrac{u\cos\varphi}{c}} = \frac{c - u\cos\varphi}{1 - \dfrac{u\cos\varphi}{c}} = c.$$

This shows that c in any direction is transformed into c, in agreement with the basic assumption of special relativity that c is the same in any direction in any inertial frame. This suggests that our answer is correct.

3021

a. A photon of energy E_i is scattered by an electron, mass m_e, which is initially at rest, as shown in Fig. 3.13. The photon has a final energy E_f. Derive, using special relativity, a formula that relates E_f and E_i to θ, where θ is the angle between the incident photon and the scattered photon.

b. In bubble chambers, one frequently observes the production of an electron-positron pair by a photon. Show that such a process is impossible unless some other body, for example a nucleus, is involved. Suppose that the nucleus has mass M and an electron has mass m_e. What is the minimum energy that the photon must have in order to produce an electron-positron pair?

(*Princeton*)

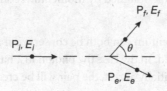

Fig. 3.13

Sol:

a. The scattering is known as the Compton effect. Conservation of energy gives

$$E_i + m_e c^2 = E_f + E_e,$$

where E_e is the energy of the electron after scattering. Conservation of momentum gives

$$\mathbf{P}_i = \mathbf{P}_f + \mathbf{P}_e,$$

where $\mathbf{P}_i$ and $\mathbf{P}_f$ are the momenta of the photon before and after scattering respectively, $\mathbf{P}_e$ is the momentum of the electron after scattering. We also have from the contraction of the momentum 4-vector of the electron

$$E_e^2 = m_e^2 c^4 + P_e^2 c^2,$$

or

$$(m_e c^2 + E_i - E_f)^2 = m_e^2 c^4 + (\mathbf{P}_i - \mathbf{P}_f)^2 c^2.$$

For the photon, $E_i = P_i c$, $E_f = P_f c$, and this becomes

$$2m_e c^2 (E_i - E_f) + (E_i - E_f)^2 = E_i^2 + E_f^2 - 2E_i E_f \cos\theta,$$

or

$$\left(\frac{1}{E_f} - \frac{1}{E_i}\right) m_e c^2 = 1 - \cos\theta.$$

b. Suppose the reaction $\gamma \to e^- + e^+$ is possible. Then the energy and momentum of the system must be conserved in all inertial frames. Consider a frame attached to the center of mass of the created pair. In this frame, the electron and positron will move in a straight line passing through the origin away from each other with the same speed v and the total momentum in zero. Conservation of momentum requires that the momentum of the original photon is also zero. However, each particle has energy $m_e \gamma c^2$, where $\gamma = \dfrac{1}{\sqrt{1 - \dfrac{v^2}{c^2}}}$, and the system has total energy $2m_e \gamma c^2$. This must also be the energy of the original photon, by energy conservation. It follows that the original photon must have a momentum $2m_e \gamma c$, contradicting the result obtained by momentum conservation. Hence the reaction is not possible.

Energy and momentum can both be conserved if another particle, say a nucleus of mass M, is involved. In the case the photon just has enough energy E to create such a pair and M is initially at rest, the pair will be created at rest, i.e.

$$E + Mc^2 = M\gamma c^2 + 2m_e c^2,$$

where $\gamma = \dfrac{1}{\sqrt{1 - \beta^2}}$ with $\beta = \dfrac{v}{c}$, v being the velocity of the nucleus after the pair creation. Momentum conservation give

$$\frac{E}{c} = M\gamma\beta c.$$

As $\gamma\beta = \sqrt{\gamma^2 - 1}$, this gives

$$\gamma^2 = 1 + \left(\frac{E}{Mc^2}\right)^2$$

and the energy equation becomes

$$(E + Mc^2 - 2m_ec^2)^2 = E^2 + M^2c^4,$$

giving

$$E = 2\left(\frac{M - m_e}{M - 2m_e}\right)m_ec^2.$$

As $M \gg m_e$, the minimum photon energy required is just slightly more than the rest energy of the created pair, $2m_ec^2$.

3022

a. A cosmic ray proton collides with a stationary proton to give an excited system moving highly relativistically ($\gamma = 1000$). In this system mesons are emitted with velocity $\bar{\beta}c$. If in the moving system a meson is emitted at an angle $\bar{\theta}$ with respect to the forward direction, at what angle θ will it be observed in the laboratory?

b. Apply the result you obtained in (a) above to mesons (rest energy 140 Mev) emitted in the moving system with momentum 0.5 GeV/c. What will θ be if $\bar{\theta}$ is 90°? What will be the maximum value of θ observed in the laboratory?

(UC, Berkeley)

Sol:

a. Let Σ, Σ' be the laboratory frame and a frame attached to the center of mass of the excited system respectively with Σ' moving with velocity βc relative to Σ in the x-direction. The velocity of a meson emitted in Σ' with velocity $\bar{\beta}c$ at angle $\bar{\theta}$ to the x'-axis is transformed to Σ as

$$u_x = \frac{u'_x + \beta c}{1 + \dfrac{u'_x \beta}{c}} = \frac{(\bar{\beta}\cos\bar{\theta} + \beta)c}{1 + \beta\bar{\beta}\cos\bar{\theta}}, \qquad u_y = \frac{u'_y}{\gamma\left(1 + \dfrac{u'_x\beta}{c}\right)} = \frac{\bar{\beta}c\sin\bar{\theta}}{\gamma(1 + \beta\bar{\beta}\cos\bar{\theta})}.$$

Hence the meson is emitted in Σ at an angle θ to the x-axis given by

$$\tan\theta = \frac{\bar{\beta}\sin\bar{\theta}}{\gamma(\bar{\beta}\cos\bar{\theta} + \beta)},$$

where

$$\gamma = 1000,$$

$$\beta = \frac{\sqrt{\gamma^2 - 1}}{\gamma} = \sqrt{1 - \frac{1}{\gamma^2}} \approx 1 - \frac{1}{2\gamma^2}$$

$$= 1 - 0.5 \times 10^{-6} = 0.9999995.$$

b. If $\overline{\theta} = 90°$, the angle of emission θ in Σ is given by

$$\tan \theta = \frac{\overline{\beta}}{\gamma \beta} = \frac{\overline{\beta}}{\sqrt{\gamma^2 - 1}}.$$

The momentum of the emitted meson is

$$\overline{p} = m\overline{\gamma}\overline{\beta}c = 0.5 \text{ GeV/c},$$

with $\overline{\gamma} = \dfrac{1}{\sqrt{1 - \overline{\beta}^2}}$, m being the rest mass of the meson. Then

$$\overline{\gamma}\overline{\beta} = \frac{0.5}{0.14} = 3.571,$$

or

$$\overline{\beta} = \frac{3.571}{\overline{\gamma}} = \frac{3.571}{\sqrt{1 + 3571^2}} = 0.963,$$

since

$$(\overline{\gamma}\overline{\beta})^2 = \overline{\gamma}^2 - 1.$$

Hence

$$\theta \approx \tan \theta = \frac{0.963}{\sqrt{10^6 - 1}} = 9.63 \times 10^{-4} \text{ rad} = 5.52 \times 10^{-2} \text{ deg} = 3.31'.$$

The maximum value of θ is given by

$$\frac{d \tan \theta}{d\overline{\theta}} = 0,$$

i.e.

$$(\overline{\beta} \cos \overline{\theta} + \beta) \cos \overline{\theta} + \overline{\beta} \sin^2 \overline{\theta} = 0,$$

or by

$$\cos \overline{\theta} = -\frac{\overline{\beta}}{\beta}.$$

Hence

$$\overline{\theta} = \arccos\left(-\frac{0.963}{0.9999995}\right) = 164.4°,$$

which gives

$$\theta = \arctan\left[\frac{0.963 \times \sin 164.4°}{1000 \times (0.963 \cos 164.4° + 0.9999995)}\right] = 0.205° = 12.3'.$$

This is obviously the maximum angle observed in the laboratory since the minimum angle is 0° for $\bar{\theta} = 0°$ and $\bar{\theta} = 180°$.

3023

An elementary particle has a half-life of τ seconds. It is travelling at a speed of 0.7 c. Calculate the observed lifetime. What distance would the particles have to travel to reduce their number to 0.75 of the initial number (assume half-life = 2 ns)?

Sol:

$$\tau' = \tau\sqrt{1 - \frac{v^2}{c^2}} = \tau\sqrt{1 - 0.49} = 7.14\tau.$$

$$N = N_0 e^{-t/\tau}h \Rightarrow t = -\tau ln.75 = 0.29\tau.$$

Distance travelled for the number to reduce to this value is

$$0.7 \times 3 \times 10^8 \times 0.29 \times 2 \times 10^{-9} = 0.12 \text{ m}.$$

3024

In a simplified model of a relativistic nucleus-nucleus collision, a nucleus of rest mass m_1 and speed β_1 collides head-on with a target nucleus of mass m_2 at rest. The composite system recoils at speed β_0 and with center of mass energy ε_0. Assume no new particles are created.

  a. Derive relativistically correct relations for β_0 and ε_0.

  b. Calculate β_0 and ε_0 (in MeV) for a ^{40}Ar nucleus impinging at $\beta_1 = 0.8$ on a ^{238}U nucleus.

  c. A proton is emitted with $\beta_c = 0.2$ at $\theta_c = 60°$ to the forward direction in the frame of the recoiling Ar + U system. Find its laboratory speed β_l and laboratory direction θ_l to within a few percent, making nonrelativistic approximations if they are warranted.

(*UC, Berkeley*)

Sol: As implied by the question the velocity of light is taken to be one for convenience

a.   For a system, $E^2 - p^2$ is invariant under Lorentz transformation. In the laboratory frame Σ, writing $\gamma_1 = \dfrac{1}{\sqrt{1 - \beta_1^2}}$,

$$E^2 - p^2 = (m_1\gamma_1 + m_2)^2 - (m_1\gamma_1\beta_1)^2.$$

In the center of mass frame Σ', $E'^2 - p'^2 = \varepsilon_0^2$. Hence

$$\begin{aligned}
\varepsilon_0^2 &= (m_1\gamma_1 + m_2)^2 - (m_1\gamma_1\beta_1)^2 \\
&= m_1^2\gamma_1^2(1 - \beta_1^2) + 2m_1m_2\gamma_1 + m_2^2 \\
&= m_1^2 + m_2^2 + 2m_1m_2\gamma_1,
\end{aligned}$$

or

$$\varepsilon_0 = \sqrt{m_1^2 + m_2^2 + \frac{2m_1m_2}{\sqrt{1 - \beta_1^2}}}.$$

In the laboratory, the system of m_1, m_2 has total momentum $m_1\gamma_1\beta_1$ and total energy $m_1\gamma_1 + m_2$. These are conserved quantities so that after the collision the composite system will move with velocity

$$\beta_0 = \frac{m_1\gamma_1\beta_1}{m_1\gamma_1 + m_2} = \frac{m_1\beta_1}{m_1 + m_2\sqrt{1 - \beta_1^2}}.$$

b.   The masses are approximately

$$m_1 = 40 \times 0.94 = 37.6 \text{ GeV},$$
$$m_2 = 238 \times 0.94 = 223.7 \text{ GeV}.$$

Then

$$\varepsilon_0 = \sqrt{37.6^2 + 223.7^2 + \frac{2 \times 37.6 \times 223.7}{\sqrt{1 - 0.64}}}$$

$$= 282 \text{ GeV} = 2.82 \times 10^5 \text{ MeV},$$

$$\beta_0 = \frac{37.6 \times 0.8}{37.6 + 223.7 \times \sqrt{1 - 0.64}} = 0.175.$$

c.   The velocity components are transformed according to

$$\beta_{lx} = \frac{\beta_{cx} + \beta_0}{1 + \beta_{cx}\beta_0} = \frac{0.2\cos 60° + 0.175}{1 + 0.2\cos 60° \times 0.175} = 0.270,$$

$$\beta_{ly} = \frac{\beta_{cy}\sqrt{1 - \beta_0^2}}{1 + \beta_{cx}\beta_0} = \frac{0.2\sin 60°\sqrt{1 - 0.175^2}}{1 + 0.2\cos 60° \times 0.175} = 0.168,$$

so the laboratory speed and direction of emission are respectively

$$\beta_l = \sqrt{0.27^2 + 0.168^2} = 0.318,$$

$$\theta_l = \arctan\left(\frac{0.168}{0.27}\right) = 31.9°.$$

Note that as

$$\frac{1}{1 + \beta_{cx}\beta_0} = \frac{1}{1 + 0.1 \times 0.175} = 0.983,$$

$$\frac{\sqrt{1 - \beta_0^2}}{1 + \beta_{cx}\beta_0} = \frac{\sqrt{1 - 0.175^2}}{1 + 0.1 \times 0.175} = 0.968,$$

both differing from 1 by less than 4%, applying nonrelativistic approximations we can still achieve an accuracy of more than 96%:

$$\beta_{lx} \approx \beta_{cx} + \beta_0 = 0.275,$$

$$\beta_{ly} = \beta_{cy} = 0.173,$$

$$\theta_l = \arctan\left(\frac{0.173}{0.275}\right) = 32.2°.$$

3025

In high energy proton-proton collisions, one or both protons may "diffractively dissociate" into a system of a proton and several charged pions. The reactions are

1. $p + p \rightarrow p + (p + n\pi)$,
2. $p + p \rightarrow (p + n\pi) + (p + m\pi)$.

Here n and m count the number of produced pions.

In the laboratory frame, an incident proton of total energy E (the projectile) strikes a proton at rest (the target). Find the incident proton energy E_0 that is

a. the minimum energy for reaction (1) to take place when the target dissociates into a proton and 4 pions,

b. the minimum energy for reaction (1) to take place when the projectile dissociates into a proton and 4 pions,

c.   the minimum energy for reaction (2) to take place when both proton dissociate into a proton and 4 pions.

$$m_\pi = 0.140 \text{ GeV}, \qquad m_p = 0.938 \text{ GeV}.$$

(Chicago

Sol: The quantity $E^2 - p^2$ for a system, where we have taken $c = 1$ for convenience is invariant under Lorentz transformation. If the system undergoes a nuclear reaction that conserves energy and momentum, the quantity will also remain the same after the reaction. In particular for a particle of rest mass m,

$$E^2 - p^2 = m^2.$$

a.   The energy for the reaction

$$p + p \rightarrow p + (p + 4\pi)$$

is minimum when all the final particles are at rest in an inertial frame, particularly the center of mass frame Σ'. Then in the laboratory frame Σ,

$$E^2 - p^2 = (E_0 + m_p)^2 - (E_0^2 - m_p^2) = 2m_p E_0 + 2m_p^2,$$

and in Σ',

$$E'^2 - p'^2 = (2m_p + 4m_\pi)^2,$$

so that

$$2m_p E_0 = 2m_p^2 + 16 m_p m_\pi + 16 m_\pi^2,$$

giving

$$E_0 = \frac{m_p^2 + 8 m_p m_\pi + 8 m_\pi^2}{m_p} = 2.225 \text{ GeV}$$

as the minimum energy the incident proton must have to cause the reaction.

b.   Since both the initial particles are protons and the final state particles are the same as before, the minimum energy remains the same, 2.225 GeV.

c.   For the reaction

$$p + p \rightarrow (p + 4\pi) + (p + 4\pi),$$

we have

$$(E_0 + m_p)^2 - (E_0^2 - m_p^2) = (2m_p + 8m_\pi)^2,$$

giving the minimum incident energy as

$$E_0 = \frac{m_p^2 + 16 m_p m_\pi + 32 m_\pi^2}{m_p} = 3.847 \text{ GeV}.$$

3026

Consider the elastic scattering of two spinless particles with masses m and μ as shown in Fig. 3.14. The Lorentz-invariant scattering amplitude (S-matrix element) may be considered as a function of the two invariant variables

$$s = (K_0 + P_0)^2 - (\mathbf{K} + \mathbf{P})^2$$

and

$$t = (K_0' - K_0)^2 - (\mathbf{K}' - \mathbf{K})^2$$

with $K^2 = K'^2 = \mu^2$ and $P^2 = P'^2 = m^2$. Obtain the physical (i.e. allowed) region in the (s, t) manifold. Compute the boundary curve $t(s)$ and make a qualitative drawing.

(*Chicago*)

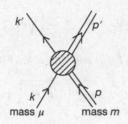

k' p'

k p

mass μ mass m

Fig. 3.14

Sol: In the elastic scattering

$$k + p \rightarrow k' + p',$$

if the system is isolated, the total energy-momentum 4-vector is conserved:

$$K + P = K' + P'.$$

In the center of mass frame of the system, the total momentum is zero:

$$\mathbf{K}' + \mathbf{P}' = \mathbf{K} + \mathbf{P} = 0.$$

Thus

$$s = (K_0 + P_0)^2 - (\mathbf{K} + \mathbf{P})^2 = (K_0 + P_0)^2$$
$$= (\sqrt{\mathbf{K}^2 + \mu^2} + \sqrt{\mathbf{P}^2 + m^2})^2$$
$$= (\sqrt{\mathbf{K}^2 + \mu^2} + \sqrt{\mathbf{K}^2 + m^2})^2,$$

as $\mathbf{P}^2 = \mathbf{K}^2$ in the center of mass frame, and

$$
\begin{aligned}
t &= (K'_0 - K_0)^2 - (\mathbf{K}' - \mathbf{K})^2 \\
&= -(\mathbf{K}' - \mathbf{K})^2 \\
&= -(\mathbf{K}'^2 + \mathbf{K}^2 - 2\mathbf{K}' \cdot \mathbf{K}) \\
&= -2\mathbf{K}^2(1 - \cos\theta),
\end{aligned}
$$

where θ is the angle of scattering of k, as the scattering is elastic.

To find the physical region in the (s, t) manifold, consider $\cos\theta$, where θ varie from 0 to π:

$$
\begin{aligned}
\theta &= 0, \quad \cos\theta = 1, \quad t = 0; \\
\theta &= \pi, \quad \cos\theta = -1, \quad t = -4\mathbf{K}^2.
\end{aligned}
$$

Hence the physical region is given by

$$
t \le 0 \quad \text{and} \quad s \ge \left(\sqrt{\mu^2 - \frac{t}{4}} + \sqrt{m^2 - \frac{t}{4}} \right)^2,
$$

the boundaries being $t = 0$ and that given by

$$
s = \left(\sqrt{\mu^2 - \frac{t}{4}} + \sqrt{m^2 - \frac{t}{4}} \right)^2,
$$

or

$$
\begin{aligned}
t &= \frac{4m^2\mu^2 - (s - m^2 - \mu^2)^2}{s} \\
&= -\frac{1}{s}[s - (m + \mu)^2][s - (m - \mu)^2].
\end{aligned}
$$

The physical region is shown as shaded area in Fig. 3.15.

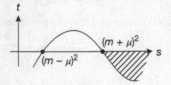

Fig. 3.15

3027

Consider the pion photoproduction reaction

$$
\gamma + p \to p + \pi^\circ,
$$

where the rest energy is 938 MeV for the proton and 135 MeV for the neutral pion.

a. If the initial proton is at rest in the laboratory, find the laboratory threshold gamma-ray energy for this reaction to "go".

b. The isotropic $3K$ cosmic blackbody radiation has an average photon energy of about 10^{-3} eV. Consider a head-on collision between a proton and a photon of energy 10^{-3} eV. Find the minimum proton energy that will allow this pion photoproduction reaction to go.

c. Speculate briefly on the implications of your result [for part (b)] for the energy spectrum of cosmic-ray protons.

(UC, Berkeley)

Sol:

a. The quantity $E^2 - P^2c^2$ is invariant under Lorentz transformation and for an isolated system is the same before and after a reaction. The threshold γ-ray energy is that for which the final state particles are all at rest in the center of mass frame. Thus

$$(E_\gamma + m_p c^2)^2 - \left(\frac{E_\gamma}{c}\right)^2 c^2 = (m_p + m_\pi)^2 c^4,$$

where E_γ is the energy of the photon and $\dfrac{E_\gamma}{c}$ its momentum, giving

$$E_\gamma = \frac{(m_\pi^2 + 2m_\pi m_p)c^4}{2m_p c^2} = 144.7 \text{ MeV}$$

as the threshold γ-ray energy.

b. That the proton collides head-on with the photon means that their momenta are opposite in direction. Then

$$(m_p \gamma c^2 + E_\gamma)^2 - \left(m_p \gamma \beta c - \frac{E_\gamma}{c}\right)^2 c^2 = (m_p + m_\pi)^2 c^4,$$

where $\gamma = \dfrac{1}{\sqrt{1 - \beta^2}}$, βc being the velocity of a proton with the minimum energy to initiate the photoproduction reaction, giving

$$\gamma(1 + \beta) = \frac{(m_\pi^2 + 2m_\pi m_p)c^4}{2E_\gamma m_p c^2} = 1.447 \times 10^{11}$$

with $E_\gamma = 10^{-9}$ MeV. As this implies $\gamma \gg 1$, we can take $\beta = 1$. Hence $\gamma = 7.235 \times 10^{10}$ and the minimum proton energy is

$$E_p = 0.938 \times 7.235 \times 10^{10} = 6.787 \times 10^{10} \text{ GeV}.$$

c.   The result implies that the part of the energy spectrum of cosmic-ray proton with $E > 6.79 \times 10^{10}$ GeV will be depleted to some degree due to interaction with the cosmic blackbody radiation.

3028

A beam of $10^6 K_l^\circ$ mesons per second with $\beta \equiv \dfrac{v}{c} = \dfrac{1}{\sqrt{2}}$ is observed to interact with a lead brick according to the reaction

$$K_l^\circ + \text{Brick} \rightarrow K_s^\circ + \text{Brick}$$

with the internal state of the lead brick identical before and after the reaction. The directions of motion of the incoming K_l° and outgoing K_s° may also be considered to be identical. (This is called coherent regeneration.)

Using

$$m(K_l) = 5 \times 10^8 \text{ eV/}c^2,$$
$$m(K_l) - m(K_s) = 3.5 \times 10^{-6} \text{ eV/}c^2,$$

give the magnitude and direction of the average force (either in dynes or in newtons) exerted on the brick by this process.

(UC, Berkeley)

Sol: Denote $m(K_l)$, $m(K_s)$ by m_l, m_s respectively. For an incoming K_l meson, the energy and momentum are respectively

$$E_l = m_l \gamma c^2 = \frac{m_l c^2}{\sqrt{1 - \frac{1}{2}}} = \sqrt{2} m_l c^2,$$

$$P_l = m_l \gamma \beta c = \sqrt{2} \cdot \frac{1}{\sqrt{2}} m_l c = m_l c.$$

Since the internal state of the lead brick remains the same after the reaction, the energies of the beam before and after the reaction must also be the same. Thus

$$E_l = E_s.$$

As

$$
\begin{aligned}
P_s^2 c^2 &= E_s^2 - m_s^2 c^4 \\
&= E_l^2 - m_s^2 c^4 = 2m_l^2 c^4 - [m_l - (m_l - m_s)]^2 c^4 \\
&\approx m_l^2 c^4 + 2m_l(m_l - m_s)c^4,
\end{aligned}
$$

or

$$P_s c \approx m_l c^2 + (m_l - m_s)c^2 = P_l c + (m_l - m_s)c^2,$$

as $m_l - m_s \ll m_l$. Hence

$$(P_s - P_l) \approx (m_l - m_s)c = 3.5 \times 10^{-6} \text{ eV/c}.$$

The change of momentum per second of the beam due to the reaction is

$$(P_s - P_l) \times 10^6 = 3.5 \text{ eV/c/s} = \frac{3.5 \times 1.6 \times 10^{-19}}{3 \times 10^8} = 1.87 \times 10^{-27} \text{ N}.$$

This is the average force exerted by the brick on the beam. As the momentum of the beam becomes larger after the interaction, this force is in the direction of the beam. Consequently the force exerted by the beam on the brick is opposite to the beam and has a magnitude 1.87×10^{-27} N.

3029

A π meson with a momentum of $5m_\pi c$ makes an elastic collision with a proton ($m_p = 7m_\pi$) which is initially at rest (Fig. 3.16).

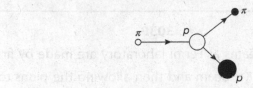

Fig. 3.16

a. What is the velocity of the c.m. reference frame?
b. What is the total energy in c.m. system?
c. Find the momentum of the incident pion in the c.m. system.

(UC, Berkeley)

Sol:

a. The system has total momentum $P = p_\pi = 5m_\pi c$ and total energy

$$E = \sqrt{p_\pi^2 + m_\pi^2 c^4} + m_p c^2 = \sqrt{26}m_\pi c^2 + 7m_\pi c^2.$$

Hence it moves with a velocity $\bar{v}$, which is also the velocity of the c.m. system, in the laboratory given by

$$\bar{v} = \frac{Pc^2}{E} = \frac{5c}{\sqrt{26} + 7} = 0.413c.$$

b. $E^2 - P^2c^2$ is invariant under Lorentz transformation, so the total energy E' in the c.m. frame is given by

$$E^2 - P^2c^2 = E'^2$$

as the total momentum in the c.m frame is by definition zero. Hence

$$E'^2 = (\sqrt{26} + 7)^2 m_\pi^2 c^4 - 25 m_\pi^2 c^4 = (14\sqrt{26} + 50) m_\pi^2 c^4,$$

or

$$E' = \sqrt{14\sqrt{26} + 50}\ m_\pi c^2 = 11.02 m_\pi c^2.$$

c. The total energy in the c.m. frame is

$$E' = \sqrt{p'^2_\pi + m_\pi^2 c^4} + \sqrt{p'^2_p + m_p^2 c^4}$$

$$= \sqrt{p'^2_\pi + m_\pi^2 c^4} + \sqrt{p'^2_\pi + 49 m_\pi^2 c^4},$$

since $|\mathbf{p}'_p| = |\mathbf{p}'_\pi|$ in the c.m. frame and $m_p = 7 m_\pi$. From (b) we have $E' = \sqrt{50 + 14\sqrt{26}} m_\pi c^2$. Substituting this in the above and solving for p'_π, we have

$$p'_\pi = \frac{35 m_\pi c}{\sqrt{50 + 14\sqrt{26}}} = 3.18 m_\pi c.$$

3030

High-energy neutrino beams at Fermi laboratory are made by first forming a monoenergetic π^+ (or K^+) beam and then allowing the pions to decay by $\pi^+ \to \mu^+ + \nu$. Recall that the mass of the pion is 140 MeV/c^2 and the mass of the muon is 106 MeV/c^2.

a. Find the energy of the decay neutrino in the rest frame of the π^+.

   In laboratory frame, the energy of the neutrino depends on the decay angle θ (Fig. 3.17). Suppose the π^+ beam has an energy of 200 GeV.

b. Find the energy of a neutrino produced in the forward direction $(\theta = 0)$.

c. Find the angle θ at which the neutrino's energy has fallen to half of its maximum energy.

(Chicago)

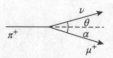

Fig. 3.17

Sol:

a. For convenience use units such that $c = 1$ (m, E, p are all in MeV). Consider the Lorentz-invariant and conserved quantity $E^2 - p^2$. In laboratory frame, before the decay

$$E^2 - p^2 = E_\pi^2 - p_\pi^2 = m_\pi^2.$$

In the rest frame of the pion, after the decay

$$\begin{aligned} E'^2 - p'^2 &= (E'_\mu + E'_\nu)^2 - (\mathbf{p}'_\mu + \mathbf{p}'_\nu)^2 \\ &= (E'_\mu + E'_\nu)^2 = p'^2_\mu + m_\mu^2 + p'^2_\nu + 2p'_\nu\sqrt{p'^2_\mu + m_\mu^2} \\ &= 2p'^2_\nu + m_\mu^2 + 2p'_\nu\sqrt{p'^2_\nu + m_\mu^2} \end{aligned}$$

as $\mathbf{p}'_\mu = -\mathbf{p}'_\nu$, and $E'_\nu = p'_\nu$ assuming the neutrino to have zero rest mass. Equating the above two expressions gives

$$E'_\nu = p'_\nu = \frac{m_\pi^2 - m_\mu^2}{2m_\pi} = 29.9 \text{ MeV}.$$

b. In the laboratory frame (Fig. 3.17), momentum conservation gives

$$p_\pi = p_\nu \cos\theta + p_\mu \cos\alpha, \qquad p_\nu \sin\theta = p_\mu \sin\alpha,$$

or

$$p_\mu^2 = p_\pi^2 + p_\nu^2 - 2p_\pi p_\nu \cos\theta,$$

and energy conservation gives

$$E_\pi = E_\nu + E_\mu.$$

As $p_\nu = E_\nu$, $p_\mu^2 = E_\mu^2 - m_\mu^2$, the last two equations give

$$p_\nu = \frac{m_\pi^2 - m_\mu^2}{2(E_\pi - p_\pi \cos\theta)}.$$

As $E_\pi \gg m_\pi$, we have

$$p_\pi = E_\pi \sqrt{1 - \frac{m_\pi^2}{E_\pi^2}} \approx E_\pi\left[1 - \frac{1}{2}\left(\frac{m_\pi}{E_\pi}\right)^2\right],$$

and hence

$$E_\nu = p_\nu \approx \frac{(m_\pi^2 - m_\mu^2)E_\pi}{2E_\pi^2(1 - \cos\theta) + m_\pi^2 \cos\theta}.$$

For neutrinos emitted in the forward direction, $\theta = 0$ and

$$E_\nu \approx \left[1 - \left(\frac{m_\mu}{m_\pi}\right)^2\right]E_\pi = 85.4 \text{ GeV}.$$

c. As

$$E_\nu \approx \frac{(m_\pi^2 - m_\mu^2)E_\pi}{2E_\pi^2 - (2E_\pi^2 - m_\pi^2)\cos\theta},$$

E_ν is maximum for neutrinos emitted at $\theta = 0$. For E_ν at half the maximum value i.e.

$$\left(\frac{m_\pi^2 - m_\mu^2}{m_\pi^2}\right)\frac{E_\pi}{2} = \frac{(m_\pi^2 - E_\mu^2)E_\pi}{2E_\pi^2 - (2E_\pi^2 - m_\pi^2)\cos\theta},$$

we have

$$\cos\theta = \frac{2(E_\pi^2 - m_\pi^2)}{2E_\pi^2 - m_\pi^2} \approx 1 - \frac{1}{2}\left(\frac{m_\pi}{E_\pi}\right)^2 \approx 1 - \frac{\theta^2}{2}$$

as θ is obviously small. Hence

$$\theta = \frac{m_\pi}{E_\pi} = 0.0007 \text{ rad} = 2.4'.$$

3031

a. A particle of mass $m_1 = 1$ g traveling at 0.9 times the speed of light collides head-on with a stationary particle of mass $m_2 = 10$ g and is embedded in it. What is the rest mass and velocity of the resulting composite particle?

b. Now suppose m_1 to be stationary. How fast should m_2 be moving in order to produce a composite with the same rest mass as in part (a)?

c. Again, if m_1 is stationary, how fast should m_2 be traveling in order to produce a composite that will have the same velocity that you found for the composite in part (a)?

(SUNY, Buffalo)

Sol:

a. Let the composite have mass m and velocity βc. Conservation of energy and of momentum give

$$m\gamma c^2 = (m_1\gamma_1 + m_2)c^2, \qquad m\gamma\beta c = m_1\gamma_1\beta_1 c.$$

where $\gamma = \dfrac{1}{\sqrt{1 - \beta^2}}$, etc. Hence

$$\beta = \frac{m_1 \gamma_1 \beta_1}{m_1 \gamma_1 + m_2} = \frac{m_1 \beta_1}{m_1 + m_2 \sqrt{1 - \beta_1^2}} = \frac{0.9}{1 + 10\sqrt{1 - 0.9^2}} = 0.168,$$

$$m^2 = (m\gamma)^2 - (m\gamma\beta)^2 = (m_1\gamma_1 + m_2)^2 - (m_1\gamma_1\beta_1)^2$$

$$= m_1^2 + m_2^2 + 2\gamma_1 m_1 m_2,$$

as $(\gamma\beta)^2 = \gamma^2 - 1$, etc., or

$$m = \sqrt{m_1^2 + m_2^2 + \frac{2m_1 m_2}{\sqrt{1 - \beta_1^2}}} = \sqrt{1 + 100 + \frac{20}{\sqrt{1 - 0.9^2}}} = 12.1 \text{ g}.$$

Thus the composite has rest mass 12.1 g and velocity 0.168c.

b. The roles of m_1 and m_2 are now interchanged so that

$$m = \sqrt{m_2^2 + m_1^2 + \frac{2m_2 m_1}{\sqrt{1 - \beta_2^2}}},$$

which is the same expression as before with $\beta_1 \to \beta_2$. Then as m_1, m_2, m remain the same, β_2 must have the value of β_1 before, that is, m_2 must move with velocity $0.9c$.

c. As in (b), we have

$$\beta = \frac{m_2 \beta_2}{m_2 + m_1 \sqrt{1 - \beta_2^2}},$$

or

$$(m_2^2 + m_1^2\beta^2)\beta_2^2 - 2m_2^2\beta\beta_2 + (m_2^2 - m_1^2)\beta^2 = 0.$$

As $m_2^2 \gg m_1^2\beta^2$, the above can be reduced to

$$m_2^2\beta_2^2 - 2m_2^2\beta\beta_2 + (m_2^2 - m_1^2)\beta^2 = 0,$$

i.e.

$$[m_2\beta_2 - (m_2 + m_1)\beta][m_2\beta_2 - (m_2 - m_1)\beta] = 0,$$

giving

$$\beta_2 = \left(1 + \frac{m_1}{m_2}\right)\beta = 0.185, \qquad \beta = \left(1 - \frac{m_1}{m_2}\right)\beta = 0.151.$$

Hence m_2 should travel at $0.185c$ or $0.151c$.

3032

A particle with mass m and total energy E_0 travels at a constant velocity V which may approach the speed of light. It then collides with a stationary particle with the same mass m, and they are seen to scatter elastically at the relative angle θ with equal kinetic energies.

a. Determine θ, relating it to m and E_0. ·

b. Find the numerical value of θ in the following limits:

   i. low energy ($V \ll c$),

   ii. high energy ($V \sim c$).

(SUNY, Buffalo

Sol:

a. As the elastically scattered particles have the same mass and the same kinetic energy, their momenta must make the same angle $\dfrac{\theta}{2}$ with the incident direction and have the same magnitude. Conservation of energy and of momentum give

$$mc^2 + E_0 = 2E, \qquad p_0 = 2p \cos\left(\frac{\theta}{2}\right),$$

where E, p are the energy and momentum of each scattered particle. Squaring both sides of the energy equation we have

$$m^2c^4 + E_0^2 + 2E_0 mc^2 = 4(p^2c^2 + m^2c^4),$$

or

$$E_0^2 + 2E_0 mc^2 - 3m^2c^4 = \frac{p_0^2 c^2}{\cos^2\left(\dfrac{\theta}{2}\right)} = \frac{E_0^2 - m^2c^4}{\cos^2\left(\dfrac{\theta}{2}\right)},$$

giving

$$\cos\left(\frac{\theta}{2}\right) = \sqrt{\frac{E_0^2 - m^2c^4}{(E_0 - mc^2)(E_0 + 3mc^2)}} = \sqrt{\frac{E_0 + mc^2}{E_0 + 3mc^2}}.$$

b. i. $V \ll c, E_0 \approx mc^2,$

$$\cos\left(\frac{\theta}{2}\right) \simeq \sqrt{\frac{2}{4}} = \frac{1}{\sqrt{2}},$$

giving

$$\theta \approx \frac{\pi}{2}.$$

ii. $V \rightarrow c, E_0 \gg mc^2$,

$$\cos\left(\frac{\theta}{2}\right) \approx 1,$$

giving $\theta \approx 0$.

3033

Of particular interest in particle physics at present are weak interactions at high energies. These can be investigated by studying high-energy neutrino interactions. One can produce neutrino beams by letting π and K mesons decay in flight. Suppose a 200 GeV/cπ meson beam is used to produce neutrinos via the decay $\pi \rightarrow \mu + \nu$. The lifetime of a π meson is $\tau_{\pi}\pm = 2.60 \times 10^{-8}$ s in its rest frame, and its rest energy is 139.6 MeV. The rest energy of the muon is 105.7 MeV, and the neutrino is massless.

a. Calculate the mean distance traveled by the pions before they decay.

b. Calculate the maximum angle of the muons (relative to the pion direction) in the laboratory.

c. Calculate the minimum and maximum momenta the neutrinos can have.

(UC, Berkeley)

Sol:

a. Let m be the rest mass of a pion. As $m\gamma\beta c^2 = 200$ GeV, we have

$$\gamma\beta = \sqrt{\gamma^2 - 1} = \frac{200}{0.1396} = 1432.7$$

and can take

$$\beta \approx 1, \qquad \gamma = 1433.$$

On account of time dilation, the laboratory lifetime of a pion is $\tau = \gamma\tau_{\pi} = 1433 \times 2.6 \times 10^{-8} = 3.726 \times 10^{-5}$s. So the mean distance traveled by the pions before they decay is

$$\tau c = 3.726 \times 10^{-5} \times 3 \times 10^8 = 1.12 \times 10^4 \text{ m} = 11.2 \text{ km.}$$

b. The total energy of the system in the rest frame Σ' of the pion is its rest energy $m_{\pi}c^2$. Conservation of energy requires that for $\pi \rightarrow \mu + \nu$,

$$m_{\pi}c^2 = E'_{\mu} + E'_{\nu},$$

the prime being used to denote quantities in the Σ' frame. As the total momentum is zero in Σ', $\mathbf{p}'_\mu = -\mathbf{p}'_\nu$ and $E'_\nu = p'_\nu c = p'_\mu c$, assuming the neutrino to have zero rest mass. Thus

$$(m_\pi c^2 - E'_\mu)^2 = p'^2_\nu c^2 = p'^2_\mu c^2 = E'^2_\mu - m^2_\mu c^4,$$

giving

$$E'_\mu = \frac{(m^2_\pi + m^2_\mu)c^2}{2m_\pi} = 109.8 \text{ MeV}.$$

Take the x'-axis along the direction of motion of the pion. Transformation equations for the muon momentum are

$$p_\mu \cos\theta = \gamma\left(p'_\mu \cos\theta' + \frac{\beta E'_\mu}{c}\right),$$

$$p_\mu \sin\theta = p'_\mu \sin\theta',$$

giving

$$\tan\theta = \frac{p'_\mu \sin\theta'}{\gamma\left(p'_\mu \cos\theta + \dfrac{\beta E'_\mu}{c}\right)}.$$

For θ to be maximum, we require

$$\frac{d\tan\theta}{d\theta'} = 0,$$

which gives

$$\cos\theta' = -\frac{p'_\mu c}{\beta E'_\mu} \approx -\frac{p'_\mu c}{E'_\mu} = -\frac{\sqrt{E'^2_\mu - m^2_\mu c^4}}{E'_\mu} = -0.271,$$

or $\theta' = 105.7°$. This in turn gives

$$\theta = 0.0112° = 0.675'.$$

Note that this is the maximum angle of emission in the laboratory since the minimum angle is 0, corresponding to $\theta' = 0$.

c. The neutrino has energy

$$E'_\nu = m_\pi c^2 - E'_\mu = 139.6 - 109.8 = 29.8 \text{ MeV}$$

and momentum $p'_\nu = 29.8$ MeV/c in Σ'. E'_ν can be transformed to the Σ frame by

$$E_\nu = \gamma(E'_\nu + \beta p'_\nu c \cos\theta').$$

As $E_\nu = p_\nu c$, $E'_\nu = p'_\nu c$, the above can be written as

$$p_\nu = \gamma(1 + \beta \cos\theta')p'_\nu.$$

Hence neutrinos emitted in the forward direction of the pion rest frame, i.e. $\theta' = 0$, will have the largest momentum in the laboratory of

$$(p_\nu)_{\text{max}} = \gamma(1 + \beta)p'_\nu \approx \frac{2\gamma E'_\nu}{c} = 8.54 \times 10^4 \text{ MeV}/c,$$

while neutrinos emitted backward in Σ' will have the smallest momentum in the laboratory of

$$(p_\nu)_{\text{min}} = \gamma(1 - \beta)p'_\nu = (\gamma - \sqrt{\gamma^2 - 1})p'_\nu \approx \frac{1}{2\gamma}\frac{E'_\nu}{c} = 1.04 \times 10^{-2} \text{ MeV}/c.$$

3034

A *K* meson of rest energy 494 MeV decays into a μ meson of rest energy 106 MeV and a neutrino of zero rest energy. Find the kinetic energies of the μ meson and neutrino into which the *K* meson decays while at rest.

(UC, Berkeley)

Sol: Conservation of energy gives

$$m_K c^2 = E_\mu + E_\nu = \sqrt{p_\mu^2 c^2 + m_\mu^2 c^4} + p_\nu c = \sqrt{p_\mu^2 c^2 + m_\mu^2 c^4} + p_\mu c,$$

as $\mathbf{p}_\mu = -\mathbf{p}_\nu$, or $p_\mu = p_\nu$, for momentum conservation. Hence

$$p_\mu = \left(\frac{m_K^2 - m_\mu^2}{2m_K}\right)c.$$

Thus

$$E_\nu = p_\nu c = p_\mu c = \left(\frac{m_K^2 - m_\mu^2}{2m_K}\right)c^2$$

$$= 235.6 \text{ MeV},$$

$$E_\mu = \sqrt{p_\mu^2 c^2 + m_\mu^2 c^4} = \left(\frac{m_K^2 + m_\mu^2}{2m_K}\right)c^2$$

$$= 258.4 \text{ MeV}.$$

Therefore the kinetic energy of the neutrino is 235.6 MeV, and that of the muon is $258.4 - 106 = 152.4$ MeV.

3035

The dot product of two four-vectors

$$A^\mu = (A^\circ, \mathbf{A}) \qquad \text{and} \qquad B^\mu = (B^\circ, \mathbf{B})$$

is here defined as

$$A^\mu B_\mu = A^\circ B^\circ - \mathbf{A} \cdot \mathbf{B}.$$

Consider the reaction shown in Fig. 3.18 in which particles of masses m_1 and m_2 are incident and particles of masses m_3 and m_4 emerge. The p's and q's are their four momenta. The variables given below are commonly used to describe such a reaction:

$$s = (q_1 + p_1)^2, \qquad t = (q_1 - q_2)^2, \qquad u = (q_1 - p_2)^2.$$

a. Show that

$$s + t + u = \sum_{i=1}^{4} m_i^2.$$

b. Assume the reaction is elastic scattering and let

$$m_1 = m_3 = \mu, \qquad m_2 = m_4 = m.$$

In the c.m. frame let the initial and final three-momenta of the particle of mass μ be $\mathbf{k}$ and $\mathbf{k}'$ respectively. Express s, t and u in terms of $\mathbf{k}$ and $\mathbf{k}'$, simplifying as much as possible. Interpret s, t and u.

c. Assume that in the laboratory frame the particle of mass m is initially at rest. Express the initial and final laboratory energies of particle μ, as well as the scattering angle, in terms of s, t and u.

(SUNY, Buffalo)

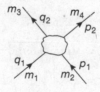

Fig. 3.18

Sol: Use units for which the velocity of light $c = 1$ for convenience.

a. q^2 is defined as $q^\alpha q_\alpha$ with $q^\alpha = (q^\circ, \mathbf{q})$, $q_\alpha = (q^\circ, -\mathbf{q})$. The quantity $q^\alpha q_\alpha$ is invariant under Lorentz transformation. Evaluating it in the rest frame of the particle:

$$q^2 = (q^\circ)^2 - \mathbf{q}^2 = E^2 - \mathbf{q}^2 = m^2.$$

Now

$$s + t + u = (q_1 + p_1)^2 + (q_1 - q_2)^2 + (q_1 - p_2)^2$$
$$= q_1^2 + q_2^2 + p_1^2 + p_2^2 + 2q_1 \cdot (q_1 - q_2 + p_1 - p_2)$$
$$= m_1^2 + m_3^2 + m_2^2 + m_4^2 + 2q_1 \cdot (q_1 - q_2 + p_1 - p_2).$$

As the 4-momenta satisfy the energy-momentum conservation law

$$q_1 + p_1 = q_2 + p_2,$$

we have

$$s + t + u = \sum_{i=1}^{4} m_i^2.$$

b. In the center of mass frame,

$$\mathbf{q}_1 + \mathbf{p}_1 = \mathbf{q}_2 + \mathbf{p}_2 = 0.$$

Hence

$$q_1^\alpha = (\sqrt{\mu^2 + k^2}, \mathbf{k}), \qquad q_2^\alpha = (\sqrt{\mu^2 + k'^2}, \mathbf{k}'),$$
$$p_1^\alpha = (\sqrt{m^2 + k^2}, -\mathbf{k}), \qquad p_2^\alpha = (\sqrt{m^2 + k'^2}, -\mathbf{k}'),$$

and

$$s = (q_1 + p_1)^2 = q_1^2 + p_1^2 + 2q_1 \cdot p_1$$
$$= \mu^2 + k^2 - k^2 + m^2 + k^2 - k^2 + 2\sqrt{(\mu^2 + k^2)(m^2 + k^2)} + 2k^2$$
$$= \mu^2 + m^2 + 2\sqrt{(\mu^2 + k^2)(m^2 + k^2)} + 2k^2,$$
$$t = (q_1 - q_2)^2 = q_1^2 + q_2^2 - 2q_1 \cdot q_2$$
$$= 2\mu^2 - 2\sqrt{(\mu^2 + k^2)(\mu^2 + k'^2)} + 2\mathbf{k} \cdot \mathbf{k}',$$
$$u = (q_1 - p_2)^2 = q_1^2 + p_2^2 - 2q_1 \cdot p_2$$
$$= \mu^2 + m^2 - 2\sqrt{(\mu^2 + k^2)(m^2 + k'^2)} - 2\mathbf{k} \cdot \mathbf{k}'.$$

Thus,

$$s = (\sqrt{\mu^2 + k^2} + \sqrt{m^2 + k^2})^2$$

is the square of the total energy of the incident particles in the center of mass frame, t is the square of the forward transfer and u is the square of the backward transfer of 4-momentum during the collision. s, t, u which are Lorentz invariant quantities are known as Mandelstam variables.

c. In the laboratory frame, we have

$$q_1^\alpha = (q_1^0, \mathbf{q}_1), \qquad q_2^\alpha = (q_2^0, \mathbf{q}_2), \qquad p_1^\alpha = (m, 0)$$

and

$$q_1^2 = q_1^\alpha q_{1\alpha} = \mu^2, \qquad q_2^2 = \mu^2,$$
$$q_1 + p_1 = q_2 + p_2.$$

Then
$$s = (q_1 + p_1)^2 = q_1^2 + p_1^2 + 2q_1 \cdot p_1$$
$$= \mu^2 + m^2 + 2q_1^0 m,$$
$$t = (q_1 - q_2)^2 = q_1^2 + q_2^2 - 2q_1 \cdot q_2$$
$$= 2\mu^2 - 2q_1^0 q_2^0 + 2\mathbf{q}_1 \cdot \mathbf{q}_2,$$
$$u = (q_1 - p_2)^2 = (q_2 - p_1)^2 = q_2^2 + p_1^2 - 2q_2 \cdot p_1$$
$$= \mu^2 + m^2 - 2q_2^0 m.$$

Hence the initial laboratory energy of particle μ is

$$q_1^0 = \frac{s - \mu^2 - m^2}{2m},$$

the final laboratory energy of μ is

$$q_2^0 = \frac{-u + \mu^2 + m^2}{2m},$$

and the scattering angle θ is given by

$$\cos\theta = \frac{\mathbf{q}_1 \cdot \mathbf{q}_2}{|\mathbf{q}_1||\mathbf{q}_2|} = \frac{\frac{t}{2} - \mu^2 + q_1^0 q_2^0}{\sqrt{[(q_1^0)^2 - \mu^2][(q_2^0)^2 - \mu^2]}}.$$

3036

The following question is a question on Newtonian gravity.

a. Calculate the radius and density of a solar-mass star ($M = 2 \times 10^{33}$ g) from which light could not escape.

b. The universe can be thought of as a sphere of gas of uniform density $\rho(t)$ and zero total energy expanding against its self-gravity. Show that if pressure can be neglected the interparticle distance increases as $t^{2/3}$.

(UC, Berkeley)

Sol:

a. By the equivalence of mass and energy, a photon of energy $E = mc^2$ has an equivalent mass m. The potential of a particle of mass m at the surface of a star of mass M and radius R is

$$V = -\frac{GMm}{R},$$

where G is the constant of gravitation. Hence for the photon to escape the star, we require $E + V \geq 0$, or $E \geq -V$. Conversely the photon will be confined to the star if $E \leq -V$, i.e.

$$mc^2 \leq \frac{GMm}{R},$$

or

$$R \leq \frac{GM}{c^2} = \frac{6.67 \times 10^{-8} \times 2 \times 10^{33}}{(3 \times 10^{10})^2} = 1.48 \times 10^5 \text{ cm} = 1.48 \text{ km}.$$

The density ρ of the sun must then be

$$\rho \geq M \left(\frac{4}{3} \pi R^3 \right)^{-1} = \left(\frac{3}{4\pi} \right) \times \frac{2 \times 10^{33}}{(1.48 \times 10^5)^3}$$

$$= 1.47 \times 10^{-1} \times \frac{10^{33}}{10^{15}} = 1.47 \times 10^{17} \text{ g/cm}^3.$$

Note that this result is consistent with the gravitational red shift. A photon of frequency ν emitted by the star will have a frequency ν' at a large distance from it, where

$$\nu' = \nu \left(1 - \frac{GM}{Rc^2} \right).$$

For the photon to *escape* the gravitational field of the star we require that $\nu' \geq 0$, or

$$R \geq \frac{GM}{c^2}.$$

b. In the expansion of a gas under the condition of uniform density, the distance between two given particles is proportional to the linear dimension of the gas and the position of any gas particle can be taken to be the center of expansion. Consider two gas particles A, B separated by a distance R. We can treat A as at the center of expansion and B as on the surface of a sphere with center at A. According to Newton's law of gravitation, B will suffer an attractive gravitational force toward A of $-\dfrac{GM}{R^2}$ per unit mass, where $M = \dfrac{4}{3}\pi R^3 \rho$, ρ being the density of the gas, is the mass of the sphere of gas. Note that the mass of the gas outside the sphere does not exert a net force on B. Neglecting pressure the equation of the motion of B is

$$\frac{d^2 R}{dt^2} = -\frac{GM}{R^2}.$$

Writing

$$\frac{d^2 R}{dt^2} = \frac{d\dot{R}}{dR}\frac{dR}{dt} = \frac{1}{2}\frac{d\dot{R}^2}{dR}$$

and noting that M does not change during the expansion, we have by integration

$$\frac{\dot{R}^2}{2} = \frac{GM}{R} + K,$$

or

$$K = \frac{\dot{R}^2}{2} - \frac{GM}{R} = T + V,$$

T, V being the kinetic and potential energies of the particle per unit mass. $K = 0$ if the total energy is zero. Hence

$$\frac{dR}{dt} = \pm\sqrt{\frac{2GM}{R}}.$$

The positive sign has to be taken for expansion. Integrating, we have, with $R = R_0$ at $t = t_0$,

$$\frac{2}{3}\left(R^{\frac{3}{2}} - R_0^{\frac{3}{2}}\right) = \sqrt{2GM}(t - t_0).$$

At large $t \gg t_0$, $R \gg R_0$ and

$$R \propto t^{\frac{2}{3}}.$$

3037

An astronaut takes an ordinary flashlight, turns it on, and leave it out in space (spin-stabilized by some rotation about its axis). What additional speed will this "photon-rocket" have gained by the time the batteries give out in two hours?

(Columbia)

Sol: Suppose the flashlight bulb is located at the focus of a paraboloid reflector so that almost the entire light output goes out in one direction. If the power of the flashlight is N watts and the time interval it is turned on is t, the total energy of the photons emitted is $E = Nt$. If the orientation of the flashlight does not change, it will gain a momentum

$$mv = \frac{E}{c} = \frac{Nt}{c},$$

or an additional speed

$$v = \frac{Nt}{mc},$$

m being the mass of the flashlight, since a photon of energy ε has a momentum $\frac{\varepsilon}{c}$.

For example, if $N = 1W$, $m = 0.3$ kg, $t = 2$ hours,

$$v = \frac{1 \times 2 \times 3600}{0.3 \times 3 \times 10^8} = 8 \times 10^{-5} \text{ m/s.}$$

3038

A hypothetical flashlight emits a well-collimated beam and is capable of converting a significant fraction of its rest mass into light. If the flashlight starts at rest with mass m_0, and is then turned on and allowed to move freely along a straight line, find its rest mass m when it reaches a velocity v relative to its original rest frame. Do not assume $c \gg v$.

(UC, Berkeley)

Sol: Let the total energy of all the photons emitted before the light reaches the velocity $v = \beta c$ be E. Then the total momentum of the photons is $\frac{E}{c}$ and is opposite in direction to v. Let the rest mass of the flashlight be m when its velocity is v. Conservation of energy gives

$$m\gamma c^2 + E = m_0 c^2,$$

and conservation of momentum gives

$$m\gamma\beta c - \frac{E}{c} = 0,$$

with $\gamma = \dfrac{1}{\sqrt{1 - \beta^2}}$. Eliminating E from the above gives

$$m\gamma(1 + \beta) = m_0,$$

or

$$m = \frac{m_0}{\gamma(1 + \beta)} = m_0 \sqrt{\frac{1 - \beta}{1 + \beta}} = m_0 \sqrt{\frac{c - v}{c + v}}.$$

3039

A particle of charge q, mass m moves in a circular orbit of radius R in the xy-plane in a uniform magnetic field $\mathbf{B} = B\hat{\mathbf{z}}$.

a. Find B in terms of q, R, m, and the angular frequency ω.

b. The speed of the particle is constant (since the $\mathbf{B}$ field does no work on it). An observer moving with uniform velocity $\beta\hat{\mathbf{x}}$ does not, however, see the particle's speed as constant. What is u_0' (the zero component of the particle's 4-velocity) as measured by this observer?

c. Calculate $\dfrac{du_0'}{d\tau}$ and, thus, $\dfrac{dp_0'}{d\tau}$. How can the energy of the particle change?

(*Princeton*)

Sol:

a. The equation of motion of the particle in the laboratory is

$$\frac{d\mathbf{p}}{dt} = q\mathbf{u} \times \mathbf{B}.$$

As $\mathbf{p}$ and $\mathbf{u}$ are parallel,

$$\mathbf{p} \cdot \frac{d\mathbf{p}}{dt} = \frac{1}{2}\frac{dp^2}{dt} = q\mathbf{p} \cdot \mathbf{u} \times \mathbf{B} = 0.$$

Hence p^2 and thus the magnitude of $\mathbf{p}$ and $\mathbf{u}$ are constant. It follows that

$$\gamma_u = \frac{1}{\sqrt{1 - \left(\dfrac{u}{c}\right)^2}}$$

is also a constant. Then, as $\mathbf{p} = m\gamma_u\mathbf{u}$, the equation of motion can be written as

$$\frac{d\mathbf{u}}{dt} = \mathbf{u} \times \omega$$

with $\omega = q\mathbf{B}/m\gamma_u$. As

$$\mathbf{u} = (\dot{x}, \dot{y}, 0), \qquad \omega = (0, 0, \omega),$$

it becomes

$$\ddot{x} = \dot{y}\omega, \quad \ddot{y} = -\dot{x}\omega, \quad \ddot{z} = 0.$$

Since the motion is confined to the xy-plane, the z equation need not be considered. The other two equations combine to give

$$\ddot{\xi} + i\omega\dot{\xi} = 0$$

by putting $x + iy = \xi$. It has general solution

$$\xi = \rho e^{-i(\omega t + \varphi)} + \xi_0,$$

where ρ, φ are real constants and ξ_0 is a complex constant. This solution is equivalent to

$$x - x_0 = R\cos(\omega t + \varphi), \qquad y - y_0 = -R\sin(\omega t + \varphi),$$

showing that the motion is circular with a radius R given by

$$u = \sqrt{\dot{x}^2 + \dot{y}^2} = R\omega,$$

ω being the angular velocity of revolution. Hence

$$B = \frac{m\gamma_u\omega}{q} = \frac{m\omega}{q} \frac{1}{\sqrt{1 - \left(\frac{u}{c}\right)^2}} = \frac{m\omega}{q} \frac{1}{\sqrt{1 - \left(\frac{R\omega}{c}\right)^2}}.$$

b. Let S, S' be respectively the laboratory frame and the rest frame of the moving observer. The zeroth component of the velocity four-vector, defined as $u^\alpha = (\gamma_u c, \gamma_u \mathbf{u})$, transforms according to

$$\gamma_u' c = \gamma(\gamma_u c - \beta\gamma_u u_1),$$

where $\gamma = \dfrac{1}{\sqrt{1 - \beta^2}}$. Thus

$$\begin{aligned}
u_0' \equiv \gamma_u' c &= \gamma\gamma_u(c - \beta\dot{x}) \\
&= \gamma\gamma_u[c + \beta\omega R \sin(\omega t + \varphi)] \\
&= \gamma\gamma_u[c + \beta u \sin(\omega\gamma_u\tau + \varphi)],
\end{aligned}$$

where τ is the proper time of the particle. Thus u_0' is not constant in S'.

c.
$$\begin{aligned}
\frac{du_0'}{d\tau} &= \gamma\gamma_u^2\beta\omega u \cos(\omega\gamma_u\tau + \varphi) \\
&= R\left(\frac{qB}{m}\right)^2 \frac{\beta}{\sqrt{1 - \beta^2}} \cos\left(\frac{qB\tau}{m} + \varphi\right).
\end{aligned}$$

If the four-momentum is defined as $p^\alpha = (mu_0, \mathbf{p})$, then, as m is a constant,

$$\frac{dp_0'}{d\tau} = m\frac{du_0'}{d\tau} = \frac{Rq^2B^2}{m} \frac{\beta}{\sqrt{1 - \beta^2}} \cos\left(\frac{qB\tau}{m} + \varphi\right),$$

which signifies a change of energy

$$\frac{dE}{d\tau} = c\frac{dp_0'}{d\tau}.$$

Note that in the S' frame, the electromagnetic field is given by

$$\begin{aligned}
E_\parallel' &= E_\parallel = 0 \\
\mathbf{E}_\perp' &= \gamma(\mathbf{E}_\perp + \mathbf{v} \times \mathbf{B}_\perp) \\
&= \gamma\mathbf{v} \times \mathbf{B}_\perp \\
&= \gamma\beta\hat{\mathbf{x}} \times B\hat{\mathbf{z}} = -\gamma\beta B\hat{\mathbf{y}},
\end{aligned}$$

so that there is also an electric field in the S' frame which does work on the particle.

3040

When two beams of protons of kinetic energy T collide head-on, the available energy for reactions is the same as for a single beam of what kinetic energy colliding with stationary protons? (Use relativistic expressions).

(UC, Berkeley)

Sol: The quantity $E^2 - p^2$ for a system is invariant under Lorentz transformation. Consider the head-on collision of two protons, each of kinetic energy T, and suppose that in the rest frame S' of one of the protons the other proton has total energy E' and momentum p'. As in the laboratory frame the total momentum of the two protons is zero, we have

$$(2mc^2 + 2T)^2 = (E' + mc^2)^2 - p'^2c^2$$
$$= (E' + mc^2)^2 - (E'^2 - m^2c^4)$$
$$= 2E'mc^2 + 2m^2c^4,$$

or

$$E' = \frac{2T^2 + 4Tmc^2 + m^2c^4}{mc^2}$$

where m is the rest mass of the proton. Hence the energy available for reactions is

$$E' - mc^2 = \frac{2T^2 + 4Tmc^2}{mc^2}.$$

3041

A photon of momentum p impinges on a particle at rest of mass m,

a. What is the total relativistic energy of photon plus particle in the center of mass frame of reference?

b. What is the magnitude of the particle's momentum in the center of mass frame?

c. If elastic backward scattering of the photon occurs, what is the momentum of the final photon in the laboratory frame?

(UC, Berkeley)

Sol:

a. Consider the quantity $E^2 - P^2c^2$ of the system which is invariant under Lorentz transformation:

$$(pc + mc^2)^2 - p^2c^2 = E'^2,$$

where E' is total energy of the system in the center of mass frame, which is by definition the inertial frame in which the total momentum vanishes. Hence

$$E' = \sqrt{2pmc^3 + m^2c^4}.$$

b. In the center of mass frame, the total momentum $P' = 0$ and p the momentum of the particle is equal and opposite to that of the photon p'. Momentum transformation

$$P' = 0 = \gamma\left(P - \frac{\beta E}{c}\right)$$

gives

$$\beta = \frac{Pc}{E} = \frac{pc}{pc + mc^2}, \qquad \gamma = \frac{1}{\sqrt{1 - \beta^2}}$$

for the center of mass frame. The particle momentum in the center of mass frame is then, using the transformation equation again,

$$p' = \gamma(0 - \beta mc) = -\gamma\beta mc = \frac{-pmc}{\sqrt{2pmc + m^2c^2}}.$$

c. Let the final momenta of the photon and the particle be $-p_1$ and p_2 respectively. Conservation of energy and of momentum give

$$pc + mc^2 = p_1c + \sqrt{p_2^2c^2 + m^2c^4},$$
$$p = -p_1 + p_2.$$

These combine to give

$$(p - p_1)^2 + 2(p - p_1)mc = (p + p_1)^2,$$

or

$$p_1 = \frac{pmc}{2p + mc}.$$

3042

We consider the possibility that one of the recently discovered particles, the $\psi'(3.7)$, can be produced when a photon collides with a proton in the reaction

$$\gamma + p \rightarrow p + \psi'.$$

In this problem we shall take the mass of ψ' to be $4M_p$, where M_p is the proton mass, which is a reasonable approximation. The target proton is initially at rest and the incident photon has energy E in the laboratory system.

a. Determine the minimum energy E that the photon must have for the above reaction to be possible. The answer can be given in units of $M_p c^2 (=938 \text{ MeV})$.

b. Determine the velocity, i.e. v/c, for the ψ' particle when the photon energy E is just above the threshold energy E_0.

(UC, Berkeley)

Sol:

a. At threshold, the final-state particles p, ψ' are stationary in the center of mass frame. Using the fact that the quantity $E^2 - P^2 c^2$ is invariant under Lorentz transformation and for an isolated system is conserved, we have, as a photon of energy E has momentum $\dfrac{E}{c}$,

$$(E_0 + M_p c^2)^2 - E_0^2 = (M_p c^2 + 4M_p c^2)^2,$$

giving

$$E_0 = 12 M_p c^2$$

as the threshold photon energy.

b. Near threshold, the ψ' is produced at rest in the center of mass frame, so its velocity in the laboratory is the velocity of the center of mass, i.e. of the system:

$$v = \frac{Pc^2}{E} = \frac{E_0 c}{E_0 + M_p c^2} = \frac{12}{13} c.$$

3043

An antiproton of energy E_0 interacts with a proton at rest to produce two equal mass particles, each with mass m_x. One of these produced particles is detected at an angle of 90° to the incident beam as measured in the laboratory. Calculate the total energy (E_s) of this particle and show that it is independent of m_x as well as of E_0.

(UC, Berkeley)

Sol: Antiproton and proton have the same mass m, say. The collision is depicted in Fig. 3.19. Momentum conservation gives

$$p_0 = p_2 \cos \theta, \qquad p_1 = p_2 \sin \theta,$$

or

$$p_2^2 = p_0^2 + p_1^2.$$

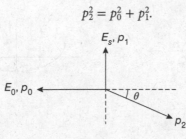

Fig. 3.19

Energy conservation gives

$$E_0 + mc^2 = E_s + \sqrt{p_2^2 c^2 + m_x^2 c^4}.$$

Combining the last two equations gives

$$(E_0 + mc^2)^2 + E_s^2 - 2(E_0 + mc^2)E_s = p_0^2 c^2 + p_1^2 c^2 + m_x^2 c^4,$$

or

$$2m^2 c^4 + 2E_0 mc^2 = 2(E_0 + mc^2)E_s,$$

since $E_0^2 = p_0^2 c^2 + m^2 c^4$, $E_s^2 = p_1^2 c^2 + m_x^2 c^4$. Hence

$$E_s = mc^2.$$

It is seen that E_s depends only on the proton mass but is independent of either m_x or E_0.

3044

a. A particle of mass m and charge e moves at relativistic speed v in a circle of radius R, the orbit being normal to a static, homogeneous magnetic field **B** as shown in Fig. 3.20. Find R in terms of the other parameters (radiation may be ignored).

b. An observer O' moving at fixed velocity v along the y-axis sees an orbit that looks like Fig. 3.21. The points a, b, c, d, e on the two figures correspond.

  i. What is the distance $y'_d - y'_b$ measured by O'?

  ii. What is the acceleration $\dfrac{d^2 x'}{dt'^2}$ of the particle at c, where it is instantaneously at rest?

  iii. What causes the acceleration at c as seen by O'?

(Princeton)

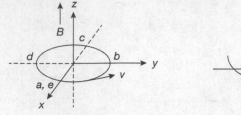

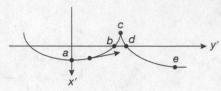

Fig. 3.20   Fig. 3.21

Sol:

a. If **p** is the momentum of the particle, we have

$$\frac{d\mathbf{p}}{dt} = e\mathbf{v} \times \mathbf{B},$$

and thus

$$\mathbf{p} \cdot \frac{d\mathbf{p}}{dt} = \frac{1}{2}\frac{dp^2}{dt} = em\gamma\,\mathbf{v} \cdot \mathbf{v} \times \mathbf{B} = 0,$$

where $\gamma = \dfrac{1}{\sqrt{1 - \dfrac{v^2}{c^2}}}$. Hence **p** and so γ and v have constant magnitudes. As shown

in Problem **3039**, the orbit is a circle of radius R given by $v = R\omega$, where

$$\omega = \frac{eB}{m\gamma}.$$

b. Let Σ, Σ' be the laboratory frame and the frame of the moving observer respectively, with Σ' moving relative to Σ in the y-direction with velocity $-v$. Lorentz transformation becomes

$$y' = \gamma(y - \beta ct) = \gamma(y + vt),$$
$$z' = z, \qquad x' = x,$$
$$ct' = \gamma(ct - \beta y) = \gamma\left(ct + \frac{vy}{c}\right),$$

as $\beta = -\dfrac{v}{c}$.

i. As

$$y_d - y_b = -2R, \qquad t_d - t_b = \frac{\pi}{\omega},$$
$$y'_d - y'_b = \gamma(y_d - y_b) + \gamma v(t_d - t_b)$$
$$= \gamma\left(-2R + \frac{v\pi}{\omega}\right)$$
$$= \frac{(\pi - 2)\gamma v}{\omega} = \frac{(\pi - 2)mv}{eB\left(1 - \dfrac{v^2}{c^2}\right)}.$$

ii. At point c, $\dfrac{dx}{dt} = 0$, $\dfrac{dy}{dt} = -v$,

$$\frac{d^2x}{dt^2} = \frac{-v^2}{R} \qquad \text{(centripetal acceleration)},$$

$$\frac{d^2y}{dt^2} = 0 \qquad \text{(tangential acceleration)}.$$

The velocity component $\dfrac{dx}{dt}$ transforms according to

$$\frac{dx'}{dt'} = \frac{dx}{dt} \Bigg/ \left(\frac{dt'}{dt}\right) = \frac{\dfrac{dx}{dt}}{\gamma\left(1 + \dfrac{v}{c^2}\dfrac{dy}{dt}\right)}.$$

In a similar way,

$$\frac{d^2x'}{dt'^2} = \frac{d}{dt'}\left(\frac{dx'}{dt'}\right) = \frac{1}{\gamma\left(1 + \dfrac{v}{c^2}\dfrac{dy}{dt}\right)} \cdot \frac{d}{dt}\left(\frac{dx'}{dt'}\right)$$

$$= \frac{1}{\gamma\left(1 + \dfrac{v}{c^2}\dfrac{dy}{dt}\right)} \cdot \frac{d}{dt}\left[\frac{\dfrac{dx}{dt}}{\left(1 + \dfrac{v}{c^2}\dfrac{dy}{dt}\right)}\right]$$

$$= \frac{\left(1 + \dfrac{v}{c^2}\dfrac{dy}{dt}\right)\dfrac{d^2x}{dt^2} - \dfrac{v}{c^2}\dfrac{d^2y}{dt^2}\dfrac{dx}{dt}}{\gamma^2\left(1 + \dfrac{v}{c^2}\dfrac{dy}{dt}\right)^3}$$

$$= \frac{-\left(1 - \dfrac{v^2}{c^2}\right)\dfrac{v^2}{R}}{\gamma^2\left(1 - \dfrac{v^2}{c^2}\right)^3}$$

$$= -\gamma^2 v\omega = \frac{-\gamma evB}{m}.$$

At point c,

$$\frac{dy'}{dt'} = \frac{\dfrac{dy}{dt} + v}{1 + \dfrac{v}{c^2}\dfrac{dy}{dt}} = \frac{-v + v}{1 - \dfrac{v^2}{c^2}} = 0.$$

As $\dfrac{dx'}{dt} = 0, \dfrac{dz'}{dt'} = 0$ also, the particle velocity $\mathbf{u}' = 0$.

iii. The transformation equations for the electromagnetic field are

$$E'_y = E_y = 0, \qquad B'_y = B_y = 0,$$
$$E'_z = \gamma(E_z - \beta c B_x) = 0,$$
$$E'_x = \gamma(E_x + \beta c B_z) = -\gamma v B,$$
$$B'_z = \gamma\left(B_z + \frac{\beta}{c}E_x\right) = \gamma B,$$
$$B'_x = \gamma\left(B_x - \frac{\beta}{c}E_z\right) = 0.$$

(In the usual geometry let y replace x, z replace y, x replace z to obtain the above). Then in Σ' the Lorentz force acting on the particle at c is

$$\mathbf{F}' = e(\mathbf{E}' + \mathbf{u}' \times \mathbf{B}') = e\mathbf{E}',$$

or

$$F' = F'_x = -\gamma e v B,$$

and the acceleration is $-\dfrac{\gamma e v B}{m}$, in agreement with (b). Hence the acceleration arises because of the presence of an electric field in Σ'.

3045

A charged particle (with charge e and rest mass m) moves in an electromagnetic field which is constant in space and time and whose components are $\mathbf{E} = (a, 0, 0)$ and $\mathbf{B} = (0, 0, b)$ in a Lorentz frame S. It is assumed that $|\mathbf{E}| \neq |\mathbf{B}|$. State the differential equations for the particle's four-vector velocity (as function of the proper time). Show that the solutions may be expressed as superpositions of exponentials, and determine the exponents. Under what conditions (on $\mathbf{E}$ and $\mathbf{B}$) are all components of the four-velocity bounded along every trajectory?

(Princeton)

Sol: The motion of the particle is described by the 4-vector equation

$$\frac{dp^\alpha}{ds} = F^\alpha,$$

where $ds = cd\tau$, τ being the proper time of the particle,

$$p^\alpha = mc^2 u^\alpha = mc^2\left(\frac{\gamma}{c}\mathbf{u}, \gamma\right),$$

$$F^\alpha = \left(\gamma\mathbf{F}, \frac{\gamma}{c}\mathbf{u} \cdot \mathbf{F}\right),$$

with $\gamma = \dfrac{1}{\sqrt{1 - \dfrac{u^2}{c^2}}}$ and $\mathbf{u}$ being the velocity of the particle.

The force acting on the particle is the Lorentz force

$$\mathbf{F} = e(\mathbf{E} + \mathbf{u} \times \mathbf{B}).$$

With $\mathbf{u} = (u_x, u_y, u_z)$, $\mathbf{E} = (a, 0, 0)$, $\mathbf{B} = (0, 0, b)$, and $\mathbf{u} \cdot \mathbf{F} = e\mathbf{u} \cdot \mathbf{E} = eau_x$, we have

$$F^\alpha = e\gamma\left(a + bu_y, -bu_x, 0, \frac{au_x}{c}\right)$$
$$= e(au_4 + cbu_2, -cbu_1, 0, au_1).$$

Hence the equations of motion are

$$m\frac{du_1}{d\tau} = \frac{e}{c}(cbu_2 + au_4),$$

$$m\frac{du_2}{d\tau} = -ebu_1,$$

$$m\frac{du_3}{d\tau} = 0,$$

$$m\frac{du_4}{d\tau} = \frac{eau_1}{c}.$$

Thus u_3 is a constant and need not be considered further. To solve the other equations, try

$$u_j = A_j e^{\lambda\tau}, \qquad j = 1, 2, 4.$$

The equations now become

$$m\lambda A_1 - ebA_2 - \frac{e}{c}aA_4 = 0,$$

$$ebA_1 + m\lambda A_2 = 0,$$

$$-\frac{e}{c}aA_1 + m\lambda A_4 = 0.$$

For a solution where not all A's vanish, we require

$$
\begin{vmatrix}
m\lambda & -eb & -\dfrac{ea}{c} \\
eb & m\lambda & 0 \\
-\dfrac{ea}{c} & 0 & m\lambda
\end{vmatrix} = 0,
$$

i.e.

$$
m\lambda\left(m^2\lambda^2 + e^2b^2 - \frac{e^2a^2}{c^2}\right) = 0.
$$

The roots are

$$
\lambda_1 = 0, \quad \lambda_2 = \frac{e}{mc}\sqrt{a^2 - c^2b^2}, \quad \lambda_3 = -\frac{e}{mc}\sqrt{a^2 - c^2b^2}.
$$

The general solution for the equation of motion is a superposition of exponentials with these exponents. For all components of the 4-velocity to be bounded along every trajectory we require that the λ's are either zero or imaginary, i.e.

$$
a \le cb, \quad \text{or} \quad |\mathbf{E}| \le c|\mathbf{B}|.
$$

3046

A particle of charge e, energy E, and velocity v moves in a magnetic field generated by a magnetic dipole of strength M located at the origin and directed along the z-axis. If the particle is initially in the xy-plane at a distance R from the origin and moving radially outward, give the minimum and maximum radii it will reach (assume the orbit is bounded).

(*Chicago*)

Sol: A particle of charge e, rest mass m and velocity $\mathbf{u}$ moving in an electromagnetic field of scalar potential Φ and vector potential $\mathbf{A}$ has Lagrangian

$$
L = -\frac{mc^2}{\gamma} - e\Phi + e\mathbf{u} \cdot \mathbf{A}.
$$

where $\gamma = \dfrac{1}{\sqrt{1 - \dfrac{u^2}{c^2}}}$. Since there is no electric field, $\Phi = 0$. The vector potential due to a magnetic dipole of moment $\mathbf{M}$ at the origin is

$$
\mathbf{A} = -\frac{\mu_0}{4\pi}\mathbf{M} \times \nabla\left(\frac{1}{r}\right) = \frac{\mu_0}{4\pi}\frac{\mathbf{M} \times \mathbf{r}}{r^3}.
$$

In spherical coordinates as shown in Fig. 3.22, we have

$$\mathbf{M} = (M\cos\theta, -M\sin\theta, 0),$$
$$\mathbf{r} = (r, 0, 0),$$

so that

$$\mathbf{A} = \frac{\mu_0}{4\pi}\frac{M\sin\theta}{r^2}\mathbf{i}_\varphi.$$

With $\mathbf{u} = (\dot{r}, r\dot{\theta}, r\dot{\varphi}\sin\theta)$, the Lagrangian is

$$L = -\frac{mc^2}{\gamma} + \frac{\mu_0}{4\pi}\frac{eM\sin^2\theta}{r}\dot{\varphi}.$$

Note that as $u^2 = \dot{r}^2 + r^2\dot{\theta}^2 + r^2\dot{\varphi}^2\sin^2\theta$, L does not depend on φ explicitly. Hence

$$p_\varphi = \frac{\partial L}{\partial\dot{\varphi}} = \left(m\gamma r^2\dot{\varphi} + \frac{\mu_0}{4\pi}\frac{eM}{r}\right)\sin^2\theta = \text{constant.}$$

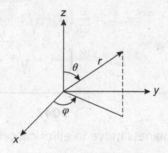

Fig. 3.22

Initially the particle is at $r = R$ and moves with velocity $v = \dot{r}$ in the xy-plane, i.e. $r = R$, $\theta = \frac{\pi}{2}$, $\dot{\varphi} = 0$ initially, giving $\frac{\mu_0}{4\pi}\frac{eM}{R}$, for the constant. Furthermore, as the only force on the particle is that due to the magnetic dipole at the origin whose magnetic lines of force at the xy-plane are perpendicular to the plane, the magnetic force is also in this plane and the motion is confined to the plane. Hence $\dot{\theta} = 0$, $\theta = \frac{\pi}{2}$ at all times. Thus

$$m\gamma r^2\dot{\varphi} + \frac{\mu_0}{4\pi}\frac{eM}{r} = \frac{\mu_0}{4\pi}\frac{eM}{R}.$$

At the maximum and minimum radii, $\dot{r} = 0$ and $\mathbf{u} = r\dot{\varphi}\mathbf{i}_\varphi$. Since magnetic force does no work as

$$\mathbf{u} \cdot \mathbf{u} \times (\nabla \times \mathbf{A}) = 0,$$

the magnitude of **u** is equal to the initial speed v, i.e. $r\dot\varphi = \pm v$, and γ is a constant. Letting

$$\alpha = \frac{\mu_0}{4\pi}\frac{eM}{m\gamma v},$$

we have

$$\pm Rr^2 - \alpha r + \alpha R = 0.$$

For the upper sign the roots are

$$r = \frac{\alpha}{2R}\left(1 \pm \sqrt{1 - \frac{4R^2}{\alpha}}\right).$$

For the lower sign the roots are

$$r = \frac{\alpha}{2R}\left(-1 \pm \sqrt{1 + \frac{4R^2}{\alpha}}\right),$$

where the positive sign is to be used since r is positive. Examining these roots we find

$$r_{max} = \frac{\alpha}{2R}\left(1 + \sqrt{1 - \frac{4R^2}{\alpha}}\right),$$

$$r_{min} = \frac{\alpha}{2R}\left(-1 + \sqrt{1 + \frac{4R^2}{\alpha}}\right).$$

3047

It is well known that planets move in elliptical orbits around the sun and the derivation of the orbit equation is a standard exercise in classical mechanics. However, if the effects of special relativity only are taken into account, the orbit is a precessing ellipse of the form

$$\frac{1}{r} = \frac{1}{r_0}\{1 + \varepsilon\cos[\alpha(\theta - \theta_0)]\},$$

where $\alpha = 1$ corresponds to the classical result of zero precession.

a. Derive this equation and express α and r_0 in terms of fundamental constants of the orbit (such as energy, angular momentum, etc.)

b. Given that the mean radius of the orbit of Mercury is 58×10^6 km and that its orbital period is 88 days, calculate the precession of Mercury's orbit in seconds of arc per century. (This effect does not, of course, account for the total precession rate of Mercury.)

(Chicago)

Sol:

a. Consider a planet of mass m and velocity $\mathbf{v}$. As it moves in an elliptical orbit, i.e. in a plane, use polar coordinates (r, θ) with the sun at the origin. The Lagrangian of the system is

$$L = -\frac{mc^2}{\gamma} + \frac{GmM}{r},$$

where $\gamma = \dfrac{1}{\sqrt{1 - \beta^2}}$ with

$$\beta^2 = \frac{v^2}{c^2} = \frac{\dot{r}^2 + r^2\dot{\theta}^2}{c^2},$$

M being the mass of the sun. As

$$\frac{\partial}{\partial \dot{r}}\left(\frac{1}{\gamma}\right) = \frac{-\gamma\dot{r}}{c^2}, \quad \frac{\partial}{\partial \dot{\theta}}\left(\frac{1}{\gamma}\right) = \frac{-\gamma r^2\dot{\theta}}{c^2}, \quad \frac{\partial}{\partial r}\left(\frac{1}{\gamma}\right) = \frac{-\gamma r\dot{\theta}^2}{c^2},$$

Lagrange's equations

$$\frac{d}{dt}\left(\frac{\partial L}{\partial \dot{q}_i}\right) - \frac{\partial L}{\partial q_i} = 0$$

give

$$\frac{d}{dt}(m\gamma\dot{r}) - m\gamma r\dot{\theta}^2 + \frac{GmM}{r^2} = 0,$$

$$m\gamma r^2\dot{\theta} = b, \text{ a constant.}$$

Letting $u = \dfrac{1}{r}$, the last two equations combine to give

$$\frac{d}{dt}\left(\frac{b}{\dot{\theta}}\frac{du}{dt}\right) + bu\dot{\theta} - GmMu^2 = 0,$$

or

$$\frac{d^2u}{d\theta^2} + u = \frac{GmMu^2}{b\dot{\theta}}, \tag{1}$$

as

$$\frac{du}{dt} = \dot{\theta}\frac{du}{d\theta}.$$

The total energy of the planet is

$$E = m\gamma c^2 - \frac{GmM}{r}.$$

Thus

$$\frac{GmMu^2}{b\dot\theta} = \frac{GmM}{b^2c^2}(E + GmM\mu)$$

and Eq. (1) becomes

$$\frac{d^2u}{d\theta^2} + \alpha^2 u = \frac{GmME}{b^2c^2},$$

(2

where

$$\alpha^2 = 1 - \left(\frac{GmM}{bc}\right)^2.$$

A special solution of Eq. (2) is

$$u = \frac{GmME}{b^2c^2\alpha^2},$$

and its general solution is thus

$$u = A\cos\left[\alpha(\theta - \theta_0)\right] + \frac{GmME}{b^2c^2\alpha^2},$$

where A and θ_0 are constants. The orbit is therefore given by

$$\frac{1}{r} = \frac{1}{r_0}\{1 + \varepsilon\cos[\alpha(\theta - \theta_0)]\}$$

with

$$r_0 = \frac{b^2c^2\alpha^2}{GmME} = \frac{(bc)^2 - (GmM)^2}{GmME}, \quad \varepsilon = Ar_0, \quad \alpha = \sqrt{1 - \left(\frac{GmM}{bc}\right)^2},$$

A, θ_0 being constants, and b, E being the angular momentum about the sun and the total energy of the planet respectively.

b. Suppose r is minimum at θ_1 and it next returns to this minimum at θ_2. Then $\alpha(\theta_2 - \theta_1) = 2\pi$. Hence the perihelion advances an angle

$$\Delta\theta = \frac{2\pi}{\alpha} - 2\pi = 2\pi\left(\frac{1}{\alpha} - 1\right)$$

in one period of revolution. Note that there is no precession if $\alpha = 1$. Since the amount of precession is small compared with 2π, α is close to unity and can be expressed as

$$\alpha \approx 1 - \frac{1}{2}\left(\frac{GmM}{bc}\right)^2,$$

and we have

$$\Delta\theta \approx \pi\left(\frac{GmM}{bc}\right)^2$$

per period of revolution. From a consideration of the gravitational attraction we have

$$\frac{GmM}{\bar{r}^2} = m\gamma\bar{r}\dot{\theta}^2,$$

where $\bar{r}$ is the mean radius of the orbit of Mercury. As

$$b = m\gamma\bar{r}^2\dot{\theta},$$

$$\frac{GmM}{bc} = \frac{\bar{r}\dot{\theta}}{c} = \frac{2\pi\bar{r}}{\tau c},$$

where $\tau = 88$ days is the period of one revolution. In a century there are

$$\frac{100 \times 365}{88} = 414.8$$

revolutions, so that the total precession per century is

$$\Theta = 414.8 \times 4\pi^3 \times \left(\frac{58 \times 10^6}{88 \times 24 \times 3600 \times 3 \times 10^5} \right)^2$$

$$= 3.326 \times 10^{-5} \text{ rad}$$

$$= 6.86 \text{ seconds of arc.}$$

This is about $\frac{1}{6}$ of the observed value, which can only be accounted for if general relativity is used for the calculation.

3048

Derive the Hamiltonian of a particle traveling with momentum $\mathbf{p} = \dfrac{m_0 v}{\sqrt{1 - \dfrac{v^2}{c^2}}}$

when it is placed in the fields defined by

$$\mathbf{E} = -\nabla\Phi - \frac{1}{c}\frac{\partial \mathbf{A}}{\partial t},$$

$$\mathbf{H} = \nabla \times \mathbf{A}.$$

(SUNY, Buffalo)

Sol: The Lagrangian of the particle, assumed to have charge q, is in Gaussian units

$$L = -\frac{m_0 c^2}{\gamma} - q\Phi + \frac{q}{c}\mathbf{v} \cdot \mathbf{A},$$

and its Hamiltonian is defined as

$$H = \sum_i \dot{x}_i p_i - L,$$

where $\dot{x}_i$ is the velocity component given, in Cartesian coordinates, by $\mathbf{v} = (\dot{x}_1, \dot{x}_2, \dot{x}_3)$ and p_i is the canonical momentum given by $p_i = \dfrac{\partial L}{\partial \dot{x}_i}$. As

$$\frac{1}{\gamma^2} = 1 - \frac{\dot{x}_1^2 + \dot{x}_2^2 + \dot{x}_3^2}{c^2}, \qquad \frac{\partial}{\partial \dot{x}_i}\left(\frac{1}{\gamma}\right) = -\frac{\gamma \dot{x}_i}{c^2},$$

and

$$\mathbf{v} \cdot \mathbf{A} = \sum_i \dot{x}_i A_i,$$

we have

$$p_i = \frac{\partial L}{\partial \dot{x}_i} = m_0 \gamma \dot{x}_i + \frac{q A_i}{c}$$

and

$$H = \sum_i \dot{x}_i p_i - L = m_0 \gamma \sum_i \dot{x}_i^2 + \frac{m_0 c^2}{\gamma} + q\Phi$$

$$= \frac{m_0 c^2}{\gamma}\left(\frac{\gamma^2 v^2}{c^2} + 1\right) + q\Phi$$

$$= m_0 \gamma c^2 + q\Phi.$$

To write H in terms of $\mathbf{p}$, we note that

$$\sum_i (m_0 \gamma \dot{x}_i)^2 = \sum_i \left(p_i - \frac{q A_i}{c}\right)^2,$$

or

$$m_0^2 \gamma^2 v^2 = \left(\mathbf{p} - \frac{q\mathbf{A}}{c}\right)^2,$$

and thus

$$m_0^2 \gamma^2 c^4 = m_0^2 \left(\frac{\gamma^2 v^2}{c^2} + 1\right) c^4 = \left(\mathbf{p} - \frac{q\mathbf{A}}{c}\right)^2 c^2 + m_0^2 c^4.$$

Therefore

$$H = \sqrt{\left(\mathbf{p} - \frac{q\mathbf{A}}{c}\right)^2 c^2 + m_0^2 c^4} + q\Phi.$$

3049

What is the velocity of a particle if its kinetic energy equals its rest energy?

(*Wisconsin*)

Sol: The kinetic energy of a particle of rest mass m_0 is

$$T = E - m_0 c^2 = m_0 c^2 (\gamma - 1),$$

where $\gamma = \dfrac{1}{\sqrt{1 - \dfrac{v^2}{c^2}}}$. As this equals $m_0 c^2$, $\gamma = 2$. Hence

$$\frac{v}{c} = \sqrt{1 - \frac{1}{\gamma^2}} = \sqrt{\frac{3}{4}},$$

or

$$v = \frac{\sqrt{3}}{2} c.$$

3050

A beam of electrons is scattered by a fixed scattering target as shown in Fig. 3.23. The electrons are elastically scattered. Each electron has an energy $E = \dfrac{5}{3} m_0 c^2$ and the beam has a flux of Q electrons per second.

a. What is the velocity of the incident electrons?

b. What are the magnitude and direction of the force on the scattering target due to the electrons?

(*Wisconsin*)

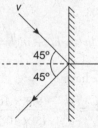

Fig. 3.23

Sol:

a. As $E = m_0 \gamma c^2 = \dfrac{5}{3} m_0 c^2$, $\gamma = \dfrac{5}{3}$ and $\dfrac{v}{c} = \sqrt{1 - \dfrac{1}{\gamma^2}} = \dfrac{4}{5}$. Hence the electron velocity is $0.8c$.

b.   Since the electrons are elastically scattered, they have the same speed before and after scattering and conservation of the momentum parallel to the target require that the incident and scattering angles are equal. Then after scattering the normal component of the momentum changes sign but remains the same in magnitude. Hence

$$\Delta p = 2p_n = 2m_0 \gamma v \cos 45° = \frac{4\sqrt{2}}{3}m_0 c.$$

The force F on the target due to the beam of electrons is equal to the impulse given to it per unit time by the beam. As Q electrons impinge on the target in unit time

$$F = 2p_n Q = \frac{4\sqrt{2}}{3}Q m_0 c,$$

and it acts vertically onto the target.

3051

The principle of equivalence asserts that gravitational and inertial masses are equal. Does a photon have nonzero gravitational mass? Explain. Suppose a photon is falling toward the earth and it falls a distance of 10 m. Calculate the effect on the frequency of the photon. What experimental technique could be used to measure this frequency change?

(Wisconsin)

Sol: The gravitational mass of a photon is not zero but is equal to the inertial mass

$$m = \frac{E}{c^2} = \frac{h\nu}{c^2},$$

in accordance with the principle of equivalence, even though its rest mass is zero. When the photon falls a distance l in a gravitational field $\mathbf{g}$, its energy increases and so does its frequency:

$$h\nu' \approx h\nu + mgl = h\nu\left(1 + \frac{gl}{c^2}\right).$$

Writing $\nu' = \nu + \Delta\nu$, we have

$$\frac{\Delta\nu}{\nu} = \frac{gl}{c^2} = \frac{9.8 \times 10}{(3 \times 10^8)^2} = 1.1 \times 10^{-15}.$$

Thus falling through a distance of 10 m in the gravitational field of the earth, the frequency of a photon will increase (blue shift) by a factor $1 + 1.1 \times 10^{-15}$. The slight increase in frequency can be detected experimentally using the Mössbauer effect.

3052

Consider a very high energy scattering experiment involving two particles with the same rest mass m_0, one initially at rest and the other incident with momentum p and total energy E.

a. Find the velocity of the center of mass $\beta^* = \dfrac{v^*}{c}$.

b. In the extreme relativistic limit $pc \gg m_0c^2$, find the total energy E^* of the system in the center of mass frame (i.e. the frame in which the total 3-momentum is zero).

(Wisconsin)

Sol:

a. The system of two particles has total energy $E + m_0c^2$ and total momentum p in the laboratory system. The velocity of the center of mass, which is the velocity of the system as a whole, in units of c, is then

$$\beta^* = \frac{pc}{E + m_0c^2}.$$

b. The quantities $E^2 - p^2c^2$ of a system is invariant under Lorentz transformation. In the laboratory frame it is

$$(E + m_0c^2)^2 - p^2c^2 = 2Em_0c^2 + 2m_0^2c^4$$

as $E^2 - p^2c^2 = m_0^2c^4$. In the center of mass frame it is $(2\bar{E})^2$, where $\bar{E}$ is the total energy of each particle. Hence

$$E^* = 2\bar{E} = \sqrt{2Em_0c^2 + 2m_0^2c^4}$$
$$\approx \sqrt{2Em_0c^2} \approx \sqrt{2pm_0c^3}$$

in the extreme relativistic limit for which $pc \gg m_0c^2$, since in this case

$$E = \sqrt{p^2c^2 + m_0^2c^4} \approx pc \gg m_0c^2.$$

3053

A particle of rest mass m and initial velocity v_0 along the x-axis is subject after $t = 0$ to a constant force F acting in the y-direction. Find its velocity at any time t, and show that $|\mathbf{v}| \to c$ as $t \to \infty$.

(Wisconsin)

Sol: The equation of motion

$$\mathbf{F} = \frac{d}{dt}(m\gamma\mathbf{v}),$$

where $\gamma = \dfrac{1}{\sqrt{1 - \dfrac{v^2}{c^2}}}$, can be written as

$$0 = \frac{d}{dt}(m\gamma\dot{x}), \qquad F = \frac{d}{dt}(m\gamma\dot{y})$$

with $\mathbf{v} = (\dot{x}, \dot{y})$, $\mathbf{F} = (0, F)$. As F is constant for $t > 0$ and initially $\dot{x} = v_0$ $\dot{y} = 0$, $F = 0$, the above integrate to give

$$m\gamma\dot{x} = m\gamma_0 v_0, \qquad m\gamma\dot{y} = Ft,$$

where $\gamma_0 = \dfrac{1}{\sqrt{1 - \dfrac{v^2}{c^2}}}$. Hence

$$\beta^2 c^2 = \dot{x}^2 + \dot{y}^2 = \frac{1}{m^2\gamma^2}(m^2\gamma_0^2 v_0^2 + F^2 t^2),$$

or

$$\gamma^2\beta^2 = \frac{\beta^2}{1 - \beta^2} = \frac{m^2\gamma_0^2 v_0^2 + F^2 t^2}{m^2 c^2},$$

giving

$$\beta^2 = \frac{m^2\gamma_0^2 v_0^2 + F^2 t^2}{m^2\gamma_0^2 v_0^2 + m^2 c^2 + F^2 t^2}.$$

As $\gamma_0^2 \dfrac{v_0^2}{c^2} = \gamma_0^2 - 1$, we have

$$v = \beta c = \sqrt{\frac{m^2\gamma_0^2 v_0^2 + F^2 t^2}{m^2\gamma_0^2 c^2 + F^2 t^2}}\, c.$$

The velocity components are

$$\dot{x} = \frac{\gamma_0 v_0}{\gamma}, \qquad \dot{y} = \frac{Ft}{m\gamma},$$

where

$$\gamma = \frac{1}{\sqrt{1 - \beta^2}} = \sqrt{\frac{m^2\gamma_0^2 c^2 + F^2 t^2}{m^2 c^2}}.$$

For $t \to \infty$, as $m\gamma_0 v_0$, $m\gamma_0 c$ remain constant we have

$$v \to \left(\frac{Ft}{Ft}\right)c = c.$$

3054

An electron of energy $E \gg mc^2$ and a photon of energy W collide.

a. What is W', the energy of the photon in the electron (e) frame of reference?

b. If $W' \ll mc^2$, the electron recoil can be neglected and the energy of the photon in the e-frame is unchanged as a result of the scattering process. What are the minimum and maximum values of the energy of the scattered photon in the laboratory (L) frame?

(Wisconsin)

Sol:

a. Suppose the photon makes angles θ, θ' with the initial direction of motion of the electron, which is taken to be the direction of the x-axis, in the L- and e-frames respectively. As $(\mathbf{p}c, E)$ forms a 4-vector, the photon energy transforms according to

$$W' = \gamma\left(W - \frac{\beta W}{c} \cdot c \cos \theta\right) = \gamma(1 - \beta \cos \theta)W,$$

where $\gamma = \dfrac{E}{mc^2}$, $\beta = \dfrac{pc}{E}$ with $p = \dfrac{1}{c}\sqrt{E^2 - m^2c^4}$ being the momentum of the electron in the L-frame. As $E \gg mc^2$,

$$pc \approx E\left(1 - \frac{m^2c^4}{2E^2}\right).$$

Hence

$$W' \approx \frac{E}{mc^2}\left[1 - \left(1 - \frac{m^2c^4}{2E^2}\right)\cos \theta\right]W$$

$$= \left[\frac{E}{mc^2}(1 - \cos \theta) + \frac{mc^2}{2E}\cos \theta\right]W. \tag{1}$$

b. In the *e*-frame, the electron is initially at rest. If its recoil can be neglected, the incident photons must be scattered back along the line of incidence with the same energy in accordance with the conservation of energy and of momentum. The transformation of energy and momentum of the photon is given by

$$W' \cos \theta' = \gamma W (\cos \theta - \beta), \qquad W' \sin \theta' = W \sin \theta,$$

or

$$\tan \theta' = \frac{\sin \theta}{\gamma (\cos \theta - \beta)}, \tag{2}$$

and

$$W = \gamma (1 + \beta \cos \theta') W' \approx \left[\frac{E}{mc^2} (1 + \cos \theta') - \frac{mc^2}{2E} \cos \theta' \right] W'. \tag{3}$$

Equation (1) shows that for W' to be maximum, $\cos \theta = -1$ or $\theta = \pi$ and

$$W'_{\text{max}} \approx \frac{2E}{mc^2}.$$

Equation (2) gives $\theta' = \pi$. The photon is scattered back so that after the collision $\theta' = 0$. Equation (3) then gives the corresponding energy in *L*-frame:

$$W_{\text{max}} \approx \frac{2E}{mc^2} W'_{\text{max}} \approx \left(\frac{2E}{mc^2} \right)^2 W.$$

Similarly, for the minimum energy, $\cos \theta = 1$, or $\theta = 0$, and

$$W'_{\text{min}} \approx \frac{mc^2}{2E} W, \ \theta' = 0.$$

After scattering $\theta' = \pi$ and

$$W_{\text{min}} \approx \frac{mc^2}{2E} W'_{\text{min}} \approx \left(\frac{mc^2}{2E} \right)^2 W.$$

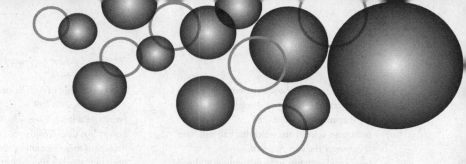

Index to Problems

Printed in the United States
By Bookmasters